PHOTOSYNTHESIS AND PHOTORESPIRATION

Proceedings of a conference held at Australian National University, Canberra, Australia, 23 November–5 December 1970

Sponsored by the Australian Academy of Sciences under the auspices of the Australia-U.S. Agreement on Scientific and Technical Cooperation

Edited by

M.D. Hatch   Division of Plant Industry
CSIRO, Canberra City, Australia

C.B. Osmond   Department of Environmental Biology
Research School of Biological Sciences
R.O. Slatyer   Australian National University, Canberra City, Australia

WILEY-INTERSCIENCE, a Division of John Wiley & Sons, Inc.
New York • London • Sydney • Toronto

FOREWORD

In October 1968 an Agreement was signed by the Governments of the United States of America and Australia to provide additional avenues for scientific and technical cooperation between the two countries.

The National Science Foundation was nominated as the Executive Agency for the United States for joint activities under this Agreement, and the Commonwealth Department of Education and Science as the counterpart Agency for Australia.

Areas of research to be encouraged under the Agreement covered a wide range of subjects in the agricultural and biological sciences, including factors affecting agricultural productivity, the management of arid and tropical lands, and the biological basis of plant responses to environmental stress.

It became apparent that a meeting concerned with recent advances in photosynthesis and photorespiration would have wide repercussions on these areas. Such a meeting would also be timely because of the discovery and evaluation, over the past few years, of the C_4 dicarboxylic acid pathway for photosynthesis. This work not only is of fundamental interest and importance, but also has considerable potential practical significance in view of the greater photosynthetic capacity of those plants that possess the C_4 pathway and related metabolic and structural attributes. These research developments have created an upsurge of interest and activity in many aspects of photosynthetic processes. Coupled with this, there have been advances in the study of photorespiration, particularly in the role of plant microbodies.

Because of this upsurge in research, there has developed a need for research workers to meet to discuss areas of agreement and disagreement and to avoid unnecessary overlap in research. Since much of the prominent work in this field is being conducted by Australian and United States scientists, there was seen to be merit in bringing together a select group of leading research workers from each country. The Agreement provided an avenue for the two Governments to give financial support for these researchers to meet for a two-week seminar in Canberra. The seminar, held between November 23 and December 5, 1970, was convened by the Australian Academy of Science, and more than fulfilled the hopes of its sponsors.

The collection of contributed papers and discussion summaries, which constitute this volume, make, I believe, a substantial contribution to contemporary understanding of problems concerned with photosynthesis and photorespiration and raise ideas for further research and for practical applications.

My predecessor, the Hon. Nigel Bowen, Q.C., M.P., was intimately concerned with the initiation of the arrangements for the Seminar, and I have much pleasure in now being associated with the final outcome of the meetings. I look forward to the possibility of further meetings in this and other fields under the auspices of the United States-Australia Agreement for Scientific and Technical Cooperation.

David Fairbairn
Minister for Education and Science
Commonwealth of Australia

PREFACE

This book reports the proceedings of a meeting held in Canberra, Australia, November 23 to December 5, 1970. The meeting was convened to consider two rapidly advancing areas of plant physiology—those centered in the C_4 pathway of photosynthesis and in photorespiration. The format of the meeting allowed generous time for formal and informal discussions as well as for workshop sessions following the presentation of the material reported here. This permitted the development of a consenus of opinion in most areas of interest or controversy and this consensus is summarized in the assessments of each major session.

The meeting was supported by the Governments of Australia and the United States of America through an agreement relating to Scientific and Technical Cooperation between the two countries. This support, and the sponsorship of the meeting by the Australian Academy of Science, is gratefully acknowledged. The conception and initial planning of the meeting was in very large measure due to the efforts of Professor R. O. Slatyer and Professor N. E. Tolbert. The local Organising Committee, convened by Professor Slatyer on behalf of the Australian Academy of Science, included Dr. M. D. Hatch, Dr. C. B. Osmond, Professor A. J. Birch, and Mr. P. D. O'Connor. This Committee wishes to acknowledge the contribution of Dr. C. B. Osmond who undertook much of the responsibility for the detailed planning and arrangements of the meeting.

We are indebted to Mr. W. Douglas, John Wiley & Sons, Inc., Australia, for advice and assistance during preparation of this book. Thanks are due to the typists Sue Osmond, Daphne Harding, and Jan Wheeler, to proofreaders Bronwyn Williams, Wendy Willett, and Remedios Rochino, and to Mr. M. Gray, CSIRO Herbarium, Canberra, for checking the list of C_4 species given in the Appendix. Permission to reproduce material previously published by the National Research Council of Canada (*Canadian Journal of Botany*), Springer-Verlag (*Planta*) and MacMillan (*Nature*) is gratefully acknowledged. Particular thanks are due to Drs. J. H. Troughton and L. A. Donaldson for providing the cover illustration, reproduced with permission from *Probing Plant Structure*, published by A. H. and A. W. Reed, Wellington, New Zealand, 1971.

M. D. Hatch
C. B. Osmond
R. O. Slatyer

Canberra, Australia
May 1971

v

CONTENTS

Section 2

Carbon Dioxide Assimilation

A.   Review Papers
     Chairman, *H. P. Kortschak*

B.   Research Papers and Assessments
     Chairmen, *M. Gibbs, D. A. Walker,* and *C. R. Slack*

Section 3

Chloroplast Structure and Function

A. Review Papers
    Chairman, D. J. Goodchild

B. Research Papers and Assessments
    Chairmen, R. G. Hiller and C. C. Black

Section 4

Photorespiration and the Role of Microbodies

A.   Review Papers
     Chairmen, C. B. Osmond and J. S. Turner

B.   Research Papers and Assessments
     Chairman, W. A. Jackson

SECTION 1

Environment, Adaptation and Evolution in Relation to Photosynthetic and Photorespiratory Gas Exchange

A. Review Papers

ADAPTIVE AND EVOLUTIONARY ASPECTS OF C_4 PHOTOSYNTHESIS

W.J.S. Downton

Research School of Biological Sciences, Australian
National University, Canberra City 2601, Australia.

SUMMARY

*Physiological characteristics related to the C_4 pathway of photo-
synthesis are reviewed. The generalizations cited should provide a
basis for further critical discussion. Plants with the C_4 pathway
are scattered taxonomically; closely related taxa may include C_3
species and plants with Crassulacean acid metabolism. The architecture of the C_4 leaf emerges as one of the most important factors in-
fluencing biochemical compartmentation, metabolite transport, and
gas exchange in the plant. Two major variations on the C_4 theme are
discussed. Aspartate or malate may carry CO_2 to the Calvin cycle in
the bundle sheath. Many species synthesizing and transporting large
quantities of malate have poorly granal bundle sheath chloroplasts.
These chloroplasts have reduced photosystem II activity and are deficient in NADPH production. The high malic enzyme in such plants
provides some degree of compensation. Species utilizing mainly aspartate have granal bundle sheath chloroplasts with apparently normal light reactions. Agranal bundle sheath chloroplasts are not likely faced with a disposal problem for excess reducing potential. This
is further evidenced by the levels of glycolate oxidase and NADPH
requiring enzymes found in agranal bundle sheath cells.*

*The Calvin cycle remains the common denominator of higher plant
photosynthesis. Studies on sugarcane tissue culture indicate that a
Calvin type cell can produce a "Kranz" type shoot. Thus mesophyll
and bundle sheath cells differ only quantitatively in their enzymology. The success of plants in fully utilizing both carboxylation
reactions depends upon adequate compartmentation of enzymes and substrates. C_4 plants and CAM plants illustrate two ways in which this
has been accomplished.*

Higher plants have been recently divided into two main groups
according to their mode of carbon fixation and associated characters
(1). Species with the C_4 dicarboxylic acid pathway of photosynthesis

3

(C_4 plants) synthesize C_4 dicarboxylic acids as initial photosynthetic products; species with the Calvin cycle (C_3 plants), mainly phosphorylated compounds. Rather than concentrate on biochemical evidence for the C_4 pathway, I will first of all discuss those physiological characteristics that have been invariably associated with the pathway. Attention will be paid to C_4 leaf architecture and its effect on certain physiological and biochemical processes. Bundle sheath chloroplast ultrastructure is related to degree of dependence of bundle sheath on mesophyll. Finally some genetic and developmental aspects of C_4 photosynthesis are considered.

PHYSIOLOGICAL CHARACTERISTICS OF C_4 PLANTS

Geographical distribution indicates that C_4 species originated primarily in the tropics, hence the synonym "tropical plants" is often used for C_4 plants (1). They seem well adapted to the extremes of light, temperature and dryness often encountered in this environment.

Response to light

Photosynthesis in young leaves of C_4 plants grown under high light intensity becomes saturated at very high light (2,3,4,5). Leaves may saturate above half of full sunlight or even fail to saturate at full sun (5). C_3 plants generally saturate at 20 to 30% of full sunlight. Some exceptions to these generalizations occur in that C_3 species such as sunflower (5) and some diploid *Triticum* spp. and *Aegilops* spp. (6) do not saturate at low light intensity. Old and shade grown leaves of *Amaranthus edulis* (C_4) saturate at levels similar to C_3 plants (2).

Response to temperature

C_4 plants perform poorly at low temperature. At less than 16° their chlorophyll is subject to photodestruction and developing leaves are chlorotic (3,7,8). The optimum temperature for C_4 photosynthesis ranges from 30° to 40° (9,10,11) and photosynthesis decreases rapidly below 15° to 20°. Panicoid and chloridoid-eragrostoid (C_4) grasses give the best relative growth rates when grown under a day/night temperature regime of 30°/25° to 36°/31° (8). C_3 grasses often become chlorotic and die at about 35° (8,12). The photosynthetic optimum of C_3 plants lies over a broad range from 10° to 25° (10,11) and they will grow as low as 5° to 10° (3). Festucoid grasses (C_3) exhibit highest relative growth rates under 21°/16° to 27°/22° or 21°/16° to 30°/25° day/night temperature regimes (8).

Photosynthetic rates

C_4 species grown and measured under their optimum conditions described above can attain photosynthetic rates of 60 to 100 mg of CO_2 assimilated $dm^{-2} hr^{-1}$ (13-15). It is therefore not surprising that C_4 species include both highly productive crops such as sugarcane, maize, and *Paspalum*, and noxious weed species like *Digitaria* (crabgrass), *Echinochloa* (barnyard grass), and *Amaranthus*. C_3 species rarely achieve this rate (*Typha latifolia* may be an exception) and

usually give values of 10 to 35 mg CO_2 dm^{-2} hr^{-1} (5,10,13,16,17)
Leaves of old and shade grown *Amaranthus edulis* have low photosynthetic rates similar to many C_3 species (2).

Water use efficiency

C₄ plants require fewer units of water to produce a unit of dry
matter than C_3 plants. In one growth study C_3 plants required an average of about 610 grams of water for the production of 1 gram of
dry matter compared to about 300 grams for equivalent production in
C₄ plants (18). Gas exchange measurements on individual leaves support these data. Under high light and temperature, maize (C₄) and
oats (C_3) have comparable transpiration (T) rates but the photosynthesis rate (P) for maize is twice that of oats. Thus maize has a
P:T ratio double that of oats (14). At low light levels, wheat (C_3)
may transpire at twice the rate of *Sorghum* (C₄). The P:T of *Sorghum*
is twice that of wheat in this instance because of transpiration
differences. The stomata of wheat open fully at low light but continue to open with increasing light intensity in *Sorghum* (19,20).

Photorespiration

C₄ species lack apparent photorespiration as judged by a low compensation point (1,21), lack of CO_2 production into CO_2 free air
(13), lack of response of photosynthesis and growth to oxygen concentration (22,23,24), and lack of a postillumination CO_2 burst (25).
The validity of these criteria as indicators of the absence of photorespiration in C₄ species will be assessed later. C_3 plants photore-
spire considerable amounts of CO_2 based on the above criteria.

TAXONOMIC DISTRIBUTION OF C₄ PHOTOSYNTHESIS

The occurrence of the C₄ pathway is taxonomically diverse. It is
present in the panicoid, chloridoid-eragrostoid, and aristidoid
lines of the Gramineae (1,26,27,28), among Cyperaceae (29,30), Amaranthaceae, Chenopodiaceae, and Portulacaceae (1,21,29,30,31), Euphorbiaceae (32), Nyctaginaceae (33), Aizoaceae (33), Compositae
(34), and Zygophyllaceae (unpublished). The genera *Atriplex*, *Kochia*,
Bassia, *Cyperus*, *Panicum*, and *Euphorbia* contain both C_3 and C₄ species (1,21,26,27,29,30,32,35). The genus *Euphorbia* also has succulent members with Crassulacean acid metabolism (CAM) (36).

The physiological characteristics associated with the C₄ pathway
(discussed above) can be used to detect C₄ species in plant surveys
(1,30). The most rapid and unambiguous means of identification is by
examination of leaf anatomy (described below). Determination of ^{13}C/
[12]C ratio is useful especially where non-viable seed or poorly preserved herbarium material is all that is available (30). The method,
however, does not allow separation of C₄ plants from plants with
Crassulacean acid metabolism (CAM) since both types of plants show
similar ^{13}C enrichment (37) as a result of CO_2 being fixed via phosphoenolpyruvate (PEP) carboxylase.

LEAF ANATOMY

All plants with the C_4 pathway have a "Kranz" type of leaf anatomy (38). The vascular bundles are surrounded by two concentric
chlorophyllous layers, an inner parenchyma bundle sheath layer and
an outer mesophyll layer. The bundle sheath layer contains specialized plastids. Some species of *Aristida* have a double bundle sheath
(39,40). A thick cell wall, which may contain a suberin layer, sepa-
rates the mesophyll from the bundle sheath (41). *Plates 1* and *2* illustrate the "Kranz" type leaf in *Chloris gayana*, a C_4 monocotyledon,
and in *Gomphrena globosa*, a C_4 dicotyledon. Many C_3 leaves have a
dorsi-ventral structure due to the presence of palisade and spongy
mesophyll layers.

Grass evolution has been viewed by some workers as having started
with a festucoid (C_3) leaf type that lacked a parenchyma bundle
sheath and had only one chloroplast type (40,42). The next advance
involved the occurrence of a thin-walled bundle sheath containing
few to no unspecialized chloroplasts and a reduction in mesophyll
airspace. This advance can be seen in the extant bambusoid and arun-
dinoid lines (C_3). Finally, the originally weakly developed bundle
sheath gave rise to thick-walled cells and the mesophyll became more
compact and radially arranged. This is evident in present day panicoid and chloridoid-eragrostoid (C_4) species. Their bundle sheath
cells house large specialized chloroplasts active in starch accumulation (1). The granal content of these chloroplasts varies with
species (30,39,42,43,44,45). Mesophyll chloroplasts are granal and
generally low in starch (1,46).

CONSEQUENCES OF "KRANZ" ARCHITECTURE

The anatomy of C_4 plants is a constant character and confers special properties to the species possessing it. It serves to compart-
mentalize biochemical events. C_4 dicarboxylic acid synthesis occurs
in the mesophyll layer (47,48,49). The acids are transported to the
bundle sheath where they are decarboxylated to provide CO_2 to the
Calvin cycle located there (46,49,50). The tight packing of the mesophyll about the bundle sheath should not allow CO_2 to pass directly from the atmosphere to the sheath. Thus the bundle sheath is
likely wholly dependent upon the mesophyll for a source of CO_2 (49,
51). The very high PEP carboxylase activity in the mesophyll layer
(47,48,49) is an effective device for trapping large quantities of
CO_2 to ensure an adequate supply of CO_2 to the Calvin cycle. It also

---

*Plate 1 The "Kranz" type leaf showing the mesophyll (M) and bundle
sheath (S) layers in Chloris gayana, a C₄ monocotyledon. The bundle
sheath chloroplasts are arranged in centrifugal position.*

*Plate 2 The two cell layers in the C₄ dicotyledon, Gomphrena globosa
showing centripetal arrangement of bundle sheath chloroplasts.*

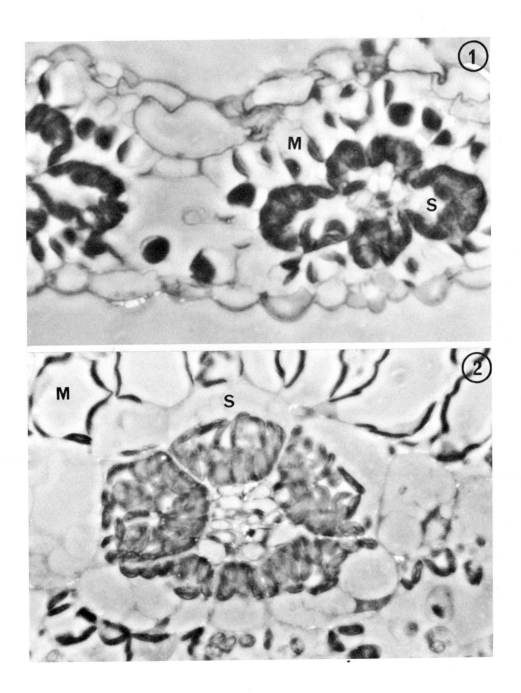

serves to recapture respiratory CO_2.

The arrangement of chlorophyllous tissue in "Kranz" leaves results in a very close association of the photosynthetic apparatus to leaf water columns (1). This may have survival value for the plant under drought conditions. The large air-liquid interfaces of the dorsi-ventral C_3 leaf would seem to place such leaves at a relative disadvantage to C_4 leaves when water is limiting. Of course many other factors such as cuticular and stomatal resistances to water vapor exchange must also be considered. The "Kranz" anatomy is associated with rapid translocation of photosynthate from synthetic sites (52). Starch accumulates in the bundle sheath when the rate of photosynthesis in the leaf exceeds rate of translocation to the phloem (1,53). Changed source-sink relations may result in extensive starch accumulation in the mesophyll (53,54). A close correlation exists between photosynthesis rate and degree of assimilate translocation among higher plants (52). If photosynthesis is subject to end-product inhibition, then the higher photosynthesis rate in C_4 leaves may result from the plant's anatomy which facilitates rapid removal of photosynthate.

EFFECTS OF PHOTOCHEMICAL AND BIOCHEMICAL INTERACTIONS

The nature of initial photosynthetic products

Most of the radioactivity fixed by C_4 plants during short periods in $^{14}CO_2$ appears in malic and aspartic acids. Repeated $^{14}CO_2$ feedings to plants revealed that each particular species consistently synthesized more of one C_4 acid than the other. The particular C_4 acid synthesized in the greatest amount correlated with levels of malic enzyme and other plant characters (50). As a consequence of this, the C_4 species studied were designated either "malate formers" or "aspartate formers" according to the major C_4 acid that the plant preferentially synthesized. ANDREWS *et al.* (55) have recently provided an enzymatic explanation for preferential C_4 acid synthesis. The designations of "malate former" and "aspartate former" should not be used in an absolute sense since it is expected that all C_4 plants will transport and decarboxylate both C_4 acids to some degree. Studies on levels of malic enzyme in various C_4 plants, however, suggest that many C_4 species utilize one particular C_4 acid almost to the exclusion of the other (50,55). For example, the low levels of malic enzyme in the "aspartate former", *Atriplex spongiosa*, (0.09 μmoles min^{-1} mg^{-1} chlorophyll) should largely preclude carboxyl transfer from malate (50). Many "malate formers", on the other hand, have malic enzyme activity well in excess of their photosynthetic rate (3-5 μmoles min^{-1} mg^{-1} chlorophyll) and would

---

Plate 3 *The dimorphic chloroplasts of Sorghum sudanense, a "malate former". Mesophyll chloroplasts (M) contain grana and have normal light reactions. Bundle sheath chloroplasts (S) lack grana (are a-granal) and are deficient in photosystem II. From (39) with permission.*

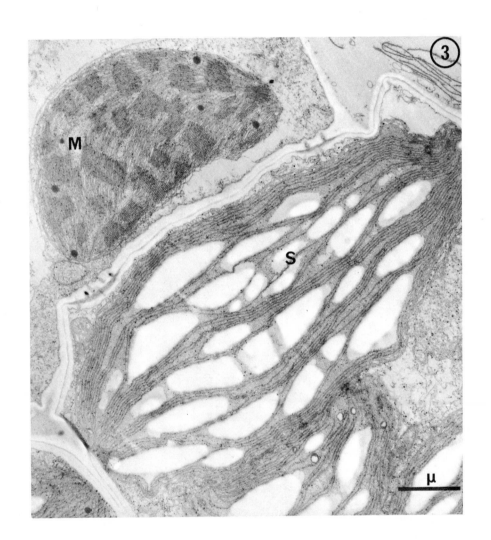

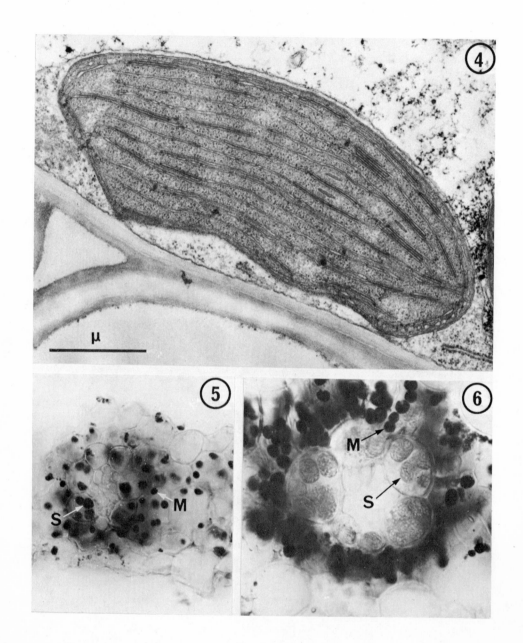

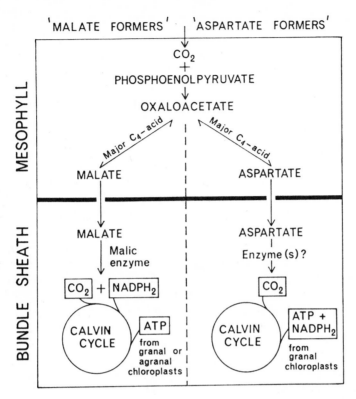

Figure 1 Reaction involved in the two modes of carboxyl transfer postulated in C_4 plants. No attempt has been made to indicate reactions necessary to maintain a carbon and nitrogen balance between the two cell layers. From (50) with permission.

seem to channel most of their fixed carbon to the Calvin cycle via malic acid (50).

Figure 1 illustrates the two modes of carboxyl transfer believed to operate in C_4 plants. "Aspartate formers" have granal mesophyll and bundle sheath chloroplasts (50). Aspartate transport serves to carry CO_2 to the bundle sheath. The mechanism of its release is

Plate 4 Immature bundle sheath chloroplast of Sorghum bicolor containing grana. Grana are lost during maturation. From (57) with permission.

Plate 5 Cross-section of immature (20% of maximum chlorophyll content) Sorghum bicolor leaf. Treatment with TNBT shows that both the granal mesophyll (M) and bundle sheath chloroplasts (S) reduce the dye to opaque diformazon. This indicates a functional photosystem II. From (57) with permission.

Plate 6 Cross-section of a mature Sorghum bicolor leaf. Only the granal mesophyll chloroplasts (M) reduce TNBT. Bundle sheath chloroplasts (S), which are agranal (as seen in Plate 3), do not reduce the dye, and lack photosystem II. From (57) with permission.

still not clear. A more complex situation is found for some of the
"malate formers". Malate transport and decarboxylation by malic enzyme delivers both CO_2 and reducing potential (NADPH) to the bundle
sheath (48,49,50). The bundle sheath chloroplasts of *some* "malate
formers" possess small grana or lack them entirely (50). *Plate 3*
shows a granal mesophyll chloroplast and an agranal bundle sheath
chloroplast in the "malate former", *Sorghum sudanense*. The agranal
chloroplast is deficient in photosystem II and cannot generate its
own NADPH by non-cyclic electron flow (46, 56). Malate transport
carries reducing potential to the sheath but only one half of the
amount required if it is assumed that each CO_2 produced by the activity of malic enzyme is refixed by the Calvin cycle (46). There is
evidence that some of the 3-phosphoglyceric acid (PGA) produced in
the bundle sheath during Calvin cycle photosynthesis may move to the
mesophyll for reduction (46,48). The loss of grana in bundle sheath
chloroplasts is a specialized feature which demands increased co-operation between the two cell layers. Bundle sheath chloroplasts in
very young tissue of certain C_4 plants such as *Sorghum bicolor* contain grana (57) and reduce tetranitro blue tetrazolium chloride
(TNBT) (*Plates 4,5*). TNBT is reduced by photosynthetic non-cyclic
electron flow which involves the co-operation of photosystem II. During bundle sheath chloroplast ontogeny, grana are lost as is the ability to reduce TNBT. *Plate 6* shows that only the granal mesophyll
chloroplasts of the mature *Sorghum* leaf reduce TNBT. The ultrastructure of this leaf is similar to that shown in *Plate 3*. Photosystem
II is lost in the mature agranal bundle sheath chloroplast, but photosystem I remains (57). These chloroplasts are likely cyclic ATP
generators. Selection for malate transport would seem a prerequisite
to loss of grana. This partially compensates for the loss of photosystem II and reducing potential (50).

The trend towards grana reduction in "malate formers" has occured within some of the more highly evolved grasses such as the Paniceae, Andropogoneae, and Maydeae (42,43,50,58) as well as certain C_4
dicotyledons such as *Gomphrena*, *Froelichia*, and *Euphorbia* (50,59,60,
61). In fact some degree of grana reduction is evident in the bundle
sheath chloroplasts of *all* "malate formers" examined to date. This
reduction can range from a case where grana consist largely of two
appressed thylakoids (the area of appression being extensive and the
frequency, very high) as is seen in *Gomphrena globosa* (59) to the
essentially agranal situation encountered in *Sorghum* and sugarcane
(39,43,59). Accompanying this decreased degree of grana formation is
a decreased emphasis on aspartate transport and an increased emphasis on malate transport and decarboxylation to provide compensatory
reducing power (50,55).

Plants with CAM can be thought of as "malate formers". The mechanism of CO_2 release from malic acid to the Calvin cycle (deacidification) is believed to involve malic enzyme in both cases. C_4 "malate formers" fix CO_2 (acidify) and decarboxylate (deacidify) simultaneously in different cell layers (49). CAM plants acidify and dea-

cidify at different times of the day and compartmentation is restricted to one cell type (62).
On the glycolate pathway and photorespiration
    Glycolate transport from the chloroplast is believed to be impor-
tant in ridding the chloroplast of excess reducing power (63). A glycolate-glyoxylate shuttle involving glycolate oxidase and NADP glyoxylate reductase apparently operates between chloroplast and peroxisome to transport hydrogen (63,64). Bundle sheath cells with agranal
chloroplasts would not require this mechanism since the production
of excess reducing potential is unlikely. Agranal bundle sheath
cells have lower activities of glycolate oxidase than granal sheath
cells (*Table 1*). Nonetheless, the glycolate oxidase levels detected
in the bundle sheaths of *Sorghum sudanense* and *Zea mays* were significant. NADP glyoxylate reductase is likely absent from agranal
sheath chloroplasts since other NADPH requiring enzymes such as NADP

TABLE I

Glycolic acid oxidase activity in extracts of leaves and of
bundle sheath cells of C_4 species (nmoles glycolate consumed
min^{-1} mg^{-1} chlorophyll). Assayed with O_2 electrode; bundle
sheath cells prepared according to (56).

Species	Leaf	Bundle sheath	Sheath chloroplasts
Atriplex hastata(C_3)	1776	–	–
Atriplex spongiosa	502	618	granal
Atriplex nummularia	500	682	granal
Sorghum sudanense	24	286	agranal
Zea mays	82	220	few, small grana
Sorghum bicolor	below level of detection		agranal

malate dehydrogenase and nitrite reductase are lacking in the bundle
sheaths of *Sorghum sudanense* and *Zea mays* (48,61,65). The glycolate
level in agranal chloroplasts may be very low if photosystem II activity is necessary for glycolate biogenesis (66,67). The mesophyll
chloroplasts of "malate formers" which are granal can dispense of excess reducing potential by malate formation and export. One might expect high levels of glycolate pathway enzymes in "aspartate formers"
especially in the granal bundle sheath cells which are probably the
most similar to C_3 plant cells (50).
    C_4 plants will give a low compensation point or lack CO_2-free air
whether or not they have high photorespiratory (glycolate pathway)
activity since mesophyll PEP carboxylase traps any photorespiratory
CO_2 produced (49). Another criterion used as evidence for the lack
of photorespiration in C_4 plants is the negligible response of photosynthesis and growth to oxygen concentration (22,23,24). High CO_2 is
known to eliminate the oxygen effect and suppress glycolate synthe-

sis in C_3 plants (68,69,70). The transport of large quantities of C_4 acids to the bundle sheath and their decarboxylation could raise the bundle sheath CO_2 concentration to a level which eliminates the oxygen effect. The oxygen sensitive postillumination CO_2 burst is another indicator of photorespiration in C_3 plants. Postillumination bursts occur in C_4 "aspartate formers" but seem to be associated with the carboxyl transfer step from C_4 acids to the Calvin cycle (50). This conclusion is based on the lack of oxygen response of the postillumination burst and the measured efflux of $^{14}CO_2$ from initial photosynthetic products (50).

Since the mesophyll PEP carboxylase system precludes the measurement of photorespiratory activity by CO_2 exchange under steady state conditions of photosynthesis, photorespiration must be investigated using other methods such as oxygen analysis. Mass spectrometry has indicated oxygen sensitive oxygen uptake by maize leaves during photosynthesis (71).

TOTIPOTENCY AND THE NEED FOR COMPARTMENTATION

All higher plants have the Calvin cycle as a common denominator of photosynthesis. Isolated parenchyma stalk tissue of sugarcane fixes CO_2 solely by the Calvin cycle (72). This tissue produces callus which has a morphologically homogeneous population of proplastids (58). These callus cultures can develop shoots with "Kranz" anatomy (58,72). Since one cell type with the Calvin cycle can generate mesophyll chloroplasts with the C_4 pathway, it suggests that both mesophyll and parenchyma sheath cells have the same genetic information and potential to form similar enzyme complements. The compartmentation of biochemical sequences in different cell layers of the "Kranz" type leaf must be a result of having some genes switched on in one layer and not in the other. Thus we find PEP carboxylase in the mesophyll and carboxydismutase in the bundle sheath (47,48,49). The same genes may be switched on to the same or varying degrees in both cell layers out of metabolic necessity; e.g. NADP glyceraldehyde-3-phosphate dehydrogenase is present in both the mesophyll and bundle sheath cells of *Sorghum sudanense* and *Zea mays*, species which are deficient in bundle sheath reduction capacity and probably transport some PGA to the mesophyll for reduction (46,48,61). Some of the glycolate pathway enzymes are present in both cell layers (73). Pyruvate,Pi dikinase(74) is therefore not likely to be an enzyme unique to mesophyll chloroplasts in C_4 plants. If a Calvin cycle type chloroplast has the inherent potential to generate a C_4 mesophyll type chloroplast then it is not surprising that the C_4 pathway often lacks taxonomic continuity. It may be a highly mutable condition susceptible to being triggered by a particular combination of environmental factors. With the apparent exception of pyruvate,Pi dikinase (74), C_4 mesophyll enzymes are also present in C_3 plants. This supports the notion that mesophyll and bundle sheath chloroplasts of C_4 plants differ only quantitatively in their enzymology and helps to explain the observation that C_3 plants often form C_4 acids under various conditions. Obviously both carboxylation reactions cannot be fully operative simultaneously in the same compartment of one cell if C_4 acids are to provide CO_2 to the Calvin cycle by decarboxylation reactions. The decarboxylated CO_2 would simply be refixed again into C_4 acids by the more efficient CO_2 trapping enzyme, PEP carboxylase. The success of plants in fully utilizing both carboxylation reactions can be attributed to adequate compartmentation. C_4 plants exploit "Kranz" anatomy; CAM plants (62) prevent short circuiting by complicated arrangements involving vacuolar storage of C_4 acids and control of PEP carboxylase activity in time by complex feedback mechanisms.

ACKNOWLEDGMENTS

Dr. T. Bisalputra and Mr. N.A. Pyliotis kindly provided the electron micrographs.

REFERENCES

1. W.J.S. DOWNTON and E.B. TREGUNNA, *Can. J. Bot.*, 46 (1968) 206.
2. M.A. EL-SHARKAWY, R.S. LOOMIS, and W.A. WILLIAMS, *J. appl. Ecol.*, 5 (1968) 243.
3. J.P. COOPER and N.M. TAINTON, *Herb. Abst.*, 38 (1968) 167.
4. T.M. CHEN, R.H. BROWN, and C.C. BLACK, *Weed Sci.*, 18 (1970) 399.
5. J.D. HESKETH and D.N. MOSS, *Crop Sci.*, 3 (1963) 107.
6. L.T. EVANS and R.L. DUNSTONE, *Aust. J. biol. Sci.*, 23 (1970) 725.
7. J.R. McWILLIAM and A.W. NAYLOR, *Plant Physiol.*, 42 (1967) 1711.
8. S. KAWANABE, *Proc. Jap. Soc. Plant Tax.*, 2 (1968) 17.
9. G. HOFSTRA and J.D. HESKETH, *Planta*, 85 (1969) 228.
10. Y. MURATA and J. IYAMA, *Proc. Crop Sci. Soc. Jap.*, 31 (1963) 315.
11. Y. MURATA, J. IYAMA, and T. HONMA, *Proc. Crop Sci. Soc. Jap.*, 34 (1965) 154.
12. R.W. DOWNES, *Proc. XI int. Grassland Cong.*, 11 (1970) 527.
13. M.A. EL-SHARKAWY, R.S. LOOMIS, and W.A. WILLIAMS, *Physiologia Pl.* 20 (1967) 171.
14. M. EL-SHARKAWY and J. HESKETH, *Crop Sci.*, 5 (1965) 517.
15. T.A. BULL, *Crop Sci.*, 9 (1969) 726.
16. J.D. HESKETH, *Crop Sci.*, 3 (1963) 493.
17. S.J. McNAUGHTON and Louise W. FULLERN, *Plant Physiol.*, 45 (1970) 703.
18. H.L. SHANTZ and L.N. PIEMEISEL, *J. agric. Res.*, 34 (1927) 1093.
19. R.W. DOWNES, *Planta*, 88 (1969) 261.
20. R.W. DOWNES, *Aust. J. biol. Sci.*, 23 (1970) 775.
21. E.B. TREGUNNA and J. DOWNTON, *Can. J. Bot.*, 45 (1967) 2385.
22. R.W. DOWNES and J.D. HESKETH, *Planta*, 78 (1968) 79.
23. J.D. HESKETH, *Planta*, 76 (1967) 371.
24. O. BJÖRKMAN, W.M. HIESEY, M. NOBS, F. NICHOLSON, and R.W. HART, *Carnegie Inst. Wash. Year Book*, 66 (1968) 228.

25. E.B. TREGUNNA, G. KROTKOV, and C.D. NELSON, *Can. J. Bot.*, 42 (1964) 989.
26. W.J.S. DOWNTON, Ph. D. Thesis (Univ. Brit. Col.) 1969.
27. J. DOWNTON, J. BERRY, and E.B. TREGUNNA, *Science*, 163 (1969) 78.
28. M.D. HATCH, C.R. SLACK, and H.S. JOHNSON, *Biochem. J.*, 102 (1967) 417.
29. H.S. JOHNSON and M.D. HATCH, *Phytochemistry*, 7 (1968) 375.
30. E.B. TREGUNNA, B.N. SMITH, J.A. BERRY, and W.J.S. DOWNTON, *Can. J. Bot.*, 48 (1970) 1209.
31. C.B. OSMOND, *Biochim. biophys. Acta*, 141 (1967) 197.
32. D.N. MOSS, E.G. KRENZER, Jr., and W.A. BRUN, *Science*, 164 (1969) 187.
33. R. CROOKSTON and D.N. MOSS, *Plant Physiol.*, 46 (1970) 564.
34. E. SCHOCH, *Z. PflPhysiol.*, (1971) in press.
35. E.B. TREGUNNA, J. DOWNTON, and P. JOLLIFFE, in Progress in Photosynthesis Research, ed. H. METZNER, I.U.B.S., Tübingen, I (1969) 488.
36. E.L. NUERNBERGK, *Planta*, 56 (1961) 28.
37. M.M. BENDER, *Phytochemistry*, in press.
38. H. MOSER, *Beih. bot. Zbl.*, 52 (1934) 378.
39. T. BISALPUTRA, W.J.S. DOWNTON, and E.B. TREGUNNA, *Can. J. Bot.*, 47 (1969) 15.
40. W.V. BROWN, *Bot. Gaz.*, 119 (1958) 170.
41. T.P. O'BRIEN and D.J. CARR, *Aust. J. biol. Sci.*, 23 (1970) 275.
42. Sr. M.C. JOHNSON, Ph. D. Thesis (Univ. Texas, Austin) (1964).
43. W.M. LAETSCH and A.J. VLITOS, *Z. PflPhysiol.*, 54 (1965) 472.
44. W.M. LAETSCH, *Am. J. Bot.*, 55 (1968) 875.
45. W.J.S. DOWNTON, T. BISALPUTRA, and E.B. TREGUNNA, *Can. J. Bot.*, 47 (1969) 915.
46. W.J.S. DOWNTON, J.A. BERRY, and E.B. TREGUNNA, *Z. PflPhysiol.*, 63 (1970) 194.
47. O. BJÖRKMAN and E. GAUHL, *Planta*, 88 (1969) 197.
48. C.R. SLACK, M.D. HATCH, and D.J. GOODCHILD, *Biochem. J.*, 114 (1969) 489.
49. J.A. BERRY, W.J.S. DOWNTON, and E.B. TREGUNNA, *Can. J. Bot.*, 48 (1970) 777.
50. W.J.S. DOWNTON, *Can. J. Bot.*, 48 (1970) 1795.
51. H.S. JOHNSON and M.D. HATCH, *Biochem. J.*, 114 (1969) 127.
52. G. HOFSTRA and C.D. NELSON, *Planta*, 88 (1969) 103.
53. M.M. RHOADES and A. CARVALHO, *Bull. Torrey bot. Club*, 71 (1944) 335.
54. J.H. HILLIARD and S.H. WEST, *Science*, 168 (1970) 494.
55. T.J. ANDREWS, H.S. JOHNSON, C.R. SLACK, and M.D. HATCH, *Phytochemistry*, in press.
56. K.C. WOO, Jan M. ANDERSON, N.K. BOARDMAN, W.J.S. DOWNTON, C.B. OSMOND, and S.W. THORNE, *Proc. natn. Acad. Sci. U.S.A.*, 67 (1970) 18.
57. W.J.S. DOWNTON and N.A. PYLIOTIS, *Can. J. Bot.*, in press.
58. W.M. LAETSCH, *Sci. Prog. Oxf.*, 57 (1969) 323.

59. K.C. WOO, N.A. PYLIOTIS, and W.J.S. DOWNTON, Z. *PflPhysiol.*, (1971) in press.
60. W.M. LAETSCH, *Plant Physiol.*, 45 (1970) S 22.
61. J.A. BERRY, *Carnegie Inst. Wash. Year Book*, 69, in press.
62. S.L. RANSON, in Plant Biochemistry, by J. BONNER and J.E. VARNER, Academic Press, New York, (1965) 493.
63. I. ZELITCH and A.M. GROTTO, *Biochem. J.*, 84 (1962) 541.
64. N.E. TOLBERT and R.K. YAMAZAKI, *Ann. N.Y. Acad. Sci.*, 168 (1969) 325.
65. G.E. MELLOR and E.B. TREGUNNA, *Can. J. Bot.*, in press.
66. Z. PLAUT and M. GIBBS, *Plant Physiol.*, 45 (1970) 470.
67. J. COOMBS and C.P. WHITTINGHAM, *Proc. R. Soc. (B.)*, 164 (1966) 511.
68. P.A. JOLLIFFE and E.B. TREGUNNA, *Plant Physiol.*, 43 (1968) 902.
69. W.A. JACKSON and R.J. VOLK, *A. Rev. Pl. Physiol.*, 21 (1970) 385.
70. I. ZELITCH, *J. biol. Chem.*, 240 (1965) 1869.
71. W.A. JACKSON and R.J. VOLK, *Nature*, 222 (1969) 269.
72. H.P. KORTSCHAK and L.G. NICKELL, *Plant Physiol.*, 45 (1970) 515.
73. D.W. REHFELD, D.D. RANDALL, and N.E. TOLBERT, *Can. J. Bot.*, 48 (1970) 1219.
74. M.D. HATCH and C.R. SLACK, *Biochem. J.*, 106 (1968) 141.

COMPARATIVE PHOTOSYNTHETIC CO_2 EXCHANGE IN HIGHER PLANTS[1]

Olle Björkman

Carnegie Institution of Washington,
Department of Plant Biology, Stanford, California, 94305, U.S.A.

SUMMARY

Studies of the influence of different environmental factors on the inhibitory effect of oxygen on the rate of net CO_2 uptake in C_3 species, and of comparative CO_2 exchange characteristics of C_3 and C_4 species, are reviewed.

C_4 species exhibit considerably higher photosynthetic rates than closely related C_3 species under conditions where the inhibitory effect of oxygen on net CO_2 uptake has a pronounced influence in C_3 species. The difference between the species is greatest at high temperatures and low CO_2 concentrations, i.e. where the inhibition caused by 21% O_2 is maximum. The inhibition by 21% O_2 is absent in C_4 species over a wide range of light intensities, temperatures, and CO_2 concentrations.

Most of the differences in photosynthetic gas exchange characteristics between C_3 and C_4 species disappear when the O_2 content is reduced to 1 or 2%. Thus the differences may largely be attributable to the presence of O_2 inhibition in C_3 but not in C_4 plants. There are no consistent differences in the rate of CO_2 uptake between C_3 and C_4 species at rate-saturating CO_2 concentrations, either under 21% or 1% O_2. However, C_4 plants have a greater intrinsic efficiency of utilizing CO_2 at low concentrations even under 1% O_2.

It is suggested that C_4 plants may have overcome the inhibitory effect of oxygen because the C_4 pathway, together with the specialized leaf anatomy associated with it, may act as an efficient CO_2 concentrating mechanism that would allow CO_2 fixation by the Calvin cycle of C_4 species to take place at an effectively higher CO_2 concentration than in C_3 species.

There is strong evidence that the higher water use efficiency of C_4 species is attributable to the high intrinsic efficiency of utilization of low CO_2 concentrations and absence of O_2 inhibition in

[1] *C.I.W.-D.P.B. Publication No. 474*

18

such species.

Thus, the available experimental evidence strongly suggest that the C₄ pathway represents an adaption to habitats with high temperatures, high irradiance, and limited water supply during the period of active growth.

During the period from the late nineteen-thirties until about five or six years ago, few laboratories were concerned with photosynthetic differentiation and adaption in higher plants. The general view among crop physiologists was that differences in photosynthetic response among higher plants were caused almost exclusively by differences in physical properties of the leaves. It became increasingly evident to those of us who were working with comparative studies of photosynthesis from widely contrasting habitats that many of the differences observed between such plants could not be adequately explained by differences in gas diffusion characteristics alone but that differentiation in the capacity of biochemical steps were also involved. There is now evidence that photosynthetic differentiation, whether genetically determined or environmentally induced, is often accomplished by differences in the relative capacities of component steps of photosynthesis.

The discovery of the C₄ pathway and the "rediscovery" of the Warburg oxygen effect and photorespiration by workers concerned with higher plant photosynthesis resulted in a renewed interest in research on photosynthetic differentiation and adaptation. This research is being carried out at different levels of organization and involves the integration of several widely different branches of plant biology. It is particularly encouraging to those of us whose primary interest lies in ecological aspects of photosynthesis that experts on the photochemical and biochemical mechanism of photosynthesis have also become interested in this field. A workshop with investigators representing such different specialities in the study of photosynthesis as those gathered in this room today, would have been an unlikely event only a few years ago.

Since the later sections deal with measurements of CO_2 exchange that were specifically designed to elucidate the biochemical mechanisms of photorespiration and C₄ photosynthesis, I will concentrate mostly on reviewing work concerning the influence upon CO_2 exchange by intact leaves of variations in light, temperature, and the gaseous composition of the atmosphere surrounding the leaf. I will also discuss the significance of these responses to photosynthetic adaptation of plants to diverse environments.

OXYGEN INHIBITION OF APPARENT PHOTOSYNTHESIS AND PHOTORESPIRATION

I would first like to review the responses of CO_2 uptake to oxygen concentration by intact leaves of C₃ species under a range of conditions. These data have been obtained by workers in a number of

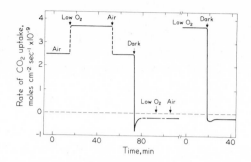

Figure 1 Time course of the rate of CO_2 exchange in response to oxygen concentration. Mimulus cardinalis, 665 nm, 100 n-einstein cm^{-2} sec^{-1}; 22°, 0.03% CO_2. After (1).

different laboratories, particularly those of Drs. KROTKOV, NELSON, and TREGUNNA in Canada, Drs. EGLE and FOCK in Germany, and in my own laboratory.

As is now well known, the rate of net CO_2 uptake of a leaf, photosynthesizing at steady-state in air of normal CO_2 and O_2 content, is markedly enhanced when the O_2 concentration is suddenly reduced (*Figure 1*). This enhancement is not temporary, but persists at least as long as anyone has followed it which probably is about a day or so. When the O_2 concentration is returned to that of normal air the rate falls and within a few minutes it returns to the previous level in air.

When the leaf is darkened a fast transient burst of CO_2 evolution is observed. This burst was first reported by DECKER, (2,3) and subsequently studies in detail by TREGUNNA *et al.*, (4,5). These workers interpreted this postillumination burst as a temporary continuation

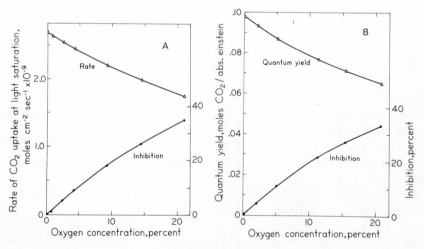

Figure 2 Rate of CO_2 uptake at light saturation (A), and slope of the linear part of the light dependence curve for CO_2 uptake (quantum yield) at 654 nm (B), as functions of O_2 concentration. Plantago lanceolata, 22°, 0.03% CO_2. After (1).

of a CO_2 production process, or photorespiration, occuring during the preceding period in the light. FOCK et al., [7] and subsequently other workers [1,6,8] found that this postillumination burst was greatly reduced if the O_2 concentration was kept low, thus suggesting that photorespiration decreases with decreasing O_2 concentration and that the inhibitory effect of O_2 on net CO_2 uptake is caused, at least in part, by photorespiration.

As illustrated in *Figure 2* my own studies [1] made it clear that the inhibitory effect of O_2 on the steady-state rate of CO_2 uptake in higher plants is present at low intensities at which the rate is a linear function of light intensity [1,9] as well as at high and moderate light intensities. In other words, the apparent quantum yield and the light-saturated rate of CO_2 uptake at normal tempera-

TABLE I

Photosynthetic CO_2 uptake (μ mole cm^{-2} sec^{-1}) in 0.2 and 21% O_2 at different wavelengths in *Solidago virgaurea*. Measurements were made at 22° and 0.03% CO_2. Light intensity at 665 nm was 6×10^3 erg cm^{-2} sec^{-1}. At other wavelengths the intensity was adjusted so that the rate of CO_2 uptake in 0.2% O_2 was approximately the same as that at 665 nm. The rates were linearly related to light intensity. After BJÖRKMAN [10].

Wavelength, nm	Rate of CO_2 uptake		Inhibition, (%) at high O_2
	0.2% O_2	22% O_2	
440	0.39	0.24	38
493	0.38	0.26	32
535	0.36	0.25	32
554	0.37	0.25	32
575	0.36	0.24	33
605	0.36	0.24	33
633	0.37	0.24	33
665	0.38	0.25	33

tures and CO_2 concentration exhibit a similar relative inhibition by O_2. As shown in *Table 1*, the inhibition of the apparent quantum yield by O_2 was found to be unaffected by the wavelength of the light and this was also found to be true for the photosynthetic rate at light intensities that were partially rate-saturating [10]. The work of BULLEY et al., [11] confirmed that the action spectrum of apparent CO_2 uptake was indeed the same in 21% and in 2% O_2. Consequently, if the O_2 inhibition were caused by a CO_2 evolving process in the light, then the action spectrum for this process must be the same as that for photosynthesis. Thus, the apparent relationship is probably a consequence of the dependency of this process on photosynthesis, rather than a direct stimulation by light.

Figure 3 illustrates the response of light-saturated CO_2 uptake

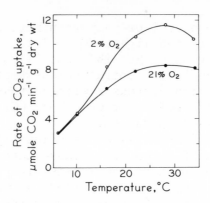

Figure 3 Rate of CO_2 uptake at light saturation as a function of temperature at 21% and 2% O_2. Marchantia polymorpha, 0.032% CO_2. After (13).

to temperature under 21% and 2% O_2. At low temperatures the O_2 inhibition is small and increases strongly with increasing temperature at least up to the point at which the optimum temperature is reached (12,13,14,15). Similarly, for a given temperature, the O_2 inhibition of light-saturated photosynthesis decreases when the CO_2 concentration is increased beyond the point where CO_2 uptake is no longer linearly dependent on CO_2 concentration (12,16,17), and may disappear entirely at sufficiently high CO_2 concentrations (17). In other words, a considerably higher CO_2 concentration is required for CO_2 saturation of photosynthesis at high than at low O_2 concentrations. It is likely that the absence of an inhibitory effect of 21% O_2 at low temperature even at normal CO_2 concentration is in part due to this level of CO_2 being rate-saturating since the CO_2 concentration required for saturation of photosynthesis decreases with decreasing temperature.

When the CO_2 concentration of the air external to the leaf is decreased far below that of normal air, a point is reached at which there is no net gas exchange, i.e. the CO_2 compensation point. If the CO_2 concentration is lowered still further the leaf will release CO_2. At low O_2 concentration it will, however, continue to take up CO_2 until the CO_2 concentration approaches zero. As was first reported by FORRESTER *et al*. (18), the CO_2 compensation point increases linearly with O_2 concentration. This has later been confirmed with other species (14,19).

It is evident from the results reported above that oxygen acts in effect, if not in fact, as a competitive inhibitor of net CO_2 uptake by intact leaves. This is in general agreement with the results obtained with isolated chloroplasts by GIBBS and coworkers (20,21). Their results also provide evidence that oxygen reacts with an intermediate of the carbon cycle, either directly, or via an oxidant produced in a Mehler reaction, resulting in the production of glycolate. The results of TOLBERT and coworkers indicate that the glycolate is in turn oxidized to CO_2 and serine outside the chloroplast via photorespiration (22). The serine may then be converted to 3-phosphoglycerate (3-PGA) whereby as much as three-fourths of the glycolate carbon would be returned to the Calvin cycle and one-fourth is lost as CO_2. However, since such a conversion requires both ATP and reduced pyridine nucleotide, it would reduce the availability of these compounds for CO_2 fixation at nonsaturating light intensities and thus reduce the amount of CO_2 fixed per light quantum absorbed. Of course, if a Mehler reaction is also involved this would result in a direct loss of reducing power which would lead to a further decreased quantum yield of CO_2 fixation. By this mechanism oxygen would inhibit net CO_2 uptake both by inhibition of CO_2 fixation and by CO_2 release in photorespiration.

Whatever the mechanism of the inhibitory effect of O_2 on net photosynthesis and the function of photorespiration may be, it is evident that they result in a substantial loss of carbon and energy to the plant under atmospheric CO_2 concentration, particularly at high temperature. Comparative growth experiments show that several species of higher plants and liverworts grow well under much lower than atmospheric oxygen concentrations; at low CO_2 concentrations dry matter production was substantially enhanced when the oxygen concentration was reduced from the usual 21% to a few percent (23,24).

COMPARATIVE CO_2 EXCHANGE OF C_3 AND C_4 SPECIES

At about the same time as it was discovered in several different laboratories that certain tropical grasses like corn, sugarcane, and pigweed, a dicotyledoneous species, had low CO_2 compensation points in the presence of 21% O_2 (25,26,27) and that they lacked a marked O_2 inhibition of CO_2 uptake (10,28,29), KORTSCHAK, HARTT, and BURR (30) and subsequently HATCH and SLACK (31) reported that sugarcane possessed a special pathway (C_4) for CO_2 fixation. Later work by HATCH and coworkers (32,33) showed that this pathway was present in many other grasses and dicots as well. The work of DOWNES and HESKETH (34) and DOWNTON and TREGUNNA (27,35) showed that a great number of species which possess the biochemical and leaf anatomical characteristics of C_4 species lacked an O_2 inhibition effect and had a low CO_2 compensation point. To date no single species among at least 70 species analyzed in many laboratories has deviated from this pattern.

It soon became apparent that several genera of higher plants and in particular the genus *Atriplex* contained C_4 as well as C_3 species (15, 36-42). The demonstration that hybrids can be obtained between C_4 and C_3 species of *Atriplex* (43) shows that such species can be closely related. Since such material is particularly well suited for studies of the functional significance of the C_4 pathway, I will, in the following, largely use the data on C_3 and C_4 species of *Atriplex* obtained in our own studies (15,44) as well as those of SLATYER (45) to illustrate the comparative CO_2 exchange characteristics of C_3 and C_4 species. The plants used in our comparative studies were grown

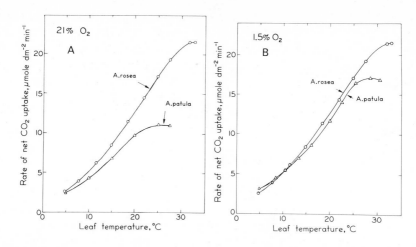

*Figure 4 Rate of CO$_2$ uptake at light saturation and 0.03% CO$_2$
as a function of temperature in Atriplex patula and A. rosea.
A: 21% O$_2$. B: 1.5% O$_2$. After (44).*

under the same controlled conditions, and the leaves used in the
comparisons were also of the same age.

As with other C$_4$ species net CO$_2$ uptake is little affected by O$_2$
concentration in *Atriplex rosea* and *A. spongiosa* whereas *A. patula*
and *A. hastata* exhibit the usual inhibition found in C$_3$ species of
higher plants (15,40,44,45). *Figure 4A* illustrates the response of
light-saturated CO$_2$ uptake to temperature in *A. rosea* and *A. patula*
in air of atmospheric CO$_2$ and O$_2$ concentration. At low temperature
the two species exhibit similar rates of CO$_2$ uptake, but as the temperature is increased *A. rosea*, the C$_4$ species, becomes increasingly
superior. At about 30° it is capable of about double the rate of *A.
patula*, the C$_3$ species.

Similarly, as shown in *Figure 5A* there is little difference in
photosynthetic rate between the species at low light intensities,

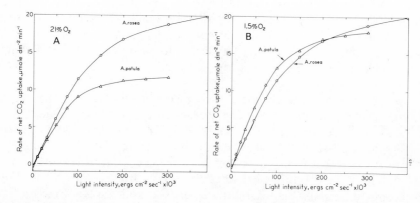

*Figure 5 Rate of CO$_2$ uptake at 27° and 0.03% CO$_2$ as a func-
tion of light intensity (400-700 nm) in Atriplex patula and A.
rosea. A: 21% O$_2$. B: 1.5% O$_2$. After (44).*

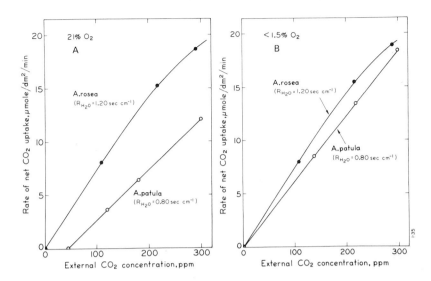

Figure 6 Rate of CO_2 uptake at light saturation and 27° as a function of external CO_2 concentration in Atriplex patula and A. rosea. A: 21% O_2. B: 1.5% O_2. After (17,44)

but as the intensity is increased light saturation occurs considerably earlier in A. patula than in A. rosea. The latter species continues to increase its photosynthetic rate up to light intensities exceeding full sunlight.

Atriplex rosea is also considerably more efficient in utilizing low CO_2 concentrations external to the leaf than A. patula (Figure 6A). In A. patula CO_2 compensation is reached at 44 ppm at 27° and increases exponentially with increasing temperature, but remains close to zero over a wide temperature range in A. rosea.

Much of the differences in CO_2 exchange between the two species disappear when the O_2 concentration is reduced to a low level. As shown in Figure 4B the temperature curves become very similar over a wide temperature range although the temperature optimum still remains somewhat higher in A. rosea than in A. patula. The light dependence curves and the light-saturated rates also become similar, but A. rosea still requires more light to reach saturation than A. patula (Figure 5B). The responses of CO_2 uptake to CO_2 external to the leaf of the two species exhibit no pronounced differences (Figure 6B).

GOLDSWORTHY (46) also reported that the CO_2 concentration required for half saturation of photosynthesis under low O_2 concentration was similar in the C_4 species sugarcane and corn and in the C_3 species, tobacco. It should be pointed out, however, that such data alone do not permit the conclusion that there are no differences in the intrinsic efficiency of the carboxylation systems between C_3 and C_4 species. The dependence of leaf photosynthesis on external CO_2 concentration most probably reflects the sum of the resistance to CO_2 diffusion and the intrinsic efficiency of the CO_2 fixing mechan-

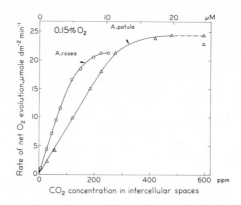

Figure 7 Rate of O_2 evolution at light saturation and 27° as a function of CO_2 concentration in the intercellular spaces in Atriplex patula and A. rosea. (After 44).

ism. Since the pathway for the diffusion of water vapor from the cell walls to the ambient air is essentially the same as that for the diffusion of CO_2 in the opposite direction the physical resistance to CO_2 diffusion can be estimated if water vapor and CO_2 exchange are measured simultaneously. In this way the CO_2 concentration in the intercellular spaces can be estimated for any given external concentration and rate of CO_2 uptake. It was found that when CO_2 uptake was expressed as a function of internal CO_2 concentration there were, indeed, significant differences between *A. rosea* and *A. patula* under low O_2 concentration (*Figure 7*).

These results are in excellent agreement with those reported by SLATYER (45) for *Atriplex spongiosa* and *A. hastata*. The efficiency of CO_2 uptake based on internal CO_2 concentration is markedly greater in the C_4 species (*A. spongiosa*) than in the C_3 species (*A. hastata*), even under low O_2 concentration. It is likely that these differences are attributable to a more efficient CO_2 fixation by β-carboxylation of phosphoenolpyruvate (PEP) than by the direct carboxylation of ribulose-1,5-diphosphate (RuDP), either because of a higher affinity of PEP carboxylase for CO_2 in comparison with RuDP carboxylase or simply because of the very high activities of the former enzyme in C_4 plants.

From the results obtained with the *Atriplex* species as well as many other species it is clear that C_4 species are capable of considerably higher maximum rates of net CO_2 uptake in normal air under conditions of high light intensities and temperatures. Such comparisons are, of course, only valid when the resistances of the stomata to CO_2 diffusion are similar. As predicted by *Figure 8A* the rate of CO_2 uptake in normal air will decrease in both types of species as the resistance to CO_2 diffusion increases, but for any given stomatal resistance the C_4 species will take up CO_2 at a faster rate than the C_3 species. The predicted difference between species would be more pronounced at resistance values (of leaf boundary layer and

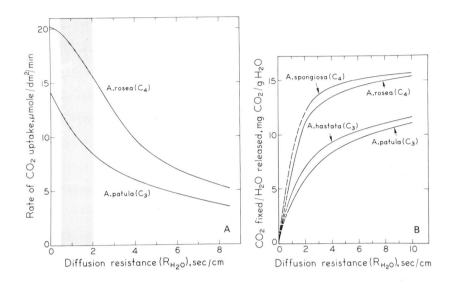

Figure 8 Calculated values for light-saturated CO_2 uptake (A), and water use efficiency of light-saturated CO_2 uptake (B), at 30°, 21% O_2, and 0.03% CO_2 as a function of leaf resistance (stomata plus leaf boundary layer) for C_3 and C_4 species of Atriplex. Shaded area indicates range of actual measurements. Experimental data from BJÖRKMAN et al. (17,44) and SLATYER (45).

stomata) usually found under natural conditions and tends to become somewhat smaller at high resistances. Conversely, for equal rates of photosynthesis the C_4 species needs to open its stomata much less than the C_3 species, i.e. it will transpire less water for a given amount of CO_2 fixed.

This is further illustrated in Figure 8B which shows the predicted ratio of the amount of CO_2 fixed to the amount of water lost. Evidently, the water use efficiency increases with increasing stomatal closure in similar manner in the C_3 and C_4 species. Thus, both types of plant can increase their water use efficiency by reducing the stomatal aperture but only at the expense of a reduced rate of photosynthesis. However, the C_4 species are capable of a higher rate of CO_2 uptake and consequently also water use efficiency than the C_3 species for any given realistic diffusion resistance value.

SLATYER (45) showed that the differences between A. spongiosa and A. hastata in photosynthetic water use efficiency corresponded to similar differences in the water use efficiency for growth. DOWNES (47) also showed that C_4 grasses have a considerably higher water use efficiency for growth than C_3 grasses when they were compared under controlled conditions (Table 2). The difference between species increased with increasing temperature. When the field data obtained by SHANTZ and PIEMEISEL early in this century (48) are grouped as C_3 and C_4 species it becomes obvious that the C_4 species,

TABLE 2

Water use efficiency (increase in dry wt (mg) per g water consumed) in C_3 and C_4 grass species at different temperatures
for growth. Values are means for three C_3 species and four C_4
species, respectively. After DOWNES (47)

Temperature	C_4 species	C_3 species	C_4/C_3
20°	4.0	2.4	1.7
25°	3.1	1.7	1.8
30°	3.6	1.1	3.3
35°	2.3	0.2	11.5

both grasses and dicots, had a higher water use efficiency than the
C_3 species (47) (*Table 3*).

It is evident from the results reviewed above that the absence of
the inhibitory effect of O_2 would be expected to have great selective advantage under conditions of high temperatures and light intensities where its presence would result in a drastic reduction of
net CO_2 uptake. The high efficiency of utilization of low CO_2 concentrations in the intercellular spaces would also result in a high
ratio of CO_2 uptake to water loss. Thus, the functional characteristics of C_4 photosynthesis would be expected to confer an advantage
in natural habitats where high light intensity, high temperature and
limited water supply prevail during much of the period of active
growth, i.e. under the conditions where C_4 species are often found
in nature.

Recent results obtained in the intense heat of the floor of Death
Valley, California, provides a striking example of how surprisingly
well adapted a higher plant can be to such extreme conditions (49).
Tidestromia oblongifolia, a C_4 perennial of the Amaranthaceae, whose
active growth takes place during the summer months, photosynthesized
at a high rate throughout the day, even though the leaf and air tem-

TABLE 3

Water use efficiency (mg dry wt g^{-1} water) in a number of C_3
and C_4 species. Data were obtained in field experiments at Ak-
ron, Colorado during the years 1911-17. After SHANTZ and PIE-
MEISEL (48), as regrouped by DOWNES (47).

Class	C_4 species	C_3 species
Dicots	3.44	1.59
Grasses	3.14	1.49

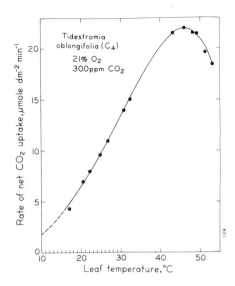

*Figure 9 Rate of CO_2 uptake at an irradiance of 1.33 cal cm^{-2}
min^{-1} (natural sunlight) as a function of leaf temperature in
Tidestromia oblongifolia. Measurements were made in the field
in Death Valley, California, on July 2, 1970 (49).*

peratures exceeded 44° most of the day and reached 50° in the early
afternoon. These temperatures proved to be close to the optimum temperature for photosynthesis which, as shown in *Figure 9*, was 46° to
47° in full sunlight. The rate of CO_2 uptake was nearly linearly related to light intensity up to full sunlight at 47°.

I would like to return briefly to the apparently similar temperature dependence of photosynthesis of C_3 and C_4 species at low oxygen
concentration and temperature where an essentially straight line is
obtained in an Arrhenius plot (logarithm of CO_2 uptake plotted a-
gainst the reciprocal of the absolute temperature of the leaf). One
point of interest is that the activation energy for photosynthesis
at low temperature is in the order of 15 to 20 kcal $mole^{-1}$ for both
the C_3 and C_4 species tested in our laboratory; this is a high activation energy for a biological reaction and suggests that the same
enzymic step may be rate-limiting in both types of species. Another
point of interest is that this high activation energy for CO_2 uptake
by intact leaves appears to be similar for CO_2 fixation catalyzed by
RuDP carboxylase *in vitro*. This may, of course, simply be a coincid-
ence, but it could mean that this enzyme is rate-limiting in both C_3
and C_4 plants. Other evidence that the carboxylation reaction is a
major rate-limiting step of light-saturated CO_2 uptake in C_3 species
in normal air gives some support to this view (9,42,50-59).

The C_4 species of *Atriplex* investigated by us had no higher CO_2
saturated rate of photosynthesis than the C_3 species. Also, EL-SHAR-
KAWAY and HESKETH (60) have shown that several C_3 species such as
sunflower and cotton can have as high CO_2 saturated rates of CO_2 uptake as C_4 species such as corn and sugarcane. The evidence that the

inhibitory effect of 21% O_2 in C_3 species is reduced as the CO_2 concentration is increased suggests that the lack of an O_2 inhibition in C_4 species may be attributable to a locally higher CO_2 concentration at the site of carboxylation of RuDP in the bundle sheath cells. Thus, the C_4 pathway would essentially function as a CO_2 concentrating mechanism which allows the carboxylation of RuDP to take place closer to CO_2 saturation than is the case in C_3 species. In such a mechanism the compartmentation that may be provided by the "Kranztypus" leaf anatomy would probably play a vital role. Recent biochemical evidence appears to be consistent with this hypothesis; I understand that Dr. HATCH will present detailed data on this matter in his review.

Clearly, more information is needed on the mechanism of both the C_4 pathway itself and of the inhibitory effect of O_2 on net CO_2 uptake, as well as on comparative performance of photosynthesis and growth in C_3 and C_4 species, before the functional and adaptive significance of the C_4 pathway can be satisfactorily assessed. Nevertheless, the information presently available from studies of CO_2 exchange in C_3 and C_4 species and the evidence that the C_4 pathway has evolved independently in a number of unrelated taxa strongly suggest that this pathway represents an adaptation that has evolved in response to selective forces operating in specific ecological habitats.

REFERENCES

1. O. BJÖRKMAN, *Physiologia Pl.*, 19 (1966) 618.
2. J.P. DECKER, *Plant Physiol.*, 30 (1955) 82.
3. J.P. DECKER, *J. Solar Energy Sci. Eng.*, 1 (1957) 30.
4. E.B. TREGUNNA, G. KROTKOV, and C.D. NELSON, *Can. J. Bot.*, 39 (1961) 1045.
5. E.B. TREGUNNA, G. KROTKOV, and C.D. NELSON, *Can. J. Bot.*, 42 (1964) 989.
6. E.B. TREGUNNA, G. KROTKOV, and C.D. NELSON, *Physiologia Pl.*, 19 (1966) 723.
7. H. FOCK, H. SCHAUB, R. ZIEGLER, and K. EGLE, *Beitr. Biol. Pflanzen*, 40 (1964) 293.
8. H. FOCK and K. EGLE, *Beitr. Biol. Pflanzen*, 42 (1966) 213.
9. O. BJÖRKMAN, *Physiologia Pl.*, 21 (1968) 84.
10. O. BJÖRKMAN, *Carnegie Inst. Wash. Yearbook*, 66 (1967) 220.
11. N.R. BULLEY, C.D. NELSON, and E.B. TREGUNNA, *Plant Physiol.*, 44 (1969) 678.
12. P.A. JOLLIFFE and E.B. TREGUNNA, *Plant Physiol.*, 43 (1968) 902.
13. O. BJÖRKMAN and E. GAUHL, *Carnegie Inst. Wash. Yearbook*, 67 (1968) 479.
14. O. BJÖRKMAN, M.A. NOBS, and W.M. HIESEY, *Carnegie Inst. Wash. Yearbook*, 68 (1970) 614.
15. O. BJÖRKMAN, E. GAUHL, and M.A. NOBS, *Carnegie Inst. Wash. Yearbook*, 68 (1970) 620.
16. H. FOCK, G. KROTKOV, and D.T. CANVIN, in Progress in Photosyn-

thesis Research, ed. H. METZNER, I.U.B.S, Tübingen, I (1969) 482.

17. O. BJÖRKMAN, unpublished data.
18. M.L. FORRESTER, G. KROTKOV, and C.D. NELSON, *Plant Physiol.*, 41 (1966) 422.
19. N.R. BULLEY, Ph. D. Thesis (Simon Frazer University, Burnaby, British Columbia) 1969.
20. P.W. ELLYARD and M. GIBBS, *Plant Physiol.*, 44 (1969) 1115.
21. M. GIBBS, *Ann. N.Y. Acad. Sci.*, 168 (1970) 356.
22. N.E. TOLBERT and R.K. YAMAZAKI, *Ann. N.Y. Acad. Sci.*, 168 (1970) 325.
23. O. BJÖRKMAN, W.M. HIESEY, M.A. NOBS, F. NICHOLSON, and R.W. HART, *Carnegie Inst. Wash. Yearbook*, 66 (1967) 228.
24. O. BJÖRKMAN, E. GAUHL, W.M. HIESEY, F. NICHOLSON, and M.A. NOBS, *Carnegie Inst. Wash. Yearbook*, 67 (1968) 477.
25. D.N. MOSS, *Nature*, 193 (1962) 537.
26. H. MEIDNER, *J. exp. Bot.*, 13 (1962) 284.
27. E.B. TREGUNNA and J. DOWNTON, *Can. J. Bot.*, 45 (1967) 2385.
28. M.L. FORRESTER, G. KROTKOV, and C.D. NELSON, *Plant Physiol.*, 41 (1966) 428.
29. J. HESKETH, *Planta*, 76 (1967) 371.
30. H.P. KORTSCHAK, C.E. HARTT, and G.O. BURR, *Plant Physiol.*, 40 (1965) 209.
31. M.D. HATCH and C.R. SLACK, *Biochem. J.*, 101 (1966) 103.
32. M.D. HATCH, C.R. SLACK, and H.S. JOHNSON, *Biochem. J.*, 102 (1967) 417.
33. H.S. JOHNSON and M.D. HATCH, *Phytochemistry*, 7 (1968) 375.
34. R.W. DOWNES and J.D. HESKETH, *Planta*, 78 (1968) 79.
35. J. DOWNTON and E.B. TREGUNNA, *Can. J. Bot.*, 46 (1968) 207.
36. J. DOWNTON, T. BISALPUTRA, and E.B. TREGUNNA, *Can. J. Bot.*, 47 (1969) 915.
37. J. DOWNTON, J. BERRY, and E.B. TREGUNNA, *Science*, 163 (1969) 78.
38. D.N. MOSS, E.G. KRENZER, Jr., and W.A. BRUN, *Science*, 164 (1969) 187.
39. C.B. OSMOND, *Biochim. biophys. Acta*, 172 (1969) 144.
40. C.B. OSMOND, J.H. TROUGHTON, and D.J. GOODCHILD, *Z. PflPhysiol.*, 61 (1969) 218.
41. O. BJÖRKMAN, *Carnegie Inst. Wash. Yearbook*, 67 (1968) 487.
42. O. BJÖRKMAN and E. GAUHL, *Planta*, 88 (1969) 197.
43. M.A. NOBS, O. BJÖRKMAN, and R.W. PEARCY, *Carnegie Inst. Wash. Yearbook*, 69 (in press).
44. O. BJÖRKMAN, R.W. PEARCY, and M.A. NOBS, *Carnegie Inst. Wash. Yearbook*, 69 (in press).
45. R.O. SLATYER, *Planta*, 93 (1970) 175.
46. A. GOLDSWORTHY, *Nature*, 217 (1968) 62.
47. R.W. DOWNES, *Planta*, 88 (1969) 261.
48. H.L. SHANTZ and L.M. PIEMEISEL, *J. Agric. Res.*, 34 (1927) 1093.
49. O. BJÖRKMAN, T. HARRISON, E. MEDINA, H. MOONEY, and R.W. PEARCY, (unpublished).

50. O. BJÖRKMAN, *Carnegie Inst. Wash. Yearbook*, 65 (1966) 454.
51. H.W. WOOLHOUSE, *Hilger J.*, 11 (1967-68) 7.
52. O. BJÖRKMAN, *Physiologia Pl.*, 21 (1968) 1.
53. P.F. WAREING, M.M. KHALIFA, and K.J. TREHARNE, *Nature*, 220 (1968) 453.
54. K.J. TREHARNE and C.F. EAGLES, in Progress in Photosynthesis Research, ed. H. METZNER, I.U.B.S., Tübingen, I (1969) 377.
55. E. GAUHL and O. BJÖRKMAN, *Planta*, 88 (1969) 187.
56. C.F. EAGLES and K.J. TREHARNE, *Photosynthetica*, 3 (1969) 29.
57. E. GAUHL, *Carnegie Inst. Wash. Yearbook*, 68 (1970) 633.
58. T.F. ANDREEVA and T.A. AVDEEVA, *Soviet Plant Physiol.*, 17 (1970) 187.
59. E. MEDINA, *Carnegie Inst. Wash. Yearbook*, 69 (in press).
60. M. EL-SHARKAWY and J. HESKETH, *Crop Sci.*, 5 (1965) 517.

SECTION 1

Environment, Adaptation and Evolution in Relation to Photosynthetic and Photorespiratory Gas Exchange

B. Research Papers and Assessment

OXYGEN EXCHANGE OF ILLUMINATED LEAVES AT CARBON DIOXIDE COMPENSATION[1]

C.L. Mulchi, R.J. Volk, and W.A. Jackson[2]

Department of Soil Science,
North Carolina State University, Raleigh, N.C. 27607, U.S.A.

SUMMARY

In the absence of net CO_2 fixation, little change occurred in the overall oxygen concentration surrounding illuminated mature soybean leaves. Nevertheless, the leaves took up and evolved considerable oxygen in nearly equimolar proportions. Both the rate of oxygen exchange and the CO_2 compensation point progressively increased as the oxygen concentration in the ambient atmosphere was raised. These increases far exceeded the stimulation of dark respiration by oxygen. At 20% oxygen, the exchange of oxygen at CO_2 compensation was enhanced by illumination, reaching a maximum at 1000-1200 ft-c. Under these conditions the rates of oxygen release and uptake were at least 400 and 350 μmoles dm^{-2} hr^{-1}, respectively.

Depleting the CO_2 concentration from compensation to essentially zero depressed oxygen release substantially but had little effect on oxygen uptake. The data suggest that a sizeable portion of the oxygen exchange at CO_2 compensation was associated with a CO_2 release and refixation process. This process appeared to be substrate limited even at 2-3% oxygen, since supplying CO_2 at subsaturating levels

---

[1] Paper No. 3334 of the Journal Series of the North Carolina State University Agricultural Experiment Station, Raleigh, N.C., U.S.A. This investigation was supported in part by Agricultural Research Service, U.S. Department of Agriculture Cooperative Agreement No. 12-12-100-9787 (34) administered by the Crops Research Division, Beltsville, Maryland. The use of trade names in this publication does not imply endorsement by the North Carolina Agricultural Experiment Station of the products named nor criticism of similar ones not mentioned.

[2] Current address: Department of Agronomy, University of Maryland, College Park, Maryland, U.S.A.

increased oxygen exchange significantly.

Experiments with pole beans have revealed that the magnitude of oxygen exchange was strongly dependent on leaf age. Rates were approximately three-fold higher in young, rapidly expanding leaves than in leaves which had ceased to enlarge.

INTRODUCTION

Increasing the ambient O_2 concentration around leaves of high compensation (C_3 photosynthesis) species commonly results in a parallel increase in Γ, the CO_2 compensation concentration (1,2,3,4). Since, at Γ, photosynthesis and photorespiration are equal, the effect of O_2 could be due to an inhibition of the former, an enhancement of the latter, or both. Current evidence in fact suggests that both processes are affected, although enhanced CO_2 exchange accompanies this increase in Γ to about 40% O_2 (5). The enhancement of photorespiration presumably results from increased glycolate synthesis and subsequent metabolism (6-12). On the other hand, oxygen may depress photosynthesis by inactivation of one or more enzymes of the carbon reduction cycle (8,13,14,15), by depletion of the pool sizes of cycle intermediates due to glycolate removal, or by diversion of electrons from the photosynthetic electron transport process to molecular oxygen (16,17).

An O_2 consuming process may occur if the ambient O_2 concentration attenuates photosynthesis or enhances photorespiration, and it must not be saturated by the O_2 evolved from photochemical sites. Moreover, at Γ, a continual release of O_2 should occur if Γ represents an equilibrium state between continual release and fixation of CO_2. This is because continual synthesis (from internally recycled CO_2) of a photorespiatory substrate (e.g. glycolate) requires a constant supply of reduced pyridine nucleotides, their turnover permitting photosynthetic electron transport to be maintained at an equivalent rate. In addition, however, O_2 release may be associated more directly with O_2 uptake as a result of diversion of electrons toward reduction of molecular O_2 as the electron carriers continue to be turned over. Such a process would result in an equimolar uptake and release of O_2.

The existence of oxygen exchange reactions by intact tissue at Γ does not establish the process whereby O_2 exerts its effect(s) on net photosynthesis. Quite sizeable rates of O_2 exchange occur at Γ with leaves of pole bean (18,19,20), but the O_2 concentrations employed were considerably below atmospheric levels. Since the magnitude of Γ is related to O_2 concentration, we considered it important to examine the pattern and intensity of O_2 exchange at higher O_2 concentrations; some of our results are reported herein. Moreover, further attention has been directed to the O_2 exchange which occurs at low O_2 concentrations where photorespiration is considered to be minimal. Soybeans (*Glycine max*, var. Dare) have been used throughout as representative of high compensation species.

EXPERIMENTAL

Plants were grown in solution culture in a controlled-environment chamber. Light at 1800 ft-c was provided for 16 hours daily and temperatures were 28 ±1 and 18 ±1°, in the light and dark periods, respectively. Leaf lengths were measured daily. Three to six days after maximal expansion, a first trifoliate leaf, or the terminal leaflet thereof, was excised under water and suspended in a Plexiglas chamber. After sealing the petiole into the potometer with a rapid-setting silicone rubber (RTV-11), the chamber cover was attached and sealed with high vacuum silicone grease.

The basic apparatus (*Figure 1*) consisted of a closed system (volume = 153 cm^3) with a diaphragm pump which provided circulation at three liters min^{-1}. The system was flushed with moist N_2 and appropriate amounts of O_2 (98 atom % ^{18}O), Ar (an internal standard) and CO_2, if desired, were introduced. Some experiments involved metering CO_2 into the system at a constant rate. For these experiments the CO_2 was generated via deposition of acid into a stirred $BaCO_3$ slurry contained in a flask which could be incorporated into, or isolated from, the system by means of a 4-way stopcock. Constant rates of acid injection into the $BaCO_3$ slurry (and hence CO_2 generation) were obtained by ejecting the acid from a variable speed syringe pump (Sage, Model 255-3). The rate of acid ejection could be very accurately adjusted, and tests without a leaf revealed constant rates of CO_2 liberation into the chamber. The source of illumination was a 750-watt incandescent lamp, the light from which was filtered through a separate 10-cm water filter and a 1-cm water jacket adjacent to the chamber. Temperature of the water circulating through the latter was regulated to obtain the desired air temperature in the chamber. Thermocouples were used to monitor leaf and air tem-

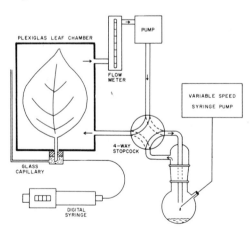

Figure 1 Closed system photosynthesis apparatus with a digital syringe potometer and a CO_2 injection flask which can be included in or excluded from the system by means of a four-way stopcock.

peratures. Cumulative water uptake was determined with a digital syringe attached to the potometer. At periodic intervals the chamber
contents were sampled and their gaseous composition determined mass
spectrometrically (21).

Much of the data presented herein simply show the changes with
time in concentration of each isotopic species of oxygen present in
the chamber. The rates so obtained are, of course, less than the actual rates of oxygen exchange. For example, the rate of disappearance of $^{18}O_2$ from the atmosphere is an underestimate of total O_2 consumption because $^{16}O_2$ is simultaneously consumed. The appearance of
$^{16}O_2$ is an underestimate of total O_2 release for the same reason. It
has been pointed out that minimal estimates of the magnitude of $^{16}O_2$
recycling can be made by assuming that each isotopic species is taken up in proportion to its concentration in the ambient atmosphere
(22,23). When $^{16}O_2$ recycling has been so estimated and its rate added to those of $^{16}O_2$ release and $^{18}O_2$ uptake, the values obtained are
called herein O_2 release and O_2 uptake. The difference between either of these estimates ($^{16}O_2$ release minus $^{18}O_2$ uptake or O_2 release
minus O_2 uptake) is called net O_2 release or net O_2 uptake, i.e. the
net change in the total oxygen content of the closed system. Rates
of O_2 release and O_2 uptake calculated from external concentrations
are also minimal (23); more correct estimates of reaction site rates
may be obtained from the ratios of isotopic oxygen species present
within the leaf (24). However, the difference between the two methods is small because of the near equilibrium existing between internal and external O_2 concentrations (24). We are aware of no procedure which provides an unequivocal measure of $^{16}O_2$ recycling within
illuminated leaves.

RESULTS

Photosynthesis at saturating CO_2 concentrations
The light response curve for O_2 release and CO_2 uptake under CO_2
saturating conditions (*Figure 2*) is presented to indicate the rates
of photosynthesis obtained at low O_2 concentration with the mature
first trifoliate leaf of soybean used in these investigations. At
each light intensity O_2 release exceeded CO_2 uptake. Attention is
directed to the rate of CO_2 uptake at 1000 ft-c ($\sim$600 μmoles dm^{-2}
hr^{-1}) and at 2500 ft-c ($\sim$1200 μmoles dm^{-2} hr^{-1}). The latter is equivalent to 52.8 mg CO_2 dm^{-2} hr^{-1}.
Oxygen uptake at CO_2 saturation vs. CO_2 compensation
Even under CO_2 saturation and low O_2 as used in the experiment
shown in *Figure 2*, a significant O_2 uptake process occurs. This has
been shown previously with leaves of pole bean by OZBUN *et al.* (18,
19,20) who also demonstrated a substantial increase in the rate of
O_2 uptake when CO_2 was depleted to Γ in a closed system. *Figure 3*
illustrates the latter phenomenon with soybeans. Depletion of CO_2 to
Γ resulted in a 2.4-fold increase in O_2 uptake. When the net CO_2 uptake rate became zero (at Γ), O_2 release diminished considerably,

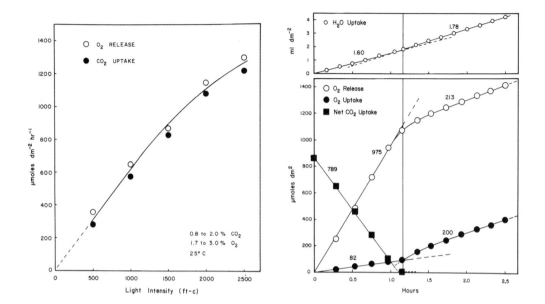

Figure 2 (left) Influence of light intensity on O_2 release and CO_2 uptake at saturating CO_2 concentrations by a soybean leaf exposed to low oxygen concentrations at 25°.

Figure 3 (right) Water uptake and oxygen exchange of a soybean leaf illuminated at 1200 ft-c in 35° environment during a period of CO_2 saturation followed by a period of CO_2 compensation at 5.4% O_2. Rates of water uptake in ml dm^{-2} hr^{-1} and gas exchange in μmoles dm^{-2} hr^{-1} are adjacent to each regression line.

but it still continued at a finite rate, somewhat in excess of the rate of O_2 uptake. An increase in the rate of water uptake by the leaf, indicating the expected increase in stomatal opening, occurred upon transition from CO_2 saturation to Γ. However, the relative increase in water uptake was small compared to that of O_2 uptake.

When a period at Γ, 4% O_2 and 1200 ft-c was followed by CO_2 saturation, the reverse sequence of events occurred (MULCHI, VOLK, and JACKSON, unpublished). With the onset of net CO_2 uptake under saturating conditions, O_2 uptake diminished from 178 to 70 μmoles dm^{-2} hr^{-1}. The concurrent decrease in water uptake (1.44 to 1.34 ml dm^{-2}) was again much smaller.

Oxygen exchange at CO_2 compensation

Rates of O_2 exchange at Γ are strongly influenced by the ambient oxygen concentration. *Figures 4, 5,* and *6* include the results of an experiment in which leaves were exposed to 2.1% O_2 at 1500 ft-c, 19.8% O_2 at 1500 ft-c, and 36.5% O_2 at 3000 ft-c. These figures depict the actual compartmental (153 cm^3) O_2 contents in μmoles and thereby illustrate the change in each isotopic species as well as the precision of measurement. In our experience, maximal rates of O_2

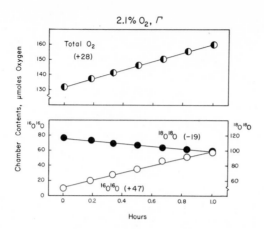

Figure 4 *Changes in concentration of* $^{16}O_2$, $^{18}O_2$ *and total oxy-gen* ($^{16}O_2$ + $^{18}O_2$) *in a closed system* (153 cm^3) *containing a soybean leaf* (0.456 dm^2) *illuminated at 1500 ft-c and exposed to 2.1% O_2 in a 30° environment at CO_2 compensation. Rate in* μ*moles* dm^{-2} hr^{-1} *are given in parentheses.*

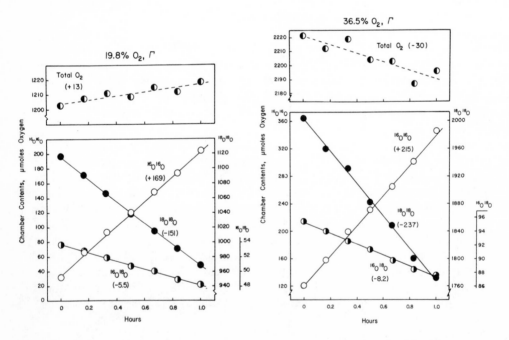

Figure 5 *(left) Changes in concentration of* $^{16}O_2$ $^{16}O^{18}O$, $^{18}O_2$, *and total* O_2 ($^{16}O_2$ + $^{16}O^{18}O$ + $^{18}O_2$) *at 19.8% O_2. Other conditions as in* Figure 4.

Figure 6 *(right) Changes in concentration of* $^{16}O_2$, $^{16}O^{18}O$, $^{18}O_2$, *and total* O_2 ($^{16}O_2$ + $^{16}O^{18}O$ + $^{18}O_2$) *at 36.5% O_2 and 3000 ft-c. Other conditions as in* Figure 4.

exchange at Γ are achieved only after periods of active CO_2 fixation. Therefore, prior to the onset of each phase, the leaf (0.456 dm^2) was exposed to saturating CO_2 ($\sim 1-2\%$) and low O_2 (2-4%) at 1000 ft-c for 0.5 hours. The chamber was then flushed and appropriate gases added to give high $^{18}O_2$ concentrations. Carbon dioxide compensation concentrations were established rapidly, remained essentially constant during each phase, and increased with increasing ambient O_2 concentration. For the phases shown in *Figures 4, 5,* and *6* the CO_2 compensation concentrations were 0.3, 0.5, and 0.9 µmoles per chamber respectively, the last value being about half the normal CO_2 concentration of air. The upper portion of each figure shows the change in total O_2 concentration ($^{16}O_2 + {}^{16}O^{18}O + {}^{18}O_2$) in the compartment; the lower portion shows the change in each oxygen isotope. Note that the ordinate increments for $^{16}O_2$ (left) and $^{18}O_2$ (right) in the lower portion of *Figures 4, 5,* and *6* are the same and that they are double those in the upper portion. For clarity, values for $^{16}O^{18}O$ have been omitted from *Figure 4*. The concentration of this isotope declined very slowly (~ 0.4 µmoles hr^{-1}) at 2.1% O_2.

The net O_2 release rates were +28, +13, and -30 µmoles hr^{-1} respectively (upper portions, *Figure 4, 5,* and *6*) showing that as the O_2 concentration increased at Γ, net O_2 declined, ultimately reaching a condition of net O_2 uptake. The rates of these net changes were rather small. At the higher O_2 concentration, they were difficult to measure with precision, but the trends are nevertheless clearly apparent.

The changes in net O_2 release with O_2 concentration reflect progressive increases in O_2 exchange. At 2.1% O_2 and 1500 ft-c there was a significant $^{18}O_2$ uptake associated with $^{16}O_2$ release (*Figure 4*). Both the uptake and release processes increased greatly as O_2 was raised to 19.8% (*Figure 5*). Further increases occurred when O_2 concentration and light intensity were increased simultaneously to 36.5% and 3000 ft-c respectively (*Figure 6*). Enhancement in O_2 uptake during illumination, resulting from increasing O_2 concentration, was considerably greater than the enhancement in darkness (*Table 1*). The light dependency of the O_2 exchange reactions at 19-20% O_2 is

TABLE I

Influence of ambient O_2 concentration (%) on O_2 uptake (µmoles $dm^{-2}hr^{-1}$) in darkness and at 1000 ft-c (at Γ) by soybean leaves in a 31° environment.

Illumination	Oxygen		
	Concentration	Release	Uptake
Darkness	2.4	—	35
	22.4	—	59
1000 ft-c, Γ	2.2	85	63
	19.0	414	400

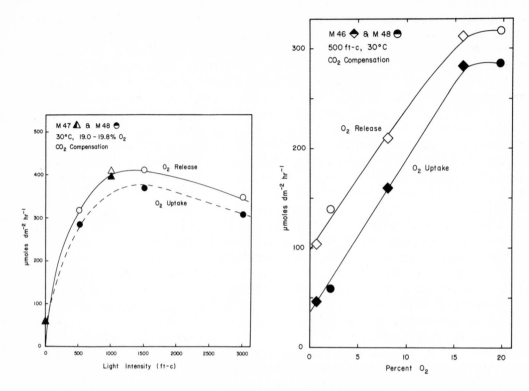

Figure 7 (left) Influence of light intensity on oxygen exchange rates of soybean leaves at CO_2 compensation in a 30° environment containing 19.0 to 19.8% O_2 (M 47 and M 48 are experiment numbers).

Figure 8 (right) Effect of O_2 concentration on oxygen exchange rates of a soybean leaf illuminated at 500 ft-c at CO_2 compensation in a 30° environment (M 46 and M 48 are experiment numbers).

summarized in Figure 7. Data have been converted to O_2 release and O_2 uptake (cf. Methods) based on a unit dm^2 leaf surface area (one side). The O_2 uptake rates measured during illumination are considerably above the dark rate and the O_2 exchange reactions saturated at 1000-1500 ft-c. Under these conditions O_2 uptake was 350-400 μmoles dm^{-2} hr^{-1}, comprising about 60% of the rate of CO_2 uptake at saturating CO_2 (cf. Figure 2). Upon increasing the light intensity to 3000 ft-c, a decline in O_2 exchange was observed at 19-20% O_2 (Figure 7). Nevertheless, when O_2 concentration was increased to 36.5% at this intensity (Figure 6), the rates of O_2 release and O_2 uptake increased considerably (to 530 and 580 μmoles dm^{-2} hr^{-1}, respectively). At 500 ft-c, O_2 exchange saturated at about 16% O_2 (Figure 8) where rates were about 280 μmoles dm^{-2} hr^{-1} for O_2 uptake. Extrapolation of the O_2 release and O_2 uptake curves to zero O_2 concentration yielded a positive intercept (Figure 8) in both instances.

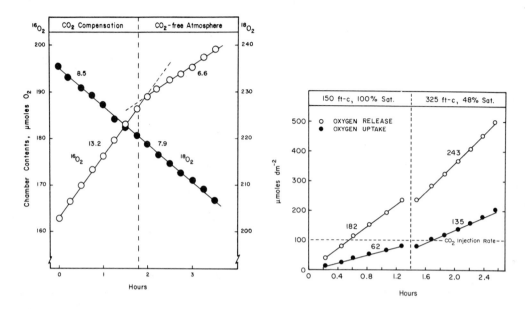

Figure 9 (left) Effect of CO_2 concentration on oxygen exchange by a soybean leaf (0.249 dm) illuminated at 300 ft-c in an environment containing 3.0% O_2 at 34°. Rates of $^{16}O_2$ release and $^{18}O_2$ uptake are presented in μmoles leaf^{-1} hr^{-1} adjacent to each regression line.

Figure 10 (right) Effect of light intensity on oxygen exchange by a soybean leaf taking up CO_2 at a constant rate: 100 μmoles dm^{-2} hr^{-1}. At 150 ft-c the CO_2 fixation capacity was saturated with CO_2; at 325 ft-c it was 48% saturated. Rates of O_2 release and O_2 uptake are presented in μmoles dm^{-2} hr^{-1} adjacent to each regression line.

Oxygen exchange at low (< 4%) oxygen concentration
    (a) The influence of a CO_2 free atmosphere
When net CO_2 measurements are employed, there appeared to be only a slow photorespiratory process at about 2% O_2 in the ambient atmosphere. Nevertheless, the foregoing experiments (*Figures 4* and *8*) and others with pole bean (18,19,20) clearly show that a significant O_2 exchange occurs. *Figure 9* illustrates the effect of a CO_2 free atmosphere on the rates of oxygen exchange. Steady state rates were first established at Γ, following which the circulating air stream was passed through Ba(OH)$_2$ to remove CO_2. The rate of $^{16}O_2$ release was depressed without a significant change in the rate of $^{18}O_2$ uptake. It thus appears that a sizeable portion of the O_2 released at Γ was associated with a CO_2 refixation process.
    (b) The influence of a low CO_2 supply
We have conducted a number of experiments to examine oxygen exchange patterns at subsaturating CO_2 concentrations. In each of

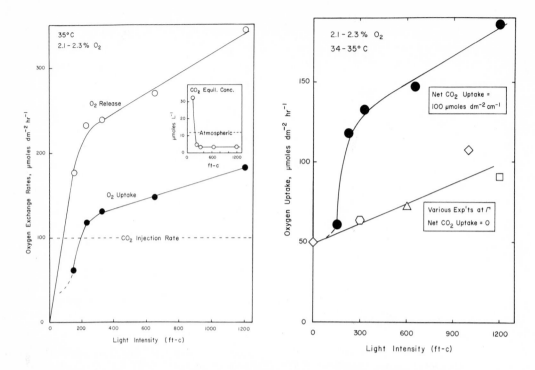

Figure 11 (left) Influence of light intensity on the oxygen exchange rate of , and CO₂ equilibrium concentration maintained by, a soybean leaf exposed to 2.2% O₂ at 35° in a closed system into which CO₂ was continuously injected at 100 μmoles dm⁻² hr⁻¹.

Figure 12 (right) Interaction between light intensity and CO₂ supply on oxygen uptake by soybean leaves exposed to 2.2% O₂ at 34 to 35°.

these experiments a leaf was enclosed in a chamber containing the selected O_2 concentration, and CO_2 was injected into the closed system at a constant rate. When this was done, there was established a static CO_2 equilibrium concentration, the magnitude of which was a function of CO_2 injection rate and light intensity. Appropriate selection of these parameters resulted in the establishment of a steady state condition wherein the CO_2 equilibrium concentration remained constant at values below normal atmospheric concentrations. *Figure 10* illustrates O_2 exchange rates at two low light intensities with a constant rate of CO_2 injection (100 μmoles dm⁻² hr⁻¹) which was just sufficient to saturate the capacity of this leaf to fix CO_2 at 150 ft-c. Immediately following the onset of CO_2 injection, the CO_2 concentration rapidly increased from Γ to 32 μmoles l⁻¹ and then remained relatively constant. Oxygen release and uptake occurred at the constant rates shown (182 and 62 μmoles dm⁻² hr⁻¹, respectively). Increasing the light intensity to 325 ft-c increased the photosynthetic capacity to 208 μmoles CO_2 dm⁻² hr⁻¹. However, since the rate of CO_2 injection remained constant at 100 μmoles dm^{-2} hr^{-1}, only 48% of the photosynthetic capacity was saturated with CO_2. Under these conditions the CO_2 concentration dropped rapidly to 3.5 μmoles l^{-1} where it again remained constant. Oxygen release increased to 243 μmoles dm^{-2} hr^{-1} and O_2 uptake to 135 μmoles dm^{-2} hr^{-1}. *Figure 11* presents the rates obtained by a sequence of such measurements at various light intensities, throughout which the CO_2 injection rate remained constant. Since the CO_2 equilibrium concentration also re-mained constant during each phase, and was independent of light in-tensity above 325 ft-c (see insert *Figure 11*), photosynthetic CO_2 fixation was essentially constant. In spite of this constancy, the effect of increased light intensity was to stimulate the rate of O_2 release and O_2 uptake in approximately parallel amounts (*Figure 11*). A comparison of the rates of O_2 uptake obtained in this experiment with those obtained at Γ in a number of other experiments reveals a significant enhancement resulting from the occurrence of CO_2 fixa-tion (*Figure 12*). It is necessary to emphasize that this result ob-tains only when the capacity for photosynthetic CO_2 fixation is not saturated by the CO_2 supply. *Figure 13* extends the pattern to higher light intensities under slightly different environmental conditions. In these experiments both the O_2 uptake at Γ and that occurring when net CO_2 fixation was 120 μmoles dm^{-2} hr^{-1} became saturated at $\sim$1200 ft-c.

(c) The influence of leaf maturity

All of the experiments reported here with soybean were conducted with fully expanded first trifoliate leaves. OZBUN *et al.* (19,20) have demonstrated that substantially greater rates may be expected with immature, rapidly expanding leaves. Some of their data taken at Γ have been replotted and are shown in *Figure 13*. Oxygen uptake was substantially greater at 1500 ft-c than in darkness at all stages of leaf development examined, and there was essential parallelism in the decline of O_2 release and O_2 uptake with increasing leaf matu-rity. The youngest leaves took up oxygen at 246 μmoles dm^{-2} hr^{-1}, a significant rate when compared to their photosynthetic capacity at 1500 ft-c, 663 μmoles CO_2 dm^{-2} hr^{-1} (19). It is emphasized that these rates of O_2 exchange at Γ were obtained at about 4% O_2. Ex-periments with immature leaves at higher O_2 concentrations have not yet been conducted.

DISCUSSION

Oxygen exchange at Γ

Substantial rates of O_2 release and O_2 uptake are exhibited by leaves illuminated at Γ (18,19,20). The present investigation re-veals that these processes were stimulated by increasing the ambient O_2 concentration at least to 36.5% (*Figures 4-6*). In contrast, much less stimulation occurred in darkness (*Table 1*). At atmospheric O_2 concentration, the O_2 exchange process was light saturated (*Figure 7*) between 1000-1500 ft-c (4.5-6.7 x 10^4 ergs cm^{-2} sec^{-1} with our

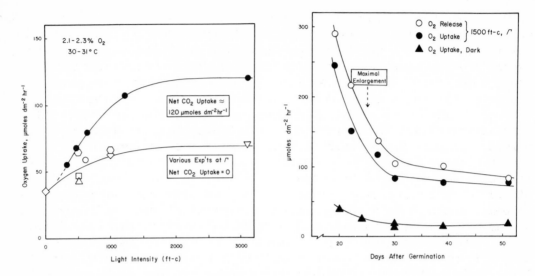

Figure 13 (left) *Interaction between light intensity and CO₂ supply on oxygen uptake by soybean leaves exposed to 2.2% O₂ at 30 to 31°.*

Figure 14 (right) *Oxygen uptake in darkness and oxygen exchange at CO₂ compensation and 1500 ft-c by pole bean leaves at various stages of maturity in an environment containing 4% O₂ at 28°. Replotted from OZBUN* et al. *(19,20).*

light source). This light intensity is considerably below that which saturated CO_2 fixation under conditions of nonlimiting CO_2 supply (*Figure 2*). As the O_2 concentration and light intensity increased, O_2 uptake at Γ was enhanced more than O_2 release, so that at 36.5% O_2 and 3000 ft-c there was a net O_2 consumption (*Figure 6*). Photoxidation of substrates in excess of those formed from internally recycled CO_2 apparently occurred under these conditions. This effect is a function both of the high oxygen concentration and high light intensity, since at 500 ft-c O_2 exchange was saturated somewhat below 20% O_2 (*Figure 8*), at which point O_2 release still exceeded O_2 uptake. Furthermore, at 20% O_2 an increase in light intensity from 1500 to 3000 ft-c depressed rather than increased the exchange rates and did so without reversing their order (*Figure 7*).

Oxygen exchange at Γ is substantial and no estimate of its magnitude can be gained from rates of net O_2 release or net O_2 uptake. At atmospheric O_2 concentrations and 1000 ft-c, O_2 uptake at Γ (*Figure 7*) was about 60% as great as the maximal capacity of the tissue for CO_2 fixation at the same light intensity (*Figure 2*). The prior experiments of OZBUN et al. (19,20; cf. *Figure 14*) show that even at ∿4% O_2, immature leaves of pole bean at Γ were able to take up O_2 at least 35% as rapidly as they could fix CO_2 at CO_2 saturation and 1500 ft-c.

The exact nature of these sizeable O_2 exchange reactions at Γ and their role in the metabolism and energy economy of plants is not obvious. Some of the O_2 uptake may be due to a blue light stimulated phenomenon (25-29). In algae, the blue light effect appears to be due to increased mitochondrial respiration (30,31). A portion of the effect may be associated with epoxyxanthophyll oxidation (32,33). Another possibility is enhanced oxidative pentose pathway activity under the restricted photosynthesis occurring at Γ (34). The dominant processes, however, are most likely due to glycolate oxidation (35) and to reaction with reductant(s) at the reducing end of photosystem I (16,17). COOMBS and WHITTINGHAM (7) have proposed a model in which light and O_2 act in concert to stimulate glycolate synthesis under CO_2 limiting conditions. Peroxide formed by reaction of O_2 with photosystem I reductant(s) is thought to enhance the formation of glycolate from a component of the photosynthetic carbon reduction cycle. Subsequent oxidation of the glycolate provides an additional O_2 uptake process. The hypothesis receives support from observations that O_2 can serve as a Hill oxidant in isolated chloroplasts (16,36, 37). Fluorescence and light scattering (17) responses to O_2 indicate that O_2 may also serve as a Hill oxidant in intact leaves. Moreover, glycolate synthesis is enhanced by O_2 and illumination in algae, chloroplasts, and leaves (6,7,8,9,11). Many of these observations can also be accommodated in the proposal of OGREN (15) who showed that O_2 competitively inhibited CO_2 incorporation by purified ribulose-1,5-diphosphate (RuDP) carboxylase from soybean leaves. A product of the oxidation of RuDP should be phosphoglycolate: hence, increased synthesis of the photorespiratory substrate would occur concomitantly with restricted photosynthesis.

The data reported here suggest that a considerable portion of the O_2 exchange observed at Γ is associated with a process involving the release and refixation of CO_2. Increased O_2 concentrations stimulated both O_2 release and O_2 uptake (*Figures 3-6* and *8*) and also increased Γ, as has been observed by others (1,2,4,5). The increase in Γ cannot be due to a general destruction of the photosynthetic mechanism in these leaves because the inhibiting effect of O_2 on net CO_2 uptake (at least at 40% O_2) can be completely eliminated if the ambient CO_2 concentration is kept at saturating levels (24,38; cf. *Figure 5* in reference 23). Depleting the ambient CO_2 concentration from Γ to essentially nil by passing the circulating atmosphere through base resulted in a decrease in O_2 release without measurably affecting O_2 uptake (*Figure 9*). The most straightforward explanation of this observation is that at Γ a significant portion of the O_2 uptake was associated with a CO_2 release process, the released CO_2 being refixed thereby permitting an equivalent O_2 release. When the atmosphere was passed through base, gradients were established such that part of the internally generated CO_2 diffused out of the leaf, thus decreasing the amount refixed photosynthetically. Unfortunately it is not possible to estimate with any certainty what portion of the respired CO_2 was released from the leaf and what portion was refixed during the base-scrubbing phase. Hence quantitative estimates of the portion of O_2 release which can be accounted for by CO_2 recycling cannot yet be made.

It is conceivable that not all of the O_2 exchange can be accounted for by CO_2 recycling. BULLEY and TREGUNNA (5) have reported that the O_2 stimulation of Γ in soybean leaves was accompanied by an accelerated exchange of CO_2. At 21% O_2, the CO_2 exchange rates at Γ were 2.7 and 3.5 mg CO_2 dm^{-2} hr^{-1} at light intensities of 6.5 x 10^4 and 28 x 10^4 ergs cm^{-2} sec^{-1}, respectively. These values (61 and 87 μmoles CO_2 dm^{-2} hr^{-1}) are considerably lower than the O_2 exchange rates which were measured at similar O_2 concentrations (*Figures 5* and *7*). If the CO_2 exchange measurements are valid, then a sizeable component of the O_2 exchange process is not associated with CO_2 exchange.

Oxygen exchange at low O_2 concentrations

Although the O_2 exchange rates at Γ are clearly enhanced as the ambient O_2 concentration is increased, significant O_2 uptake and release was observed at quite low O_2 concentrations. Values at least as high as dark respiration rates were observed at 1% O_2 or less (*Figure 8*), and extrapolation of the O_2 uptake rates to zero % O_2 yielded a positive intercept clearly different from zero. This positive intercept has been confirmed in other experiments in which O_2 uptake was measured at Γ between zero and 5% O_2, and the same pattern is also apparent in maize (39). Moreover, the significant depression in O_2 release resulting from maintenance of a CO_2 free atmosphere (*Figure 9*) occurred under low light (300 ft-c) and low O_2 ($\sim$3%), both conditions being conducive to minimal photorespiratory rates as determined from CO_2 exchange measurements. Hence there is reason to suspect that respiratory activity does exist at low O_2 concentrations in illuminated leaves. Results to be presented elsewhere show that even at 2% O_2, $^{13}CO_2$ is taken up considerably faster than $^{12}CO_2$, relative to their molar concentrations. At least one explanation for this difference is the maintenance of a higher $^{12}CO_2/$$^{13}CO_2$ ratio within the leaf than in the atmosphere (23). If this explanation is correct, a finite rate of respiration occurs at low O_2 concentrations during illumination, at least at saturating ambient CO_2.

It is also significant that a substantial enhancement in O_2 uptake at low O_2 concentrations occurs when the ambient CO_2 concentration is increased from Γ to a higher but subsaturating level (*Figures 12* and *13*). The data suggest that a substrate for the O_2 uptake process is synthesized. Furthermore, the parallel increase in O_2 release and O_2 uptake during subsaturating CO_2 uptake (*Figure 11*) can be interpreted as a light-mediated enhancement of the substrate pool for the respiratory process. Whether this is a consequence of increased diversion of immediately fixed CO_2 to formation of the substrate or of diversion of previously accumulated photosynthate into the substrate pool is not known.

Ambient CO_2 concentrations have a significant regulatory influence on O_2 uptake during illumination. A subsaturating rate of CO_2

fixation enhances the process over the rate existing at Γ (*Figure 12 and 13*) while saturating CO_2 concentrations restrict it (*Figure 3*). Although water uptake was suppressed by the saturating CO_2 concentrations compared to the rate at Γ, the depression was not as great as that of O_2 uptake (*Figure 3*). Similar observations of enhancement in O_2 uptake upon CO_2 depletion have been made with algae (40,41). The indication is that even at relatively low O_2 concentrations, O_2 and CO_2 essentially compete as oxidants for light-generated reductants, and that there is also an additional component of O_2 uptake associated with continual oxidation of a respiratory substrate. The nature of this substrate is not known. The conditions under which the observations can be made ($\sim$2% O_2 and 300 ft-c) are such that glycolate synthesis is likely to be minimal. The possibility therefore exists that a vestige of normal dark respiratory processes continues during illumination.

REFERENCES

1. M.L. FORRESTER, G. KROTKOV, and C.D. NELSON, *Plant Physiol.*, 41 (1966) 422.
2. E.B. TREGUNNA, G. KROTKOV, and C.D. NELSON, *Physiologia Pl.*, 19 (1966) 723.
3. O.V.S. HEATH and B. ORCHARD, *J. exp. Bot.*, 19 (1968) 176.
4. J. POSKUTA, *Physiologia Pl.*, 21 (1968) 1129.
5. N.R. BULLEY and E.B. TREGUNNA, *Can. J. Bot.*, 48 (1970) 1271.
6. J.A. BASSHAM and M. KIRK, *Biochem. biophys. Res. Commun.*, 9 (1962) 376.
7. J. COOMBS and C.P. WHITTINGHAM, *Proc. R. Soc. (B)*, 164 (1966) 511.
8. P.W. ELLYARD and M. GIBBS, *Plant Physiol.*, 44 (1969) 1115.
9. Z. PLAUT and M. GIBBS, *Plant Physiol.*, 45 (1970) 470.
10. N.E. TOLBERT, in Photosynthetic Mechanisms of Green Plants, NAS-NRC Publ. 1145, Washington, D. C., (1963) 648.
11. N.P. VOSKRESENSKAYA, Y.A. WIIL, G.S. GRISHINA, and T.R. PARNIK, *Photosynthetica*, 4 (1970) 1.
12. I. ZELITCH, *Plant Physiol.*, 43 (1968) 1829.
13. M. GIBBS, P.W. ELLYARD, and E. LATZKO, in Comparative Biochemistry and Biophysics of Photosynthesis, eds. K. SHIBITA, *et al.*, Univ. Tokyo Press, Tokyo, (1968) 387.
14. J.S. TURNER and E.G. BRITTAIN, *Biol. Rev.*, 37 (1962) 130.
15. W.L. OGREN, *Nature*, (1971) in press.
16. U. HEBER and C.S. FRENCH, *Planta*, 79 (1968) 99.
17. U. HEBER, *Biochim. biophys. Acta*, 180 (1969) 302.
18. J.L. OZBUN, R.J. VOLK, and W.A. JACKSON, *Plant Physiol.*, 39 (1964) 523.
19. J.L. OZBUN, R.J. VOLK, and W.A. JACKSON, *Crop Sci.*, 5 (1965) 69.
20. J.L. OZBUN, R.J. VOLK, and W.A. JACKSON, *Crop Sci.*, 5 (1965) 497.
21. R.J. VOLK and W.A. JACKSON, *Crop Sci.*, 4 (1964) 45.

22. W.A. JACKSON and R.J. VOLK, *Nature*, 222 (1969) 269.
23. W.A. JACKSON and R.J. VOLK, *A. Rev. Pl. Physiol.*, 21 (1970) 385.
24. C.L. MULCHI, Ph. D. Thesis, (North Carolina State University, Raleigh, N.C.) 1970.
25. N.P. VOSKRESENSKAYA and G.S. GRISHINA, *Fiziol. Rast.*, 8 (1961) 579.
26. N.P. VOSKRESENSKAYA and G.S. GRISHINA, *Dokl. Akad. Nauk SSSR.*, 144 (1962) 625.
27. N.P. VOSKRESENSKAYA, G.S. GRISHINA, S.N. CHMORA, and N.M. POYAR-KOVA, *Can. J. Bot.*, 48 (1970) 1251.
28. J. POSKUTA, *Experientia*, 24 (1968) 796.
29. N.R. BULLEY, C.D. NELSON, and E.B. TREGUNNA, *Plant Physiol.*, 44 (1969) 678.
30. W. KOWALLIK and H. GAFFRON, *Nature*, 215 (1967) 1038.
31. U. KOWALLIK and W. KOWALLIK, *Planta*, 84 (1969) 141.
32. K.H. LEE and H.Y. YAMAMOTO, *Photochem. Photobiol.*, 7 (1968) 101.
33. C.A. TAKEGUCHI and H.Y. YAMAMOTO, *Biochim. biophys. Acta*, 153 (1968) 459.
34. G.H. KRAUSE and J.A. BASSHAM, *Biochim. biophys. Acta*, 172 (1969) 553.
35. I. ZELITCH, *A. Rev. Pl. Physiol.*, 15 (1964) 121.
36. A.H. MEHLER, *Archs Biochem. Biophys.*, 33 (1951) 65.
37. G. FORTI and A.T. JAGENDORF, *Biochim. biophys. Acta*, 54 (1961) 322.
38. P.A. JOLLIFFE and E.B. TREGUNNA, *Plant Physiol.*, 43 (1968) 902.
39. R.J. VOLK and W.A. JACKSON, Manuscript in preparation, (1971).
40. A.H. BROWN and D. WEIS, *Plant Physiol.*, 34 (1959) 224.
41. J.S. BUNT, *Biochem. biophys. Res. Commun.*, 35 (1969) 748.

THE LIGHT RESPONSE OF CO_2 EXCHANGE: ON THE SOURCE OF DIFFERENCES BETWEEN C_3 AND C_4 SPECIES

Roger M. Gifford

Department of Agronomy, Cornell University, Ithaca, N.Y. 14850, U.S.A. and Division of Plant Industry, CSIRO, Canberra City 2601, Australia

SUMMARY

The results of various gas exchange studies of C_4 plants (Zea mays, Pennisetum spp., Panicum maximum) and of C_3 plants (Brassica campestris, Calopogonium mucunoides) are examined. The conclusion is drawn that the apparent nonsaturation of C_4 net CO_2 exchange at full sunlight intensities is probably largely a stomatal response rather than a response of the mechanism of CO_2 fixation. The assumptions made in the analysis which yields this conclusion are discussed.

The continued response of net CO_2 exchange by maize leaves to increasing light intensity up to full sunlight was documented by HESKETH and MUSGRAVE (1). This phenomenon did not occur with every leaf studied but later studies indicated that it is usually true not only for maize but also for a range of species (2-8) which are now regarded as exhibiting the C_4 dicarboxylic acid pathway of CO_2 fixation (9,10) and the associated "syndrome". Species lacking the C_4 pathway usually show light saturation at one fifth to one quarter full sunlight intensity (11,3). Despite some exceptions (12,13,9) to the above generalizations, the correlation between the possession of the high light saturating characteristic and the C_4 syndrome has been sufficiently close that the question emerges as to whether the saturation characteristics *per se* of C_4 plants bears any "cause and effect" relationship with the typically high absolute rates of net photosynthesis and the biochemical characteristics of such species.

I can see three possibilities. Data of SLACK (14) and HATCH and SLACK (15) suggest the possibility that rapid changes in the activities of some enzymes of the C_4 carbon fixing system in response to the current light intensity may tend to match the rates of dark reactions to the potential rate of light reactions. However, the en-

zyme which was observed to be activated increasingly with light intensity (pyruvate,Pi dikinase) did so with a half-time of 10-15 minutes (9) which is slower than the commonly found time of equilibration of net photosynthesis to new light intensity (e.g. 8). A second
possibility involves the suggested greater requirement for cyclic
phosphorylation of the C_4 system than the C_3 system (16). Cyclic
phosphorylation by isolated chloroplasts does not saturate in bright
light whereas noncyclic photophosphorylation does (17). Thus a
greater dependence on "nonsaturating" cyclic photophosphorylation
may be invoked as an hypothesis to explain the saturation characteristic of C_4 plants. A third possible explanation is that the extent
of stomatal opening as a function of light intensity might be involved. The study on maize (cv. Pa83) described below was done in
R.B. MUSGRAVE'S laboratory to examine the stomatal hypothesis on
material grown under field conditions similar to those used by HES-
KETH and MUSGRAVE (1).

Leaves of mature plants in the ear-filling stage were studied
with a double leaf chamber which was operated simultaneously as an
assimilation chamber and nitrous oxide diffusion porometer of the
SLATYER and JARVIS (18) type (for details see 19,20). Artificial
illumination was from "Cool-beam" incandescent spotlights shining
equally on *both* leaf surfaces. Intensity was varied with layers of
cheesecloth. Temperature was controlled at 30°. Ambient CO_2 concentration, C_a, was maintained constant at 280 ppm.

The diffusion resistance analysis was carried out separately for
each side of the leaf. Hence

$$\Sigma r_u = (C_a - \Gamma)/NCE_u \tag{1}$$

$$\Sigma r_l = (C_a - \Gamma)NCE_l \tag{2}$$

(19), where Σr is the total resistance to CO_2 uptake, C_a is the ambient CO_2 concentration, Γ is the CO_2 compensation point, and NCE is
the net CO_2 exchange rate per unit surface area. Subscripts u and 1
refer to upper and lower leaf surfaces. Σr is considered to be composed of three additive parts, the boundary layer resistance r_a, the
stomatal resistance r_s, and the "residual resistance" r_r comprising
the gamut of diffusional and chemical limitations to CO_2 uptake not
accounted for by $(r_a + r_s)$. The entirely extracellular resistance to
diffusion of N_2O diffusion through the leaf blade, r_n, is considered
to be made up of five additive terms

$$r_n = r_{a,u} + r_{s,u} + r_z + r_{s,l} + r_{a,l} \tag{3}$$

where r_n is the intercellular space resistance to N_2O diffusion
through the mesophyll tissue. Combining equations (1), (2), and (3)
and assuming $r_{r,u} = r_{r,l}$ we get

$$(\Sigma r_u + \Sigma r_l) = (2r_r - r_z) + r_n \tag{4}$$

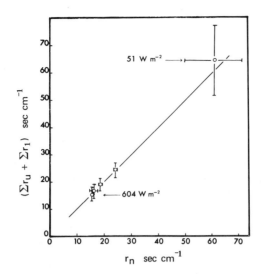

*Figure 1 Correlation between total resistance to CO_2 uptake
and stomatal resistance to N_2O diffusion over a range of visible light intensities. Each point is the mean of three leaves
(cv. Pa83). 2 x S.E. for each parameter indicated by horizon-
tal and vertical bars. For significance of the diagonal line
see the text.*

The results of a study involving 3 maize leaves are summarized in
Figure 1 by plotting $(\Sigma r_u + \Sigma r_l)$ against r_n. The diagonal line is
the theoretical curve predicted by equation (4) if variation of r_n
as a function of light intensity accounted completely for variation
of Σr, and $(2r_r - r_z)$ is constant. The data suggest that stomatal
resistance is the major resistance controlling the shape of the
light response curve of NCE.

We are interested in what the light response curve would look
like if the stomatal resistance had remained constant. The following
procedure has been used to approximate such a curve. To adjust a
measured NCE to some other value (NCE') appropriate to another r_s
(r_s') we define $\Delta r_s = (r_s - r_s')$. Then assuming it is valid to leave
r_r constant while adjusting r_s, NCE' = $(C_a - \Gamma)/\Sigma r'$ where $\Sigma r' = (\Sigma r -
\Delta r_s)$ and $\Sigma r = (C_a - \Gamma)/NCE$. The assumption concerning holding r_r
constant is not necessarily strictly valid and we shall return to
further discussion of that later. Using this procedure the mean
light response curve for the 3 maize leaves was adjusted to a constant $(r_a + r_s) = 2$ s cm^{-1} (4 s cm^{-1} for each side of the leaf). r_a
was o.7 s cm^{-1} for each side of the leaf. In making the calculations
it was assumed $r_z = 2$ s cm^{-1}. This value of r_z was obtained by calculations based on anatomical studies of similar maize leaves (20).
The results plotted in *Figure 2* show that the adjusted NCE saturated
rather abruptly at about 100 Wm^{-2} (visible). Also the data at low
light intensity get very close to the 8 quantum efficiency line
(corresponding to about 30% efficiency for the light source used).
However, it is at such low light intensity that the assumption con-

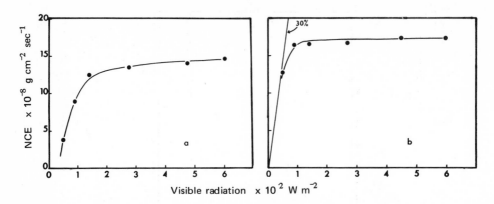

Figure 2 *Light response curve of net CO$_2$ exchange rate for Zea mays. (a) NCE as measured (b) NCE after adjustment (NCE' at $r_a + r_s = 2$ s cm^{-1}).*

cerning r_r vs. C_w (CO$_2$ concentration at the mesophyll cell wall) is most likely to be in error (see below), so the last observation should be viewed with caution.

Drs. M.M. LUDLOW and H.G. McPHERSON have provided me with data which may also be treated in this way. r_s was estimated from transpiration measurements in both sets of data. Although these data do not separate the gas exchange of each side of the leaf, any error resulting will probably have little effect on the saturating light intensity for the adjusted curves. After adjustment of the data to a condition of ($r_a + r_s$) constant at 2 s cm^{-1}, both sets of data indicate saturation at less than full sunlight intensity (*Figures 3* and *4*). The bulrush millet (*Pennisetum typhoides*) leaves studied by McPHERSON saturate at around 350 W m^{-2} (visible) after the adjustment. The *Pennisetum purpureum* leaves of LUDLOW responded differently to each other one saturating very abruptly in between 100 and 150 W m^{-2} the other not saturating completely. LUDLOW'S *Panicum maximum* data saturate at about 250 W m^{-2}.

A similar adjustment was made to light response curves of two C$_3$ species, turnip (*Brassica campestris*) (21) and a legume (*Calopogonium mucunoides*) (LUDLOW, personal communication). Adjustment of the curves to ($r_a + r_s$) = 2 s cm^{-1} for all light intensities had little effect on the saturating light intensity for these species. It remained at about 200 W m^{-2} (visible) for both species.

Returning to the assumption that r_r may be left constant while ($r_a + r_s$) is adjusted; this is equivalent to assuming that r_r is independent of C_w. Such was true for the maize cultivar studied, when exposed to 604 W m^{-2} of visible radiation (19). It was not studied at lower light intensities. It seems to me to be likely that, for light intensities below some low value, r_r may increase with increasing C_w. This would cause the true curves adjusted to constant ($r_a + r_s$) to deviate below the ones presented here at low light intensity, and could conceivably cause the saturating light intensity of the adjusted curves to be greater than indicated in this presen-

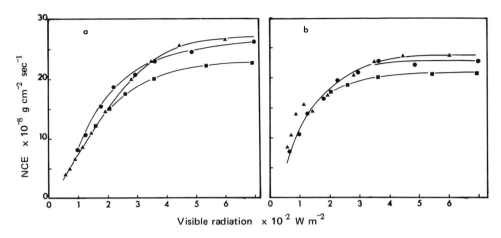

Figure 3 *Light response curves of net CO_2 exchange rate for Pennisetum typhoides leaves of various maturities. C_a = 0.03%. Data provided by H.G. McPHERSON. (a) NCE as measured (b) NCE after adjustment (NCE' at $r_a + r_s$ = 2 s cm^{-1}).*

tation. But surprisingly TROUGHTON and SLATYER (1969) found contrary evidence. For cotton leaves exposed to 110 W m^{-2} of visible radiation r_r *decreased* rather than increased with C_w. Hence it is not possible to be too categorical about the precise validity of the adjusted curves in the present study. The stomata must, however, be playing a major role.

In conclusion it seems that the light saturating characteristic of C_4 species may not be associated directly with the special C_4

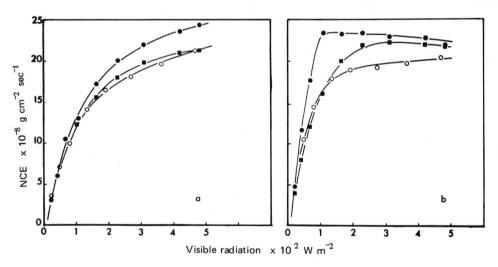

Figure 4 *Light response curves for leaves of Pennisetum purpureum (o———o and ●———●), and of Panicum maximum (■———■). C_a = 0.03%. Data provided by M.M. LUDLOW. (a) NCE as measured (b) NCE after adjustment (NCE' at $r_a + r_s$ = 2 s cm^{-1}).*

carbon fixation mechanisms. An indirect effect through stomatal aperture in combination with typically low residual resistances in C_4 species is probably the major effect. More precise quantitative evaluation of the stomatal control awaits experiments which enable CO_2 exchange at constant internal CO_2 concentration (C_w) to be plotted against light intensity.

ACKNOWLEDGMENTS

I am greatly indebted to Dr. R.B Musgrave with whose facilities the maize study was performed, and to Drs. M.M. Ludlow and G.L. Wilson and Drs. H.G. McPherson and R.O. Slatyer who generously furnished data prior to its appearance in press.

REFERENCES

1. J.D. HESKETH and R.B. MUSGRAVE, *Crop Sci.*, 2 (1962) 311.
2. J.D. HESKETH, *Crop Sci.*, 3 (1963) 493.
3. J.D. HESKETH and D.N. MOSS, *Crop Sci.*, 3 (1963) 107.
4. M. EL-SHARKAWY and J.D. HESKETH, *Crop Sci.*, 5 (1965) 517.
5. J.D. HESKETH and D. BAKER, *Crop Sci.*, 7 (1967) 285.
6. D.N. MOSS, *Conn. Agric. Exp. Sta. Bull.*, 664 (1964) 86.
7. J.P. COOPER and M.M. TAINTON, *Herb. Abs.*, 38 (1968) 167.
8. T.A. BULL, *Crop Sci.*, 9 (1969) 726.
9. M.D. HATCH, C.R. SLACK, and T.A. BULL, *Phytochemistry*, 8 (1969) 697.
10. C.C. BLACK, T.M. CHEN, and R.H. BROWN, *Weed Sci.*, 17 (1969) 338.
11. R.H. BOHNING and C.A. BURNSIDE, *Am. J. Bot.*, 43 (1956) 557.
12. L.T. EVANS and R.L. DUNSTONE, *Aust. J. biol. Sci.*, 23 (1970) 725.
13. J.E. BEUERLEIN and J.W. PENDLETON, *Agron. Abs.*, 62 (1970) 28.
14. C.R. SLACK, *Biochem. biophys. Res. Commun.*, 30 (1968) 483.
15. M.D. HATCH and C.R. SLACK, *Biochem. J.*, 112 (1969) 549.
16. C.C. BLACK and B.C. MAYNE, *Plant Physiol.*, 45 (1970) 738.
17. T.M. CHEN, R.H. BROWN, and C.C. BLACK, *Plant Physiol.*, 44 (1969) 649.
18. R.O. SLATYER and P.G. JARVIS, *Science*, 151 (1966) 574.
19. R.M. GIFFORD and R.B. MUSGRAVE, *Physiologia Pl.*, 23 (1970) 1048.
20. R.M. GIFFORD, Ph. D. Thesis (Cornell Univ., Ithaca, N.Y.), *Diss. Abstr.*, 31 (1970) 70-14, 383.
21. P. GAASTRA, *Meded. LandbHoogesch. Wageningen*, 59 (1959) 1.

ADAPTATION OF SORGHUM PLANTS TO LIGHT INTENSITY: ITS EFFECT ON GAS EXCHANGE IN RESPONSE TO CHANGES IN LIGHT, TEMPERATURE, AND CO_2

R.W. Downes

Division of Plant Industry, CSIRO, Canberra City, 2601, Australia

SUMMARY

Plants of Sorghum sudanense were grown under conditions of low light intensity and rates of photosynthesis and transpiration were observed simultaneously in attached leaves under a range of conditions of leaf temperature, light intensity, and CO_2 concentration. Net rate of photosynthesis and stomatal opening saturated at relatively low light intensity. Photosynthetic rate was maximal near normal ambient CO_2 concentrations but stomatal resistance increased as CO_2 concentration in the air increased. When comparable levels of stomatal resistance were induced in sorghum and wheat, net rate of photosynthesis was higher in sorghum suggesting an adaptive advantage of C_4 species under conditions of moisture stress. The differential effect of light intensity on stomata of C_3 and C_4 species was discussed in relation to rates of assimilation and transpiration in the crop situation.

INTRODUCTION

The net rate of photosynthesis (P) increases with light intensity in C_4 grasses up to full sunlight (1), this being associated with decreasing stomatal resistance to CO_2 diffusion (2). However, in shade adapted C_4 *Amaranthus* plants, P does not increase beyond moderate levels of radiation (3,4).

To define the effects of adaptation to low radiation in *Sorghum sudanense*, seedlings were grown in glasshouses in winter and gas exchange was measured when leaf temperature, light intensity, and CO_2 in the air were varied. It was anticipated that if P was modified by light adaptation there would be an effect on an associated parameter such as resistance to CO_2 diffusion. Since stomata provide a resistance to both CO_2 and water exchange, and stomatal aperture is modified by environmental conditions, the relationship between stomatal movements and rates of gas exchange received particular attention.

EXPERIMENTAL

Seedlings of *Sorghum sudanense* (Stapf.) cv. Greenleaf were grown at day/night temperatures of 30/25° with a 16 hour photoperiod of normal winter radiation (300 cal cm^{-2} day^{-1}) extended by 25 ft-c light from incandescent sources. Using procedures described previously (2), net photosynthetic rate (P) and transpiration rate (T) were measured simultaneously on young attached leaves.

Leaf temperature was varied between 15 and 40° while leaves were exposed to 1 of 3 levels of photosynthetically active radiation (0.09, 0.27, 0.45 cal cm^{-2} min^{-1}, referred to as levels 1, 2, and 3 respectively). With leaf temperature constant at 28°, the CO_2 concentration in the air stream was varied between 0 and 130 x 10^{-8} g cm^{-3} at the chamber outlet, for each of the 3 light intensities. The CO_2 response curve was also determined for high radiation adapted (summer grown) leaves at 28° and light level 3.

In a subsidiary experiment, sorghum plants adapted to low light and wheat plants, grown at 21/16° in low light, were exposed to low root temperatures which increased stomatal resistance. Values of P and T were recorded simultaneously while leaf temperature was 28° and radiation was constant at level 2.

Components of P and T were derived from the equations (5):

$$P = (C_a - C_w)/(r_a + r_s)_{CO_2}$$

$$T = (W_w - W_a)/(r_a + r_s)_{water}$$

where C_a and C_w represent the CO_2 concentrations in the external air and at the surfaces of the mesophyll cell walls respectively, and W_a and W_w represent the water vapor concentrations at these respective sites; r_a and r_s represent the resistances provided by the boundary layer and by stomata and intercellular spaces of the leaf respectively, to CO_2 and water vapor, as defined by subscripts. Constants (6) were used to relate $(r)CO_2$ and (r)water values.

RESULTS AND DISCUSSION

The photosynthetic response to temperature (*Table 1*) was similar to that of plants grown at high light intensity (2). However, adaptation to low light intensity reduced the photosynthetic rate under high radiation in *Sorghum* as it did in *Amaranthus* (3,4). In plants adapted to low light, the stomata were unable to open fully under high light conditions, unlike those of high light conditioned plants (2). Since old leaves are shaded in the crop canopy, low P values in old leaves (3) may be due, in part, to shade effects.

Although phosphoenolpyruvate carboxylase activity is lower in shade adapted than in high light adapted leaves (4), values of

TABLE I

Effect of leaf temperature and light intensity on P ($\times 10^{-8}$ g cm^{-2} sec^{-1}), ($r_a + r_s$) (sec cm^{-1}), and ($C_a - C_w$) ($\times 10^{-8}$ g cm^{-3}) of sorghum plants adapted to low light intensity.

Light intensity (cal cm^{-2} min^{-1})		Leaf temperature (°C)					
		15	20	25	30	35	40
0.09	P	5.1	7.1	7.8	7.8	7.9	7.6
	($r_a + r_s$)CO_2	7.8	7.6	6.8	7.1	7.1	7.2
	($C_a - C_w$)	41	53	53	55	57	54
0.27	P	7.6	10.0	12.8	15.1	15.1	15.3
	($r_a + r_s$)CO_2	4.8	4.3	3.6	3.7	3.5	3.7
	($C_a - C_w$)	37	44	47	55	54	57
0.45	P	8.5	11.1	14.2	16.7	17.4	16.9
	($r_a + r_s$)CO_2	4.9	4.1	3.5	3.2	3.1	3.4
	($C_a - C_w$)	41	45	49	53	54	56

($C_a - C_w$), which among other things reflect activity of carboxylating enzymes (2), were constant near their physical maximum in low light adapted plants (*Table 1*).

Effects of light intensity and C_a on P and ($r_a + r_s$)CO_2 are shown in *Figure 1*. Irrespective of light adaptation or test light intensity, P was maximal near normal ambient CO_2 concentrations (60 $\times$ 10^{-8} g cm^{-3}). At low levels of C_a, P and ($r_a + r_s$)CO_2 were almost linearly related to C_a. When C_a was about 30 $\times$ 10^{-8} g cm^{-3}, the slope of the ($r_a + r_s$)CO_2 response increased but remained linear at higher C_a values. Increasing ($r_a + r_s$)CO_2 was associated with a flattening and downturn in the P response to C_a. This P response differs from that reported in maize (7) and a nonlinear relationship between P and C_a between C_a values of 30 and 60 $\times$ 10^{-8} g cm^{-3} contrasts with findings in *Pennisetum* (8).

Since P is apparently zero when C_a is zero, metabolism of newly fixed carbon compounds is unlikely to be providing the energy for an ion pump (9) inducing the maximal stomatal opening observed under these conditions. It appears possible that respiration in guard cells, perhaps involving glycolate metabolism (10), provides the energy for stomatal opening when C_a is low. The change in slope of the ($r_a + r_s$) - C_a relationship (*Figure 1B*) may be associated with a change from respiration-powered to photosynthesis-dependent stomatal opening. It may be significant that C_4 species have evolved a spatial isolation of the interdependent C_3 and C_4 pathways in different cells (11) yet processes which control stomatal movements in C_4 species may all operate within the one guard cell.

When stomatal opening is modified by water availability, P is inversely related to ($r_a + r_s$)CO_2 in both C_3 and C_4 species (*Figure 2*). When ($r_a + r_s$)CO_2 is constant, P in sorghum is 2.5 times that in

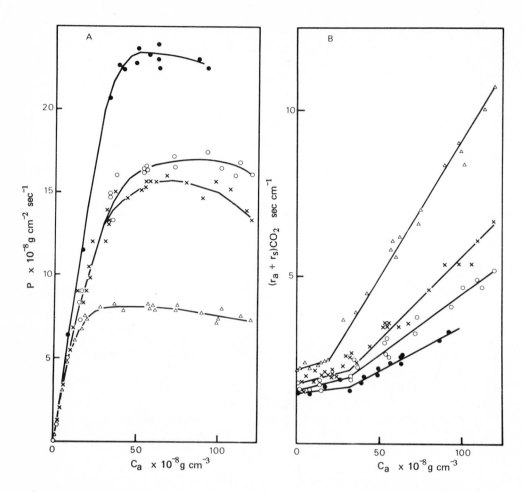

*Figure 1 Photosynthetic rate (Figure 1A) and CO_2 diffusion re-
sistance (Figure 1B) in sorghum leaves in response to ambient
CO_2 concentration and light intensity. (●) high light adapted,
(o) low light adapted (both tested at 0.45 cal cm^{-2} min^{-1} P.A.
R.), (x) low light adapted (tested at 0.27 cal cm^{-2} min^{-1} P.A.
R.), (△) low light adapted (tested at 0.09 cal cm^{-2} min^{-1} P.A.
R.).*

wheat and when P is constant, $(r_a + r_s)CO_2$ in sorghum is 2.5 times
that in wheat. Because T is also modified by resistance, the data
indicate that sorghum is more efficient in using water, efficiency
being related to $(C_a - C_w)$ (2). In addition, when moisture stress
induces comparable partial stomatal closure in C_3 and C_4 species,
the C_4 species are capable of maintaining higher P values and are
thus more able to meet the respiratory demands necessary to sustain
life. Although this suggests an adaptive advantage of C_4 over C_3
species in dry environments, other strategies like deep roots and
annual habit are also available to minimize the adverse effects of
moisture stress.

The effects of light adaptation and light sensitivity of sorghum

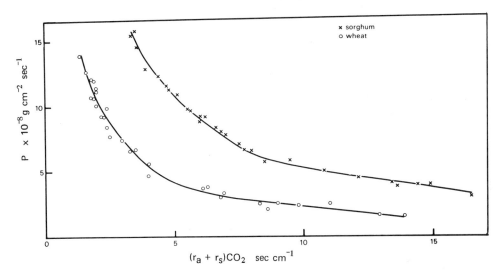

Figure 2 Photosynthetic rate in sorghum and wheat in response to changes in CO_2 diffusion resistance induced by low root temperature.

stomata are important in the crop situation. Stomatal resistance is likely to be high in sorghum leaves deep in the crop canopy so that lower leaves should have little effect on total P and T, whereas stomatal resistance is low in C_3 leaves at low light intensity (12). It appears that these phenomena should not detract substantially from the superior water use efficiency of C_4 species (2) in the crop situation, although changes in r_a may cause some modification (13). However, differential stomatal behavior of C_3 and C_4 species in response to light is likely to account for similar net assimilation rates in stands of C_3 and C_4 species (14), especially under low radiation conditions (12).

ACKNOWLEDGMENT

The able technical assistance of Mr. R. Davis in operating and maintaining the equipment was invaluable.

REFERENCES

1. J.D. HESKETH and D.N. MOSS, *Crop Sci.*, 3 (1963) 107.
2. R.W. DOWNES, *Aust. J. biol. Sci.*, 23 (1970) 775.
3. M.A. EL-SHARKAWY, R.S. LOOMIS, and W.A. WILLIAMS, *J. appl. Ecol.*, 5 (1968) 243.
4. M.D. HATCH, C.R. SLACK, and T.A. BULL, *Phytochemistry*, 8 (1969) 697.
5. P. GAASTRA, *Meded. LandbHoogesch. Wageningen*, 59 (1959) 1.
6. J. GALE and A. POLJAKOFF-MAYBER, *Physiologia Pl.*, 21 (1968) 1170.
7. J.D. HESKETH, *Crop Sci.*, 3 (1963) 493.
8. H.G. McPHERSON, Ph. D. Thesis (Australian National University,

Canberra) 1970.

9. R.A. FISCHER and T.C. HSIAO, *Plant Physiol.*, 43 (1968) 1953.

10. I. ZELITCH, *A. Rev. Pl. Physiol.*, 20 (1969) 329.

11. C.R. SLACK, M.D. HATCH, and D.J. GOODCHILD, *Biochem. J.*, 114 (1969) 489.

12. R.W. DOWNES, *Planta*, 88 (1969) 261.

13. R.O. SLATYER, *Planta*, 93 (1970) 175.

14. J.P. COOPER, *Herb. Abst.*, 40 (1970) 1.

ANALYSIS OF THE DIFFERENCE BETWEEN MAXIMUM LEAF NET PHOTOSYNTHETIC RATES OF C_4 GRASSES AND C_3 LEGUMES

M.M. Ludlow

Division of Tropical Pastures, CSIRO,
St Lucia, Qld., 4067, Australia

SUMMARY

The higher leaf net photosynthetic rate of C_4 grasses compared with C_3 legumes was associated with a lower intracellular resistance because stomatal resistances were similar. The lower intracellular resistance appeared to be associated with a lower "transfer" rather than "chemical" resistance. Evidence for the relative importance of the chemical and transfer components of the intracellular resistance is discussed.

INTRODUCTION

It has been shown that higher relative growth rates of tropical grasses compared with tropical legumes were associated with higher net assimilation rates (1). These in turn were associated with higher photosynthetic rates despite higher respiration rates. Confirmation of higher rates of photosynthesis and reasons for the difference between grasses and legumes were sought by studying the carbon dioxide and water vapor exchange of attached leaves at near-optimum leaf temperatures, ambient carbon dioxide concentrations, and high irradiance. Differences in rates of net photosynthesis were analyzed in terms of carbon dioxide transfer resistances, and differences in the intracellular resistances were further analyzed in terms of its transfer and chemical components. Data for only one grass (*Sorghum almum*) and one legume (*Calopogonium mucunoides*) are presented to illustrate the characteristics of grass and legume groups, except in the analysis of the intracellular resistance when mean data for a number of grasses and legumes are presented.

EXPERIMENTAL

Conditions of plant growth, the gas exchange apparatus, and the methods used to calculate the carbon dioxide transfer resistances have been described (2). The intracellular resistance (r_i) is the series sum of the resistances between the mesophyll cell wall and the site of carboxylation where the carbon dioxide concentration is zero. It has a transfer component ($r_{transfer}$) which may not be purely diffusional if there are enzyme-mediated transport mechanisms (e.g. carbonic anhydrase) or if cytoplasmic streaming is involved, and a chemical component ($r_{chemical}$) which is composed of photochemical (or excitation) and biochemical (or carboxylation) resistances. Transfer and chemical resistances were estimated using CHARTIER'S method which is based on a modified form of RABINOWITCH'S equation for photosynthesis at the biochemical level and GAASTRA'S carbon dioxide diffusion equation (3). The assumptions made in the calculations will be discussed later.

RESULTS AND DISCUSSION

The net photosynthetic rate of *S. almum* was twice that of *C. mucunoides* and the difference in rate between them was associated with a difference in the intracellular resistance (*Table 1*).

TABLE I

Net photosynthetic rate (mg CO_2 dm^{-2} hr^{-1}) and carbon dioxide transport resistances (sec cm^{-1}) of *Sorghum almum* and *Calopogonium mucunoides*. (Visible irradiance 380 W m^{-2}, CO_2 concentration 300 ± 5 µl l^{-1}, leaf temperature 30 ± 1°, leaf-air vapor pressure difference 17 ± 3 mm Hg.

Parameter	*S. almum* C_4	*C. mucunoides* C_3
Net photosynthetic rate	76	37
Resistances		
boundary layer (r_a)	0.76	0.56
stomatal (r_s)	0.94	0.84
intracellular (r_i)	1.10	3.00

The difference in the intracellular resistance between C_4 grasses and C_3 legumes appears to be associated with a difference in transfer rather than a difference in chemical resistance (*Table 2*). Comparable results have been obtained for the C_4 grass, maize, and the C_3 legume, *Phaseolus vulgaris* (3,4). These results are contrary to the more widely held view, arising from other methods of analysis, that the chemical resistance is the most important component of the intracellular resistance in C_3 plants and probably also in C_4 plants (5), and that it probably differs between C_3 and C_4 plants because of the different biochemical pathways of carbon dioxide fixation

(6). Because of this apparent conflict, some of the assumptions in CHARTIER'S method will now be examined and the evidence for the importance of the chemical resistance discussed.

TABLE 2

Transfer and chemical components of the intracellular resistances (sec cm^{-1}) of (a) *C. mucunoides* and the mean of three grasses (*Pennisetum purpureum, Panicum maximum,* and *Cenchrus ciliaris*) and (b) the mean of four legumes (*C. mucunoides, Glycine wightii, Phaseolus atropurpureus,* and *Vigna luteola*) and four grasses (*S. almum, Panicum maximum, Pennisetum purpureum,* and *Brachiaria ruziziensis*). Values were calculated using CHARTIER'S method and assuming that the inhibition of net photosynthesis by 21% oxygen is a valid estimate of photorespiration rate and that all the respiratory flux enters the intercellular spaces.

		$r_{transfer}$	$r_{chemical}$	$\dfrac{r_{chemical}}{r_{transfer}}$
(a)	grasses	0.34 ± 0.040	0.27 ± 0.032	0.75
	legumes	2.21 ± 0.067	0.24 ± 0.035	0.11
(b)	grasses	0.74	0.69	0.80
	legumes	3.25	0.10	0.03

Two important assumptions in CHARTIER'S method are that the transfer resistance is independent of the carbon dioxide concentration inside the cell, and that transfer and chemical resistances are constant over a small range of irradiances approaching light saturation and over which data are obtained. At our present state of knowledge these seem reasonable assumptions and will not be discussed further. In addition, values had to be given to the rate of photorespiration and the proportion of respiratory carbon dioxide entering the intercellular spaces of the leaf. In the absence of evidence to the contrary, I have made the usual assumption that the photorespiration rate was zero in C_4 grasses. Furthermore, I have assumed that the inhibition of net photosynthesis by 21% oxygen is a measure of the photorespiration rate of C_3 legumes, and that all the respiratory flux enters the intercellular spaces.

Whether all or none of the respiratory carbon dioxide enters the intercellular spaces has virtually no effect on the ratio of chemical to transfer resistance (*Figure 1*). Similarly, any error resulting from the inhibition of net photosynthesis by 21% oxygen being an overestimate of photorespiration rate (7) is negligible. The real value of photorespiration rate must lie within the range indicated because the carbon dioxide efflux into carbon dioxide free air is an

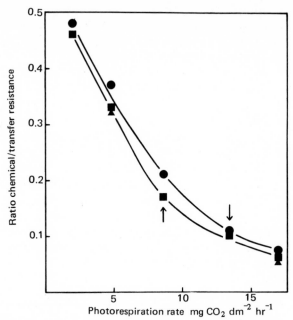

Figure 1 Effects of photorespiration rate and the proportion
of the respiratory flux entering the intercellular space of
the leaf (● = all, ■ = half, ▲ = none) upon the relative sizes
of the transfer and chemical components of the intracellular
resistance of C. mucunoïdes. The ascending and descending arrows indicate the rates of photorespiration as measured by
carbon dioxide efflux into carbon dioxide free air and the inhibition of net photosynthesis by 21% oxygen, respectively.
The other three rates are hypothetical values.

underestimate.

Therefore, the conclusions that the transfer resistance is greater than the chemical resistance in C_3 legumes and of similar size in
C_4 grasses seems justified on the basis of the data presented here
and at the present state of knowledge. Other, though not unequivocal,
data on the influence of mesophyll cell size (8) and decenylsuccinic
acid (9) on net photosynthesis suggest that the transfer resistance
is an important component of intracellular resistance of C_3 plants.
Although C_3 and C_4 plants have different leaf anatomy (10), there is
no information which could explain the possible difference in transfer resistance.

Evidence for the importance of the chemical resistance is circum-
stantial. It is based on "correlations" between the activity of photosynthetic (mainly carboxylation) enzymes and net photosynthetic
rate during aging, between leaves grown at different irradiances,
between plants with the C_4 or the C_3 pathway, and following various
defoliation and growth substance treatments (6). Alternatively, it
is based on low values of the transfer resistance calculated from
subcellular dimensions and liquid phase diffusion coefficients (6,
11). However, before this evidence can be properly evaluated a num-

ber of questions must be answered:

(a) Is the relationship between net photosynthetic rate and enzyme activity causal, or is it merely a correlation? If the relationship is causal, does the enzyme activity determine the rate of photosynthesis or vice versa?

(b) Is it reasonable to suppose that the activity of the carboxylation enzyme limits photosynthesis, or is it more likely that some other enzyme activity or the rate of a particular reaction is limiting?

(c) What effect does carbonic anhydrase, cell size, cell wall, and membrane permeability, and spatial arrangement of organelles have on the transfer resistance?

It is evident that the measurement of the transfer or chemical components of the intracellular resistance both within and between C_3 and C_4 plants is not yet entirely satisfactory, and that a critical analysis of CHARTIER'S method and answers to the above questions are needed.

ACKNOWLEDGMENTS

The assistance of P. Chartier and A. Laisk is gratefully acknowledged. Part of this work was done at the Botany Department, University of Queensland.

REFERENCES

1. M.M. LUDLOW and G.L. WILSON, *Aust. J. agric. Res.*, 21 (1970) 183.
2. M.M. LUDLOW, *Planta*, 91 (1970) 285.
3. P. CHARTIER, M. CHARTIER, and J. CATSKY, *Photosynthetica*, 4 (1970) 48.
4. P. CHARTIER, personal communication.
5. M.D. HATCH and C.R. SLACK, *A. Rev. Pl. Physiol.*, 21 (1970) 141.
6. J.A. RAVEN, *Biol. Rev.*, 45 (1970) 167.
7. M.M. LUDLOW and P.G. JARVIS, in Plant Photosynthetic Production, a Manual of Methods, eds. Z. SESTAK, J. CATSKY, and P.G. JARVIS, Dr. W. Junk, The Hague, in press.
8. D. WILSON and J.P. COOPER, *Nature*, 214 (1967) 989.
9. P.J.C. KUIPER, *Plant Physiol.*, 40 (1965) 915.
10. W.M. LAETSCH, *Am. J. Bot.*, 55 (1968) 875.
11. A. LAISK, *Fiziol. Rast.*, 17 (1970) 40.

THE C₄ PATHWAY RELATED TO GROWTH RATES IN SUGARCANE

T.A. Bull

David North Plant Research Centre, Toowong, Qld. 4066, Australia.

SUMMARY

High photosynthetic rates in recently expanded leaves of sugar-cane were characteristic only of young plants growing in summer. Mature leaves of summer grown plants and young leaves of winter grown plants gave P_{max} values at high light and CO_2 levels similar to those of C_3 plants. In both cases the reduced efficiency and lower rates at P_{max} were attributable to increased stomatal resistance associated with lower light regimes.

Growth analysis showed that maximum rates of dry matter production in sugarcane occurred at the time of year when photosynthetic activity was similar to that of C_3 plants. Comparison with growth in C_4 and C_3 species failed to reveal any inherent superiority of the C_4 pathway and suggested that high growth rates in sugarcane were largely associated with an extended growing season.

INTRODUCTION

The rapid growth rates recorded in some tropical grasses have been linked with the ability to utilize full sunlight and to photosynthesize at rates significantly higher than most temperate species (1,2). Several physiological and morphological characters have been associated with this ability, including the possession of the C_4 photosynthetic pathway (2).

SLATYER (3) compared growth in two species of *Atriplex* which differ in photosynthetic pathway. He showed that while the C_4 species displayed high early growth rates the C_3 species developed a larger leaf area and its growth rate eventually surpassed that of the C_4 species.

The present investigation of photosynthetic responses in sugarcane (*Saccharum* L. hybrid) attempts to assess the importance of the C_4 pathway in making this plant what BURR (1) has termed "one of the most efficient and productive of all crops".

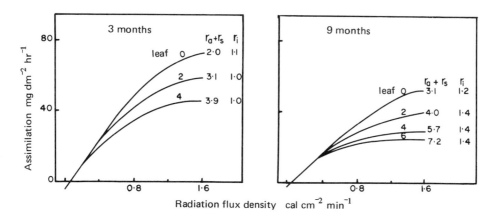

Figure 1 Light response of photosynthesis in leaves from young and old sugarcane (cv. Pindar). Boundary layer plus stomatal resistance ($r_a + r_s$) and internal leaf resistance (r_i) values (sec cm^{-1}) are included. Leaves were numbered successively down the stalk from the most recently expanded (0).

EXPERIMENTAL

The conditions of growth and method of measuring photosynthesis, transpiration, and leaf resistances have been outlined elsewhere (4). All radiation flux density values refer to total radiant flux, not just the 400-700 nm portion of the spectrum.

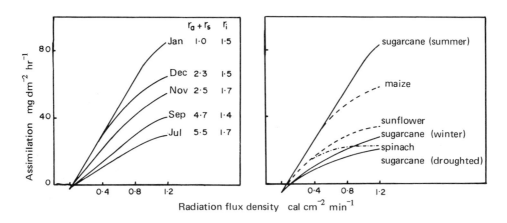

Figure 2 Light response curves for photosynthesis in leaves of (a) sugarcane as a function of time of year and (b) young, old and droughted sugarcane plants compared with maize (C_4), spinach (C_3), and sunflower (C_3).

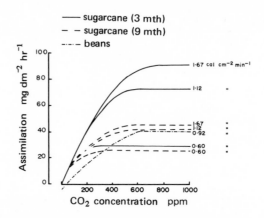

Figure 3 CO₂ response curves for photosynthesis in young and old sugarcane plants measured at three light intensities and compared with the response in French bean (Phaseolus vulgaris).

RESULTS AND DISCUSSION

Light response

Potential photosynthesis rates in sugarcane declined in the older leaves of young and mature plants (*Figure 1*). Little change in the intracellular leaf resistance (r_i) could be measured and the reduction in P_{max} was largely attributable to increased stomatal resistance.

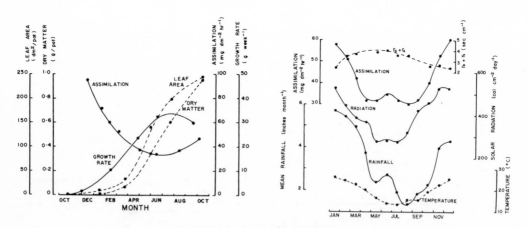

Figure 4 Growth analysis and photosynthesis measurements in sugarcane over the normal growth period.

Figure 5 Seasonal variation in mean monthly assimilation rate, radiation receipt, rainfall, temperature, and leaf resistance ($r_a + r_s$).

Assimilation rates in young fully expanded leaves varied with the time of year due to irreversible changes in stomatal resistance (*Figure 2*). During winter or after severe droughting, P_{max} values in sugarcane leaves were similar to those of C_3 plants but the efficiency at low light intensities was lower.

CO_2 response

Photosynthesis in leaves from young plants responded to CO_2 up to about 700 v.p.m. at 1.67 cal cm^{-2} min^{-1}. In older plants, however, photosynthesis saturated for CO_2 at about 500 ppm and was not responsive to the higher light intensities (*Figure 3*). Although P_{max} in older plants was similar to that for a C_3 species the efficiency at lower CO_2 levels remained unaltered.

Seasonal variation in photosynthesis

Concurrent measurements of photosynthesis rate and whole plant growth rates were made over a normal growing season (*Figure 4*). Potential photosynthetic rate declined from high values (>90 mg dm^{-2} hr^{-1}) early in the season to about 30-40 mg dm^{-2} hr^{-1} from April to September. Dry matter and leaf area production did not show any significant increase until photosynthetic rates in single leaves were at their minimum.

Mean monthly values for assimilation rate and leaf resistance obtained from several years results were found to correlate with seasonal changes in radiation flux density, rainfall, and temperature (*Figure 5*). Although photosynthesis rates in leaves from 12-15 month old plants did not reach the high rates measured in young plants they were almost double the rates prevailing during winter. A less

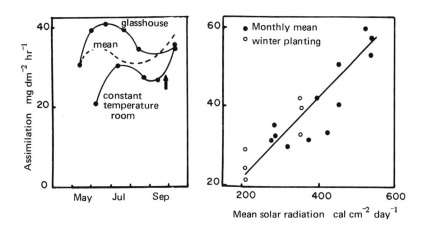

Figure 6 (a) Photosynthesis rates in young plants grown over winter in a glasshouse (ca 350 cal cm^{-2} day^{-1}) or constant temperature room (200 cal cm^{-2} day^{-1}) compared with monthly mean values. The arrow indicates transfer from constant temperature room to glasshouse.
(b) General correlation of mean monthly assimilation rate and daily radiation receipt.

marked, 24 month cycle has been reported from detached leaf measurements in Hawaii (5). The difference in response may arise from the more constant climate of Hawaii coupled with the practice of growing cane for two or more years.

To investigate whether the high rates were characteristic of young plants or related to some component of the environment several varieties were planted in April and grown over winter. All plants were grown in a phytotron at a temperature of 26° and given adequate water and nutrients. Some were grown in a glasshouse and received natural sunlight at an average of 350 cal cm^{-2} day^{-1}. Others were grown in a constant temperature room and given artificial light at an intensity of 0.3 cal cm^{-2} min^{-1} (200 cal cm^{-2} day^{-1}).

Photosynthetic rates in these plants were consistent with the daily radiation level and fitted a correlation between photosynthesis and radiation regime derived from monthly means (*Figure 6*). In neither treatment did the photosynthetic rates approach those measured in similar aged plants grown during summer. The results suggest that photosynthetic ability in sugarcane is largely regulated by seasonal changes in solar radiation receipt.

Photosynthesis and growth

Although the final yield of sugarcane varieties and their wild relatives differed, the photosynthetic capacities of single leaves did not (*Table 1*). The difference found between varieties was of the same order as that found within a single variety even under the most rigorously controlled growth conditions. However, yield was highly correlated with leaf area over the first 5 months of growth indicat-

TABLE I

A comparison of photosynthetic rate (P = mg dm^{-2} hr^{-1}) in leaves of 6 week old plants with dry matter production (g/pot) and leaf area (dm^2/pot), both measured after 5 months growth, in varieties of *Saccharum* spp.

Species	Variety	P*	Dry matter	Leaf area
S. officinarum	Chittan	46	415	150
	HQ409	53	350	116
S. robustum	US 57-86-3	53	291	128
S. spontaneum	SES 327	57	497	173
	SES 356	50	208	86
Commercial	B49-119	48	423	144
	Pindar	52	431	136
	Trojan	47	322	116
	L.S.D. 1%	10	82	28

* Radiation flux density, 1.12 cal cm^{-2} min^{-1}

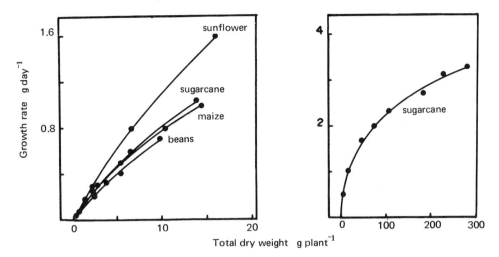

Figure 7 (a) Growth rates in two C₄ plants (maize and sugar-cane) and two C₃ plants (bean and sunflower) grown at 28°. Single leaf assimilation rates measured at 1.12 cal cm^{-2}min^{-1} were, sugarcane 67, maize 61, sunflower 34, and bean 19 mg dm^{-2}hr^{-1}.
(b) Growth rate as a function of leaf area.

ing that rate of leaf area production is of more importance than differences in photosynthetic rate.

Since the high assimilation rates in young sugarcane did not cor-relate with the period of rapid growth a comparison was made with growth in some species. Beans (*Phaseolus vulgaris* L.), maize (*Zea mays* L.), sunflower (*Helianthus annuus* L.), and sugarcane were grown in a controlled temperature glasshouse at 28°. Although these species differ in growth pattern any advantage of the C₄ pathway would be ex-pected to become evident during the early phase of growth. Over the first 40 to 50 days sugarcane exhibited the lowest growth rates de-spite potentially high assimilation rates in single leaves (*Figure 7*). Plotting growth rate against leaf area also failed to demonstrate any superior performance by the C₄ plants.

Similarly when growth rate was expressed against size of plant (*Figure 8*) only sunflower stood out as being more efficient. Subse-quent growth rates in sugarcane exceeded those of the other species but were associated with larger plant size. Apparently sugarcane owes at least part of its reputation as an efficient crop to an ex-tended growing season rather than an inherently superior photosyn-thetic pathway.

Implication of the C₄ pathway

The C₄ pathway was not directly implicated in the rapid growth rates measured in sugarcane. Growth rates were fastest when photo-synthetic rates were similar to those measured in a C₃ plant like sunflower. SLATYER (3) has also reported that growth rates in a C₃ species of *Atriplex* may exceed those of a C₄ species and WARREN WIL-SON (6) measured similar net assimilation rates in C₃ and C₄ plants

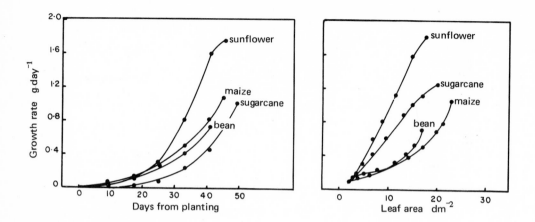

Figure 8 Growth rate as a function of plant size in two C₄ and two C₃ plants.

grown under a high radiation regime.

LAETSCH (7) and DOWNES (8) have suggested that the superior efficiency of water use in C₄ plants may have endowed them with a selective advantage over C₃ plants when water stress limits stomatal aperture. Although this may be true for some species it appears to be of minor importance in sugarcane. The ratio of photosynthesis to transpiration (P/T) was highest (0.07) in sugarcane during summer when water stress was low. During the cool, dry winter P/T dropped to about 0.025 and was not greatly different from the value of 0.02 measured in several C₃ plants.

Present evidence shows that maximum C₄ activity in sugarcane occurs when daily radiation receipt and leaf temperatures are highest. Conceivably temperature tolerance may be important in providing a selective advantage for some C₄ species. Sugarcane can maintain full photosynthesis at leaf temperatures of 40° and can recover from several hours at 55°. Such temperatures are lethal to many C₃ species. On the other hand leaf temperatures of 18° or less greatly inhibit photosynthesis in sugarcane but do not adversely affect most C₃ plants. However, examples of C₃ species tolerant of high temperatures can be cited and suitable selection pressures may produce tolerant strains from most species.

Present evidence indicates that the C₄ pathway confers no advantage to plant growth which cannot be circumvented by C₃ species should the required selection pressure arise.

REFERENCES

1. G.O. BURR, C.E. HARTT, H.W. BRODIE, T. TANIMOTO, H.P. KORTSCHAK, D. TAKAHASHI, F.M. ASHTON, R.E. COLEMAN, *A. Rev. Pl. Physiol.*, 8 (1957) 275.

2. M.D. HATCH and C.R. SLACK, in Progress in Phytochemistry, ed. L. REINHOLD and Y. LIWSCHITZ, Interscience, London, 2 (1970) 35.
3. R.O. SLATYER, *Planta*, 93 (1970) 175.
4. T.A. BULL, *Crop Sci.*, 9 (1969) 726.
5. H.P. KORTSCHAK and A. FORBES, in Progress in Photosynthesis Research, ed. H. METZNER, I.U.B.S., Tübingen, 1 (1969) 383.
6. J. WARREN WILSON, *Ann. Bot.*, 31 (1967) 41.
7. W.M. LAETSCH, *Am. J. Bot.*, 55 (1968) 875.
8. R.W. DOWNES, *Planta*, 88 (1969) 261.

RELATIONSHIP BETWEEN PLANT GROWTH AND LEAF PHOTOSYNTHESIS IN C$_3$ AND C$_4$ SPECIES OF *Atriplex*

R.O. Slatyer

Research School of Biological Sciences,
Australian National University, Canberra City, 2601, Australia.

SUMMARY

Photosynthesis, growth, and transpiration of C$_3$ and C$_4$ species of Atriplex were compared over a period of 23 days.

The results provided clear evidence that, although rates of leaf photosynthesis in the C$_4$ species (A. spongiosa) were always as high as, or higher than, in the C$_3$ species (A. hastata), overall plant growth rates were controlled primarily by leaf area. At the beginning of the experiment, growth was more rapid in A. spongiosa but by the end of the period growth was more rapid in A. hastata, because this species put consistently more of its photosynthate into new leaf production.

Transpiration rates were consistently lower in A. spongiosa than in A. hastata and the ratio declined slightly during the experiment. In consequence, water use efficiency, both on a single leaf and whole plant basis, was much greater in the C$_4$ species.

INTRODUCTION

Considerable interest has been evoked in recent years by the demonstration of higher rates of photosynthesis in leaves of plants exhibiting C$_4$ metabolism (1). This interest has been reinforced by the demonstration that transpiration rates of the leaves of C$_4$ plants may be lower than those of C$_3$ species (2,3,4).

If these phenomena apply to whole plants, and to plant communities, as well as to single leaves, it is clear that they have far-reaching ecological and agronomic implications. Accordingly, it was decided to examine these relationships, and to compare rates of leaf photosynthesis and transpiration with rates of total plant growth and water use, over an extended period of time.

For plant material, the genus *Atriplex* was selected since it contains species with both C$_3$ and C$_4$ metabolism (3,5,6). A C$_4$ species

76

endemic to high radiation arid environments (*A. spongiosa*) and a C$_3$
species endemic to humid coastline regions (*A. hastata*) were used. A
full account of these experiments is reported elsewhere (7).

MATERIALS AND METHODS

Preparation

Seedlings of *A. spongiosa* and *A. hastata* were grown in water culture in a naturally lit glasshouse in early summer (November-Decem-
ber).

Temperature regulation of the glasshouse prevented day temperatures from rising above 32° and night temperatures from falling below 20°. Radiation levels during the pre-experimental preparatory
phase were relatively low, but sunny conditions prevailed throughout
most of the experimental phase and incident radiation levels averag-
ed 570 cal cm^{-2} day^{-1}. The experimental stage commenced when the
plants were about 3 weeks old and had developed approximately 8
leaves. At this stage 12 uniform plants of each species were selected from the large number originally sown, and grouped into 6 pairs.

General experimental procedure

On day 0 of the experimental period, one pair of plants was harvested and fresh weight, dry weight, dry weight distribution (leaf,
stem and fruit and root), and leaf area were determined. On days 7,
12, 17, and 23, four more pairs were harvested and the same data
collected.

Prior to day 7, selected leaves on all plants were tagged as
"standard" young leaves. These were well formed young leaves, at
morphologically reproducable locations on each plant, which aged
progressively as the experiment continued.

On each of the subsequent sampling occasions, leaf chamber measurements of the primary CO$_2$ and water vapor exchange characteristics were made on one plant of each pair prior to harvest. These
measurements were made on the aging tagged leaves, and also on
leaves assessed as "standard" young leaves on each occasion. By this
procedure it was possible to distinguish between the effects of
aging of a plant as a whole from that of an individual leaf. For
further details see (7).

RESULTS

In *Figure 1* the patterns of change in total growth rate, leaf
area, and net assimilation rate, obtained from the whole plant responses, are depicted for both species.

At the beginning of the experimental period the total growth rate
and leaf area of *A. spongiosa* was substantially greater than that of
A. hastata (x 2.2 and x 1.7 respectively). The net assimilation
rate, expressed per unit leaf area was also greater (x 1.3).

As the experiment continued, however, a progressive decline in
the ratio of total growth rate and leaf area occurred until, by the

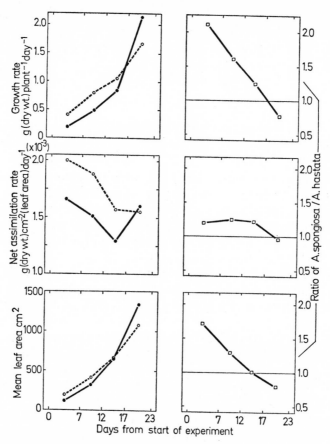

Figure 1 Changes in growth rates, net assimilation rates, and leaf areas, of A. spongiosa (o-----o) and A. hastata (●————●) during the experimental period. The ratios of each of these parameters (□————□) are given in the three diagrams on the right hand side of the Figure. From (7) with permission.

final sampling occasion, both were less in *A. spongiosa* than in *A. hastata*. Although the ratio of the net assimilation rates also declined gradually, to a level of about unity, it is clear that the most important factor causing a change in the pattern of total growth rate was the change in leaf area. During the experimental period, therefore, a consistently higher proportion of photosynthate was used to produce new leaf material in *A. hastata* than was the case in *A. spongiosa*.

In *Figure 2*, data for total water use per plant, transpiration rate, and water use efficiency are plotted. Initially, water use in *A. spongiosa* was also significantly higher (x 1.4) than in *A. hastata* and, as was the case with growth rate, a progressive decline occurred during the experiment, final values being about 0.7 x *A. hastata*. Transpiration rate, per unit leaf area, remained at a relatively constant level throughout the experiment, and was substantially lower in *A. spongiosa* than *A. hastata*. In consequence, water use

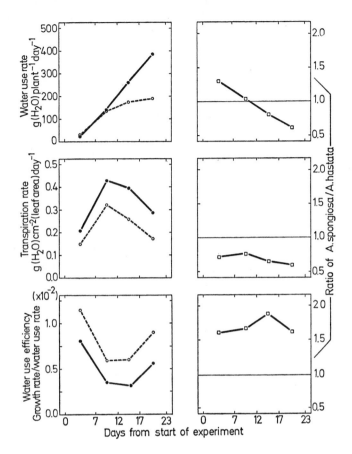

Figure 2 *Changes in water use rates, transpiration rates, and water use efficiency of A. spongiosa (o-----o) and A. hastata (●————●) during the experimental period. The ratios of each of these parameters (□————□) are given in the three diagrams on the right hand side of the Figure. From (7) with permission.*

efficiency (growth rate/water use rate, or net assimilation/transpiration rate) was substantially higher in A. *spongiosa* than in A. *hastata* and remained relatively constant during the experimental period.

In *Figure 3* gas exchange data from single leaves is presented for both species. The pattern of net photosynthesis, measured in young and aging leaves under light saturated conditions and ambient CO_2 concentrations, closely paralleled the equivalent data (net assimilation rate) from the whole plants; the initial levels in A. *spongiosa* being substantially higher than in A. *hastata*; the final levels being similar, and a progressive decline in A. *spongiosa* being associated with relatively constant values in A. *hastata*.

The data for total resistance to CO_2 transfer reflect these trends, and the data for intracellular resistance, r_i, and stomatal resistance, r_s, explain the patterns observed. It is apparent that r_i levels in A. *spongiosa* were always lower than in A. *hastata*,

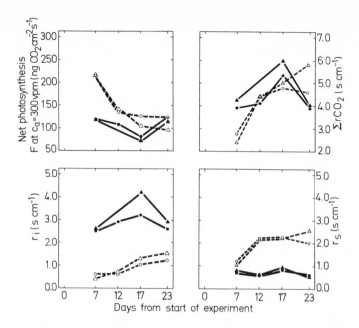

Figure 3 Changes in primary characteristics of CO_2 exchange and water vapor exchange of young (o-----o) and aging (Δ-----Δ) leaves of A. spongiosa, and young (●———●) and aging (▲———▲) leaves of A. hastata during the experimental period.

associated with the more efficient carboxylation in this species. The levels of r_i were initially very low (about 0.5 sec cm^{-1}) and comparable in both young and aging leaves. During the experimental period they increased to final levels of about 1.5 sec cm^{-1}. In *A. hastata* the initial and final levels were about 2.5 and 3.0 sec cm^{-1} respectively, although some intermediate points showed higher values.

By comparison, stomatal resistance in *A. spongiosa* was always higher than in *A. hastata* and increased during the experimental period from about 1.0 to 2.5 sec cm^{-1}, whereas they remained relatively constant in *A. hastata* at about 0.7 sec cm^{-1}.

DISCUSSION

The primary objective of this experiment was to determine whether or not a species possessing C_4 metabolism was necessarily more productive than one with C_3 metabolism. The results have provided clear yet simple evidence that the interaction between metabolism and morphogenesis can yield a range of productivity patterns, with a key determinant being the degree to which contemporary photosynthate is utilized in the development of additional photosynthetic surface as compared with non-photosynthetic material.

Therefore, even though the C_4 species, *A. spongiosa*, maintained higher net assimilation rates (and higher leaf net photosynthesis rates) than the C_3 species, *A. hastata*, until the final sampling occasion, the fact that the ratio of its leaf area to that of *A. has-*

tata declined from 1.7 to 0.8 during the experimental period was the main reason why the total growth rate of *A. spongiosa* dropped from a level more than twice as high as *A. hastata* at the beginning of the experiment to a value less than *A. hastata* by the end of the measurement period.

It is of interest that, despite the lower r_i levels, net leaf photosynthesis rates (and whole plant net assimilation rates) in *A. spongiosa* did actually decline to levels comparable to those in *A. hastata* by the end of the experiment. This was caused by a progressive and pronounced increase in stomatal diffusive resistance r_s, (particularly in the older leaves), so that the total diffusive resistance (ΣrCO_2) in *A. spongiosa* finally exceeded that in *A. hastata*.

In applying these results to field conditions, great caution must be exercised. In the first place, community structure and the distribution of leaf area become relatively more important in determining total productivity than specific leaf photosynthesis. Secondly, where economic yield, of grain for example, is under consideration many additional factors are involved and these may be only slightly determined by rates of leaf photosynthesis.

The transpiration data must also be applied to field conditions with caution because the magnitude of the external resistance, r_a, can vary widely depending on air movement and on community, as well as plant structure, and is frequently a much more important determinant of community transpiration than r_s.

In consequence, the water use efficiency values obtained for single plants may not be reproduced in the field, particularly when it is appreciated that, apart from the community effects already referred to, the relative and absolute values of r_s and r_i can be expected to fluctuate widely depending on such factors as heat, water, or nutrient stress, as well as endogenous factors such as age. However, the tendency of *A. spongiosa* to have a higher r_s value and a lower r_i value than *A. hastata* appears to be genotypic in origin and may therefore be reflected in some aspects of field behavior.

REFERENCES

1. M.D. HATCH and C.R. SLACK, in Progress in Phytochemistry, eds. L. REINHOLD and M. LIWSCHITZ, Wiley, Interscience, N.Y., 2 (1970) 35.
2. J. HESKETH, *Planta*, 76 (1967) 371.
3. G. HOFSTRA and J.D. HESKETH, *Planta*, 85 (1969) 228.
4. R.W. DOWNES, *Planta*, 88 (1969) 261.
5. W.J.S. DOWNTON, T. BISALPUTRA, and E.B. TREGUNNA, *Can. J. Bot.*, 47 (1969) 915.
6. C.B. OSMOND, *Biochim. biophys. Acta*, 172 (1969) 144.
7. R.O. SLATYER, *Planta*, 93 (1970) 175.

FEEDBACK CONTROL OF PHOTOSYNTHESIS IN SUGARCANE

K.T. Glasziou and T.A. Bull

David North Plant Research Centre, Toowong, Qld. 4066. Australia.

SUMMARY

End-product repression of photosynthesis by sucrose has been reported in detached leaves of sugarcane (1). In repeating this work source-sink relationships for photosynthesis were modified in several ways. Detached leaves were dipped into osmotic solutions or osmotic solutions were pumped through leaves. Leaves attached to the plant were girdled with steam or wax to kill the phloem and prevent translocation without affecting water transport. Whole plants were moved from growing to non-growing environments to induce ripening and produce increased sugar storage. In all cases any reduction in photosynthesis was accompanied by a concomitant increase in stomatal resistance. No conclusive evidence was obtained to support the existence of direct end-product repression of photosynthesis in sugarcane.

INTRODUCTION

Saccharum species and commercial sugarcane varieties differ widely in their morphology and yield of dry matter. Measurements of photosynthesis in a range of varieties grown under controled conditions have shown that photosynthetic capacities of single leaves are not sufficiently different to account for the variation in final yield (2). During the early phases of growth yields generally correlated with leaf area but once a full canopy has been obtained factors other than leaf area become limiting and varietal susceptibility to stress may prevent full photosynthetic capacity being achieved under field conditions.

One feature common to all sugarcane varieties is their ability to store high levels of sugar in the stalk. Sucrose levels in mature internodes may commonly exceed 20% fresh weight and, during ripening, leaf sugars may rise from <1% to about 2% and sugar levels in the apical stalk tissue increase from 1% to about 5%. Although photosynthesis declines over the year it tends to remain relatively constant over the ripening period despite the increasing sugar levels in the remainder of the plant (2). The current investigation was undertaken to obtain quantitative data on the role of feedback or end-product repression in causing yield variation in sugarcane.

EXPERIMENTAL

Growth conditions and methods of measuring photosynthesis, transpiration, and leaf resistances have been presented elsewhere (3,4).

RESULTS AND DISCUSSION

HARTT(1) has reported that rates of photosynthesis in detached sugarcane leaves decline after several hours illumination and that this reduction in photosynthesis was accompanied by a build up in sucrose in the leaf. Initial rates were completely restored following a dark period. HARTT interpreted these results as indicating an end-product repression of photosynthesis.

Measurements were made on detached leaves by WALDRON *et al.* (4) using a photosynthesis chamber similar to that used by HARTT and the phenomenon she reported was readily reproduced. However, although the decline in photosynthesis was accompanied by an increase in leaf sugar level, stomatal resistance also increased. Interposing a dark period allowed photosynthesis rates and stomatal apertures to recover but no change could be measured in leaf sugar level.

Initial rates of photosynthesis could be restored in leaves with depressed rates by treating them with ammonia vapor to cause stomatal opening. This suggests that inhibition of photosynthesis arose from stomatal closure rather than direct end-product repression. Consequently the inhibition of photosynthesis is not a simple function of sucrose concentration in the whole leaf but it may correlate with sucrose levels in a particular compartment.

We have continued our investigations using leaves still attached to the stalk so that water stress effects common in detached leaves can be avoided. Shading or removing other leaves on the plant caused a slight but significant increase in photosynthesis but this correlated with an improved water status in the leaf due to reduced transpiration by the remainder of the plant.

However, the following evidence suggests that photosynthesis may vary in a manner unrelated to transpirational losses and therefore indicates that controls other than stomatal aperture may operate.

Firstly, 42 large plants were placed in an airconditioned glasshouse which could be used as a giant photosynthesis chamber. The plants were grown at a low temperature (17° day/10° night) for three weeks and then temperatures were increased to 24°/19°. Little change in photosynthesis occured over the next 2 weeks despite a doubling in the transpiration rate. However, over the following 11 days photosynthesis doubled while transpiration remained unaltered. Since no change in leaf area or canopy structure occured over this period the

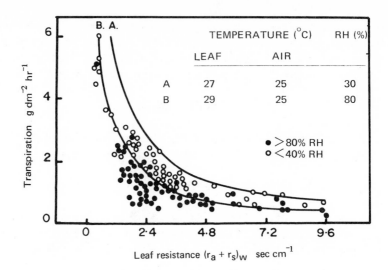

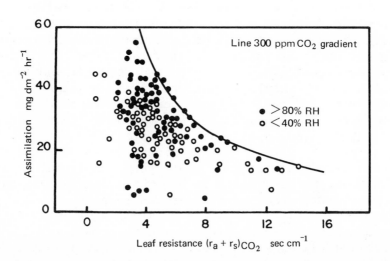

*Figure 1 Transpiration and photosynthesis measured at two re-
lative humidity levels and related to leaf resistance. Lines
A and B were calculated from leaf and air temperatures common
during actual measurements at each humidity. The line in the
assimilation curve was calculated assuming a 300 ppm gradient from ambient air to the sites of photosynthesis and represents P_{max}.*

increase in photosynthesis occured in those leaves already present.
   Secondly, measurements of photosynthesis, transpiration, and leaf
resistance were made on single attached leaves at humidities below
40% and above 80% RH. Transpiration rates fell close to the theoretical curves obtained by calculation from prevailing leaf and air

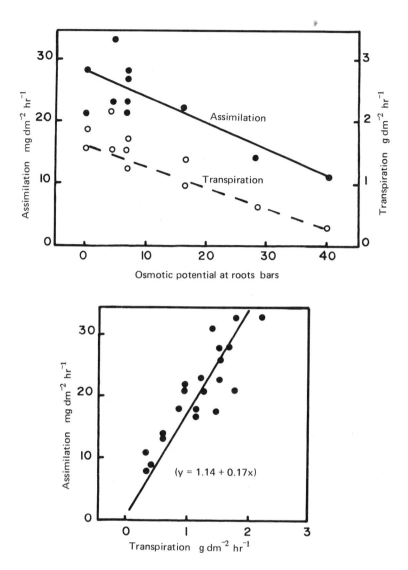

Figure 2 *Depression of photosynthesis and transpiration by the application of mannitol solutions to the roots.*

temperatures (*Figure 1*). On the other hand photosynthesis rates filled a large part of the area beneath the P_{max} curve calculated by assuming a 300 ppm CO_2 gradient from the ambient air to the site of photosynthesis. Many of the very low values fell below what would be expected from known sources of variation such as age of plant, time of year, *etc.* and may indicate controls of photosynthesis which operate independently of stomatal resistance.

One possible cause of depressed photosynthesis would be end-product inhibition caused by sucrose build up in the xylem. In mature stalks xylem sugar levels can be as high as 0.2 to 0.3 M (5). Girdling a leaf with steam or hot wax caused leaf sugars above the girdle to increase from 0.5 to over 4.5% fresh weight. Although marked

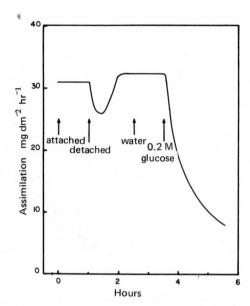

Figure 3 *Photosynthesis measurements on a leaf of an apical shoot of sugarcane. The shoot was detached under water and water, followed by 0.2 M glucose, was pumped through the shoot and leaf by a peristaltic pump connected to the base of the shoot.*

repression of photosynthesis occured during the experiment it was accompanied by an increase in leaf resistance. Similarly, placing the cut ends of detached leaves in 0.2 M osmotic solutions or apply-ing osmotic solutions to the roots also caused a rapid depression of photosynthesis *(Figure 2)*. However, the depression was directly cor-related with decreased transpiration rates and was proportional to the osmotic potential applied, again indicating regulation by sto-mates.

Apical buds of a mature sugarcane stalk can be induced to develop into a side shoot which physically resembles a young sugarcane plant. TANIMOTO (personal communication) showed that photosynthesis in these shoots is similar to the rates prevailing in leaves in the main stalk and not to the high rates measured in young plants. Such a result would be expected from HARTT'S hypothesis due to repression by sugar levels in the stalk tissue.

A series of measurements was made on the leaf of a side shoot *(Figure 3)*. An initial photosynthetic rate was established and the shoot was then detached under water and allowed to re-establish a steady rate which was slightly higher than the original. A peristal-tic pump was attached to the base of the shoot and water was pumped up. The end of the leaf protruding from the chamber was trimmed so that water could drip out of the xylem vessels of the cut end. This treatment rapidly flushed out all shoot and leaf xylem sugars but did not alter the photosynthetic rate. When the water was replaced by 0.2 M glucose or sucrose then photosynthesis was rapidly inhibited and leaf resistance increased from 4 to 25 sec cm^{-1}.

The possibility that depressed rates of photosynthesis resulted from increased sucrose in leaf xylem was checked by direct sampling of xylem vessels. Microcapillary tubes were inserted into xylem vessels of leaves in which the maximum rate of photosynthesis had fallen to about 10-15 mg dm^{-2}hr^{-1} (GAYLER, personal communication). Sugar levels were below the limit of detection (0.0006 M).

These results leave us with no evidence that sugar in the leaf xylem is responsible for reduced rates of photosynthesis. Osmotic solutions applied to the roots or introduced into the xylem appear to act primarily on stomatal resistance and do not provide evidence of feedback control on photosynthetic processes.

TABLE I

Effect of an extended period of low temperature (16°) on photosynthesis and leaf resistance in recently expanded leaves of sugarcane. Light intensity 1.12 cal cm^{-2}min^{-1}, 80% RH

Time (Weeks)	Treatment	Photosynthesis (mg dm^{-2}hr^{-1})	$r_a + r_s$ (sec cm^{-1})
0		35	4
6		37	5
12		14-20*	11-17*
17	Plant growing slowly	29	6
"	Plant not growing	1-11*	10-90*
"	Mature leaf - above plant	13-20*	8-16*
21		6	38
"	Roots @ 30° for 3 days	6	37
"	Leaf @ 24° for 3 days	2-9*	32-80*
"	Water pumped up leaf	17	9
"	Plant @ 24° for 3 days	38	3

*Variable results due to oscillating stomatal apertures

The effect of transfering plants from a growing to a non-growing environment was followed in large plants placed at 16° for an extended period. Photosynthetic rates in already expanded leaves were little affected but leaves which developed at the low temperatures exhibited rates below 10 mg dm^{-2}hr^{-1} (Table 1). These low rates were associated with a severe depression of apical internode elongation, increased sugar levels in the apical internodes and a marked increase in stomatal resistance. However, a leaf attached to an older internode with higher sugar levels was less affected by the cold treatment and had potential rates about double those in young leaves.

Raising the root or leaf temperature separately did not affect the inhibited rate of photosynthesis.

An increased leaf water content partially reversed the inhibition but normal photosynthesis rates were only obtained when the whole plant was moved to a higher temperature (24°) for 3 days, i.e. when growth resumed. Although these changes are consistent with a feedback repression by stored sucrose all changes in photosynthesis were accompanied by changes in stomatal resistance. Consequently these results provide no evidence to link sucrose level with photosynthetic repression other than via an effect on leaf water potential.

In summary we conclude that HARTT'S hypothesis of sucrose repression of photosynthesis in sugarcane was based on experiments in which no satisfactory evidence was provided that the detached leaves did not develop water deficits on exposure to high light intensity for long periods. We have induced low rates of photosynthesis in detached leaves, in steam girdled leaves, and in whole plants but these were always accompanied by increased leaf resistance. Repression of photosynthesis induced by environmental stress was reversed by placing plants under favorable environments for relatively short periods but in some cases the recovery of photosynthesis rate lagged behind increased transpiration.

These results may be explicable in terms of either end-product feedback or of a lesion arising from extreme environments but we have no evidence to separate these phenomena. Sucrose phosphate phosphatase in sugarcane leaves is inhibited by sucrose (6) and it is probable that sucrose repression of photosynthesis may occur by this or other biochemical mechanisms. However, experience with field and phytotron measurements leads us to think that the major control over photosynthesis is provided by stomatal response to light, carbon dioxide concentration, and in the present context, to leaf water potential.

REFERENCES

1. C.E. HARTT, *Naturwissenschaften*, 21 (1963) 666.
2. T.A. BULL, This volume, (1971).
3. T.A. BULL, *Crop Sci.*, 9 (1969) 726.
4. J.C. WALDRON, K.T. GLASZIOU, and T.A. BULL, *Aust. J. biol. Sci.*, 20 (1967) 1043.
5. J.S. HAWKER, *Aust. J. biol. Sci.*, 18 (1965) 959.
6. J.S. HAWKER, *Biochem. J.*, 102 (1967) 401.

A RELATIONSHIP BETWEEN NET PHOTOSYNTHESIS, DIFFUSIVE RESISTANCE, AND CARBOXYLATING ENZYME ACTIVITY IN BEAN LEAVES

T.F. Neales, K.J. Treharne,[1] and P.F. Wareing[2]

Botany Department, University of Melbourne,
Parkville, Vic., 3052, Australia

SUMMARY

Leaves of Phaseolus vulgaris of widely differing rates of net photosynthesis were obtained by using a technique of partial defoliation. The predominant part of the differences in net photosynthesis could not be attributed to changes in leaf stomatal resistance. The activities of ribulose-1,5-diphosphate carboxylase in these leaves were positively and significantly correlated with net photosynthesis ($r^2 = 0.77$). The activities of other enzymes were less well correlated with net photosynthesis. It is concluded that the predominant rate-limiting factor in net photosynthesis, at saturating light intensities and ambient CO_2 concentrations of about 360 ppm, was the rate of carboxylation, and not the rate of diffusive transfer of CO_2 to the photosynthetic sites.

INTRODUCTION

MONTEITH (1) introduced the term carboxylation resistance (r_x'), when discussing the resistances to CO_2 transport towards the photosynthetic sites in a leaf. If this resistance assumes rate-limiting values, then "enzymic reactions and other processes may limit the rate of photosynthesis when both light and CO_2 are abundant" (2). Since 1965 experimental evidence has been published which indicates that carboxylating enzyme activity may indeed be a determinant of the rate of net photosynthesis (P) (3-8). In this paper we also establish a positive correlation between the rates of P and the activity of ribulose-1,5-diphosphate (RuDP) carboxylase.

---

[1] *Welsh Plant Breeding Station, Aberystwyth, U.K.*

[2] *Botany Department, University College of Wales, Aberystwyth, U.K.*

89

EXPERIMENTAL

Plant material and sampling procedure

Plants of *Phaseolus vulgaris* (Var. Canadian Wonder) were grown
for 20 days in soil from seed in a controlled environment (25°, 2000
ft-c, 16 hr photoperiod). We used the established phenomenon (7,9,
10) that partial defoliation reduces the apparent senescence rate of
the remaining leaves, the gas exchange and enzyme content of which
were periodically measured.

Gasometric methods

The water vapor and CO_2 fluxes of leaves were measured psychro-
metrically and by infra red gas analysis respectively. The methods
were similar to those of SLATYER (11) and CHARTIER *et al.* (12).

Enzyme assay procedures

The preparation of enzyme extracts from leaves was similar to
that of TREHARNE and COOPER (13), as was also the reaction mixture
of the assay of RuDP carboxylase. No evidence of the presence of
partial inhibitors in the leaf extracts was obtained. Enzyme reaction rates were linear over a period of 15 minutes; a 5 minute incubation time was used. An incubation mixture without RuDP was assayed on each occasion, giving a control value of acid-stable radioactivity. Other enzymes assayed, and the methods used, are indicated
below. Phosphoenolpyruvate (PEP) carboxylase (13), alkaline fructose
1,6 diphosphatase (FDPase) (14), ribose-5-phosphate (R-5-P) isomerase
(15), phosphoribulokinase (16), phosphoglycerate kinase (17).

RESULTS AND DISCUSSION

*The effect of partial defoliation on the rates of net photosynthesis
and on RuDP carboxylase activity*

The time-course of the effects of partial defoliation on CO_2 and
water vapor exchange was followed over 4 days. Three replicates were
separately measured at each sampling time. The corresponding leaves
of similarly treated plants were assayed for RuDP carboxylase activity.

Figure 1 shows that in the control (undefoliated) plants, there
were reductions to 41% of the initial values, in both P and RuDP
carboxylase activity over 3 days. The values on day 4 were similar.

However, in the leaves of the partially defoliated plants, P increased to 125% of the initial value on day 3 and the mean value of
RuDP carboxylase activity was unaltered (*Figure 1*). On the fourth
day after defoliation, P declined to 115% of the initial value and
the individual measurements of RuDP carboxylase activity were highly
variable. Partial defoliation thus delays the decay, with time, of
the photosynthetic capacity of bean leaves and the decrease in P in
the control plants is accompanied by a similar decrease in the activity of RuDP carboxylase activity. These observations accord with
those of WOOLHOUSE (6,10) using *Perilla* leaves and indicate that one
of the determinants of P at high irradiance and an ambient CO_2

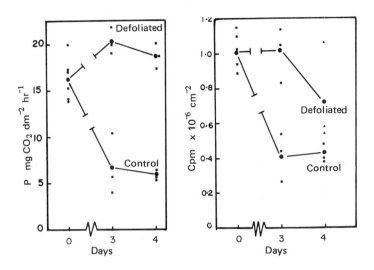

Figure 1 *The time-course of P and RuDP carboxylase activity in the monofoliate leaves of bean plants after partial defolia-tion on Day 0. (For conditions of experiment, see Table 2A).*

concentration of c.360 ppm is the activity of RuDP carboxylase, thus supporting the evidence cited above.

The correlation between P and the activity in the leaf of six en-zymes involved in photosynthesis

The strong association between net photosynthetic rate and the activity of RuDP carboxylase in bean leaves has already been shown (*Figure 1*). However, there remains the possibility that this is a spurious correlation: not indicative of a casual association. There-fore, in the same experiment, we also measured the activities of five other photosynthetic enzymes (listed in *Table 1A*) and the sol-uble protein concentration in the supernatant assay mixture. Three independent measurements were made on each leaf. The net photosyn-thetic rate of similarly treated leaves was also measured (*Figure 1*).

The correlation coefficient (r) relating the mean values of P and enzyme activities (*Table 1A*) are shown in *Table 1B*. P is most close-ly (r = 0.88 and significant at P = 0.05) correlated (*Table 1B*) with RuDP carboxylase activity, followed (in order of value of r^2) by positive correlations with soluble protein (r^2 = 0.62, P = 0.10), PEP carboxylase (r^2 = 0.59, P > 0.10) and malate dehydrogenase (r^2 = 0.05, P > 0.10). There was no association between P and the other enzymes assayed.

These results reinforce the previous observation that, within the measurements made, the correlation between P and RuDP carboxylase is the strongest. The association between P and the soluble protein concentration may be a reflection of the high concentration of "Fraction I" protein in the leaf extracts (18,19). The correlation between P and PEP carboxylase and between P and malate dehydrogenase

TABLE IA

The changes in net photosynthesis (P, mg CO_2 dm^{-2} hr^{-1}) and enzyme activities in control (cont.) and defoliated treatments (defol.) for 4 days after partial defoliation (six replicates at day 0 and three each sampling thereafter).

	Day 0	Day 3		Day 4	
		Cont.	Defol.	Cont.	Defol.
Net photosynthesis(P)	16.3	6.7	20.3	5.9	18.7
Enzyme activity					
RuDP carboxylase*	1.01	0.40	1.01	0.43	0.72
PEP carboxylase*	0.11	0.07	0.11	0.10	0.14
Alkaline FDPase**	4.70	3.24	1.62	3.24	3.42
Phosphoribulokinase**	3.80	1.44	0.90	-	-
R-5-P isomerase***	1.12	1.47	1.00	2.81	3.10
Malate dehydrogenase#	1.40	1.11	1.84	1.14	3.00
Soluble protein##	0.31	0.32	0.43	0.30	0.44

  * Acid stable radioactivity, $\times 10^{-6}$ cpm cm^{-2} min^{-1}
 ** Pi released or incorporated, $\times 10^{-1}$ µmoles cm^{-2} min^{-1}
*** $\times 10^{-2}$ ΔOD (540 nm) cm^{-2} min^{-1}
  # $\times 10^{-2}$ ΔOD (340 nm) cm^{-2} min^{-1}
mg cm^{-2}

TABLE IB

The correlation between P and enzyme activities from Table IA

Enzyme	Correlation coefficient(r)	r^2	Probability
RuDP carboxylase	0.88	0.77	0.05
PEP carboxylase	0.77	0.59	0.20
Alkaline FDPase	-0.22	-	-
Phosphoribulokinase	-	-	-
R-5-P isomerase	-0.23	-	-
Malate dehydrogenase	0.71	0.50	0.20
Soluble protein	0.79	0.62	0.10

may be insignificant but, if real, could be interpreted to indicate that photosynthetic carboxylations by PEP carboxylase, in a C_3 leaf, are significant but of smaller magnitude than those effected by RuDP carboxylase.

Effects of partial defoliation on the leaf diffusive resistances

    Measurements of the transpiration rate of the leaves of control and defoliated plants indicated (*Table 2A*) that the leaves of the control plants had lower transpiration rates, as well as lower

TABLE 2A

Effect of partial defoliation on net photosynthesis (P, mg CO_2 dm^{-2} hr^{-1}) and transpiration (mg H_2O dm^{-2} hr^{-1}). (Irradiance at leaf surface 0.49 cal cm^{-2} min^{-1}; leaf temperatures, 26.4°-27.4°; mean CO_2 concentration at inlet to leaf chamber, 364±10 ppm.)

Sampling time	P	P(adjusted)*	Transpiration rate
Day 0	16.3	-	2634
Day 3			
Control	6.7	10.7	2130
Defoliated	20.3	-	3183
Day 4			
Control	5.9	9.6	1723
Defoliated	18.7	-	2925

* P(adjusted) = P of control plants when "donated" with the r_1 values of defoliated plants; calculated from new estimates of C_w and the actual estimates of r_r.

TABLE 2B

Effect of partial defoliation on leaf diffusive resistances to CO_2 transfer (sec cm^{-1}). Conditions as in Table 2A.

Sampling time	Diffusive resistances			
	r_a	r_1	r_r	Σr
Day 0	0.7	2.3	11.7	14.7
Day 3				
Control	0.7	3.2	32.2	36.1
Defoliated	0.7	1.4	9.6	11.7
Day 4				
Control	0.7	4.4	35.1	40.2
Defoliated	0.7	1.8	10.1	12.6

values of P, than those of the treated plants. Hence it is necessary to apportion the observed decreases in P in the control plants (*Figure 1*) to that which can be attributed to increases in the gas-phase resistances, r_a and r_1 (common to both water vapor and CO_2 transport) and that attributable to increases in the residual resistance (r_r) (20). When the residual resistance is measured in the linear part of a CO_2 response curve it equals the intracellular resistance (r_i) of other workers. The "adjusted" P values (*Table 2A*) are the mean net fluxes of CO_2 that the control plants would have if these leaves were given the smaller values (*Table 2B*) of r_1 of the

defoliated plants. These adjusted values of P of the control plants
on day 3 and day 4 are 53% and 51% respectively of the values of the
defoliated plants (*Table 2A*). In other words, the predominant increase of the total diffusive resistance to CO_2 transfer in the
leaves of the control plants (*Table 2B*) which caused the large decreases in P (*Figure 1*), is not in the stomatal resistance component,
but in the residual (nongas phase) resistance (21). Also the correlation coefficient of P with $1/r_r$ was 0.996 (P<0.001), indicating
that the effects of changes in stomatal resistance on P are negligible.

*Which component of the residual resistance to CO_2 transfer is affect-
ed by defoliation*

It has been shown that the large decreases in P in defoliated
plants (*Figure 1*) can be attributed to large increases (*Table 2B*) in
the residual resistance (r_r). MONTEITH (1) proposed that r_r can be
apportioned into components as follows:

$$r_r = r_m + r_x + r_e \qquad (1)$$

where r_m = liquid diffusion resistance, r_x = carboxylation resistance, r_e = excitation resistance. r_x and r_e are not diffusive resistances and have been designated (21) "chemical resistances".
Hence there remains the problem of apportioning the observed increases in r_r (*Table 2B*) to one of the three components in equation
(1).

Figure 2 shows that, at limiting light intensities, the photochemical efficiencies of the leaves of control and treated plants
are similar. Since r_e has been defined (1) as a resistance inversely
proportional to irradiance (I), and because at limiting light intensity P $\propto$ I, it follows that $r_e \propto 1/P$. Thus the similarity of the response of P to low irradiance levels in the leaves of plants with
different values of r_r (control and treated plants) indicates that
the values of r_e in the leaves of both types is also similar.

We have not been able to get separate estimates of the liquid
diffusion resistance (r_m) and the carboxylation resistance (r_x),
thus it cannot be certain that the observed differences in r_r (*Table
2B*) are not due to large changes in r_m. However, as these differences in r_r are correlated with large differences in the RuDP carboxylase activity (*Figure 1, Table 1A*), it seems unlikely that changes in r_m account for the predominant part of changes in r_r. Furthermore, CHARTIER *et al.* (12) have shown in young bean leaves that,
although widely different estimates of r_x are obtained (0.12 - 1.18
sec cm^{-1}), the variation in r_m (6.90 - 7.06 sec cm^{-1}) is small.
RACKHAM (22) concluded that r_m was a small proportion of the total
resistance to CO_2 transfer and therefore not normally rate-limiting.

We tentatively conclude that, in the leaves of the plants we used,
the carboxylation resistance (r_x) increases with senescence and that
this increase correlates with the observed decreases (*Figure 1*) in
activity of RuDP carboxylase and in net photosynthesis. This

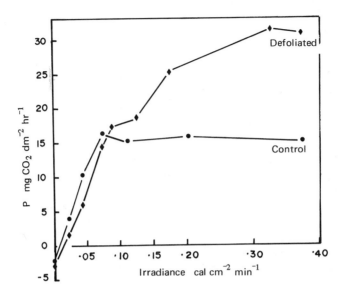

Figure 2 The effect of irradiance on P of bean leaves from control and partially defoliated plants, 3 days after partial defoliation. [Conditions of experiment: inlet (CO_2) = 390 ppm; leaf temperature, 21.6 - 23.2°.]

conclusion disagrees with that of CHARTIER *et al.* (12), using young bean leaves. They indicated (their *Table 2*) that r_x, at its highest estimated value, was only 17% of $(r_m + r_x)$.

CONCLUSION

More important than the above discussion of attempts to apportion the observed increases r_r into its components, is the general conclusion that, in these experiments the net photosynthetic rate of leaves (in saturating light and c. 360 ppm CO_2) was closely correlated with the activity of the primary carboxylating enzyme, RuDP carboxylase (*Figure 1, Tables 1A and 1B*). These experiments thus support the statement (8) that "resistance to CO_2 diffusion may not be the only rate-limiting step in apparent photosynthesis even at high light intensities and atmospheric CO_2 concentrations".

ACKNOWLEDGMENTS

One of us (T.F.N.) acknowledges the scientific hospitality and help of the Director and staff of the Welsh Plant Breeding Station, Aberystwyth, where this experimental work was done. My thanks are especially due to Dr. John Cooper and other members of the Department of Development Genetics at the W.P.B.S. for help, advice, and encouragement.

REFERENCES

1. J.L. MONTEITH, in Environmental Control of Plant Growth, ed. L.T. EVANS, Academic Press, London and New York, (1963) 95.
2. J.L. MONTEITH, *Field Crop Abstr.*, 18 (1965) 213.
3. O. BJÖRKMAN, *Carnegie Inst. Wash. Yearbook*, 65 (1966) 454.
4. O. BJÖRKMAN, *Physiologia Pl.*, 21 (1968) 1.
5. O. BJÖRKMAN, *Carnegie Inst. Wash. Yearbook*, 68 (1969) 614.
6. H.W. WOOLHOUSE, *Hilger J.*, 11 (1967-68) 15.
7. P.F. WAREING, M.M. KHALIFA, and K.J. TREHARNE, *Nature*, 220 (1968) 453.
8. W.A. JACKSON and R.J. VOLK, *A. Rev. Pl. Physiol.*, 21 (1970) 385.
9. K. MOTHES and C. ENGELBRECHT, *Flora*, 139 (1952) 1.
10. H.W. WOOLHOUSE, *Symp. Soc. exp. Biol.*, 21 (1967) 179.
11. R.O. SLATYER, *Planta*, 93 (1970) 175.
12. P. CHARTIER, M. CHARTIER, and J. CATSKY, *Photosynthetica*, 4 (1970) 48.
13. K.J. TREHARNE and J.P. COOPER, *J. exp. Bot.*, 20 (1969) 170.
14. E. RACKER and J. SCHROEDER, *Archs. Biochem. Biophys.*, 74 (1958) 326.
15. B. AXELROD, *Methods Enzymol.*, I (1959) 363.
16. E. RACKER, *Archs. Biochem. Biophys.*, 69 (1957) 300.
17. A. PETERKOFSKY and E. RACKER, *Plant Physiol.*, 36 (1967) 409.
18. S.G. WILDMAN and J. BONNER, *Archs. Biochem.*, 14 (1947) 381.
19. M.W. STEER, B.E.S. GUNNING, T.A. GRAHAM, and D.J. CARR, *Planta*, 79 (1968) 254.
20. P. GAASTRA, *Meded. LandbHoogesch. Wageningen*, 59 (1959) 1.
21. H. MEIDNER, *Nature*, 222 (1969) 876.
22. O. RACKHAM, in Light as an Ecological Factor, ed. R. BAINBRIDGE *et al.*, Blackwell, Oxford, (1967) 167.

THERMAL RESPONSES OF THE PRIMARY CARBOXYLATING ENZYMES FROM C$_3$ AND C$_4$ PLANTS ADAPTED TO CONTRASTING TEMPERATURE ENVIRONMENTS

P.J. Phillips and J.R. McWilliam

Division of Plant Industry, CSIRO, Canberra City, 2601, Australia

SUMMARY

In species adapted to contrasting thermal environments, the maximum specific activities for phosphoenolpyruvate carboxylase and ribulose-1,5-diphosphate carboxylase were found at temperatures close to those optimal for growth and photosynthesis. The apparent Km for ribulose-1,5-diphosphate in extracts of C$_3$ and C$_4$ species was independent of temperature. Activation energies for this enzyme were lower in species adapted to low temperatures. The ribulose-1,5-diphosphate carboxylase of the alpine psychrophile Caltha was more thermolabile at elevated temperatures than that of thermophiles such as Atriplex. The Atriplex enzyme was no more heat stable than that from mesophilic species. At reduced temperatures, chilling sensitive C$_4$ species showed a sudden increase in activation energies for phosphoenolpyruvate carboxylase, suggestive of a conformational change in a membrane-bound enzyme due to a phase change occurring in the membrane lipids.

INTRODUCTION

The effects of temperature on plants are primarily exerted through the influence of heat on synthesis, stability, and activity of controlling molecules such as enzymes, and on molecular complexes in membranes which in turn produce changes in cellular metabolism. Variations in enzyme response to temperatures outside the optimum for growth could provide a biochemical basis for adaptation to different thermal environments.

In this study the specific activities and kinetic properties of the two primary carboxylating enzymes, ribulose-1,5-diphosphate (RuDP) carboxylase and phosphoenolpyruvate (PEP) carboxylase have been studied in an attempt to relate thermal responses to the growth and photosynthesis of plants adapted to contrasting temperature regimes.

97

EXPERIMENTAL

The study was based on three species, chosen because of their contrasting temperature adaptations. These were *Caltha intraloba*, an alpine psychrophile (a C_3 plant); wheat (*Triticum aestivum*), a temperate mesophile (a C_3 plant); and saltbush (*Atriplex nummularia*), a semi-arid thermophile (a C_4 plant). In addition three C_4 species were included, millet (*Pennisetum typhoides*) and maize (*Zea mays*), chilling sensitive tropical species, and Johnson grass (*Sorghum bicolor* var. *halepense*), a sub-tropical perennial weed, naturalized in continental U.S.A., (1) which is chilling resistant .

Growth rates were measured in the Canberra phytotron in controlled temperatures under artificial light and 16-hour photoperiods. In the enzyme and photosynthesis studies, plants were grown under similar conditions, in natural daylight. Johnson grass was sampled from the field in the spring.

Temperature response curves for photosynthesis were obtained for *Caltha*, wheat, and saltbush from 0 to 52°. Net photosynthesis (P) and transpiration of attached leaves were measured in an open system containing an infra red gas analyzer and differential psychrometers (2). High flow rates of air (315 ppm CO_2) were maintained and the photosynthetically active radiation from a quartz-iodine lamp was 0.4 g cal cm^{-2} min^{-1}. All values of P were corrected for changes in stomatal resistance (3).

Crude enzyme extracts were prepared by grinding young leaves with buffer in a mortar at 4° and filtering through "mira-cloth". Enzymes were assayed spectrophotometrically, RuDP carboxylase and PEP carboxylase being assayed as previously described (4,5). RuDP carboxylase from C_4 plants was assayed by the method of ANDREWS and HATCH (6). Extracts were preincubated for 5 min in absence of substrate, assayed at 25°, and specific activity measured on a protein base (7). Enzyme half-life was measured in detached leaves, incubated at the desired temperature in the dark with cycloheximide to prevent new protein synthesis (8). Half-lives were calculated from the time-dependent decrease of enzyme activity over the temperature range of 25° to 50°.

RESULTS AND DISCUSSION

Growth and photosynthesis

The contrasting temperature optima for growth and photosynthesis for *Caltha*, wheat, and saltbush (*Table 1a,b*) reflect the close adaptation to the temperatures experienced in the natural environments. *Caltha* grew actively and gave relatively high rates of photosynthesis below 10°, under conditions similar to alpine environments during spring. Wheat gave a typical temperate plant response (9), giving optimal rates near 20°. Saltbush, although capable of growth and photosynthesis over the entire temperature range, showed maximum performance above 30° (*Table 1a,b* and *Figure 1*). Unlike wheat, which

TABLE Ia

Optimum temperatures (°C) for relative growth rate (RGR, $mg \; mg^{-1} \; day^{-1}$) and net photosynthesis (P, $\times 10^{-8} \; g \; cm^{-2} \; sec^{-1}$) in three species adapted to contrasting temperature environment.

Parameter	*Caltha*	Wheat	Saltbush
Optimum temperature			
for RGR	10*	15-20**	30-35
for P	10.5	23.5	36.5
P at optimum	6.9***	16.5	12.0

*   Optimum for absolute growth
**   FRIEND *et al.* (10)
*** Based on whole plant tops

TABLE Ib

The effect of temperature on net photosynthesis (expressed as % maximum).

Temperature	*Caltha*	Wheat	Saltbush
5	94	46	23
15	98	82	65
30	68	95	95
40	0	62	98
50	0	9	56

was permanently damaged by long exposure to temperatures much above 35°, saltbush grew and photosynthesized actively at these temperatures which are experienced during summer in semi-arid environments. Rates of P for wheat and saltbush were comparable at their optimum temperatures and although C_4 species have been reported as having rates higher than C_3 species under high light (9), the high reflectance of saltbush leaves and high stomatal resistance (2.5 to 3.0 sec cm^{-1}) compared with that for wheat (0.4 sec cm^{-1}) provide a partial explanation.

Figure 1 shows the upward shift in optimum temperature for maximum photosynthesis which parallels the temperature at which saltbush plants were grown prior to measurement. This correlation between the environmental temperature and the temperature optimum for photosynthesis suggests that saltbush retains phenotypic flexibility, and is able to modify its photosynthetic rate to adjust to changing temperatures.

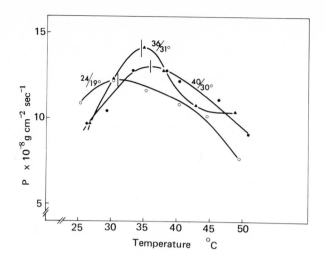

*Figure 1 Temperature response curves for net photosynthesis
(corrected for stomatal resistance) for leaves of saltbush
(Atriplex nummularia) grown at the three day/night temperatures shown.*

Specific activities of enzymes

Maximum specific activities of RuDP carboxylase from *Caltha*,
wheat, and saltbush, and also PEP carboxylase from saltbush were obtained from plants growing at temperatures near the optimum for

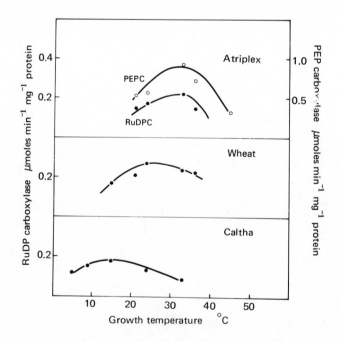

*Figure 2 Specific activities of RuDP carboxylase and PEP carb-
oxylase from C_3 and C_4 plants grown at different temperatures.*

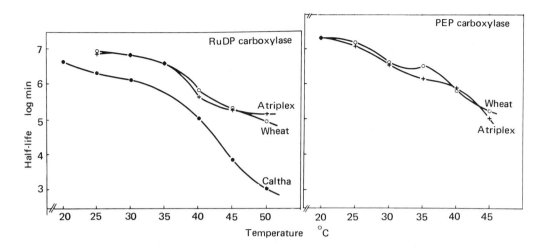

Figure 3 Half-lives (log_e minutes) for RuDP carboxylase and PEP carboxylase.

growth and photosynthesis for the species (*Figure 2*). There is no
suggestion that increased quantities of enzymes are maintained to
offset the decelerating effect on reaction rate of lowered temperatures. The ability to maintain photosynthesis at high temperatures,
is not readily explained but evidence for increased enzyme production offsetting the increased rate of thermal denaturation is discussed later. The results do indicate however that the optimum temperature for photosynthesis in these species is also the temperature
at which the plants maintain maximum available carboxylating enzyme.
Protein and activity levels per gram fresh weight (not presented)
for these plants indicate that the increased specific activities are
not due to increased quantities of protein.
Thermal responses
    (a) High temperature
    The decreased quantity of carboxylating enzymes above the optimum
for specific activity suggests that the ability to photosynthesize
at these temperatures may be dependent on the formation of enzymes
with increased thermal stability. Evidence to support this view based largely on work with thermophilic bacteria, has been presented by
LANGRIDGE (11). In the present study of the *in vivo* thermal stability of RuDP carboxylase and PEP carboxylase, there were no differences between wheat and saltbush in the rate of denaturation of
either enzyme with increasing temperature (*Figure 3*). However RuDP
carboxylase from *Caltha*, was much more unstable over the entire temperature range. *Caltha* may possess a more thermolabile RuDP carboxylase which could be responsible for its failure to photosynthesize
much above 35°. The similarity of the denaturation curves for the
carboxylating enzymes of wheat and saltbush, however, indicates that
increased stability cannot provide an explanation for the increased
high temperature tolerance of saltbush. It is possible that

increased rates of protein synthesis may occur in saltbush at high
temperatures to offset enzyme lost by denaturation in order to maintain the quantity of enzyme measured from plants grown at different
temperatures.

(b) Median temperatures

Adaptation to temperatures within the range from 15° to 40° was
investigated in four species, including the two C_3 representatives,
Caltha and wheat, and two C_4 species, saltbush and millet (*Table 2*).

TABLE 2

Activation energies (Ea, K cal mole^{-1}) for RuDP carboxylase
and apparent Km at 25° for RuDP (mM) from C_3 and C_4 species
with contrasting temperature optima for photosynthesis.

Species	Photosynthesis optimum (°C)	Ea	Km RuDP
Caltha (C_3)	10.5	5.94	0.30
Wheat (C_3)	23.5	7.34	0.41
Saltbush (C_4)	36.5	7.85	0.021
Millet (C_4)	35-40*	10.96	0.046

* (H.D. MCPHERSON, personal communication)

Activation energies (Ea) for RuDP carboxylase derived from changes
in Vmax tended to decrease with decreased temperature of adaptation,
which suggests that psychrophiles such as *Caltha* may be able to photosynthesize at low temperatures by possessing enzymes with low sensitivity to a decrease in temperature. Conversely, tropical species
such as millet, which do not experience low temperatures during
growth in their natural habitat, can operate with a higher Ea value.

The apparent Km for RuDP did not change significantly with temperature, thus substrate affinity has no adaptive significance within this temperature range. The apparent Km for the two C_4 plants
was about one tenth the value obtained for the C_3 species, as reported earlier (6).

(c) Low temperatures

Tolerance to chilling temperatures between 15° and 2° was studied
by comparing activation energies of PEP carboxylase in four species
which varied in their chilling sensitivity. With the chilling-sensitive species maize and millet, the reaction rate deviates from the
Arrhenius formulation below 12° (*Figure 4*). An abrupt discontinuity
in the Arrhenius plot, and the associated three-fold increase in the
activation energy commencing at about 10° to 12° (approximately 3.5
on the absolute scale) has been previously reported by LYONS and
RAISON (13) for succinate oxidation by mitochondria from chilling-
sensitive plants. They suggested that the increase in Ea is due to a
conformational change in the membrane-bound oxidative enzymes,

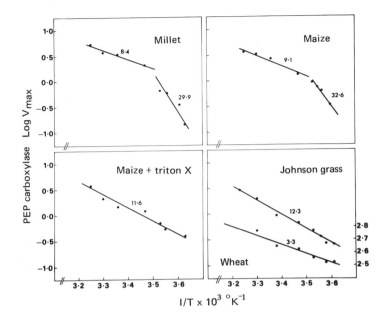

Figure 4 Arrhenius plots for PEP carboxylase from chilling-sensitive (millet and maize) and chilling-resistant (Johnson grass and wheat) species.

induced by a phase change in the mitochondrial membrane lipids (14). Treatment of maize chloroplast extracts with 0.5% Triton-X eliminated any discontinuity (*Figure 4*) and resulted in a higher but constant activitation energy over the temperature range. This provides evidence for an association between PEP carboxylase and a lipid component of a membrane, and raises the possibility, as suggested by SLACK *et al.* (15) and BALDRY *et al.* (16), that PEP carboxylase could be associated with the outer chloroplast membrane.

No discontinuity was observed for PEP carboxylase from wheat, which is consistent with the low temperature tolerance of this C_3 species, however, the lack of any discontinuity in the plot for Johnson grass is of interest, as this is a C_4 species and occurs, with other species in the *Sorghum* genus, mainly in tropical environments. The strain of Johnson grass used in this study was obtained from Oklahoma, U.S.A. where it is a naturalized perennial weed, and its ability to maintain a constant activation energy over a wide range of temperatures may explain its chilling tolerance.

ACKNOWLEDGMENTS

We wish to thank Mr. R. Davis for his assistance and for the provision of photosynthesis equipment, and Drs. J. Langridge, C.B. Osmond and M.D. Hatch for advice and suggestions made during this study.

REFERENCES

1. A.S. HITCHCOCK, Manual of the Grasses of the United States, Miscell. Publ. 200, U.S. Gov. Printing Office, Washington p 1051
2. R.O. SLATYER and J.F. BIERHUIZEN, *Plant Physiol.*, 39 (1964) 1051.
3. R.M. GIFFORD, this volume (1971).
4. E. RACKER, *Archs. Biochem. Biophys.*, 69 (1957) 300.
5. D.A. WALKER, *Biochem. J.*, 67 (1957) 73.
6. T.J. ANDREWS and M.D. HATCH, *Phytochemistry*, (1970) in press.
7. O.H. LOWRY, N.J. ROSEBROUGH, A.L. FARR, and R.J. RANDALL, *J. biol. Chem.*, 193 (1951) 265.
8. D. GRAHAM, M.D. HATCH, C.R. SLACK, and R.M. SMILLIE, *Phytochemistry*, 9 (1970) 521.
9. M.A. EL-SHARKAWY and J.D. HESKETH, *Crop Sci.*, 4 (1964) 514.
10. D.J. FRIEND, V.A. HELSON, and J.E. FISHER, *Can. J. Bot.*, 40 (1962) 939.
11. J. LANGRIDGE, *A. Rev. Pl. Physiol.*, 14 (1963) 441.
12. J. LANGRIDGE, *Molec. gen. Genet.*, 103 (1968) 116.
13. J.M. LYONS and J.K. RAISON, *Plant Physiol.*, 45 (1970) 386.
14. J.K. RAISON and J.M. LYONS, *Archs. Biochem. Biophys.*, (1970) in press.
15. C.R. SLACK, M.D. HATCH, and D.J. GOODCHILD, *Biochem. J.*, 114 (1969) 489.
16. C.W. BALDRY, C. BUCKE, and J. COOMBS, *Biochem. biophys. Res. Commun.*, 37 (1969) 828.

CHARACTERISTICS OF HYBRIDS BETWEEN C_3 AND C_4 SPECIES OF *Atriplex*[1]

Olle Björkman, Malcolm Nobs, Robert Pearcy,
John Boynton, and Joseph Berry

Carnegie Institution of Washington,
Department of Plant Biology, Stanford, Calif., 94305, U.S.A.

SUMMARY

Atriplex rosea has all of the photosynthetic and biochemical characteristics of the C_4 pathway of photosynthesis, whereas A. patula ssp. hastata has those of the C_3 pathway. In spite of the difference in photosynthetic pathway, these species are closely related, as we have successfully hybridized them with A. rosea as the female parent. We report here studies with the F_1 and F_2 hybrid generations. The F_1 hybrids are intermediate between the parents in leaf morphology and anatomy. They have intermediate levels of phosphoenolpyruvate carboxylase and initial CO_2 fixation occurs via carboxylation of both phosphoenolpyruvate and ribulose-1,5-diphosphate. In spite of the intermediate nature of the hybrid in these traits, photosynthetic CO_2 exchange characteristics are not intermediate. The hybrids achieve about the same light-saturated rate of CO_2 uptake as the parents at CO_2 saturation; however, they require a higher concentration of CO_2 to saturate than even A. patula, and the inhibitory effect of O_2 on net photosynthesis is quantitatively similar to that in A. patula.

The leaf anatomy of the individual F_2 hybrids analyzed varied from the "Kranz-typus" of the C_4 parent to the conventional type of the C_3 parent. The activity of phosphoenolpyruvate carboxylase varied more than ten-fold. All photosynthetic gas exchange characteristics except the CO_2 compensation point (which varied from 19 to 46 ppm) are similar to those of the F_1 hybrid. These results imply that "Kranz-typus" leaf anatomy and high phosphoenolpyruvate carboxylase do not in themselves lead to the efficient photosynthetic characteristics normally associated with C_4 photosynthesis, and show that a lowered CO_2 compensation point need not be correlated with C_4 photosynthetic characteristics. Apparently the component steps not only

---

[1] *C.I.W.-D.P.B. Publication No. 473*

must be present, but also be appropriately coordinated in time and
space.

INTRODUCTION

The group of plants now known to possess the C_4 pathway of photosynthesis is diverse, containing members from several families of
higher plants. In spite of the taxonomic diversity of this group,
these plants hold in common a remarkable set of characteristics
which show absolute correlation. These characteristics include: the
production of C_4 acids as the initial products of CO_2 fixation; the
presence of high activities of phosphoenolpyruvate (PEP) carboxylase; the lack of an inhibition of photosynthesis by the oxygen content of normal air; the occurrence of a very low CO_2 compensation
point; the presence of "Kranz-typus" anatomy with dimorphic chloroplasts in the bundle sheath and mesophyll tissue; and a different
carbon isotope discrimination pattern from C_3 plants. This group of
characteristics has recently been referred to as the L-syndrome (1).
The occurrence of the syndrome in unrelated taxa and the division of
several taxa into groups possessing the syndrome and others lacking
it, raises important questions concerning its inheritance and evolution. The existence of the syndrome itself also raises questions
concerning the relationship of its various traits to the operation
of the C_4 pathway of photosynthesis. Several causal relationships
have been suggested, but the apparent absence of natural variability
in the expression of the syndrome has hindered critical tests. Our
studies on artificially produced hybrids between *Atriplex patula*
ssp. *hastata* Hall and Clem., a normal C_3 species (2), and *Atriplex*
rosea L., a C_4 species (2,3, see also 4), have provided material
uniquely suited to studying these relationships.

   The genus *Atriplex* is of world-wide distribution and contains o-
ver 200 species. Basing their classification on embryonic characters
alone HALL and CLEMENTS, in 1923, divided the genus into two major
subgenera: *Euatriplex* and *Obione*. All species of the subgenus *Obione*
that have been analyzed thus far possess the C_4 pathway or at least
the associated photosynthetic or anatomical characteristics. *Atri-*
plex patula, and all other *Atriplex* species that have been found to
lack the C_4 pathway or associated characteristics, belong to subgenus *Euatriplex* (5,6,7,8). HALL and CLEMENTS also placed *Atriplex ro-*
sea in this same subgenus; moreover, they considered that the *rosea*
group was closely related to the *patula* group. This was a major consideration in our choosing these species for our comparative studies
and hybridization experiments.

EXPERIMENTAL

   All plant materials used in the physiological, biochemical, and
anatomical investigations were grown in controlled growth cabinets
under a light intensity of 1.1×10^5 erg cm^{-2} sec^{-1} (400-700 nm) for

16 hours alternating with 8 hours darkness. The temperature during the light and dark periods was 25° and 20°, respectively. The seedlings were grown in perlite with nutrient solution added daily.

PEP carboxylase and ribulose-1,5-diphosphate (RuDP) carboxylase activities were assayed according to BJÖRKMAN and GAUHL (9). Carbonic anhydrase was assayed according to RICKLI *et al.* (10). Feedings with $^{14}CO_2$ and chromatographic separations were conducted as described by PEARCY *et al.* (11) and photosynthesis measurements as described by BJÖRKMAN *et al.* (12). Electron micrographs were prepared as described by BOYNTON *et al.* (13).

RESULTS AND DISCUSSION

Atriplex rosea served as the female parent in the hybridization experiments. Emasculation of the highly self-fertile plants was achieved by removal of the bisexual capitate inflorescences leaving the predominantly pistillate axial inflorescences. Plants were inspected, inflorescences containing staminate flowers removed, and the pistillate axial inflorescences pollinated with A. *patula* pollen at least twice daily during the flowering period.

Approximately 10% of the axile flowers that were repeatedly pollinated with *Atriplex patula* pollen yielded seeds. Nearly all seeds were viable, germinated rapidly, and gave rise to strong F₁ seedlings. These were uniform and morphologically and anatomically intermediate between the parents. The reciprocal cross with A. *patula* as the female parent has not been successful. The absence of unisexual pistillate inflorescences in this species makes it extremely difficult, if not impossible, to obtain controlled cross-pollination.

The F₁ hybrid is diploid with the same chromosome number as its parents. The karyotypes of the parents are very similar and both have 9 pairs of chromosomes. The karyotype of the F₁ hybrid is indistinguishable from those of the parental species. Meiotic pairing of the chromosomes in the hybrid is, however, irregular and seed-set is poor. We have, nevertheless, been successful in obtaining enough F₂ seeds for experimental purposes.

Figure 1 shows the distribution of chromosome numbers obtained with the F₂ seedlings investigated thus far. These numbers range from slightly above diploid to more than pentaploid. Many of these F₂ hybrids are essentially sterile, but several of them have excellent seed-set and may well prove to be self-perpetuating. Leaf shapes of the parents, the F₁ hybrid, and some F₂ hybrids are shown in *Figure 2A*. The F₁ is clearly intermediate between the parents. Its growth habit (not shown) is also intermediate. In spite of its polyploid nature the F₂ population shows considerable segregation in a number of morphological characteristics, in fact as much as one would expect from a diploid F₂ population. Leaf shape, for example, varies from that of A. *patula* to that of A. *rosea* and a complete recombination of the parental leaf shapes has been obtained.

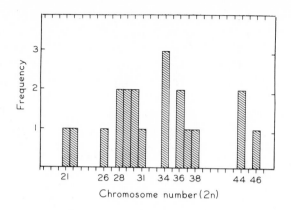

Figure 1 Frequency distribution of chromosome numbers for the F2 seedlings investigated.

The leaf anatomy (*Figure 2B*) also shows a very wide range of variation. *A. rosea* has the typical "Kranz-typus" anatomy of other C_4 plants whereas *A. patula* completely lacks this structure. The F_1 hybrid has a distinct bundle sheath, but a rather poorly developed radiate mesophyll and no differentiation of hypodermal tissue. Among the F_2 hybrids, no. 7735-3, (to the extreme left in the upper row of the F_2 hybrids), shows essentially complete expression of the "Kranz-typus" anatomy of *A. rosea* whereas no. 7735-2 (lower right corner) closely resembles *A. patula*. Other F_2 hybrids show many of the possible recombinations of the basic anatomies of the parents.

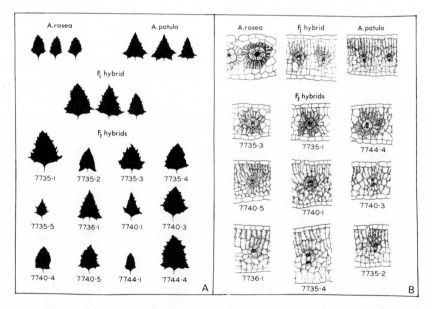

Figure 2 Leaf shapes (2A) and camera lucida drawings of living leaf sections (2B) of Atriplex patula, A. rosea, and their first and second generation hybrids.

The intermediate nature of the F_1 hybrids and the segregation of morphological and anatomical traits in the second generation leave absolutely no doubt that these are hybrid derivatives. The success of this hybridization justifies HALL and CLEMENTS' contention that these species are closely related, and shows that the minimum genetic divergence between C_3 and C_4 plants is much less than one previously would have expected. Recent biochemical evidence that the pathway of carbon in C_4 plants is fundamently similar to that in C_3 plants (14, see also 15) is consistent with this genetic divergence. Conceivably, there may have been interconversion between C_4 and C_3 photosynthesis several times during the evolution of the genus *Atriplex*, and probably many other genera.

Feedings of radioactive $^{14}CO_2$ to the parents, *A. rosea* and *A. patula*, showed the patterns typical of the C_4 and C_3 pathways, respectively. The F_1 hybrid was clearly intermediate. The percent of C_4 products fell within the range of 21 to 48%. The fraction of C_4 products in the hybrid could be manipulated by the changing of the CO_2 concentration of the atmosphere surrounding the leaf. The higher value was obtained at low CO_2 (100 ppm), the lowest value at 800 ppm CO_2. The proportions of C_4 products in the F_2 hybrids also fell within this range. In other studies, we have shown that, indeed, both carboxylation systems operate in parallel in the F_1 hybrid (11).

TABLE I

PEP carboxylase activity in leaf extracts of hybrids from a cross between *Atriplex rosea* and *A. patula*. Activities expressed in μmole CO_2 min^{-1} g^{-1} fresh weight.

Plant	Activity
A. rosea (mean ± S.D.)	24.4 ± 1.9
A. patula "	1.4 ± 0.1
F_1 hybrid "	6.4 ± 1.5
F_2 hybrid 7735-1	3.21
" 7735-2	0.99
" 7735-3	5.59
" 7735-4	1.59
" 7736-1	2.05
" 7740-1	12.90
" 7740-3	10.53
" 7740-4	1.98
" 7740-5	8.68
" 7744-4	8.48

The activity of PEP carboxylase in leaf extracts of the parents and hybrids is shown in *Table 1*. The activity of this enzyme in *A. rosea* and *A. patula* is similar to that found in other C_4 and C_3 plants, respectively. The F_1 is clearly intermediate while the individual F_2 hybrids show great variation. Other enzymes involved in the C_4 pathway as adenylate kinase, pyrophosphatase, and pyruvate,Pi dikinase, have also been detected in the F_1 hybrid.

TABLE 2

RuDP carboxylase activity in leaf extracts of hybrids from a cross between *Atriplex rosea* and *A. patula*. Activity expressed in µmole CO_2 min^{-1} g^{-1} fresh weight.

Plant	Activity
A. *rosea* (mean ± S.D.)	4.09 ± 0.53
A. *patula* "	12.64 ± 0.06
F_1 hybrid "	11.55 ± 1.53
F_2 hybrid (10 individuals)	11.43 ± 2.1

Table 2 shows the activity of RuDP carboxylase which is very probably involved in photosynthesis by both C_3 and C_4 plants (9). The activities are high in *A. rosea* but are still lower than in *A. patula*. The levels of RuDP carboxylase in the F_1 hybrid and in all of the F_2 hybrids do not differ from that in *A. patula*.

TABLE 3

Light-saturated rate of CO_2 uptake and inhibitory effect of 21% O_2 at 27° and 300 ppm CO_2 of hybrids from a cross between *Atriplex rosea* and *A. patula*. Rates expressed in µmole CO_2 min^{-1} dm^{-2} leaf area.

Plant	Net rate of CO_2 uptake 21% O_2	1.3% O_2	Ratio 21% / 1.3% O_2
A. *rosea* (mean ± S.D.)	21.8 ± 1.0	21.8 ± 1.0	1.00 ± 0.00
A. *patula* "	11.8 ± 0.6	18.5 ± 1.1	0.63 ± 0.01
F_1 hybrid "	6.5 ± 0.7	11.3 ± 1.3	0.57 ± 0.04
F_2 hybrids "	6.8 ± 0.7	10.5 ± 1.0	0.64 ± 0.03

In spite of the intermediate nature of the F_1 hybrids with regard to biochemical and anatomical characteristics, photosynthetic gas exchange characteristics are clearly not intermediate between the parental species. *Table 3* shows that the light saturated photosynthetic rate at 300 ppm CO_2 external to the leaf is considerably lower than in either parent. This lower rate is not caused by higher leaf resistances to gas diffusion. Remarkably, the F_2 hybrids show little variation but are very similar to the F_1 hybrid in spite of the wide variation in leaf anatomy and PEP carboxylase activity among them.

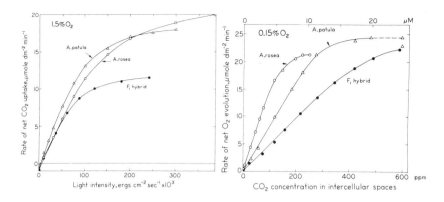

Figure 3 (left) The rate of photosynthetic CO$_2$ uptake as a function of light intensity for Atriplex patula, A. rosea, and their F$_1$ hybrid, measured at 27°, 300 ppm CO$_2$, and 1.5% O$_2$.

Figure 4 (right) Photosynthetic O$_2$ evolution as function of CO$_2$ concentration in the intercellular spaces under 0.15% O$_2$ and 27° leaf temperature in the F$_1$ hybrid and the parental species. Saturating light intensities were used for each plant. The CO$_2$ concentration in the intercellular spaces was calculated from the measured resistance to water vapor transfer, assuming a value of 1.56 for the ratio of the diffusion coefficient for CO$_2$ to that of water.

Another notable point is the inhibitory effect of O$_2$ on photosynthesis. *Table 3* also shows the ratio of the net rate of CO$_2$ uptake in 21% and 2% O$_2$ at 300 ppm CO$_2$. There is no effect of O$_2$ on *A. rosea*, but the inhibitory effect of O$_2$ is quantitatively similar in all the other plants.

The photosynthetic rate at 300 ppm CO$_2$ was lower in the F$_1$ and F$_2$ hybrids than in the parents even under low O$_2$ concentration. As shown in *Figure 3*, the initial slope of the curve for CO$_2$ uptake as a function of light intensity of the F$_1$ hybrid is close to that of the parents, indicating that the lower photosynthetic rate of the F$_1$ hybrid found at higher light intensities is not due to a lower efficiency of the photoacts.

Figure 4 shows that the CO$_2$ saturated rates of photosynthesis are similar in both parents and the F$_1$ hybrid, but the hybrid requires much higher CO$_2$ concentration in the intercellular spaces to reach saturation, and is considerably less efficient in utilizing low CO$_2$ concentrations than either parent. The F$_2$ hybrids are similar to the F$_1$ hybrids in this respect.

One gas exchange characteristic which does show some segregation is the CO$_2$ compensation point, shown in *Table 4*. The range of the F$_2$ hybrids analyzed is from 19 to 46 ppm. The parents with less than 2 ppm and 44 ppm are at the extremes.

It seemed conceivable that the high requirement for CO$_2$ exhibited by the hybrids was due to low carbonic anhydrase activities. *Table 5*

TABLE 4

CO_2 compensation point at 27° and 21% O_2 in hybrids from a cross between *Atriplex rosea* and *A. patula*.

Plant	Compensation point, ppm CO_2
A. rosea	<2
A. patula (mean ± S.D.)	44.2 ± 1.4
F_1 hybrid "	37.0 ± 3.5
F_2 hybrid 7735-1	37
" 7735-2	43
" 7735-3	29
" 7735-4	34
" 7736-1	46
" 7740-3	40
" 7740-4	35
" 7740-5	19
" 7744-4	27

TABLE 5

Carbonic anhydrase activity in leaf extracts. Activities expressed according to RICKLI *et al.* (1964).

Plant	Enzyme units
A. rosea	385
A. patula	275
F_1	294
Sorghum sudanense	very low

shows that this was not the case. The F_1 hybrid had as much carbonic anhydrase as *A. patula*, the C_3 parent. Thus it is certainly adequate. Surprisingly, *A. rosea* had equally high activities of this enzyme in contrast to what has been reported for other C_4 species (16,17). Our measurements of carbonic anhydrase activity in *Sorghum* and some other C_4 species gave values similar to those previously reported. These were much lower than in *A. rosea*.

The testing of F_2 individuals for photosynthetic characteristics, leaf anatomy, and enzyme activities were carried out on the same leaf in order to search for correlations among the various segregating characteristics. *Table 6* reports comparative data for several F_2 individuals. The anatomy of these leaves can be determined by cross reference to *Figure 2B*.

F_2 hybrid 7740-3 is an individual with high PEP carboxylase activity. In this case, it exceeds the activity of RuDP carboxylase.

TABLE 6

The segregation of PEP carboxylase, RuDP carboxylase and some photosynthetic characteristics in selected F_2 hybrids. Enzyme activities and photosynthetic rates are expressed in μmole CO_2 min^{-1} dm^{-2} leaf area. Photosynthetic rates were measured at light-saturation, 1.3% O_2, 300 ppm CO_2, and 27°. CO_2 compensation point was measured at 21% O_2 and 27°.

	7740-3	7735-1	7735-3	7740-5
RuDP carboxylase	22.2	26.1	30.4	37.5
PEP carboxylase	25.9	7.9	12.7	22.0
Photosynthetic rate	8.6	10.2	9.7	10.9
Inhibition by 21% O_2	34%	39%	36%	35%
CO_2 compensation point	40 ppm	37 ppm	29 ppm	19 ppm

Evidently, high PEP carboxylase had no effect on CO_2 compensation, photosynthetic rate or inhibitory effect of O_2. This illustrates that high PEP carboxylase per se is not sufficient for these C_4 characteristics. The anatomy of this leaf resembles the F_1 hybrid.

F_2 hybrid 7735-1 has a leaf anatomy similar to *A. rosea*. There is also no expression of the gas exchange characteristics associated with it in C_4 species, indicating that "Kranz-typus" anatomy is also not enough to bring about these characteristics. The PEP carboxylase of this leaf is low.

F_2 hybrid 7735-3 has a similar leaf anatomy to 7735-1, but has a somewhat higher PEP carboxylase activity. In this case, the CO_2 compensation point is significantly lower than the *A. patula* parent (29 ppm and 44 ppm, respectively).

F_2 hybrid 7740-5 has yet more PEP carboxylase and a well developed radiate mesophyll. In this case the compensation point is correspondingly lower, 19 ppm. This may support the hypothesis that refixation of respiratory CO_2 by PEP carboxylase in the radiate mesophyll leads to the low compensation point of C_4 plants. These conditions are, however, apparently not sufficient to result in high photosynthetic rate and lack of O_2 inhibition. We are now attempting to select F_2 and F_3 hybrids from large segregating populations for efficient photosynthesis by using selective growth conditions, i.e., low CO_2 concentration.

Dr. J. BOYNTON has conducted ultrastructural studies of the two parents and the hybrid derivatives. *Plate 1* shows an *A. patula* leaf cut in cross-section to the vascular bundle. The cells surrounding this bundle are similar in size, wall thickness, and structural organization to palisade and spongy mesophyll cells elsewhere in the leaf. There is no evidence of bundle sheath development.

Plate 2 shows a similar region of an *A. rosea* leaf. There is a well-developed bundle sheath with thick-walled cells containing abundant chloroplasts and mitochondria, and a very small periferal vacuole. Higher magnification micrographs indicate that the chloroplasts of A. *rosea*, in contrast to all other C_4 species investigated, apparently lack a periferal reticulum.

Plate 3 shows a similar region of the F_1 hybrid. The bundle sheath is evident, but the cells are smaller, more vacuolate, and have thinner walls than those of A. *rosea*. The structural organization of the F_1 bundle sheath cells resembles more closely that of mesophyll cells of A. *patula*. Bundle sheath cells of all F_2 hybrids thus far examined resemble those of the F_1 shown above.

Conceivably, the absence of efficient photosynthesis in the hybrids could be related to the incomplete ultrastructural differentiation of the bundle sheath tissue. However, it should be pointed out that it is unlikely that the C_4 characteristics are passed by chloroplast inheritance; the chloroplasts of all of the hybrids were most probably derived from the maternal parent which was A. *rosea*, the C_4 plant.

The studies presented here show that the genetic divergence between C_3 and C_4 species need not be great. However, the studies also show that at least in *Atriplex*, the inheritance of the high efficiency of utilization of low CO_2 concentrations and the absence of oxygen inhibition of net CO_2 uptake is probably complex. The segregation of individual characteristics such as high PEP carboxylase or "Kranz-typus" leaf anatomy seem to be determined by relatively few genes. However, the high photosynthetic efficiency normally associated with the C_4 pathway evidently requires that several different components be present and probably also properly coordinated, i.e., the genes determining the component steps and their coordination must be inherited together. These do not, in this case, appear to be

---

Plate 1 Electron micrograph of a thin section through part of vascular bundle with chloroplast-containing phloem parenchyma (P), two adjacent vacuolate border parenchyma cells (BP) and a portion of a vacuolate mesophyll cell (MP) in a fully expanded leaf of A. patula. x 5400.

Plate 2 Electron micrograph of a thin section through parts of two thick-walled bundle sheath cells (BS) and three adjacent thin-walled vacuolate mesophyll cells (MP) in a fully expanded leaf of A. rosea. Numerous plasmodesmata (PD) traverse the wall between the two cell types. x 5400. The cell walls of the bundle sheath cells are even thicker when grown at higher light intensities (100,000 vs. 50,000 ergs cm^{-2} sec^{-1} used in this case.)

Plate 3 Electron micrograph of a thin section through a comparable leaf area of the F$_1$ hybrid, A. rosea x A. patula. Moderately well developed bundle sheath cells with central vacuoles (BS) surround the vascular bundle and the remainder of the leaf is occupied by vacuolate mesophyll cells (MP). x 5400.

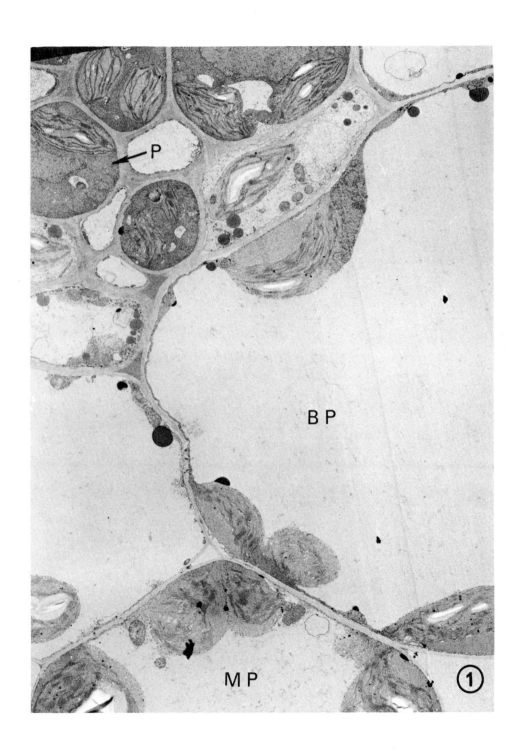

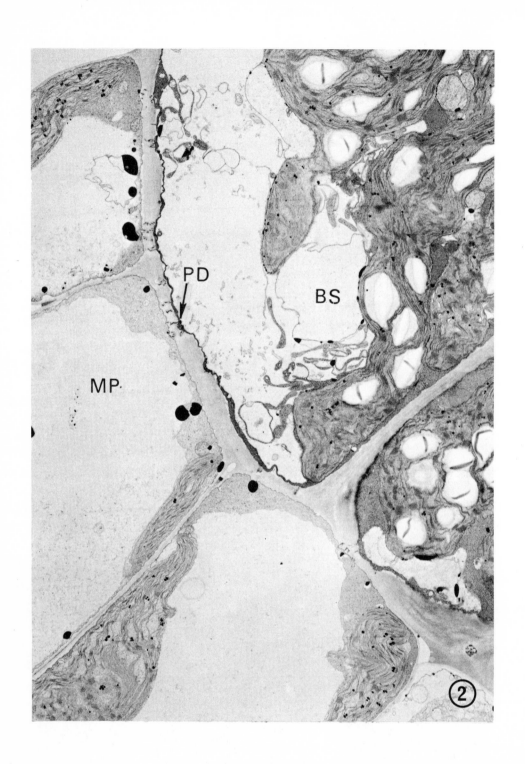

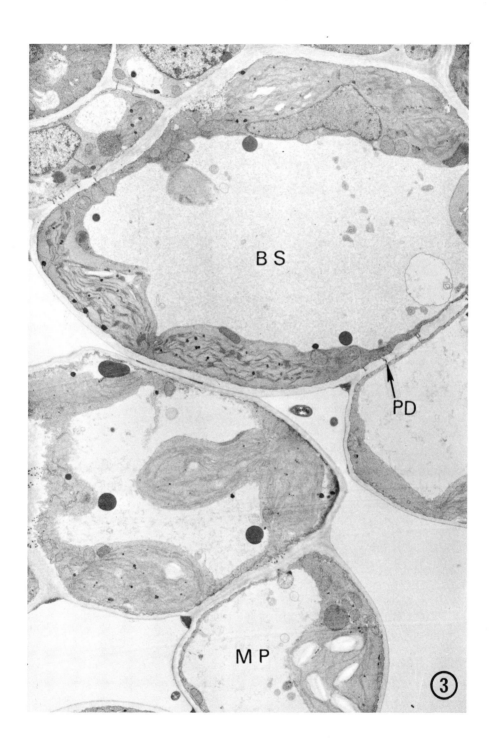

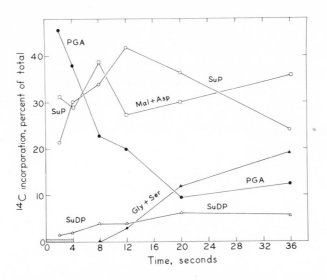

Figure 5 *Distribution of radioactivity after 2 or 4 seconds of photosynthesis in air containing $^{14}CO_2$ followed by photosynthesis in air containing $^{12}CO_2$ for different lengths of time. Light intensity during preillumination and feedings was 8.8 x 10^4 ergs cm^{-2} sec^{-1}; final CO_2 concentration was approximately 400 ppm. Abbreviations are: Asp, aspartic acid; Gly, glycine; Mal, malic acid; PGA, 3-phosphoglyceric acid; Ser, serine; SuP, sugar phosphates; SuDP, sugar diphosphates.*

linked.

Pulse-chase feedings of $^{14}CO_2$-$^{12}CO_2$ (*Figure 5*) provide additional evidence that proper coordination or compartmentation may be lacking in the hybrids. In the F_1 hybrid initial fixation of CO_2 occurs both via β-carboxylation of PEP and by direct carboxylation of RuDP forming both 3-phosphoglyceric acid (3-PGA) and C_4 acids. The decline in label in 3-PGA during the chase indicates that it serves as a precursor of sugar phosphates and products of the glycolate pathway as in C_3 plants. However, the label in C_4 acids does not decline during the chase unlike the situation in normal C_4 plants (18), indicating that the C_4 acids may not be properly metabolized in the hybrid. Ironically, in this case, the ability to fix CO_2 by β-carboxylation of PEP is correlated with a reduced rather than an enhanced efficiency of photosynthetic CO_2 uptake. The reason for this is not yet clear, but it is interesting that the level of RuDP carboxylase in the F_1 and all F_2 hybrids was higher than in the *A. rosea* parent. This was the only biochemical characteristic, which like the photosynthetic gas exchange characteristics did not segregate. If a considerable portion of this enzyme is located in the same cells as the PEP carboxylase it may lead to a detrimental interaction between the two carboxylation systems.

REFERENCES

1. E.B. TREGUNNA, B.N. SMITH, J.A. BERRY, and W.J.S. DOWNTON, *Can. J. Bot.*, 48 (1970) 1209.
2. O. BJÖRKMAN, E. GAUHL, and M.A. NOBS, *Carnegie Inst. Wash. Yearbook*, 68 (1969) 620.
3. W.J.S. DOWNTON, T. BISALPUTRA, and E.B. TREGUNNA, *Can. J. Bot.*, 47 (1969) 915.
4. O. BJÖRKMAN, this volume, (1971).
5. W.J.S. DOWNTON, Ph. D. Thesis, (University of British Columbia, Vancouver, Canada.) 1969.
6. E.B. TREGUNNA, W.J.S. DOWNTON, and P. JOLLIFFE, in Progress in Photosynthesis Research, ed. H. METZNER, I.U.B.S., Tübingen, I (1969) 488.
7. W.M. LAETSCH, *Am. J. Bot.*, 55 (1968) 875.
8. C.B. OSMOND, *Z. PflPhysiol.*, 62 (1970) 129.
9. O. BJÖRKMAN and E. GAUHL, *Planta*, 88 (1969) 197.
10. E.E. RICKLI, S.A.S. GHAZANFAR, B.H. GIBBONS, and J.T. EDSALL, *J. biol. Chem.*, 239 (1964) 1065.
11. R.W. PEARCY and O. BJÖRKMAN, *Carnegie Inst. Wash. Yearbook*, 69 (1971) in press.
12. O. BJÖRKMAN, R.W. PEARCY, and M.A. NOBS, *Carnegie Inst. Wash. Yearbook*, 69 (1971) in press.
13. J.E. BOYNTON, M.A. NOBS, O. BJÖRKMAN, and R.W. PEARCY, *Carnegie Inst. Wash. Yearbook*, 69 (1971) in press.
14. J.A. BERRY, *Carnegie Inst. Wash. Yearbook*, 69 (1971) in press.
15. M.D. HATCH, this volume (1971).
16. E.R. WAYGOOD, R. MACHE, and K.C. TAN, *Can. J. Bot.*, 47 (1969) 1455.
17. R.G. EVERSON and C.R. SLACK, *Phytochemistry*, 8 (1968) 581.
18. M.D. HATCH and C.R. SLACK, *Biochem. J.*, 101 (1966) 103.

CARBON DIOXIDE COMPENSATION IN PLANTS WITH C_4 CHARACTERISTICS

Dale N. Moss

Department of Agronomy & Plant Genetics,
University of Minnesota, St. Paul, Minnesota, 55101.

SUMMARY

Measurement of CO_2 compensation points of more than 800 species show that low compensation point is invariably associated with a starch-forming green bundle sheath. Studies with 158 genotypes of corn, 100 of wheat, and 25 of barley showed no genotypic variation in CO_2 compensation point. A screening technique was developed which permitted assay for C_4 traits in large populations of wheat and oats after treatment of seeds with ionizing radiation. No plants with C_4 traits were found. Crosses between C_3 and C_4 species of Panicum and Euphorbia were unsuccessful. Other surveys established that a green bundle sheath is not, in itself, a positive indication of C_4 photosynthesis.

We have worked in several areas concerning traits which differ in C_3 and C_4 species. I shall discuss briefly four of these areas.

COMPENSATION POINTS OF SPECIES

We have measured the CO_2 compensation points of more than 800 species during the course of various experiments in recent years. This has included about 500 species of Gramineae, a few other monocots and about 300 dicots. Summaries of parts of this extensive sampling were published previously (1,2,3). Of interest at this conference are the following: I. The C_4 traits in Gramineae are found in clearly defined evolutionary patterns essentially as suggested by DOWNTON and TREGUNNA (4). An exception, within present classification schemes, occurs in the genus *Panicum* where both C_3 and C_4 species are found (3,5). II. An erratic pattern is found in the occurrence of C_4 species among dicots. C_4 traits occur in several species in the Amaranthaceae, Chenopodiaceae, Portulacaceae, and Euphorbiaceae (3,4) and, recently were reported, in the Aizoaceae, Nyctagenaceae, and Zygophyllaceae (1). III. Both C_3 and C_4 species have been found in the *Euphorbia* and *Atriplex* genera (3). In all cases where low compensation points have been found, the species have also had starch-forming green bundle sheaths. Our studies and evidence from other laboratories on different traits, suggest that the C_4 condition consists of a "package" of several traits which appear to be mutually dependent. Wherever the C_4 condition exists, all of these traits occur.

VARIATION OF COMPENSATION POINTS WITHIN SPECIES

The CO_2 compensation points of healthy leaves are basically different for C_4 and C_3 species; in our studies, C_4 species have compensation points near 0 ppm and the C_3 species have compensation points near 50 ppm (at 25°).

HEICHEL and MUSGRAVE (6) reported that compensation points varied in maize (*Zea mays* L.) genotypes between 9 and 26 ppm and were inversely correlated with differences in leaf photosynthesis with a range between 85 and 25 mg $CO_2 dm^{-2} hr^{-1}$. Our experience with maize is strikingly different than that reported by HEICHEL and MUSGRAVE. In 1969 we grew 33 maize genotypes in the field, including some inbreds which were also used by HEICHEL and MUSGRAVE. We measured photosynthesis and CO_2 compensation points on excised leaves at about the time the genotypes flowered. All had compensation points near 0 ppm and photosynthesis in intense light near 50 mg $CO_2 dm^{-2} hr^{-1}$ (7). In 1970, we planted 125 genotypes in the field and again measured CO_2 compensation points. Duplicate measurements were made on each genotype . All had low compensation points with a mean of 1.3 ppm and a standard error of ± 1.2 ppm. The maximum for any genotype was 5 ppm.

Similarly, little variation occurs in compensation points within C_3 species. We found little variation in compensation points of 100 wheat (*Triticum aestivum* L.) genotypes (52 ± 2 ppm at 23°) or 25 barley (*Hordeum vulgare* L.) genotypes (55 ± 2 ppm at 23°) selected from the USDA World Collection to represent a diverse genetic background (3). Thus, our experience leads us to believe that CO_2 compensation points are essentially invariable. The reasons why they should not vary are not clear. The compensation point is often assumed to be a balance between two independent opposed CO_2 fluxes, respiration and photosynthesis. Thus to have a compensation point of 0 ppm, C_4 species should have no respiration or have some system to recapture internal CO_2. Even if green cells in C_4 species had a trapping mechanism which prevented respiratory CO_2 from escaping into the atmosphere, it is difficult to visualize how this system could trap CO_2 from epidermal or other non-green cells. Likewise, it is difficult to visualize how opposing fluxes of CO_2 in C_3 plants could result in the same compensation point for different plants within a species, even when the plants are raised in different environments which result in strikingly different morphologies. Furthermore, the fact that many C_3 species have nearly identical compensation points is difficult to explain.

Many factors, such as leaf temperature (8,9,10) or water stress

(10) affect the CO_2 compensation point of C_3 species. These factors have been carefully controlled in our work. We have also found that significant errors in measurement can result from systems components. Even the use in a system of a few centimeters of "Tygon" or rubber tubing, in which CO_2 readily dissolves, can give erroneous results and introduce striking variability to compensation point measurements. Also, CO_2 will reversibly adsorb on the surface of copper tubing. Thus, extreme care must be taken to insure accuracy if variability is to be sought among genotypes.

We have checked our system by introducing air with known concentrations of CO_2 into the system, closing it, and determining whether the concentration is maintained for long periods of time and whether a rapid change to a new concentration results in either adsorption or desorption from the system components into the atmosphere. When these precautions are not taken, or when leaves are not in optimum condition, then variability in measurements readily occurs. This variability may even be indirectly related to genotype because leaf size and intensity of photosynthesis both affect the equilibrium concentration in a closed system, if there are any internal CO_2 sources or sinks. We feel that it is probable that our results differ from some others found in the literature because of differences in measuring systems or of physiological status of the tissue being measured.

SCREENING FOR COMPENSATION POINT DIFFERENCES

We have been interested in evaluating the probability of finding individual plants within C_3 species which might have C_4 characteristics. We developed a method for screening large populations of C_3 plants for C_4 characteristics based on our evidence that all C_4 plants have CO_2 compensation points near 0 ppm (11). Of C_3 and C_4 plants placed in a closed, illuminated system, the C_4 plants are better competitors for the limited amount of CO_2 available. The result is that the CO_2 concentration of the atmosphere falls below the CO_2 compensation concentration for C_3 plants. At this low CO_2 concentration the C_4 plants photosynthesize at the expense of respiratory CO_2 from the C_3 plants, which soon die. If any plant in a C_3 population had C_4 characteristics, it should survive.

We have used this method of screening C_3 population for C_4 traits to evaluate large populations of wheat (*Triticum aestivum* L.) and oats (*Avena sativa* L.) grown from seed treated with up to 7 cycles of ionizing radiation. We have screened several thousand oat seedlings and well over 50,000 wheat seedlings and not found a single one with the C_4 characteristics.

THE RELATION OF CO_2 COMPENSATION AND CHLORENCHYMATOUS VASCULAR BUNDLE SHEATHS

Dr. BJÖRKMAN reported at this conference the results of crossing C_3 and C_4 species in the genus *Atriplex*. We tried similar crosses

with *Panicum* and *Euphorbia* species without success. We also sought
other species among the dicots which might be closely enough related
to cross but which differ in respect to C_3 and C_4 traits. Our approach was to examine species which have green vascular bundle
sheaths to determine whether they were C_4 species. We collected and
grew 88 species (from 22 families) which have chlorenchymatous bundle sheaths (1). Of these, only three species, *Tribulus terrestris*
L., *Boerhaavia paniculata* L.C. Rich, and *Trianthema portulacastrum*
L., had low CO_2 compensation points. These also were the only three
species which formed starch in the bundle sheaths and in which the
bundle sheath was composed of the prominent, thick-walled cells typical of C_4 species. Thus, a green bundle sheath alone is not a positive identification of C_4 photosynthesis.

REFERENCES

1. R.K. CROOKSTON and D.N. MOSS, *Plant Physiol.*, 46 (1970) 564.
2. E.G. KRENZER, Jr. and D.N. MOSS, *Crop Sci.*, 9 (1969) 619.
3. D.N. MOSS, E.G. KRENZER, Jr., and W.A. BRUN, *Science*, 164 (1969)
   187.
4. W.J.S. DOWNTON and E.B. TREGUNNA, *Can. J. Bot.*, 46 (1968) 207.
5. W.J.S. DOWNTON, J. BERRY, and E.B. TREGUNNA, *Science*, 163 (1969)
   78.
6. G.H. HEICHEL and R.B. MUSGRAVE, *Crop Sci.*, 9 (1969) 483.
7. D.N. MOSS, in Prediction and Measurement of Photosynthetic Productivity, Pudoc, Wageningen (1970) 323
8. O. BJÖRKMAN, E. GAUHL, and M.A. NOBS, *Carnegie Inst. Year Book*,
   68 (1969) 620.
9. O.V.S. HEATH and H. MEIDNER, *J. exp. Bot.*, 18 (1967) 746.
10. H. MEIDNER, *J. exp. Bot.*, 13 (1962) 284.
11. K.M. MENZ, D.N. MOSS, R.Q. CANNELL, and W.A. BRUN, *Crop Sci.*, 9
    (1969) 692.

ASPECTS OF THE EVOLUTION OF THE PHOTOSYNTHETIC CARBOXYLATION REACTION IN PLANTS

John H. Troughton

Physics and Engineering Laboratory, DSIR,
Private Bag, Lower Hutt, New Zealand

SUMMARY

The close correlation between the $\delta^{13}C$ value of higher plant tissue and the carbon pathway of photosynthesis of the plants was used to trace the evolutionary development of the carbon pathways during geological time. The $\delta^{13}C$ value of carbon from geological samples, mainly coal, from a range of periods from the Pliocene to the Permian, was measured and compared with the $\delta^{13}C$ value of present day plants. Difficulties associated with the interpretation of this data are discussed and the interpretation is made that the C_3 type carbon metabolism was present during the periods represented by the samples. The lack of evidence for the presence of the C_4 pathway cannot be taken as indicating that it was not present as it may have been present in other regions. Results from other workers were used to extend the general conclusion back to the Cambrian Period and to regions in both the northern and southern hemispheres.

INTRODUCTION

The evolutionary development of photosynthesis has been difficult to trace because of the lack of sufficient and appropriate information within the geological records. The evolution of the photochemical reactions of photosynthesis have been discussed recently (1) and this paper is restricted to aspects of photosynthesis involving carbon dioxide uptake, carboxylation, and photorespiration.

The study of the evolution of the carbon pathways in plants is facilitated by the phenomenon that higher plants preferentially take up the lighter of the two naturally occurring isotopes of carbon (^{12}C and ^{13}C) during photosynthesis (2,3,4). Consequently the ratio of the carbon isotopes in geological specimens can be used to indicate possible mechanisms involved in the derivation of the carbon.

EXPERIMENTAL

The $^{13}C/^{12}C$ ratio was measured by mass spectrometry after the method of CRAIG (5,6). This carbon isotope ratio is normally expressed as a $\delta^{13}C$ value where

$$\delta^{13}C\permil = \left[\frac{^{13}C/^{12}C \quad \text{sample}}{^{13}C/^{12}C \quad \text{P.D.B. standard}} - 1\right] \times 1000$$

RESULTS AND DISCUSSION

Fractionation of carbon isotopes in plants

Since higher plants discriminate against the heavier isotope of carbon, ^{13}C, they therefore have $\delta^{13}C$ values more negative than air (7,8). However, the extent of discrimination by plants is variable and recent results show that angiosperms can be divided into two major groups on the basis of their $\delta^{13}C$ values (9.10). This feature is illustrated in *Figure 1* which is prepared from the results of TROUGHTON, HENDY, and CARD (unpublished). In these results, which included assays of more than 100 species, the $\delta^{13}C$ value was

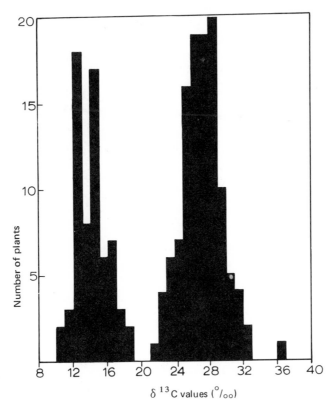

Figure 1 Histogram depicting the bimodal distribution of $\delta^{13}C$ values for a range of higher plants.

directly correlated with the particular pathway of carbon metabolism. Without exception, species known to be C_4 type plants were in
the group with less negative $\delta^{13}C$ values (mean of $-14^\circ/_{oo}$) while all
the C_3 type plants were in the group with more negative $\delta^{13}C$ values
(mean of $-29^\circ/_{oo}$). This technique of distinguishing between C_3 and
C_4 type species can be used to screen higher plants for the carbon
pathway. For example, *Atriplex buchananii*, which is indigenous to
New Zealand has a $\delta^{13}C$ value consistent with the presence of a C_4
type of carbon metabolism.

It has not been possible to quantitatively account for this fractionation of carbon isotopes found in plants. The fractionation is
likely to be associated with the low molecular weight compound CO_2
and the reactions involved include the processes of transport of CO_2
into the plant, the carboxylation reaction (including the carboxylating species), and photorespiration. That the difference in $\delta^{13}C$
value between the two major groups is likely to be associated with
biochemical differences is supported by data from succulent plants.
Succulents exhibiting Crassulacean acid metabolism (CAM) have $\delta^{13}C$
values similar to C_4 type plants in spite of differences in the resistances to transfer of CO_2 into the plants and in the timing of
the subsequent metabolic reactions involved in photosynthesis.

Although the sites of the fractionation of carbon have not been
identified, it is still possible to use the correlation between the
$\delta^{13}C$ value and pathway of carbon metabolism to trace the evolutionary development of the two pathways of photosynthesis in plants.

Comparison of present day plants

An initial comparison of some present day members of the plant
kingdom revealed that, apart from the angiosperms where both monocotyledons and dicotyledons contain C_3 and C_4 plants, all other
plants that were measured had $\delta^{13}C$ values consistent with a C_3 type
metabolism. This included examples from the Coniferophyta which a-
rose in the carboniferous era ($\delta^{13}C$ of $-24.0^\circ/_{oo}$), the Filicophyta
from the Devonian era ($\delta^{13}C$ of $-26.5^\circ/_{oo}$) and the Bryophyta which
are considered to have arisen during the Cambrian period (mosses
$-31.4^\circ/_{oo}$; liverwort $-32.3^\circ/_{oo}$; *Selaginella* $-30.8^\circ/_{oo}$).

Carbonaceous material deposited in geological time

The natural carbon isotope ratio values of carbon deposited during geological time are difficult to interpret because they may be
influenced by many other factors apart from fractionation associated
with photosynthesis. These factors include changes in the $\delta^{13}C$ value
of the environment, variation in $\delta^{13}C$ values within a plant, large
scale variations in the physical and chemical environment and the
possibility of exchange of carbon between deposits of different
ages.

Available evidence suggests that there were variations in the
$\delta^{13}C$ value of the atmosphere during geological time but that these
variations were less than $-3^\circ/_{oo}$ (11). The preferential uptake of
^{12}C by vegetation has led to some variation in the $\delta^{13}C$ value of the
atmosphere during periods when land plants were increasing (11).

Substantial changes in the environment and in particular in the O_2, CO_2, and temperature levels have occurred during geological time and these changes may have affected the $\delta^{13}C$ values of the plants. The temperature effect has only been studied over the range 20° to 30° and for maize, cotton, *Atriplex hastata*, and *Atriplex spongiosa* but there was only minor variation in the $\delta^{13}C$ value (TROUGHTON, HENDY, and CARD, unpublished). Similarly, CO_2 concentration has little effect, as tomato plants grown in 0.015% CO_2 had $\delta^{13}C$ values of -21.6°/oo and in 1.5% CO_2, -25°/oo. Furthermore our studies show there was little effect (less than -3°/oo) on the $\delta^{13}C$ value of *Dactylis glomerata* and *Festuca arrundinaceae* grown in a wide range of ecological sites from Finland to Israel. The oxygen level would be expected to influence the $\delta^{13}C$ value of the plants through an effect on photorespiration but the magnitude of this effect is, as yet, unknown.

The $\delta^{13}C$ value of a carbon deposit would primarily refer to carbon assimilated during periods of rapid photosynthesis. Variation in the $\delta^{13}C$ value between plant components is likely to be small as it has been shown for maize that the leaf (-11.4°/oo), seed (-11.2°/oo) and pollen (-12.1°/oo) have similar $\delta^{13}C$ values as also are the values for a pine needle (-25.7°/oo), young wood (-25.5°/oo), and even paper derived from pine trees (-25.3°/oo) (TROUGHTON, HENDY, and CARD, unpublished). It has often been pointed out that the lipid component of plants has more negative $\delta^{13}C$ values (-35°/oo) than the cellulose component (2,3,12), but, at least for higher plants, the proportion of the total carbon in the lipid fraction is small and is unlikely to significantly affect the $\delta^{13}C$ value for the plant.

These results indicate that the variation in the $\delta^{13}C$ value of carbon derived from plants during geological time is likely to be small, or at least significantly less than the difference in the $\delta^{13}C$ value between plants with the different carbon pathways of photosynthesis.

To trace the evolution of the $\delta^{13}C$ values a series of coal samples from the New Zealand region was chosen and this series contained representatives from seven geological periods from the Pliocene to the Permian. As shown in *Table 1* there may have been a trend toward less negative $\delta^{13}C$ values with age but the dominant feature of the results is that the $\delta^{13}C$ value of all the samples is within the range of values measured for plants known to have the C_3 type carbon metabolism.

Similar results have been reported for samples of carbon from a similar range of geological periods from America and Australia. CRAIG (5) reports $\delta^{13}C$ values more negative than -18°/oo and a mean of about -27.0°/oo for a range of coal and shale samples extending back to the Cambrian period. For samples of coal from Australia, JEFFERY and associates (13) measured $\delta^{13}C$ values within the range -19.7°/oo to -26.9°/oo and with a mean value of about -24.0°/oo. The results of specialized coal-types, carbonaceous shales, and graphites analyzed by JEFFERY *et al.* (13) were of interest because of

TABLE I

Natural isotopic carbon composition of some geological samples

Geological period	Sample	$\delta^{13}C$ value (‰)
Pliocene	coal	-27.9
Miocene	"	-27.9
Oligocene	"	-27.5
Eocene	"	-27.3
Cretaceous	"	-25.9
Triassic	"	-25.1
Permian	"	-25.9
	"	-25.1
	wood (from Antartica)	-22.5
	pollen	-25.1

some values in the range -12‰ to -16.2‰ but the interpretation of these results is difficult because the carbon is suggested to be of marine origin. Similarly, although the $\delta^{13}C$ values from the carbonaceous rocks of the Soudan Iron Formation are suggestive that C_3 type carbon metabolism was operative 2.7 billion years ago there are doubts about interpretation because at least for the extractable fraction of carbon it seems likely that the carbon migrated into the rock subsequent to deposition (14).

Although it is difficult to interpret the precise meaning of the $\delta^{13}C$ values, these results suggest that the C_3 type carbon metabolism has been operative as a major mechanism for the fixation of atmospheric CO_2 since plants came on land. These results do not exclude the possibility that the C_4 type metabolism has also been in existence during this same period even though it has not been uncovered here. The technique of measuring the natural isotopic carbon ratio in geological specimens would, however, be useful in tracing the evolutionary development of the C_4 pathway and for investigating its geographical distribution during geological time.

ACKNOWLEDGMENTS

The author sincerely acknowledges the co-operation and interest of Dr. T.A. Rafter, Institute of Nuclear Science, DSIR, New Zealand, in this work and the assistance of Mrs. K.A. Card and Dr. C.H. Hendy with the experimental aspects. The samples of coal that were used were supplied by Mr. T. Grant-Taylor of the N.Z. Geological Survey.

REFERENCES

1. J.M. OLSON, *Science*, 168 (1970) 438.
2. P.H. ABELSON and T.C. HOERING, *Proc. natn. Acad. Sci. U.S.A.*, 47

(1961) 623.

3. R. PARK and S. EPSTEIN, *Geochim. Cosmochim. Acta*, 21 (1960) 110.
4. M. CALVIN, Chemical Evolution, Oxford University Press, (1969).
5. H. CRAIG, *Geochim. Cosmochim. Acta*, 3 (1953) 53.
6. H. CRAIG, *Geochim. Cosmochim. Acta*, 12 (1957) 133.
7. F.E. WICKMAN, *Geochim. Cosmochim. Acta*, 2 (1952) 243.
8. H. CRAIG, *J. Geol.*, 62 (1954) 115.
9. M. BENDER, *Radiocarbon*, 10 (1968) 468.
10. B. SMITH, quoted by W.M. LAETSCH, *Sci. Prog. Oxf.*, 57 (1969) 323.
11. J.N. WEBER, *Geochim. Cosmochim. Acta*, 31 (1967) 2343.
12. S.R. SILVERMAN, *J. Am. Oil Chem. Soc.*, 44 (1967) 691.
13. P.M. JEFFERY, W. COMPSTON, D. GREENHALGH, and J. de LAETER, *Geochim. Cosmochim. Acta*, 7 (1955) 255.
14. P.E. CLOUD, J.W. GRUNER, and H. HAGEN, *Science*, 148 (1965) 1713.

EVOLUTIONARY, ADAPTIVE, AND ENVIRONMENTAL ASPECTS OF THE PHOTOSYNTHETIC PATHWAY: ASSESSMENT

L.T. Evans

Division of Plant Industry, CSIRO, Canberra City, 2601, Australia

EVOLUTION

The taxonomic distribution of succulents, of plants with Crassulacean acid metabolism (CAM), and of those with CO_2 fixation by the C_4 dicarboxylic acid pathway (C_4 plants), as indicated in *Figure 1*, suggests that both CAM and the C_4 pathway have arisen polyphyletically among several of the more advanced orders of both dicots and monocots. The Caryophyllales, Euphorbiales, and Asterales include both CAM and C_4 plants. There are several orders which, on present knowledge, include CAM but no C_4 plants (e.g. Rosales, Liliales, Bromeliales), or C_4 but no CAM plants (e.g. Cyperales, Poales).

The taxonomic distribution also suggests that the Calvin cycle is the more primitive pathway of photosynthesis. This conclusion is supported by TROUGHTON'S report that all the lower plant remains which he examined displayed the greater discrimination against ^{13}C characteristic of the Calvin pathway *vis-a-vis* both C_4 and CAM (6) plants. Moreover, all plants apparently rely on the Calvin cycle for the ultimate steps in CO_2 fixation, the CAM and C_4 pathways operating essentially as mechanisms for prefixing and concentrating CO_2 in the photosynthetic tissue. During ontogeny, tissues initially with C_3 characteristics can develop C_4 characteristics, as in the cultured sugarcane mentioned by DOWNTON, or in the loss of coupling between photosystems I and II with age described by BISHOP *et al*.

All these findings favor the concept of the CAM and C_4 pathways being recent addenda to the more primitive Calvin cycle.

Figure 1 Taxonomic distribution of succulent, CAM, and C_4 plants. The evolutionary arrangement of orders, and their family composition, is that given by TAKHTAJAN (1). The distributions are provisional and based on the following sources: succulents (2), CAM (3,4,5), C_4 plants (this volume).

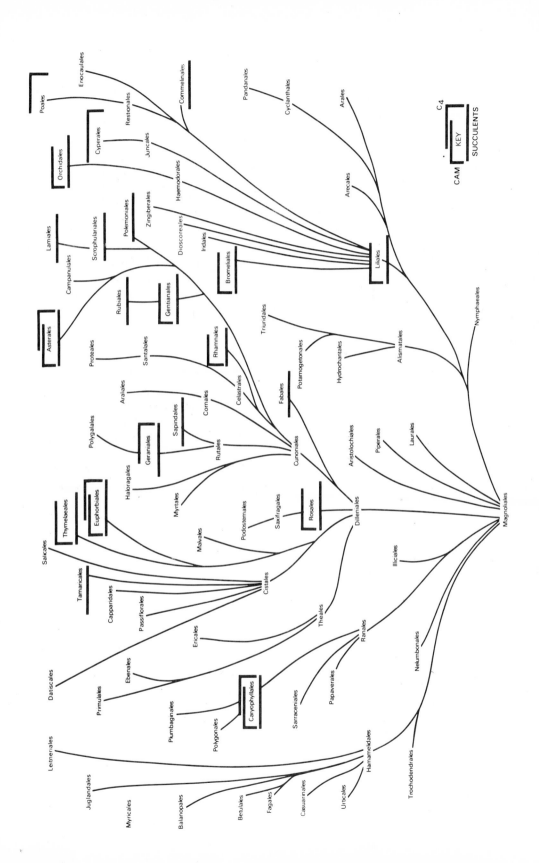

ADAPTIVE ASPECTS: GROWTH AND PHOTOSYNTHESIS

Before we consider under what conditions the CAM and C_4 pathways
may confer an adaptive advantage, the point should be made that a
high photosynthetic rate is not necessarily advantageous in terms of
either species survival or crop yields. Even with long established
and intensively selected crops, where one might expect such an advantage, there is no evidence that the photosynthetic rate per unit
leaf area has increased during evolution and selection. BULL mentioned that the wild sugarcanes have rates as high as the modern hybrids, and we (7) found that the maximum photosynthetic rate has
fallen quite markedly during the evolution of wheat. Practitioners
of growth analysis have long been aware that differences in relative
growth rates are far more strongly determined by differences in the
extent of investment into new leaf growth than by those in net assimilation rate, as SLATYER (8) found in a comparison of C_3 and C_4
species of *Atriplex*. In terms of survival and adaptation, performance under extreme conditions is probably more important than relative photosynthetic rate under optimal conditions.

COOL, LOW LIGHT ENVIRONMENTS

Under primitive atmospheres, with low O_2 and high CO_2 levels, the
Calvin cycle would have been as effective as the C_4 pathway, and
this still appears to be the case with the present atmosphere in
cool, low light environments. The structure of the light-trapping
photosynthetic units seems to be similar in C_3 and C_4 plants, but
the latter may require additional ATP to run the C_4 cycle, and perhaps also for shuttling 3-phosphoglycerate (3-PGA) and for maintaining the very rapid and massive transfer of malate etc. to the bundle
sheath cells. Thus, in spite of the advantage for the carbon balance
they derive from their lack of photorespiration, C_4 plants may have
quantum requirements which are no lower than those of C_3 plants even
in air, as BJÖRKMAN'S results (9) with *Atriplex patula* (C_3) and *A.
rosea* (C_4) indicate. At low oxygen concentration the quantum requirement was 35-40% higher in the C_4 than in the C_3 species. These
measurements were made only in red light, and need to be extended to
other C_3/C_4 comparisons, under more balanced light sources, but they
suggest that the C_4 pathway should not be referred to as "efficient
photosynthesis"; in terms of light requirement it may be less so
than the Calvin cycle. Moreover, the stomatal mechanism in C_4 plants
may require higher light intensities to reach full opening, as GIF-
FORD showed, indicating that the stomatal resistance of C_4 plants
may be higher than that of C_3 plants under low light.

HIGH LIGHT ENVIRONMENTS

In environments with high insolation and high day temperatures,
the carbon loss by photorespiration in C_3 plants becomes substantial,

and CAM and C_4 plants derive a marked advantage from their more efficient internal trapping of CO_2. Along with the higher photosynthetic rates of the C_4 plants under high insolation, there is their more efficient use of water, long known from the data of SHANTZ and PIEMEISEL (10). The lower transpiration ratio in C_4 grasses has been confirmed by several participants in this symposium. Comparing the data of SLATYER (8) and DOWNES (11) with those of NEALES *et al.* (12) for some CAM plants, the ratio of water lost to CO_2 gained is similar in CAM (in the range 30 to 116) and C_4 (50 to 100) plants, and much lower than the ratio of 150 to 250 common among C_3 plants. However, as SLATYER (8) has emphasized, one cannot extrapolate directly to the crop or plant community from these single leaf studies, since crop structure and other features can modify this difference, and in a community r_1 tends to become small relative to r_a. BULL, for example, found that sugarcane had no advantage over several C_3 plants in either growth rate or efficiency of water use in some conditions.

THE ROLE OF NIGHT TEMPERATURE

The main differentiating factor in relation to the adaptive advantages of CAM and C_4 plants may be night temperature. CAM operates very efficiently at low night temperatures, and TING pointed out that the nocturnal CO_2 fixing mechanism, first recognized for that it is by WARBURG in 1886, becomes pronounced with marked diurnal fluctuation of temperature. Under these conditions we get the required compartmentation between CO_2 fixation and reduction in time, diurnally, rather than in space as in the morphologically differentiated C_4 leaves. But this diurnal compartmentation is effective only when there is available a massive CO_2 fixing pool, hence the succulence of CAM plants. The data of NEALES *et al.* (12) emphasize just how massive this pool is; *Agave* plants, for example, could sustain an average dark fixation rate of 8 mg CO_2 dm^{-2} hr^{-1} for 8 hours.

In the case of the C_4 plants, it is not altogether clear why they grow as poorly as they do at low temperatures. BJÖRKMAN pointed out that there is no obvious enzymatic basis for this. It may be due in part to greater photolability of the chloroplast membranes at low temperatures, accentuated in a maize mutant (13), or to some failure in the mechanism of translocation. HILLIARD and WEST (14,15) found that instead of the usual nocturnal removal of starch that occurs during warm nights, the starch accumulated in *Digitaria* leaves at a night temperature of 10° to the point where chloroplast functions, such as the Hill reaction and nucleic acid synthesis, were impaired. In this case the starch accumulated in the mesophyll cells, whereas it is usually in the bundle sheath of C_4 plants. Other C_4 plants also showed a parallelism between starch accumulation and sensitivity to low temperatures. Another mechanism which may be involved in this sensitivity, and which may prove to be more general, could be the impairment of chloroplast membrane permeability suggested by PHILLIPS and McWILLIAM from their Arrhenius plots for phosphoenolpyruvate (PEP) carboxylase activity.

GENETIC ASPECTS

The C_4 syndrome has many components: β-carboxylation, anatomical compartmentation, apparent lack of photorespiration and O_2 inhibition, low CO_2 compensation points, high maximum rates of photosynthesis and growth, high light saturation, and high optimum temperatures. Some of these features can be found among C_3 plants, for example the very high photosynthetic rates at high intensities found in the primitive wheats (7), in some soybeans, and in *Typha* (16). However, the full syndrome suggested a complex and integrated evolutionary change. So long as it was found only in the tropical grasses this posed no problem, but when found in several other orders, its polyphyletic origin seemed difficult to account for at first. Now it appears that the syndrome may follow from only a few central changes, and BJÖRKMAN'S data with the *Atriplex* crosses suggest that only a few genes are involved.

As LAETSCH put it, it is simply "CAM mit Kranz", the evolution of the β-carboxylation CO_2-concentrating mechanism plus the morphological compartmentation given by the Kranz anatomy. The latter alone is not enough, as MOSS showed. Of course, more than this is involved, as consideration of the traffic required between mesophyll and bundle sheath in the latest Hatch-Slack scheme emphasizes. Some of BJÖRKMAN'S F_2 hybrids had β-carboxylation plus Kranz anatomy, yet apparently lacked the necessary coordination and synchrony for normal C_4 operation.

Moreover, functional differentiation among plants with the C_4 pathway seems to have occurred in several orders, between those biased towards aspartate labeling, and those biased towards malate, as DOWNTON indicated. Here too biochemical differentiation appears to be accompanied by anatomical differentiation, such as the centrifugal arrangement of the bundle sheath chloroplasts, the greater reduction of granal development and the loss of photosystem II activity in the bundle sheath chloroplasts of plants with malate bias.

PHYSICAL ASPECTS: RESISTANCE ANALYSES

Minimum stomatal resistances appear to be somewhat lower in C_3 than in C_4 plants, but this is more than counterbalanced by the noticeably lower residual resistances in the C_4 plants (8). The residual resistance comprises a physical term for diffusion of CO_2 in the liquid phase from the cell walls through the cytoplasm and across the chloroplasts to the sites of carboxylation, and a chemical term for the carboxylation resistance.

A possible explanation of the much lower residual resistance of the C_4 plants would be a much lower carboxylation resistance, because of the greater affinity for CO_2 of PEP carboxylase compared with ribulose-1,5-diphosphate (RuDP) carboxylase, and of the lower photorespiratory loss. However, the first attempts by CHARTIER *et al.* (17) to partition the residual resistance in a C_3 plant suggested that the carboxylation resistance was extremely small (0.1 - 0.6 sec cm^{-1}), the physical resistance being the dominant one (4.9 - 9.1 sec cm^{-1}). RACKHAM'S (18) calculations, based on chloroplast and wall dimensions, also suggested that the diffusional resistance could be substantial (0.3 - 4 sec cm^{-1}). Using CHARTIER'S methods, LUDLOW concluded that the carboxylation resistance was low (0.2 - 0.3 sec cm^{-1}) in both C_3 and C_4 plants, and that the higher residual resistance of the C_3 plants was due to their much higher diffusional resistance. SLATYER queried the validity of CHARTIER'S method of analysis on several grounds, and new methods clearly need to be developed. Moreover, the partitioning into diffusional and carboxylation resistances is not altogether logical. On the one hand, the carboxylation resistance may well have a diffusional component at the molecular level as GIFFORD pointed out. On the other, the diffusional term ends with the initial β-carboxylation in the mesophyll of C_4 plants, and takes no account of diffusion to the bundle sheath cells, as LUDLOW noted.

In spite of the low carboxylation resistances derived by the CHARTIER analysis, there are many reasons for believing that, at least in C_3 plants, RuDP carboxylase activity constitutes a major limitation to photosynthetic rate. In the paper by NEALES *et al.*, the highest correlation was between photosynthetic rate and RuDP carboxylase activity, and BJÖRKMAN mentioned many instances where these parameters were highly related in comparisons between varieties, between sun and shade forms, and within varieties under varying conditions of light intensity, temperature, nutrient, disease and age, as in the experiments of WOOLHOUSE (19).

Since RuDP carboxylase comprises such a high proportion of total leaf protein, these correlations suggest that its activity is a major limitation on photosynthetic rate. Hence the evolutionary advantage of mechanisms increasing its CO_2 fixing ability, such as the C_4 pathway.

REFERENCES

1. A. TAKHTAJAN, Die Evolution der Angiospermen, Fischer, Jena, (1959).
2. H. JACOBSEN, A Handbook of Succulent Plants, Blandford, London, (1954).
3. J. BRUINSMA, *Acta bot. neerl.*, 7 (1958) 531.
4. E.L. NUERNBERGK, *Planta*, 56 (1961) 28.
5. K.H. SCHUTTE, R. STEYN, and M.v.D. WESTHUIZEN, *J. S. Afr. Bot.*, 33 (1967) 107.
6. M.M. BENDER, *Phytochemistry*, in press.
7. L.T. EVANS and R.L. DUNSTONE, *Aust. J. biol. Sci.*, 23 (1970) 725.
8. R.O. SLATYER, *Planta*, 93 (1970) 175.

9. O. BJÖRKMAN, E. GAUHL, and M.A. NOBS, *Carnegie Inst. Wash. Yearbook*, 68 (1969) 620.

10. H.L. SHANTZ and R.L. PIEMEISEL, *J. agric. Res.*, 34 (1927) 1093.

11. R.W. DOWNES, *Aust. J. biol. Sci.*, 23 (1970) 775.

12. T.F. NEALES, A.A. PATERSON, and V.J. HARTNEY, *Nature*, 219 (1968) 469.

13. A. MILLERD, D.J. GOODCHILD, and D. SPENCER, *Plant Physiol.*, 44 (1969) 567.

14. J.H. HILLIARD and S.H. WEST, *Science*, 168 (1970) 494.

15. S.H. WEST, in Plant Response to Climatic Factors, UNESCO, Paris, (1971) in press.

16. S.J. McNAUGHTON and L.W. FULLEM, *Plant Physiol.*, 45 (1970) 703.

17. P. CHARTIER, M. CHARTIER, and J. CATSKY, *Photosynthetica*, 4 (1970) 48.

18. O. RACKHAM, in Light as an Ecological Factor, eds. R. BAINBRIDGE, G.C. EVANS, and O. RACKHAM, Blackwells, Oxford, (1966) p. 167.

19. H.W. WOOLHOUSE, *Symp. Soc. exp. Biol.*, 21 (1967) 179.

SECTION 2

Carbon Dioxide Assimilation

A. Review Papers

MECHANISM AND FUNCTION OF THE C_4 PATHWAY OF PHOTOSYNTHESIS

M.D. Hatch

Division of Plant Industry, CSIRO, Canberra City, 2601, Australia

SUMMARY

The key data relating to the mechanism and function of the C_4 pathway is reviewed. The kinetics of labeling of compounds in leaves provided with $^{14}CO_2$ indicate that radioactivity can enter the C-1 of 3-phosphoglycerate via the C-4 of C_4 dicarboxylic acids and that this is the major or sole route of 3-phosphoglycerate formation. From these studies and knowledge of the activity and distribution of enzymes it would appear that CO_2 is incorporated initially into C_4 acids in the mesophyll cells. Depending upon the species, either malate or aspartate or both acids are then transported to the bundle sheath chloroplasts where the C-4 carboxyl is released as CO_2 and refixed by ribulose-1,5-diphosphate carboxylase. The remaining C_3 compound, pyruvate in the case of malate decarboxylation and alanine in the case of aspartate decarboxylation, is then returned to the mesophyll cells to serve as a precursor of phosphoenolpyruvate. It is proposed that the reactions unique to the C_4 pathway serve as a mechanism for concentrating CO_2 at the site of action of ribulose-1, 5-diphosphate carboxylase in the bundle sheath cells.

INTRODUCTION

As I see it, my task at this stage is to draw together and interpret the information available prior to this meeting on the mechanism and function of the C_4 dicarboxylic acid pathway, henceforth refered to as the C_4 pathway. Species which utilize the Calvin cycle will be refered to as C_3 pathway species. The situation as of the latter part of last year has been summarized in recent reviews (1,2) and at the International Botanical Congress in Seattle. I will therefore only try to highlight some of the key evidence available to that time with particular emphasis on those observations which I think ought to be kept in mind during the subsequent discussions of this symposium. I will then move on to consider in some detail the more recent developments and interpretations.

RADIOTRACER STUDIES

Until recently at least, information about the mechanisms of the C_4 pathway has been derived mainly from time-course and pulse-chase radiotracer studies with leaves, combined with comparative studies on the activity and location of enzymes. Amongst the radiotracer

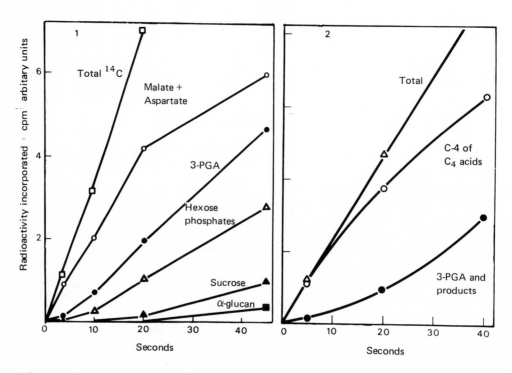

Figure 1 Rate of labeling of compounds in sugarcane leaves exposed to $^{14}CO_2$ under steady-state conditions. [Data from HATCH and SLACK (4)]

Figure 2 Rate of labeling of C-4 of C_4 acids and of 3-PGA and its products in maize leaves at lower light intensity (2000 ft candles). [Data from JOHNSON and HATCH (6)]

studies that remain critical in the present context are· the simple time-course studies in which the incorporation of $^{14}CO_2$ is followed under steady-state conditions for photosynthesis. With sugarcane and maize leaves radioactivity is incorporated at a steady rate from zero time and appears initially in the C_4 dicarboxylic acids oxalo-acetate, malate, and aspartate (3-6). As shown in *Figure 1*, for su-garcane, the rate of labeling of these acids is maximal from zero time. The next compound to be rapidly labeled is 3-phosphoglycerate (3-PGA) but it may be worth noting that under some conditions (5) several seconds may elapse before any label is detected in compounds other than C_4 acids. Label appears initially in the C-4 of the C_4 acids; the half time for saturation of this carbon is about 10 sec compared with 30 sec for the C-1 of 3-PGA (1,4). With increasing time greater proportions of the fixed radioactivity appears in the various sugar phosphates and the photosynthetic end products sucrose and starch.

The kinetics of the labeling of the C_4 dicarboxylic acids is totally consistent with these acids being first products of CO_2 fixa-

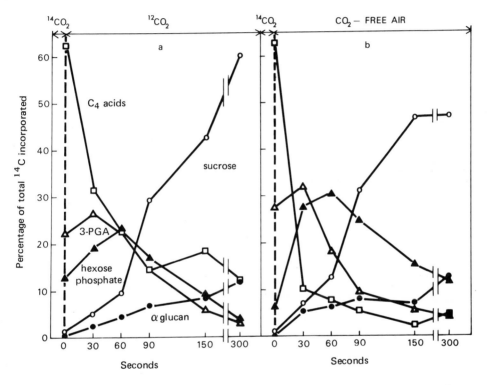

Figure 3 Changes in the distribution of radioactivity after *Sorghum* leaves were exposed to 0.035% $^{14}CO_2$ for 15 sec and then transfered to (a) air or (b) CO_2-free air. [Data from JOHNSON and HATCH (6)]

tion and intermediates in 3-PGA formation. In saying this it should be particularly kept in mind that these changes in rates of labeling occur while there is a linear increase in total radioactivity incorporated. To further emphasize this point *Figure 2* shows the kinetics of labeling of the C_4 acids and other compounds expressed in their simplest form. In this study with maize leaves the rate of labeling of the C-4 of the C_4 acids is compared with the rate of labeling of 3-PGA and its products (6). In accordance with the label moving through the C-4 carbon to 3-PGA, the initial rate of labeling of this carbon is comparable to the rate of total incorporation but subsequently declines. This decline is compensated for by an increase in the rate of labeling of 3-PGA and its products in such a way that the rate of total incorporation remains essentially constant.

The results of pulse-chase studies provide more direct information about the transfer of label from C_4 acids to 3-PGA. For instance the radioactivity residing in the C_4 dicarboxylic acids of *Sorghum* leaves after 15 sec in $^{14}CO_2$ is rapidly mobilized into 3-PGA, triose phosphates and hexose phosphates during a chase in $^{12}CO_2$ (6) *(Figure 3a)*. After 150 sec in $^{12}CO_2$ it is mostly located in the end products sucrose and starch. By degrading the C_4 acids it was possible to show that essentially all the C-4 label is transfered by 90 sec, the residual radioactivity in the C_4 acids at later times being in the other carbons of these compounds. When leaves were transfered to CO_2 free air instead of normal air almost 90% of the radioactivity in the C_4 acids was transfered to other compounds in 30 sec *(Figure 3b)*. These studies clearly demonstrate how rapidly this C_4 acid pool involved in photosynthesis is turning over.

In another experiment in which time-course and pulse-chase data were obtained simultaneously for maize leaves it was possible to quantitate the transfer of radioactivity from the C_4 acids to 3-PGA (6). In essence this experiment showed that the radioactivity entering 3-PGA and its products during various periods of exposure to $^{14}CO_2$ was almost entirely accounted for by the radioactivity moving from the C-4 of the C_4 acids. The pulse-chase data was used to estimate the rate constant for loss of radioactivity from the C-4 of dicarboxylic acids. With this rate constant the radioactivity lost from the C-4 of the C_4 acids could be calculated for any time during the period of exposure of leaves to $^{14}CO_2$. The calculated values for the loss of ^{14}C from the C-4 of the C_4 acids corresponded closely with the amount of ^{14}C observed to appear in 3-PGA and its products.

I should mention just one more radiotracer study which I will refer to again later. This experiment showed that the radioactivity in the C_4 acids of leaves exposed to $^{14}CO_2$ was transfered to 3-PGA and subsequent products at essentially the same rate during a chase in normal air and in air containing 5% CO_2, that is nearly 200 times that in normal air (4). At that time, we concluded from this observation that there was no free CO_2 pool involved in the transfer of radioactivity from the C-4 of the C_4 acids to 3-PGA.

ENZYMES OF THE C_4 PATHWAY

Turning to enzyme studies, *Table 1* summarizes the earlier data on the comparative activity and location of the enzymes implicated in the pathway. A feature of all plants showing the labeling characteristics I have just described is their very high phosphoenolpyruvate (PEP) carboxylase activity. This activity is around 5 times that for the photosynthesis rates for these species and about 60 times that present in leaves of species with the C_3 pathway. Leaves of C_4 path-

way species also contain adequate quantities of an NADP malate dehydrogenase and aspartate aminotransferase to account for malate and
aspartate formation from oxaloacetate during photosynthesis.

TABLE I

Summary of the activity (μmole min^{-1} mg^{-1} chlorophyll) and location of enzymes implicated in photosynthesis in C_4 pathway
plants*

Enzyme and location**	Activity***	Average ratio for C_4 to C_3 species#
Photosynthesis rate	3-5	2
Mesophyll		
PEP carboxylase	16-21	60
NADP-malate dehydrog. ##	2-14	10
Aspartate aminotransf.##	4-40	10
Pyruvate,Pi dikinase	3-7	∞
Adelylate kinase	17-45	75
Mesophyll and bundle sheath		
Pyrophosphatase	15-60	15
3-PGA kinase	50	Similar
NADP-triose-P dehdrog.	5-10	Similar
Bundle sheath		
FDP aldolase	8	Similar
Alkaline FDPase	1	Similar
R 5-P isomerase	15	Similar
Ru 5-P kinase	10-25	Similar
RuDP carboxylase	2-4	Similar
Malic enzyme	$\begin{bmatrix} 10\text{-}12 \\ 0.1\text{-}1 \end{bmatrix}$	$\begin{bmatrix} 50 \\ 2 \end{bmatrix}$

* Abstracted from HATCH and SLACK (1) and more recent data
  (18).
** Designated as mesophyll or bundle sheath when at least 90%
   of the activity was located in these cells.
*** Enzyme activity in leaf extracts. The ranges shown include
    the values for several species.
The approximate ratio of average activities recorded for
  species with the C_4 pathway and the C_3 pathway.
In some species this enzyme is about equally distributed
   between the two cell types (18).

One of the predictions from radiotracer studies was that the C-1,
C-2, and C-3 carbons of the C_4 acids must provide the precursor for
PEP regeneration (1). This was supported by the discovery of the en-
zyme pyruvate,Pi dikinase that converts pyruvate to PEP and that

apparently occurs only in the leaves of species with the C_4 pathway. The other products of this reaction are PPi and AMP and it was significant that the C_4 pathway species also contained comparatively high activities of the enzymes pyrophosphatase and adenylate kinase which would be likely to be involved in the metabolism of these products.

Information about the intra- and intercellular location of these enzymes was initially provided by examining their association with mesophyll and bundle sheath chloroplasts isolated in non-aqueous media (7,8). These studies indicate that the first group of enzymes listed in *Table 1*, which are directly or indirectly involved in the formation and interconversion of dicarboxylic acids, are largely or solely associated with mesophyll chloroplasts. Several enzymes operative in the Calvin cycle, including fructose 1,6-diphosphate (FDP) aldolase, FDPase, ribose phosphate (R 5-P) isomerase, and ribulose 5-phosphate (Ru 5-P) kinase, had comparable activities in the leaves of C_4 pathway and C_3 pathway species. These enzymes were associated with bundle sheath chloroplasts as also was ribulose-1,5-diphosphate (RuDP) carboxylase and malic enzyme. The studies of BLACK and colleagues, reported in this volume, substantially support these conclusions about the intercellular distribution of enzymes.

Earlier estimates suggested that the activity of RuDP carboxylase in C_4 pathway leaves was much lower than the maximum photosynthesis rates and less than that in leaves of C_3 pathway species (1). We now know this is not correct and this aspect will be considered in more detail below. Some species contained very high activities of malic enzyme but in others the activity was little above that of Calvin cycle plants and considerably less than the maximum photosynthesis rates (1,9,10). In contrast to the other enzymes, the two enzymes involved in the conversion of 3-PGA to triose phosphates, 3-PGA kinase, and NADP glyceraldehyde 3-phosphate dehydrogenase, were about equally distributed between the two types of chloroplasts in maize at least. The possible significance of this observation will be considered later.

INTERPRETATIONS AND RECENT DATA

Other than some very interesting data on the regulation of some of the key enzymes of the process (1,2,11) this was, informationwise, about the "state of the art" towards the latter part of last year. It seemed from this data that:-

(i)   the C_4 acids were formed in the mesophyll chloroplasts,

(ii) the C-4 of these acids was at least the major source of the C-1 of 3-PGA,

(iii) the conversion of 3-PGA to triose phosphates occured in both types of chloroplasts,

(iv) and that the conversion of triose phosphates to hexose phosphates and the reactions leading to RuDP formation were operative in the bundle sheath chloroplasts.

Until recently the balance of evidence appeared to be inconsistent with the transfer of carbon from the C_4 dicarboxylic acids to the C-1 of 3-PGA proceeding via a free CO_2 pool, and against the involvement of RuDP carboxylase in this process. As an alternative we had proposed the operation of a transcarboxylation reaction purely on the basis of the negative evidence for the former possibility. Briefly, the key points of evidence against decarboxylation and refixation and against the involvement of RuDP carboxylase were:-

(i)   the pulse-chase data with the 5% CO_2 chase, refered to above, which was at least inconsistent with the existence of freely diffusable intermediate pool of $^{14}CO_2$,

(ii)   the evidence that the activity of RuDP carboxylase in C_4 pathway species was apparently too low for its integral operation in the overall process of photosynthesis,

(iii) and the data showing that a considerable portion of the early-labeled 3-PGA was located in the mesophyll chloroplasts while the RuDP carboxylase was located in the bundle sheath chloroplasts (8).

The possibility that RuDP carboxylase was involved in the transfer of carbon from the C_4 acids to 3-PGA was re-opened by the observations of BJÖRKMAN and GAUHL (12) reported late last year. They showed that the level of RuDP carboxylase in several species with the C_4 pathway was in fact similar to those in C_3 pathway species. These activities were just sufficient to account for the integral operation of the enzyme in photosynthesis providing the concentration of CO_2 at its site of action was near saturating. It turned out that the lower activities reported earlier were due in part to incomplete extraction of the enzyme from bundle sheath cells and in part to the use of assay conditions which were non-optimal for the enzyme from C_4 pathway plants (12,13). Incidently, our recent studies (13) have shown the RuDP carboxylase from C_4 pathway species has a sharp pH optimum at 7.7, is inhibited by concentrations of HCO_3 above 20 mM, and has a much lower Km for RuDP than that reported for the enzyme obtained from C_3 pathway species.

When it was established that the activity of RuDP carboxylase was sufficient to account for refixation of CO_2 released from the C_4 acids, assuming transfer of the C-4 to 3-PGA occurs by decarboxylation and refixation, several other previously anomalous observations seemed to fit into place. You will recall that at least some C_4 pathway species contained high levels of malic enzyme and that this enzyme was located in the bundle sheath chloroplasts along with RuDP carboxylase *(Table 1)*. BERRY, DOWNTON, and TREGUNNA (14) arrived at a similar conclusion about the distribution of these enzymes using a different technique. As a consequence of this kind of data it was apparent that malic enzyme was ideally situated to release CO_2 from the dicarboxylic acids right at the site of action of RuDP carboxylase. It followed that at least in those species with high malic enzyme, the major route of carbon flow could involve formation of malate in the mesophyll chloroplasts followed by transfer of this acid

to the bundle sheath chloroplasts where it is metabolized by malic
enzyme and RuDP carboxylase. In this regard it should be noted that
there is qualitative evidence from microradioautographic studies indicating that C_4 acids must move rapidly into the bundle sheath
chloroplasts. Furthermore, I should also mention the high frequency
of plasmodesmata connecting the two types of cells which could facilitate the rapid intercellular transport of metabolites. This was
originally noted by LAETSCH and PRICE (15) in sugarcane and has now
been observed in several other C_4 pathway species.

We had proposed earlier (4) that following the transfer of the
C-4 of the C_4 acids to the C-1 of 3-PGA the residual C_3 compound
would have to give rise to pyruvate and thence PEP in order that the
cycle be sustained. In the scheme just outlined this pyruvate would
be formed directly by the action of malic enzyme and we would have
to propose that it is transported back to the mesophyll chloroplasts.

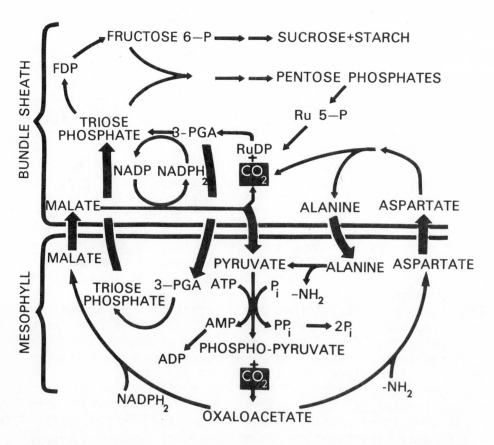

*Figure 4 Schematic representation of the reactions and inter-
cellular movements (heavy arrows) of metabolites proposed to
occur during the operation of the C_4 pathway. Alternative processes whereby CO_2 is released by transport of malate (left-
hand side) or aspartate (right-hand side) are shown. The
scheme is a modification of that presented earlier (18).*

This interpretation is depicted schematically in *Figure 4*.

The particular species which have high malic enzyme are also characterized by having relatively few bundle sheath chloroplasts and only about 30% of the total leaf chlorophyll and few if any grana in these chloroplasts (PYLIOTIS, WOO, and DOWNTON, this volume). The earlier suggestion (8) that these chloroplasts may be deficient in the capacity to photoreduce NADP has been supported by recent studies (16,17). In view of this it is not without significance that the operation of the scheme involving malic enzyme would result in the transfer not only of carbon, but also of reducing power, to the bundle sheath cells *(Figure 4)*. However, stoichiometrically, only half of the $NADPH_2$ necessary for the reduction of 2 molecules of 3-PGA to triose phosphates would be provided by malic enzyme. If there is insignificant photoreduction of NADP in the bundle sheath chloroplasts this deficiency could be overcome if about half of the 3-PGA formed in these chloroplasts is transported to the mesophyll chloroplasts for reduction to triose phosphates. Since FDP aldolase and FDPase are apparently specifically located in the bundle sheath chloroplasts the triose phosphates would then have to be returned to the latter chloroplasts for conversion to sugar phosphates *(Figure 4)*. Whether the complexity of this process, in terms of transport, worries the leaf as much as it does the observer, remains to be seen. However, this proposition does make sense of two other observations:- (i) that a substantial part of the early labeled 3-PGA appears to be located in the mesophyll chloroplasts (8) and, (ii) that about half the 3-PGA kinase and NADP glyceraldehyde 3-P dehydrogenase activities of these leaves are located in the mesophyll chloroplasts (8).

Thus, for the particular group of C_4 pathway species with high malic enzyme activity and other associated characteristics, we are at least able to make a reasonable and complete proposal for the route of CO_2 fixation. However, as I indicated earlier, there is another group of C_4 pathway plants which have only low activities of malic enzyme, more comparable to the activities present in C_3 pathway plants. It is probably significant that these species, in contrast to the species with high malic enzyme, contain a major part of the leaf chlorophyll in their bundle sheath chloroplasts and have a normal frequency of grana in these chloroplasts.

A possible clue as to the nature of the alternative biochemical process operating in these species was provided when it was found that they contain much higher activities of aspartate and alanine aminotransferases (18 and JOHNSON *et al.*, this volume). Comparative information on the activity of these enzymes is summarized in *Table 2* in which species are divided into 3 groups according to their enzyme activities. Group 1 includes those species with relatively high malic enzyme activity. This activity considerably exceeds the maximum photosynthesis rates of these species of 3 to 5 μmoles $min^{-1}mg^{-1}$ chlorophyll. The species of this group contain comparable activities of NADP malate dehydrogenase. Group 3 includes species of the grasses *Chloris* and *Eragrostis* and of the dicotyledons *Amaranthus* and *Atriplex* which have relatively low malic enzyme activity, in many cases about the activity present in leaves of C_3 pathway species. These species also have considerably less NADP malate dehydrogenase. DOWNTON (10) has also demonstrated this kind of a division with respect to malic enzyme amongst a wide selection of C_4 pathway species.

TABLE 2

Malic enzyme activity (μmoles min^{-1} mg^{-1} chlorophyll) in C_4 pathway species, and its relation to other enzyme activities*

	Malic enzyme	NADP malate dehydrogenase	Alanine aminotrans-ferase	Aspartate aminotrans-ferase
Group 1 *Zea* *Saccharum* *Sorghum*	10-12	9-14	0.3-0.7	4.5-5.7
Group 2 *Gomphrena*	5.5	2.3	16	15
Group 3 *Chloris* *Eragrostis* *Amaranthus* *Atriplex*	0.1-1	0.6-3	16-32	22-41

* Data from ANDREWS, JOHNSON, SLACK, and HATCH (18). The ranges of activity are for the different species examined.

The point I want to particularly emphasize in the data of *Table 2* is that the species of Group 3 contained very high levels of alanine and aspartate aminotransferases relative to the activities in Group 1 species. Assayed with α-ketoglutarate as the acceptor there was about 30 to 60 times more alanine aminotransferase and 5 to 10 times more aspartate aminotransferase in these species. *Gomphrena celosoides* was clearly intermediate between groups 1 and 3 in respect to both the activities of malic enzyme and aminotransferases. This could be taken as a warning that while such a division of this C_4 pathway species may be convenient it is mechanistically almost certainly a quantitative rather than qualitative division.

We have additional information on the intercellular distribution of these aminotransferases (18 and JOHNSON *et al.*, this volume). However, at this point I simply want to suggest that the presence of high aspartate and alanine aminotransferase activities in the Group 3 species would be consistent with aspartate replacing malate as a CO_2 donor. Aspartate aminotransferase would be required for the synthesis of aspartate and alanine aminotransferase for the regeneration of pyruvate, and hence PEP, from the C_3 compound remaining after decarboxylation. This alternative route is also depicted schematically in *Figure 4*.

FUNCTION OF THE C_4 PATHWAY

If the C_4 pathway does operate in the manner proposed it might be useful to conclude by considering the possible advantages or function of such a process. Superficially at least, fixing CO_2 only to release it and refix it again would appear to be a fairly fruitless exercise. However, you will recall I said earlier that the levels of RuDP carboxylase in C_4 pathway species was just adequate to account for its integral role in photosynthesis *provided* it operates in the cell with near-saturating concentrations of CO_2. This proviso, I believe, provides a clue to the role of the unique reactions of the C_4 pathway.

TABLE 3

Characteristics of PEP carboxylase and RuDP carboxylase

Parameter	RuDP carboxylase	PEP carboxylase
Ratio maximum activity in extracts to maximum photosynthesis rate	1-3[*]	Approx. 5[**]
K_m for CO_2[***]	450 μM	7 μM
Concentration CO_2 required to give maximum rates of photosynthesis[#]	At least 450 μM	Approx. 2 μM

[*]   For C_3 pathway and C_4 pathway species (9,12,13).
[**]   For C_4 pathway species (1,2).
[***] See ref. 19 and 20.
[#]   Estimated from the data provided above.

To develop this point it would be useful to consider the quantitative information summarized in *Table 3* for the carboxylase with which we are concerned. Firstly, it can be readily calculated that the concentration of CO_2 in solution, in equilibrium with the CO_2 in air, is about 8 μM at physiological temperatures and irrespective of the pH. It should also be noted that the concentration of a CO_2 pool in the leaf derived by diffusion from air, and serving photosynthesis, must be less than this concentration under steady state conditions. We now know that CO_2 rather than HCO_3^- is the substrate for RuDP carboxylase, and that the K_m for CO_2 is about 450 μM (19). Also, the activity of RuDP carboxylase isolated from leaves of C_4

pathway plants and C_3 pathway plants, assayed under optimal conditions, is about equal to or at the most about twice the maximum photosynthesis rate for these species. Assuming that the properties and performance of isolated RuDP carboxylase represents those of the enzyme within the leaf we can consider this information in one of two ways. If the enzyme is served by CO_2 pools derived by simple diffusion of CO_2 from the air then the internal concentration of CO_2 will be less than 8 μM, that is less than 1/50 of the K_m for CO_2. The turnover of the enzyme in leaves will therefore be only about 1/50 to 1/100 of that necessary to account for photosynthetic CO_2 fixation. Alternatively, if the enzyme has an integral role in photosynthesis it must operate at CO_2 concentrations of about 450 μM or more.

The evidence is strongly in favor of the latter proposition and hence we must have mechanisms operating to concentrate CO_2 against a gradient. The C_4 pathway provides all the elements for such a CO_2 concentrating mechanism. Firstly, PEP carboxylase would appear to be perfectly capable of fixing CO_2 at adequate rates from pools formed by diffusion of CO_2 from air. The potential activity of the enzyme in leaves of C_4 pathway plants is about 5 times the photosynthesis rate (Table 3). We also know from the recent studies of WAYGOOD and colleagues (20) that CO_2 is also the substrate for this enzyme and that the K_m for CO_2 is about 7 μM. Hence, if other conditions are optimal this enzyme could account for the observed rates of photosynthesis with an internal CO_2 concentration of only about 1/4 of that in equilibrium with air, that is with an internal CO_2 concentration of about 2 μM.

At least for those species with high malic enzyme activity most of the other requirements for a CO_2 concentrating mechanism are also fulfilled. Thus, provided the malate formed via PEP carboxylase in the mesophyll chloroplasts is transported at an adequate rate to the bundle sheath chloroplasts, then CO_2 could be accumulated in these chloroplasts right at the site of action of RuDP carboxylase. We can only speculate at the moment that mechanisms which are similar in principle are operative in the species with low malic enzyme activity but high aminotransferase activities.

Thus, following this argument to its logical conclusion we could regard the unique reactions of the C_4 pathway simply as a complex mechanism for concentrating CO_2 for fixation by RuDP carboxylase. Of course, RuDP carboxylase may also fix CO_2 diffusing from the air to the bundle sheath cells. However, I would hasten to emphasize that for many C_4 pathway species at least, there is considerable evidence they have adapted themselves to rely largely or solely on the route via C_4 acids for providing RuDP carboxylase with CO_2. In support of this view are the observations that:-

    (i)   the label entering 3-PGA and its products in maize leaves exposed to $^{14}CO_2$ is entirely accounted for by label lost from the C-4 of C_4 acids (6).

    (ii)   leaves incorporate $^{14}CO_2$ at a steady rate from zero time

and that the initial rate of labeling of C_4 acids is comparable to this rate (4,6).

(iii) inhibitors of phosphopyruvate carboxylase strongly inhibit
CO_2 fixation by C_4 pathway species but not by C_3 pathway
species (21).

(iv) growth conditions which reduce phosphoenolpyruvate carboxylase and pyruvate,Pi dikinase to about 10% of maximum levels reduce the maximum photosynthesis rates but do not alter the patterns of labeling when $^{14}CO_2$ is supplied (22).

In addition, the fact that ^{14}C moves from C_4 acids into 3-PGA at the
same rate during a chase in normal air and air containing 5% $^{12}CO_2$
(4) would be consistent with very restricted diffusion of atmospheric CO_2 to the pool in the bundle sheath cells.

Clearly, much remains to be established or confirmed from the
purely mechanistic point of view. The proposed intermediate pools of
CO_2 in bundle sheath cells must be characterized and information is
required about how and where aspartate might be decarboxylated in
the bundle sheath cells. Obtaining clear and direct evidence for the
proposed intercellular transport of metabolites may prove to be inherently difficult.

A second generation of experiments using the tried methods such
as radiotracer studies with leaves could in some instances provide
additional information. Studies with aqueously-extracted chloroplasts could yield much useful information provided these notorious-
ly intractable organelles can be coaxed to behave themselves properly after isolation. However, probably the most rewarding studies in
the immediate future will be those with cell preparations made by
differential grinding of leaves. The art of the preparation of these
cells has been brought to a high state of refinement by Clanton
BLACK and his colleagues and a report on progress with these studies
appears in the following paper.

REFERENCES

1. M.D. HATCH and C.R. SLACK, in Progress in Phytochemistry, ed.
    L. REINHOLD and Y. LIWSCHITZ, Interscience, London, 2 (1970)
    35.
2. M.D. HATCH and C.R. SLACK, *A. Rev. Pl. Physiol.*, 21 (1970) 141.
3. H.P. KORTSCHAK, C.E. HARTT, and G.O. BURR, *Plant Physiol.*, 40
    (1965) 209.
4. M.D. HATCH and C.R. SLACK, *Biochem. J.*, 101 (1966) 103.
5. M.D. HATCH, C.R. SLACK, and H.S. JOHNSON, *Biochem. J.*, 102
    (1967) 417.
6. H.S. JOHNSON and M.D. HATCH, *Biochem. J.*, 114 (1969) 127.
7. C.R. SLACK, *Phytochemistry*, 8 (1969) 1387.
8. C.R. SLACK, M.D. HATCH, and D.J. GOODCHILD, *Biochem. J.*, 114
    (1969) 489.
9. C.R. SLACK and M.D. HATCH, *Biochem. J.*, 103 (1967) 660.
10. W.J.S. DOWNTON, *Can. J. Bot.*, 48 (1970) 1795.

11. H.S. JOHNSON and M.D. HATCH, *Biochem. J.*, 119 (1970) 273.

12. O. BJÖRKMAN and E. GAUHL, *Planta*, 88 (1969) 197.

13. T.J. ANDREWS and M.D. HATCH, *Phytochemistry*, (1970) in press.

14. J.A. BERRY, W.J.S. DOWNTON, and E.B. TREGUNNA, *Can. J. Bot.*, 48 (1970) 777.

15. W.M. LAETSCH and I. PRICE, *Am. J. Bot.*, 55 (1969) 77.

16. W.J.S. DOWNTON, J.A. BERRY, and E.B. TREGUNNA, *Z. PflPhysiol.*, 63 (1970) 194.

17. K.C. WOO, Jan M. ANDERSON, N.K. BOARDMAN, W.J.S. DOWNTON, C.B. OSMOND, and S.W. THORNE, *Proc. natn. Acad. Sci. U.S.A.*, 67 (1970) 1825.

18. T.J. ANDREWS, H.S. JOHNSON, C.R. SLACK, and M.D. HATCH, *Phytochemistry*, (1971) in press.

19. T.G. COOPER, D. FILMER, M. WISHNICK, and M.D. LANE, *J. biol. Chem.*, 244 (1969) 1081.

20. E.R. WAYGOOD, R. MACHE, and C.K. TAN, *Can. J. Bot.*, 47 (1969) 1455.

21. C.B. OSMOND and P.N. AVADHANI, *Plant Physiol.*, 45 (1970) 228.

22. M.D. HATCH, C.R. SLACK, and T.A. BULL, *Phytochemistry*, 8 (1969) 697.

PHOTOSYNTHESIS IN MESOPHYLL CELLS AND BUNDLE SHEATH CELLS ISOLATED FROM *Digitaria sanguinalis* (L.) SCOP. LEAVES

G.E. Edwards[1] and C.C. Black[2]

Department of Biochemistry, University of Georgia,
Athens, Georgia, 30601, U.S.A.

SUMMARY

Fully developed crabgrass leaves were used to isolate mesophyll cells and bundle sheath cells to study the metabolism of these distinct types of photosynthetic cells. Data is presented on: identification of the isolated cell types with various types of microscopy; the yields of intact cells; the localization of phosphoenolpyruvate carboxylase and NADP malate dehydrogenase in mesophyll cells and ribulose-1,5-diphosphate carboxylase and malic enzyme in bundle sheath cells; the stimulation of CO_2 fixation with the mesophyll cells by pyruvate and with the bundle sheath cells by ribose-5-phosphate and ribulose-1,5-diphosphate; the primary formation of dicarboxylic acids in the mesophyll cells and 3-phosphoglycerate in the bundle sheath cells; the almost equal distribution of glyceraldehyde-3-phosphate dehydrogenase activity in both cell types; the presence of starch synthesizing enzymes in both cell types; and the preferential localization of peroxisomal and mitochondrial activity in the bundle sheath strands. A scheme is proposed for photosynthesis in crabgrass leaves which integrates the available data on leaf anatomy, CO_2 fixation, subsequent carbon reduction and metabolism, and the light reactions.

In 1965 and 1966 H.P. KORTSCHAK and coworkers formally reported some of their extensive photosynthetic experiments with sugarcane which had been in progress for a number of years (1,2,3). In these experiments, the very unexpected observation was recorded that photosynthetic fixation of $^{14}CO_2$ by sugarcane leaves resulted in a primary labeling of the organic acids, malic and aspartic, following a short (less than 5 sec) exposure to $^{14}CO_2$. The expected first stable

---

[1] *U.S. Public Health Service Postdoctoral Fellow*
[2] *Supported by NSF grant GB7772.*

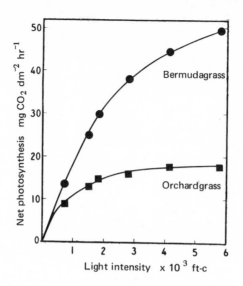

Figure 1 *Response of net photosynthesis to light intensities with bermudagrass and orchardgrass (13).*

product of photosynthesis, phosphoglyceric acid (4), only contained 10-15% of the ^{14}C while 70-80% of the ^{14}C was in malic plus aspartic acid (1). The pioneering experiments of KORTSCHAK and coworkers were confirmed by M.D. HATCH and C.R. SLACK, who in a series of very perceptive reports, presented a scheme to account for photosynthetic carbon metabolism in sugarcane and a variety of other plants (5,6,7, 8). However, a key enzyme, transcarboxylase, in the operation of their theoretical pathway has not been demonstrated.

Late in 1967, workers from at least four laboratories submitted work for publication correlating the presence of the C_4 pathway of CO_2 fixation in certain plant species with a specific leaf anatomy and chloroplast ultrastructure, high net rate of photosynthesis, low rate of photorespiration, low CO_2 compensation point, and little effect of oxygen levels on net photosynthesis (9,10,11,12).

In this laboratory we were studying the light intensity response curves of various photosynthetic reactions with isolated chloroplasts and intact organisms. Thus we realized that the light intensity response curve for a plant which possessed the C_4 pathway (*Figure 1*, bermudagrass) was not similar to the typical light intensity curves of electron transport and noncyclic photophosphorylation with isolated chloroplasts (13,14,15). However, cyclic photophosphorylation [phenazine methosulfate (PMS) catalyzed] in chloroplasts does respond in a fashion similar to bermudagrass leaf photosynthesis (*Figure 1*) as the light intensity increases (15). So we examined a variety of photochemical reactions in C_4 plants and these studies are presented in more detail in this volume (see MAYNE, EDWARDS, and BLACK).

Our studies of carbon metabolism in C_4 plants were initiated to determine the pathway of CO_2 fixation in bermudagrass. When we exposed bermudagrass leaves to $^{14}CO_2$ for 5 seconds the percent distribution of radioactivity in malate and aspartate was 33% and 54% while in sugarcane the respective distribution was 62% and 22% (16). In an experiment in which bermudagrass leaves were exposed to $^{14}CO_2$ for 5 sec and then transferred to $^{12}CO_2$ under continuous illumination the radioactivity in malate remained relatively constant while aspartate rapidly lost radioactivity. The enriched labeling of aspartate over malate and the rapid loss of aspartate labeling relative to malate in $^{12}CO_2$ with bermudagrass were directly opposite to the pattern reported in sugarcane (5). In bermudagrass we assayed the levels of ribulose-1,5-diphosphate (RuDP) carboxylase and phosphoenolpyruvate (PEP) carboxylase and found substantial levels of both enzymes. In *Table 1* these enzyme activities are compared in several plants. A number of laboratories now have reported the detection of substantial levels of RuDP carboxylase in C_4 plants (17, 18,19,20,21). Thus the early reports of low levels of RuDP carboxylase in C_4 plants (5,6,7) are probably invalid. In the original pathway of HATCH and SLACK (5) RuDP carboxylase was not given a definite role in CO_2 fixation in C_4 plants.

TABLE I

Carboxylase enzyme levels in leaves of plants with differing photosynthetic pathways (activity, μmoles CO_2 hr^{-1} mg^{-1} chlorophyll)

| | C_4 plants | | | C_3 plants | |
	Crab-grass	Sugar-cane	Bermuda-grass	Barley	Tall Fescue
PEP carboxylase	236	830	680	14	21
RuDP carboxylase	132	130	170	296	380

In 1968-1969 it was clear to workers in numerous laboratories that C_4 plant leaves contained substantial levels of two carboxylases, and related anatomical studies also indicated that two types of chloroplasts and photosynthetic cells were present in the same leaf (10,12,22,23). Thus in order to understand leaf photosynthesis in C_4 plants it was imperative that the two types of chloroplasts be isolated. The initial report of these separation efforts was an enrichment of the bundle sheath chloroplast fraction using aqueously prepared sugarcane chloroplasts separated on a step-sucrose gradient (24). Then a differential grinding technique to localize RuDP and PEP carboxylase in leaves was presented by BJÖRKMAN and GAUHL (17). BERRY *et al.* (19) used the same method to obtain an indication of the distribution of the carboxylases, malic enzyme, and NADP malic dehydrogenase between the two photosynthetic cell types. The separation of mesophyll and bundle sheath chloroplasts by density fractionation in nonaqueous media also has been achieved (25,26). All of these experimental procedures for isolating chloroplasts to localize enzymes and to study photosynthesis in isolated chloroplasts have definite limitations, a discussion of which is beyond the limits of this manuscript. Being familiar with many of these problems, and the subsequent uncertainties which arise in trying to interpret the experimental results, we decided to approach the problem by isolating the two distinct photosynthetic cell types from a C_4 plant leaf.

In the isolation of distinct cell types from leaves we reasoned that plant cell walls generally are much stronger than internal membranes such as those surrounding chloroplasts. With most plant tissue the exertion of sufficient force to break the cell wall and liberate the cell contents also could result in membrane rupture and possibly a subsequent loss of metabolic activity. Hence, if one could isolate cell types from a fully differentiated tissue, it should be possible to study distinct cell types and gain a clearer understanding of metabolism in the intact tissue.

In preliminary studies we surveyed leaves from numerous plants to determine the ease of cell release by various types of grinding or enzymatic digestion. We discovered that whole cells could be isolated from *Digitaria sanguinalis* (crabgrass) leaves simply by mashing the leaves in a mortar. Thus we concentrated upon crabgrass and developed isolation techniques for mesophyll cells, bundle sheath strands, and bundle sheath cells (27). Basically the technique involves the gentle application of mechanical force to release cells from the tissue and then a separation of cell types by filtration based on cell size and shape (20,27).

Plate 1 is a light micrograph illustrating the general features of each cell type which are observed in a light microscope. The distinguishing features of bundle sheath cells in the light microscope are: chloroplasts are packed into the cells, often seeming to completely fill the cell; cells have angular edges; and the cells appear to be shaped as rectangulars, cylinders, or squares. The distinguishing features of mesophyll cells in the light microscope are: a tendency for the chloroplasts to be scattered in each cell; cells generally have rounded edges; and the cells often have a rounded or egg-shaped appearance.

The crabgrass mesophyll cell preparations appear microscopically to be over 93% intact and essentially free from other contaminating whole cells (*Table 2*). Mesophyll cells are our best preparation. The bundle sheath cell preparations are about 60-75% intact bundle sheath cells with 5-10% contamination by mesophyll cells. The contamination by mesophyll cells is considered in interpreting metabolic studies. Many of our studies are photosynthetic, light minus dark, so the contamination of bundle sheath cell preparations by broken cells or vascular tissue is not considered a serious problem. Bundle sheath strands are used in many photosynthetic studies instead of isolating bundle sheath cells. The yields of each preparation are

TABLE 2

Cell counts in typical preparations of bundle sheath cells and mesophyll cells of *Digitaria sanguinalis*.

Preparation	Exp.	Total cells counted	% intact bundle sheath cells	% intact mesophyll cells	% cells broken or without chloroplasts
Bundle sheath cells	1	1312	73.4	6.4	20.2
	2	1505	59.0	8.5	32.5
	3	1222	74.0	4.4	21.6
Mesophyll cells	1	1490	0.9	93.0	6.1
	2	1176	1.0	93.2	5.8
	3	1568	0.9	93.5	5.6

given in *Table 3* on the basis of total leaf chlorophyll. These are typical yields which can vary somewhat in each experiment due to factors in the isolation procedures such as force or time of grinding a particular fraction; particularly when isolating bundle sheath cells. We routinely isolate fresh cells for each experiment and use as soon as possible (within 1 to 2 hours) since the entire isolation procedure requires only 20 to 30 minutes.

Since C_4 plants have a definite cell morphology in leaf cross sections (12,22,23) which may be related to the pathways of CO_2 fixation we have investigated the shape and arrangement of cells in the crabgrass leaf with the scanning electron microscope. *Plate 2* is a view of a bundle sheath strand illustrating bundle sheath cells tightly arranged around the vascular tissue with the arrows indicat-

TABLE 3

Typical yields of various fractions, on a chlorophyll basis, when isolating cells from *Digitaria sanguinalis* leaves.

Preparation	Total chl (μg)		Yield (%)	
	#1	#2	#1	#2
Whole leaf (4 g)	8384	8960	100.0	100.0
Mesophyll cells	1335	1000	15.9	11.2
Bundle sheath strands	664	1630	7.9	18.2
Bundle sheath cells	198	496	2.4	5.5

ing areas formerly occupied by bundle sheath cells.

A light microscope view of a longitudinal section from the middle of an intact crabgrass leaf shows that the vascular tissue and the bundle sheath cells are surrounded by mesophyll cells with distinct air spaces between the mesophyll cells (*Plate 3*). We were able to gently tease complete vascular tissue sections, with bundle sheath and mesophyll cells still attached, out of crabgrass leaves and obtain scanning electron micrographs of the internal leaf morphology of crabgrass (*Plate 4*). In this three dimensional view, the vascular tissue is surrounded tightly by bundle sheath cells, and the bundle sheath cells are surrounded in loose circles by mesophyll cells. A comparison of the scanning electron micrograph (*Plate 4*) with the light micrograph (*Plate 3*) strengthens this interpretation. Electron microscopy studies of intact leaves and of these isolated leaf preparations also supports this interpretation (H.H. MOLLENHAUER and G.E. EDWARDS, unpublished data). We propose that a comparative study of the micrographs supports the view that the internal morphology of the crabgrass leaf is a series of parallel veins, each of which is surrounded tightly by bundle sheath cells which are surrounded more sparsely by mesophyll cells (as shown in *Plate 4*), with this entire assembly sandwiched between an upper and a lower layer of epidermal cells (27).

Upon the successful isolation of the two cell types from fully differentiated crabgrass leaves we undertook a characterization of their metabolic activities and have reported some of these results (20,27,28,29). In order to relate metabolic activities to intact leaves it was necessary to determine the distribution of chlorophyll and protein between bundle sheath cells and mesophyll cells in the intact leaf. In *Table 4* the distribution of chlorophyll between the cell types is given as determined by two separate methods which suggest that there is an approximately equal distribution of chlorophyll between bundle sheath cells and mesophyll cells in the whole leaf. When protein:chlorophyll ratios were determined for bundle sheath strands and mesophyll cells and compared with the data from

---

Plate 1 Light micrograph of a crabgrass bundle sheath cell and two mesophyll cells. x 940.

Plate 2 Scanning electron micrograph of an isolated crabgrass bundle sheath strand. The arrows denote areas where vascular tissues are exposed due to the removal of bundle sheath cells. x 900.

Plate 3 Longitudinal sections through the middle of a crabgrass leaf. Arrows denote air spaces. Light micrograph. x 645.

Plate 4 Scanning electron micrograph of a vein from a crabgrass leaf with the bundle sheath cells packed around the vascular tissues and the mesophyll cells loosely surrounding the bundle sheath cells. Arrows denote areas formally occupied by mesophyll cells. x 900.

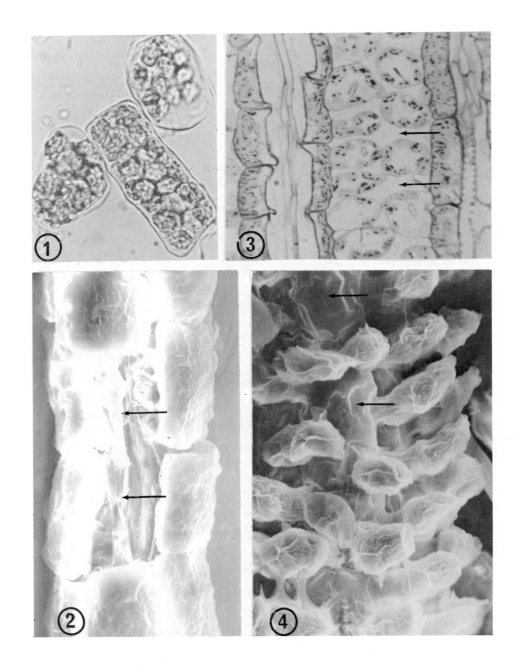

TABLE 4

Distribution of chlorophyll between the two photosynthetic
cell types in leaves of *Digitaria sanguinalis*.

Method	Chlorophyll distribution in leaf (%)	
	Bundle sheath cells	Mesophyll cells
I Chlorophyll $a:b$ ratio*	52	48
II Chlorophyll per cell and ratio of cells**	43	57

* Chlorophyll $a:b$ was determined in 96% ethanol extracts for
the whole leaf, the mesophyll cells, and the bundle sheath
cells, giving values of 4.4, 2.8, and 5.9 respectively. The
percentage distribution between the two cell types was determined by solving simultaneous equations.

** Cell counts were made on preparations of the two cell types
and chlorophyll per cell was determined giving 3.5×10^{-5} µg
chlorophyll per mesophyll cell and 9.0×10^{-5} µg chl per bundle sheath cell. From the ratio of cells determined in longitudinal and cross sections the ratio of mesophyll cells to
bundle sheath cells in the whole leaf was found to be 3.3 to
1. From this data the chlorophyll distribution was determined.

method II, *Table 4*, about equal distribution of protein between bundle sheath strands and mesophyll cells in the intact leaf was indicated. We conclude that to express a biochemical activity on a chlorophyll or protein basis gives a good indication of the relative
distribution of the activity between bundle sheath strands and mesophyll cells in a crabgrass leaf.

TABLE 5

Levels of RuDP carboxylase and PEP carboxylase in mesophyll
and bundle sheath cells of *Digitaria sanguinalis* (activity,
µmoles CO_2 hr^{-1} mg^{-1} chlorophyll).

Preparation	RuDP carboxylase	PEP carboxylase
Total leaf	150, 115	250, 223
Mesophyll cells	12, 14	635,740
Bundle sheath cells	290,225	55, 54

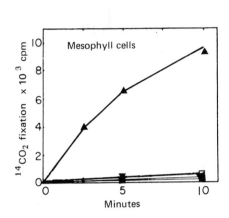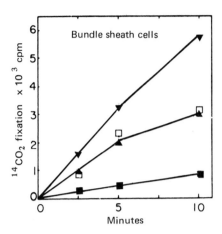

Figure 2 (left) Photochemical $^{14}CO_2$ fixation by isolated mesophyll cells. Upper curve, plus 5 mM pyruvate. Lower curves,
dark, plus 5 mM R-5-P, and control (20).

Figure 3 (right) Photochemical $^{14}CO_2$ fixation by isolated bundle sheath cells. Upper curve, plus 5 mM R-5-P. Middle curve,
plus 5 mM pyruvate and control. Lower curve, dark (20).

Assays for RuDP and PEP carboxylase with mesophyll and bundle
sheath cells and with whole leaf extracts show that the carboxylases
are differentially distributed in the cell types (Table 5). Such
differences in carboxylase distribution suggest that CO_2 assimilation during leaf photosynthesis is different in mesophyll and bundle
sheath cells (20).

The mesophyll cell preparation of crabgrass lacked the capacity
to fix substantial quantities of CO_2 in the absence of other substrates. However, pyruvate induced a substantial rate of CO_2 fixation with illuminated mesophyll cells. Ribose-5-phosphate (R-5-P)
had little effect on the CO_2 fixation by mesophyll cells in light or
in darkness (Figure 2). Bundle sheath cells fix some CO_2 during
illumination without the addition of other substrates (Figure 3) although this endogenous capacity is variable (1-15 µmoles of CO_2 fixed hr^{-1} mg^{-1} chlorophyll. R-5-P enhanced photosynthetic CO_2 fixation
by bundle sheath cells while pyruvate had no effect on the CO_2 fixation (Figure 3).

Table 6 shows the effect of several exogenous substrates on CO_2
fixation by mesophyll and bundle sheath cells of crabgrass. The mesophyll cells have a very high capacity to fix CO_2 when provided
with PEP whereas R-5-P and RuDP enhance the CO_2 fixation in bundle
sheath cells. The enhancement of CO_2 fixation by PEP in the bundle
sheath cell preparation (Table 6) indicates that part of the PEP
carboxylase may be located in bundle sheath cells. However, we propose that the bundle sheath cells which have high RuDP carboxylase
activity and possibly the pentose cycle of photosynthesis may be

TABLE 6

Effect of various substrates on CO_2 fixation (μmoles CO_2 hr^{-1} mg^{-1} chlorophyll) by mesophyll and bundle sheath cells of *Digitaria sanguinalis* and mesophyll cells of spinach*.

| Treatment and substrate | *Digitaria sanguinalis* | | Spinach |
	Mesophyll cells	Bundle sheath cells	Mesophyll cells
light, no addition	0.4	1.2	51.2
dark, no addition	0.1	1.0	0.1
light, PEP	1078	198	107.0
dark, PEP	968	183	72.0
light, R-5-P	0.4	32.2	53.2
dark, R-5-P	0.3	8.1	5.8
light, R-5-P, ADP	4.6	67.0	−
dark, R-5-P, ADP	3.3	28.7	−
light, RuDP	1.3	34.3	−
dark, RuDP	1.7	17.0	−
light, ADP	0.6	11.0	−
dark, ADP	0.3	1.8	−
light, 2-PGA	50.5	42.8	−
light, aspartate	0.3	1.7	−
light, alanine	0.2	1.9	−

* Concentrations of additives were 2 mM ADP, 4 mM PEP, 5 mM R-5-P, 2 mM RuDP, 5 mM 2-PGA, 10 mM aspartate, and 10 mM alanine.

analogous to C_3 plant cells. Spinach mesophyll cells photoassimilate CO_2 predominantly by the Calvin cycle and contain PEP carboxylase activity comparable to crabgrass bundle sheath cells (*Table 6*). PEP enhances CO_2 fixation in isolated spinach mesophyll cells, either in the dark or light (*Table 6*). Comparable PEP carboxylase activity has been reported in leaf extracts from numerous C_3 plants (30).

Table 7 shows the products of $^{14}CO_2$ fixation with mesophyll cells supplemented with specific substrates. The formation of malate is light dependent (presumably light is utilized to form NADPH). The data with bundle sheath cells show little malate production in light or dark with PEP and show some 3-phosphoglycerate (3-PGA) formation. While 3-PGA was not detected as an end product in mesophyll cells with the addition of various substrates (*Table 7*), when 3-PGA, RuDP,

TABLE 7

Products formed by isolated cells of *Digitaria sanguinalis* during photosynthesis in $^{14}CO_2$ with various substrates (% of total fixation).

	Malate	Oxalo-acetate	3-PGA	Others
Mesophyll cells				
10 mM pyruvate, light 20 min	55	35	*	10
10 mM 2-PGA, light 20 min	56	31	*	13
1 mM PEP, light 20 min	56	44	*	*
1 mM PEP, dark 20 min	*	95	*	5
Bundle sheath cells				
1 mM PEP, light 20 min	5	67	20	8
1 mM PEP, dark 20 min	*	78	11	11

* Not detected

or R-5-P was added to bundle sheath cells with a 15 min incubation in the light 60-70% of the ^{14}C was in 3-PGA.

We consider that these experiments with cell types, e.g. carboxylase localization, differential CO_2 fixation with exogenous substrates, and the products of CO_2 fixation, indicate that β-carboxylation is the primary pathway of CO_2 assimilation in mesophyll cells while the bundle sheath cells use the Calvin cycle for CO_2 fixation (20).

In order to assist in establishing the routes of carbon flow following CO_2 fixation and to integrate photosynthesis with other cellular activities we have assayed numerous enzymes in the cell types. Some of this data is presented in *Table 8* and *9*. Several points emerge from these studies which assist in understanding cell and leaf metabolism. First is the roughly equal distribution of NAD and NADP glyceraldehyde 3-P dehydrogenase which indicates that the carbon reduction step of photosynthesis occurs in both cell types. Second is the localization of NADP malate dehydrogenase in the mesophyll cells where the reduction of oxaloacetate (OAA) to malate should occur to operate the C_4 pathway of CO_2 fixation and conversely the localization of malic enzyme in the bundle sheath cells for decarboxylation and the production of NADPH to occur. Third, the enzymes of starch metabolism are present in both cell types (although there may be different routes of synthesis) which agrees well with previous work showing starch accumulation in both cell types (23). Fourth is the cytochrome *c* oxidase concentration in bundle sheath strands which also agrees with a high concentration of mitochondria in bundle sheath cells previously observed with the electron microscope (23).

TABLE 8

Distribution of enzymes in mesophyll cells and bundle sheath cells of *Digitaria sanguinalis* (activity, μmoles hr^{-1} mg^{-1} chlorophyll).

Enzyme	Whole leaf	Mesophyll cells	Bundle sheath cells
A. Enzymes located in both cells			
Pyrophosphatase (pH 8.9)	2460	1820	2660
Adenylate kinase	1800	1900	1500
Glyceraldehyde 3-P dehydrogenase (NADP)	–	206	284
(NAD)	72	92	101
B. Enzymes with preferential localization			
Malic enzyme	195	26	543
Fructose 1, 6-DP aldolase	153	50	588
Cytochrome *c* oxidase	22	4	45
Malate dehydrogenase (NADP)	35	103	*
C. Enzymes of starch metabolism			
UDPG-transglucosylase	5.3	*	4.4
ADPG-transglucosylase	5.1	*	8.8
Phosphorylase	53	28	92
Sucrose-6-P synthetase	39	51	26

* Not detected

Finally, the levels of peroxisomal enzymes in crabgrass leaves and the concentrating of activity in the bundle sheath strands (*Table 9*) are pertinent to understanding photorespiration in C$_4$ plants. In some unpublished experiments (A. LIU) we have shown that internally the bundle sheath and mesophyll cells cycle the CO$_2$ derived from glycolate metabolism which partially explains the difficulties encountered in studying photorespiration with intact leaves. Finally, we can obtain a quantitative understanding of leaf metabolism on a cellular level as a result of these studies. In *Table 10* a general summary of some leaf metabolic activities are presented. Unfortunately, it is not within the scope of this manuscript to give a tho-

TABLE 9

Activity and localization of peroxisomal enzymes (activity, m moles hr^{-1} mg^{-1} catalase, μmoles hr^{-1} mg^{-1}for others) .

	Catalase		Glycolate oxidase		Hydroxypyruvate reductase	
	Chl.*	Prot.**	Chl.	Prot.	Chl.	Prot.
Spinach leaves	153	12	33	2	645	108
Digitaria leaves	27	3	6	0.5	34	6
Digitaria mesophyll cells	13	1	3	0.2	18	3
Digitaria bundle sheath strands	55	4	9	0.5	109	16

* Chlorophyll basis
** Protein basis

rough discussion of each topic so the remainder of the discussion must be limited to photosynthesis.

We can now propose a scheme for photosynthesis in crabgrass leaves in which we integrate the data on leaf anatomy, CO_2 fixation, subsequent carbon metabolism including reduction, and light reactions (also see MAYNE, EDWARDS and BLACK in this volume). In outline the scheme is given in *Figure 4*. As atmospheric CO_2 enters the leaf substomatal cavity it encounters continuous intercellular gaseous regions and can diffuse freely to the surface of both the mesophyll cells and bundle sheath cells (*Plates 3,4*). The very active PEP carboxylase in the mesophyll cells fixes most (about 85% in crabgrass) of the CO_2 as quickly as it is available at atmospheric con-

TABLE 10

General summary of metabolic activity in cells of *Digitaria sanguinalis* leaves

Activity	Mesophyll cells	Bundle sheath cells (including vascular tissue)
CO_2 fixation	β carboxylation	Calvin cycle
Glycolate respiration	20-30%	70-80%
Mitochondrial respiration	10%	90%
Starch synthesis	yes	yes

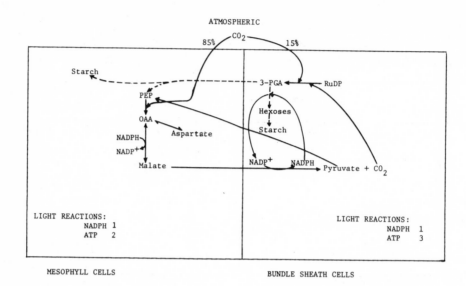

Figure 4 Scheme for photosynthesis in a crabgrass leaf.

centrations. There seems to be little doubt that an atmospheric CO_2 concentration of 0.03% is not saturating for photosynthesis at light intensities above 2,000 or 3,000 ft-c. In C_4 leaves CO_2 fixation at rates of 300 to 450 µmoles of CO_2 hr^{-1} mg^{-1} chlorophyll are observed (11,15) and we have observed similar rates with crabgrass leaves in midsummer. PEP carboxylase activity in isolated crabgrass mesophyll cells is as high as 1800 µmoles of CO_2 fixed hr^{-1} mg^{-1} chlorophyll (27,28) and several other reports indicate PEP carboxylase activity in the range of 600 to 1100 µmoles of CO_2 fixed hr^{-1} mg^{-1} chlorophyll (11,17) in crude leaf extracts of C_4 plants. Second, the bundle sheath cells also fix atmospheric CO_2 (about 15% in crabgrass) via the Calvin cycle. The reasons for reaching this conclusion are as follows:

(i)   There is an active RuDP carboxylase in extracts from isolated bundle sheath cells.

(ii)   The isolated bundle sheath cells fix CO_2.

(iii) In the original experiments, and in most subsequently, on C_4 plants, 10-12% of the ^{14}C appeared in 3-PGA from [14]CO_2 fed to sugarcane leaves in the shortest times of exposure, 0.6 sec, to [14]CO_2 (1). We have found that crabgrass leaves, in a 5 sec exposure to [14]CO_2, fix about 10-15% of the ^{14}C into 3-PGA.

(iv)   From studies on the internal leaf morphology we have calculated the surface area of mesophyll cells and bundle sheath cells which may be exposed to gases from the atmosphere, bundle sheath cell surface being about 10% of the total exposed cell surface area. Assuming that both cells are equally permeable to CO_2 these calculations agree well with the data cited above on the initial products of atmospheric [14]CO_2 fixation with intact leaves [*Plates 3,4,* and (27)].

In the bundle sheath cells the atmospheric CO_2 fixed directly could enter the classical Calvin cycle and go directly to starch. Thus the bundle sheath cells have a complete cycle of photosynthesis which is independent of the mesophyll cells. However the bundle sheath cells are partially dependent upon the mesophyll cells for CO_2 and perhaps some NADPH (*Figure 4*). The mesophyll cells in contrast do not appear capable of operating a complete cycle of CO_2 fixation, including reduction, independent of the bundle sheath cells. The primary fixation of CO_2 is into a 4-C compound and there is no evidence to date that this carbon can be reduced and utilized to synthesize hexoses or starch in the mesophyll cells. Thus in the scheme (*Figure 4*) we envision the mesophyll cells as "CO_2 harvesting cells" transporting the CO_2 "trapped" in dicarboxylic acids to the bundle sheath cells. Seemingly this could be an efficient mechanism for concentrating CO_2 in the bundle sheath cells and thus allowing RuDP carboxylase to operate at its maximum capability. A discussion integrating carbon metabolism, light reactions, and energy requirements in mesophyll cells and bundle sheath cells is given in this volume (MAYNE, EDWARDS and BLACK) and in (29).

Finally it should be noted clearly that this is a minimal scheme and does not allow for the diversion of compounds into other pathways within a cell or the movement of compounds between cells. For example the pyruvate in bundle sheath cells seemingly should be subject to many other fates particularly in view of the high mitochondrial concentration there. The 3-PGA and 2-PGA formed in the bundle sheath cells also are compounds which appear as likely candidates to move to mesophyll cells to produce PEP or reduced trioses. Reduced trioses, pentoses, and hexoses also may move between cells. Certainly many of these branch pathways appear feasible and some are supported by the experiments reported here such as the effects of various exogenous substrates on CO_2 fixation by cell types (*Table 6*); but we propose that the scheme given (*Figure 4*) best accomodates the available anatomical and biochemical data.

ACKNOWLEDGMENTS

We are grateful to a number of colleagues for actively participating in some of these experiments and for lively discussions. These include Drs. R.H. Brown, B.C. Mayne, H.H. Mollenhauer, T.M. Chen, S.S. Lee, J. Travis, V. Gracen, and Miss Ann Liu.

REFERENCES

1. H.P. KORTSCHAK, C.E. HARTT, and G.O. BURR, *Plant Physiol.*, 40 (1965) 209.
2. H.P. KORTSCHAK and C.E. HARTT, *Naturwissenschaften*, 53 (1966) 1.
3. G.O. BURR, C.E. HARTT, H.W. BRODIE, T. TANIMOTO, H.P. KORTSCHAK, D. TAKAHASHI, F.M. ASHTON, and R.E. COLEMAN, *A. Rev. Pl. Physiol.*, 8 (1957) 275.

4. A.J. BASSHAM and M. CALVIN, The Path of Carbon in Photosynthesis, Prentice Hall, Englewood Cliffs, New Jersey, (1957) 104 pp.
5. M.D.HATCH and C.R. SLACK, *Biochem. J.*, 101 (1966) 103.
6. M.D. HATCH, C.R. SLACK, and H.S. JOHNSON, *Biochem. J.*, 102 (1967) 417.
7. C.R. SLACK and M.D. HATCH, *Biochem. J.*, 103 (1967) 660.
8. M.D. HATCH and C.R. SLACK, *Biochem. J.*, 106 (1968) 141.
9. R.W. DOWNES and J.D. HESKETH, *Planta*, 78 (1968) 79.
10. W.J.S. DOWNTON and E.B. TREGUNNA, *Can. J. Bot.*, 46 (1968) 207.
11. H.S. JOHNSON and M.D. HATCH, *Phytochemistry*, 7 (1968) 375.
12. W.M. LAETSCH, *Am. J. Bot.*, 55 (1968) 875.
13. R.H. BROWN, R.E. BLASER, and H.L. DUNTON, *Proc. X int. Grassland Cong.*, 10 (1966) 108.
14. J.F. TURNER, C.C. BLACK, and M. GIBBS, *J. biol. Chem.*, 237 (1962) 577.
15. T.M. CHEN, R.H. BROWN, and C.C. BLACK, *Plant Physiol.*, 44 (1969) 649.
16. T.M. CHEN, R.H. BROWN, and C.C. BLACK, *Plant Physiol.*, 47 (1971) in press.
17. O. BJÖRKMAN and E. GAUHL, *Planta*, 88 (1969) 197.
18. C.B. OSMOND, J.H. TROUGHTON, and D.J. GOODCHILD, *Z. PflPhysiol.*, 61 (1969) 218.
19. J.A. BERRY, W.J.S. DOWNTON, and E.B. TREGUNNA, *Can. J. Bot.*, 48 (1970) 777.
20. G.E. EDWARDS, S.S. LEE, T.M. CHEN, and C.C. BLACK, *Biochem. biophys. Res. Commun.*, 39 (1970) 389.
21. T.J. ANDREWS, H.S. JOHNSON, C.R. SLACK, and M.D. HATCH, *Phytochemistry*, (1970) in press.
22. W.M. LAETSCH, *Sci. Prog. Oxf.*, 57 (1969) 323.
23. C.C. BLACK and H.H. MOLLENHAUER, *Plant Physiol.*, 46 (1970) 828.
24. C.W. BALDRY, J. COOMBS, and D. GROSS, *Z. PflPhysiol.*, 60 (1968) 78.
25. C.R. SLACK, *Phytochemistry*, 8 (1969) 1387.
26. C.R. SLACK, M.D. HATCH, and D.J. GOODCHILD, *Biochem. J.*, 114 (1969) 489.
27. G.E. EDWARDS and C.C. BLACK, *Plant Physiol.*, 47 (1971) 18.
28. G.E. EDWARDS, T.M. CHEN, S.S. LEE, and C.C. BLACK, *Plant Physiol.*, 46 (1970) Suppl., p. 6.
29. B.C. MAYNE, G.E. EDWARDS, and C.C. BLACK, *Plant Physiol.*, 47 (1971) in press.
30. D.A. WALKER, *Biol. Rev.*, 37 (1962) 215.

NONAUTOTROPHIC CO$_2$ FIXATION AND CRASSULACEAN ACID METABOLISM

Irwin P. Ting

Department of Life Sciences,
University of California, Riverside, 92502 , U.S.A.

SUMMARY

The dark or nonautotrophic CO$_2$ fixation pathway is described in terms of its carbon flow, subcellular regulation, subcellular localization and relationship to the C$_4$ photosynthetic pathway. It is concluded that the general carbon flow is: phosphoenolpyruvate + CO$_2$ → oxaloacetate → malate → pyruvate + CO$_2$. The products, viz., malate and other organic and amino acids, are highly compartmented and separated from mitochondrial, microbody, and chloroplast metabolism. The pathway is thought to be localized in the cytosol. In C$_4$ photosynthesis and Crassulacean acid metabolism, the nonautotrophic CO$_2$ fixation pathway is coupled to the C$_3$ photosynthetic pathway at the level of malate. In plants with Crassulacean acid metabolism, carboxylation to form malate in the dark or decarboxylation to form CO$_2$ and pyruvate in the light are separated in time whereas in C$_4$ photosynthesis the two phases are separated spatially; carboxylation in the mesophyll cells and decarboxylation in the bundle sheath cells.

Nonphotosynthetic CO$_2$ fixation appears to be ubiquitous throughout the Plant and Animal Kingdoms. Almost all organisms investigated have some capacity to incorporate CO$_2$ or bicarbonate into organic constituents. The general phenomenon was first observed with propionic acid bacteria by WOOD and WERKMAN in 1935 (1). From subsequent work, the concept of heterotrophic CO$_2$ fixation developed. The term heterotrophic was apparently applied because the process was known only from heterotrophic organisms, i.e., those organisms which required preformed substrates to obtain energy. For plants, the term nonautotrophic may be somewhat more descriptive since the process is dependent on energy derived from carbon compounds previously synthesized by the plant autotrophically. The term dark fixation is somewhat misleading in that it occurs in light as well as dark. Hence,

169

the term nonautotrophic will be used to distinguish it from photosynthetic or autotrophic CO_2 fixation.

Crassulacean acid metabolism (CAM) is a phrase used to describe a
specific series of biochemical events which occur in many succulent
plants. In general, CAM refers to the relatively massive nonautotrophic CO_2 fixation occurring in the dark in green tissues with the
concomitant accumulation of significant quantities of organic acids.
The metabolism is characterized by a distinct diurnal fluctuation of
organic acids. During the first portion of a dark period followed by
a sufficient light period, organic acids tend to increase (2). In
the subsequent light period, organic acids decrease until the following dark period at which time the cycle is repeated. Almost without exception, the fluctuating organic acid is l-malate. A major
difference between nonautotrophic CO_2 fixation by green succulent
tissue and nongreen, nonsucculent root tissue appears to be the lack
of demonstrable organic acid fluctuation in the latter. Nonautotrophic or dark CO_2 fixation including CAM has been reviewed by several
workers (see for example 3-7).

THE METABOLIC PATHWAY

Products

The metabolic pathway for nonautotrophic CO_2 fixation appears to
be relatively well established. THURLOW and BONNER (8) showed that
$^{14}CO_2$ was incorporated into organic acids in the dark. In crassulacean plants, THOMAS and RANSON (9) demonstrated the relationship between CO_2 fixation and organic acid synthesis. SALTMAN and coworkers
extensively investigated the metabolic pathway for CO_2 fixation and
organic acid biosynthesis in succulent and nonsucculent plants.
Their work led to the firm conclusion that the first stable products
of CO_2 fixation were malate and aspartate (10,11) and that there
were no fundamental differences between succulent and nonsucculent
plants. The ultimate product accumulating, however, depends on the
species. Most investigators have concluded that essentially no neutral or carbohydrate compounds are synthesized from nonautotrophical
ly fixed carbon dioxide. An obvious exception seems to be seedlings
which are converting lipoidal material to carbohydrates (12). Because of the unique labeling pattern of certain plants (see later),
there is some indication that organic phosphates, e.g., 3-phospho-
glycerate (3-PGA), should be labeled as a first product. There is,
however, little published evidence for the latter.

WEBB and BURLEY (13) and JOSHI *et al.* (14) showed that salts
could alter the products of CO_2 fixation; amino acids were predominant in the presence of NaCl and organic acids in the absence of
NaCl.

The products of CO_2 fixation, *viz.*, organic and amino acids, are
in active metabolic pools (15-18). The release of CO_2 by decarboxylation follows first order kinetics. The two phase nature of the decarboxylation has led to the suggestion that there are at least two

active intracellular pools of products (see later).

Labeling pattern in malate

VARNER and BURRELL (19) determined that there was more ^{14}C label in the C-4 position of malate than in the C-1 position after a period of nonautotrophic [14]CO_2 fixation by *Bryophyllum* leaves. BRADBEER *et al.* (20) in similar experiments, concluded that the distribution of label from [14]CO_2 in the C-4 and C-1 carbons of malate maintained a 2 to 1 ratio. AVADHANI (21) showed that in *Kalanchoe* the label was mostly in the C-4 position under anaerobic conditions. In pea leaves and pea cotyledons, the C-4 to C-1 labeling was also about 2 to 1 (22) as it was in carrot root tissue (23). In fumitory leaves (20), potato tuber tissue (24), sunflower leaves (20), and beet root tissue (23) most of the label was in the C-4 position after dark fixation of [14]CO_2. The latter labeling pattern is similar to that of C_4 plants in which the label is initially in the C-4 position of malate, other carbons being labeled only secondarily.

It seems reasonably clear that labeling in the C-4 position is consistent with β-carboxylation of a three carbon compound such as phosphoenolpyruvate (PEP). The asymmetric labeling in succulents is difficult to explain. BRADBEER *et al.* (20) suggested an initial carboxylation by a ribulose carboxylase type reaction to form one C-1 labeled 3-PGA and one unlabeled one. Subsequent β-carboxylation of both would result in the 2 to 1 labeling of C-4 to C-1. Further experimental evidence for the double carboxylation is needed.

Alternative hypotheses to account for asymmetric labeling involve unequal randomization through other enzymes. As indicated, the patterns which seem to exist are either mostly C-4 labeling, equal or randomized labeling, and asymmetric labeling with an approximate 2 to 1 distribution in C-4 to C-1. One could visualize that the latter pattern could occur if the malate or other organic acid products formed were all shunted into a pool with no possibility of randomization (C-4 label only), or shunted into two pools; one with complete randomization and one without. If exactly half of the product went into each pool, a 3 to 1 (C-4 to C-1) would result. If all of the product entered a pool with complete randomization possibilities, the equal distribution would result. Depending on the portion of product shunted to a closed pool and that to an open pool with label randomization, the label distribution could vary accordingly. Regardless of the mechanism, compartmentation of products is necessary to maintain unequal distribution of ^{14}C within the malate molecule.

The steady state metabolic pathway of CO_2 fixation as envisioned here is shown in *Figure 1*. In nonautotrophic or dark CO_2 fixation, the initial substrate, PEP, apparently comes from carbohydrate precursors. In the case of the unequal labeling of malate outlined by BRADBEER *et al.* (20), exactly one-half of the PEP would be labeled in a [14]CO_2 experiment and the carboxylation of PEP (reaction I) shown in *Figure 1* would represent the second in the sequence. Oxaloacetate, the first product, is either reduced immediately to malate

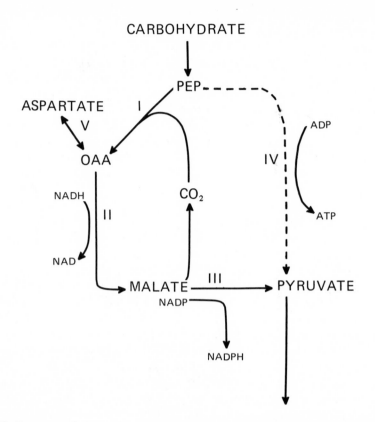

Figure 1 Presumed steady state flow of carbon through the non-autotrophic CO_2 fixation pathway.

via a soluble-malate dehydrogenase (reaction II) or converted to aspartate, perhaps via a soluble-aspartate aminotransferase (reaction V). We visualize the former reaction, i.e., the malate reaction, to represent the steady state flow of carbon, whereas the aspartate forming reaction is presumed to be into a side pool. The relative amount of aspartate formed would be a function of the enzymic activities. Since the sequence forming malate from PEP is generally non-reversible, the utilization of malate is assumed to be by decarboxylation via the malate enzyme (reaction III). The pyruvate generated is presumably used in a variety of metabolic reactions. The direct route for pyruvate synthesis from PEP via pyruvate kinase (reaction IV) would compete with the CO_2 fixation pathway. The ratio of carboxylation (reactions I and II) to decarboxylation (reaction III) would govern the extent to which malate accumulates.

INTRACELLULAR LOCALIZATION OF PATHWAY

Compartmentation of metabolites
    It is known that there are multiple compartments of organic acids within the cell (26-29). In *Bryophyllum* leaves MacLENNAN *et al.* (26)

reported data indicating that less than 30% of the malate synthesized from acetate was in an active turnover pool and less than 50% of the evolved CO_2 came from citric acid cycle reactions. LIPS and BEEVERS (27) in a similar study with corn roots, showed that the malate synthesized from CO_2 was in a pool different from that synthesized from acetate. Evidence for a cytoplasmic pool of malate was presented by LIPS and BEEVERS (1966b) and by OSMOND and LATIES (29). In the latter study, data were obtained which indicated that malate synthesized from CO_2 fixation in beet discs was present in a cytoplasmic pool with a relatively short half-life and a vacuolar pool with a longer half-life. In our own work with *Opuntia*, other succulents, and root tips (15-18,30) evidence was found for a two phase direct decarboxylation of malate synthesized from CO_2. These data were taken to indicate two pools of organic acids (18).

During nonautotrophic fixation of CO_2, generally no carbohydrate products are formed from the incorporated carbon; only organic and amino acids are significantly labeled. For example, in cactus root tips, after a 2 hour dark fixation period, 80% of the ^{14}C was in organic acids (16). Of the organic acids, 66.4% was in malate, 11.4% in citrate, and trace amounts in fumarate, succinate, and isocitrate. Because the tissue was actively respiring (about 200 μl hr^{-1} g^{-1} fresh wt), the data were taken to indicate compartmentation of the products outside of the mitochondria.

It seems reasonable to conclude that malate and other products of nonautotrophic CO_2 fixation are generally not in equilibrium with all other metabolic processes within the cell.

Compartmentation of enzymes

  (a) Phosphoenolpyruvate carboxylase (orthophosphate:oxaloacetate carboxy-lyase (phosphorylating), EC 4.1.1.31)

PEP carboxylase activity was reported by BANDURSKI and GREINER (31) to be in extracts of spinach leaves. Later investigators, e.g., MAZELIS and VENNESLAND (32), ROSENBERG *et al.* (33), and JOHNSON and BRUFF (34) reported activity in chloroplast fractions isolated from spinach. GARNIER-DARDART (35) reported that PEP carboxylase was localized in the chloroplasts of the succulent plant, *Bryophyllum daigremontianum*. SLACK *et al.* (36) suggested that in maize PEP carboxylase may be bound to the envelope of mesophyll chloroplasts. In our own work, we observed that PEP carboxylase was associated with nonaqueously prepared chloroplasts of *Opuntia*, but not with aqueously prepared chloroplasts (37). BRANDON (38) reported PEP carboxylase localization in mitochondrial fractions of *Bryophyllum*, and MAZELIS and VENNESLAND (32) reported it in mitochondrial fractions of cauliflower and pea seedlings. In corn root sections and in *Opuntia* green tissue, we could not obtain evidence for mitochondrial compartmentation (39,40). The published data taken together seem to suggest some particulate localization of PEP carboxylase.

In nonaqueously prepared chloroplasts from the succulent, *Opuntia ficus-indica*, NADP triose phosphate dehydrogenase maintained almost a constant ratio of activity to chlorophyll (37). PEP carboxylase

was not associated with chlorophyll on a constant basis. These data
suggest that PEP carboxylase may not be associated with chloroplasts
and that it is not localized in the same subcellular compartment
with NADP-triose phosphate dehydrogenase.

In cotton leaf tissue (Deltapine Smoothleaf), there are three
known isoenzymes of PEP carboxylase (40). One of the isoenzymes is
associated with both aqueously and nonaqueously prepared chloroplast
fractions; however, sucrose density gradient centrifugation did not
suggest that the particulate isoenzyme was localized in intact chloroplasts, mitochondria, or microbodies (40). When nonaqueously prepared chloroplasts from cotton were layered at the bottom of a linear carbon tetrachloride-hexane gradient and centrifuged to equilibrium, the chlorophyll and NADP-triose phosphate dehydrogenase activity showed a close correlation, whereas the PEP carboxylase was
displaced toward a lighter fraction. A similar experiment in which
nonaqueously prepared chloroplasts were pelleted in a series of increasing density gradients (CCl_4-hexane), showed much the same. The
close correspondence of chlorophyll and NADP-triose phosphate dehydrogenase in chloroplast fractions prepared by a variety of methods
seems to suggest localization in chloroplasts. PEP carboxylase, however, does not seem to be localized with NADP-triose phosphate dehydrogenase and there is no valid evidence that it is a chloroplast
stroma enzyme.

(b) Malate dehydrogenase (L-malate:NAD oxidoreductase, EC
1.1.1.37)

All plant tissues investigated have at least two isoenzymes of
malate dehydrogenase, a particulate mitochondrial form and a soluble
or nonparticulate form.

Many green tissues have an additional malate dehydrogenase isoenzyme. In spinach leaf tissue the microbodies or leaf peroxisomes
have a specific isoenzyme of malate dehydrogenase (41,42). Other C_3
plant leaves investigated in our laboratory also have a microbody-
malate dehydrogenase. The exact function is not understood, but it
probably has little direct function in CO_2 fixation. Green tissues
of the succulent, *Opuntia ficus-indica*, also have three malate dehydrogenase isoenzymes (43); two particulate forms and a soluble
form. Our investigations with green leaf tissue of corn have indicated two mitochondrial NAD malate dehydrogenases and one soluble
form.

Recently, investigations concerning an NADP specific malate dehydrogenase have appeared (44,45). The activity appears to be higher
in C_4 plants than in C_3 plants. Nonaqueous preparations of maize
leaf chloroplasts indicated that it is localized with NADP triose
phosphate dehydrogenase in chloroplasts. Our qqueously purified
chloroplasts from spinach leaves also indicated unique localization
of NADP malate dehydrogenase in intact spinach chloroplasts. The
latter observation was verified during this conference by Hilary
JOHNSON and myself.

Because of the fairly conclusive evidence that malate synthesized from nonautotrophic CO_2 fixation is not in equilibrium with malate synthesized in the mitochondria, it seems reasonable to conclude that the soluble-malate dehydrogenase functions to a large extent in the nonautotrophic CO_2 fixation pathway. One would predict, therefore, that insofar as PEP carboxylase mediated synthesis of malate takes place in chloroplasts that the NADP malate dehydrogensae would be the coupling enzyme and that nonchloroplastic malate synthesis would be coupled with the soluble-NAD malate dehydrogenase.

(c) Malic enzyme (L-malate:NADP oxidoreductase (decarboxylating), EC 1.1.1.40)

Malic enzyme, as well as the other enzymes of the pathway, is known to exist in multiple forms. In apple fruit, DILLEY (46) isolated and purified a form which ran as a single band on acrylamide gels. In corn root tissue, we were able to locate a single isoenzyme (by starch gel electrophoresis) and reported evidence that it was a soluble or nonparticulate protein (39). Hence it was concluded that nonautotrophic CO_2 metabolism insofar as it is mediated by this particular malic enzyme isoenzyme is in the cytosol.

In green tissue of *Bryophyllum*, BRANDON (38) reported malate enzyme activity localized in mitochondria. In green cactus tissue, we were able to demonstrate three different proteins with malate enzyme activity apparently localized in three different subcellular compartments; *viz*., cytosol, mitochondria, and chloroplast fractions (47).

(d) Aspartate aminotransferase (L-aspartate:α-oxoglutarate aminotransferase, EC 2.6.1.1)

Aspartate aminotransferase or glutamate-oxaloacetate transaminase is classically considered to be a mitochondrial enzyme. In higher plant tissues, multiple molecular forms are known (48). Spinach leaf tissue has at least two isoenzymes which are separable on DEAE-cellulose and by starch gel electrophoresis (unpublished results). Organelle purification on sucrose gradients has indicated that at least one of these isoenzymes is located in purified mitochondria and one other is most likely soluble.

(e) Summary

All of the data in the literature taken together strongly suggest that the products of dark or nonautotrophic CO_2 fixation are compartmentalized and separate from mitochondrial, microbody, and probably chloroplast metabolism. It is generally suggested that nonautotrophic CO_2 metabolism is localized in the cytosol. The main enzymes mediating the reactions, *viz*., PEP carboxylase, malate dehydrogenase, malic enzyme, and aspartate aminotransferase, all exist as isoenzymes or in multiple molecular forms. It is clearly shown that these multiple molecular forms are localized in different subcellular compartments. If dark CO_2 fixation metabolism is composed of cytosol reactions, then one expects that the soluble isoenzymes mediate the flow of carbon. The other isoenzymes would participate only to the extent that the isoenzymic reactions are in equilibrium and that carbon flows from one compartment to another.

REGULATION OF THE METABOLIC PATHWAY

Phosphoenolpyruvate carboxylase
PEP carboxylase catalyzes the following reaction:

$$PEP + CO_2 \xrightarrow{Mg^{++}} OAA + Pi \tag{1}$$

The phosphate acceptor is water and the reaction takes place with a large loss in free energy (49). For this reason, the carboxylation reaction is essentially nonreversible. Coupled with malate dehydrogenase,

$$OAA + NADH \rightarrow MALATE + NAD \tag{2}$$

the oxaloacetate is quickly reduced to malate. The thermodynamic equilibrium constant for the malate dehydrogenase reaction is about 1.7×10^{12} (50) and hence the overall reaction (equations 1 and 2) is generally not reversible. It is not surprising, therefore, that tissues with these enzymes tend to accumulate malate.

In microbial systems, PEP carboxylase is an allosteric enzyme activated by acetyl-CoA and severely inhibited by aspartate and malate (51,52). There is no evidence that acetyl-CoA affects plant PEP carboxylases (40). We have reported the inhibition of corn root PEP carboxylase by 1-malate and to a lesser extent D-malate, aspartate, and glutamate (53). QUEIROZ (54) reported that the PEP carboxylase of *Kalanchoe* leaves was sensitive to 1-malate; however, IZUI *et al.* (55) did not find an effect of malate on a spinach leaf enzyme. Of the three cotton leaf isoenzymes, only one was particularly sensitive to organic acid metabolites (40). At concentrations of 4 to 30 mM 1-malate, however, all were inhibited.

In crude homogenates, the corn root carboxylase is completely inhibited by ADP (39). The cotton isoenzymes are also inhibited by ADP and to a lesser extent by ATP, AMP, and 3'-5'-cyclic AMP (40).

A PEP carboxylase isolated from potato tissue was inhibited by high bicarbonate concentrations when in the presence of manganese, but not when in the presence of magnesium or cobalt (56). WALKER and BROWN (57) reported that carbon dioxide at high concentrations acted as a noncompetitive inhibitor with respect to PEP. All of the cotton leaf isoenzymes were inhibited by high concentrations of bicarbonate (2.5 mM) (40).

The plant PEP carboxylases have Km's for PEP in the range of 0.04 to 0.4 mM and for bicarbonate in the range of 0.05 to 0.22 mM.
Malate dehydrogenase
As previously stated, the malate dehydrogenase mediated reaction strongly favors reduction of oxaloacetate. The Michaelis constants for oxaloacetate vary with pH, but range from about 0.01 to 0.03 at pH 7.5. The Michaelis constant for 1-malate is on the order of 8-15 mM (43).

Relatively little is known about the regulation of NAD-malate de-hydrogenases in plant tissues. Some evidence suggests that the ma-late dehydrogenase of the Enterobacteriaceae is regulated by NADH in an allosteric manner (58). In plant tissues, the malate dehydrogen-ase isoenzymes are inhibited by the substrate, oxaloacetate. The three known isoenzymes, soluble, mitochondrial, and microbody, are affected differentially; however, the soluble and microbody proteins are more sensitive to pH changes being inhibited more by oxaloace-tate at low pH than at high pH. The mitochondrial form is inhibited about equally at all pH values (41,42,59).

A variety of small, naturally occurring metabolic substrate and product molecules inhibit or otherwise affect the activity of plant malate dehydrogenases (43). For example, α-ketoglutarate, cis-aconi-tate, isocitrate, succinate, fumarate, pyruvate, aspartate, gluta-mate, and asparagine tend to inhibit all isoenzymes. The affect is much greater in the direction of malate oxidation than oxaloacetate reduction. Inhibitor concentrations had to be relatively high (10 mM or more) and no obvious differential effects were noted. The physio-logical significance of these effects is not clear at this time.

Malic enzyme

Malic enzyme catalyzes the following reversible reaction:

$$MALATE + NADP \overset{Mg^{++}}{\underset{}{\rightleftarrows}} PYRUVATE + CO_2 + NADPH \qquad (3)$$

Although the equilibrium constant (0.05 1 mole^{-1}) favors malate for-mation, the CO_2 release tends to drive the reaction toward decarb-oxylation (60). Hence, there is the suggestion that the *in vivo* re-action is toward malate utilization rather than malate synthesis (5, 15). Michaelis constants for 1-malate are in the range of 0.1 to 0.7 mM (46,47).

In a discussion of the diurnal variation of organic acids in suc-culents, QUEIROZ (54) indicated that 1-malate activated the malate enzyme. Kinetic studies of the apple fruit enzyme did not suggest substrate activation; however, D-malate enhanced the activity when in the presence of 1-malate (49). The cactus isoenzymes were inhi-bited by D-malate (47). Carefully controlled kinetic experiments de-signed to determine if there was any substrate (1-malate) activation were negative (47).

WALKER (61) reported that high CO_2 tended to inhibit malate de-carboxylation by malate enzyme and enhance pyruvate carboxylation.

In vivo control of CO_2 metabolism

Because the cytosol-localized nonautotrophic CO_2 fixation pathway is at a branched metabolic point (*Figure 1*), one might expect that the flow of carbon from PEP to pyruvate would be under strict meta-bolic control. Some outward evidence for the control is apparent in the obvious and dramatic diurnal fluctuation of organic acids in CAM (2). It is known that starch and organic acids (e.g., malate) fluc-tuate; malate is low and starch is high in the light (2). Under conditions in which malate is high in the light, there is little net CO_2 exchange (62).

The diurnal variation of organic acids in green succulent tissue can be accounted for by several regulatory processes. Firstly, carboxylation in the dark would tend to generate malate. Steady state metabolism of the malate by decarboxylation would tend to limit the amount of malate (and other products) formed. In cactus stem tissue, of the $^{14}CO_2$ fixed into malate and other stable compounds during a two hour period, a subsequent two hour period resulted in a 50% turnover of products (18). In corn root tips, we estimated from enzymic activities that the rate of decarboxylation could be about 50% of the rate of carboxylation (39). Indeed, the turnover of CO_2 fixation products was about 44.5% during a two hour chase period following a two hour carboxylation period (15). It is relatively certain, therefore, that both carboxylation and decarboxylation occur concurrently in both green and nongreen tissues. The observed diurnal fluctuation of organic acids could result, therefore, from alterations in the relative rates of carboxylation and decarboxylation.

In the simplest sense, one can visualize autotrophic CO_2 fixation in the light competing with nonautotrophic CO_2 fixation. Hence, in the light, relatively less CO_2 would go into nonautotrophic fixation products than into autotrophic fixation products, e.g., carbohydrates. This should be true regardless of whether malate is a major product in the light or not. In the light, carbon eventually appears in carbohydrate fractions. If steady state decarboxylation of nonautotrophic fixation products continues, then the products will tend to disappear faster than they are formed. Thus a diurnal fluctuation will occur as a function of light (18). With succulent plants that have stomates closed during the day (63,64), carbon or CO_2 from malate will be largely rechanneled into photosynthetic reactions and the diurnal fluctuation may be magnified.

MUKERJI (65) reported that the enzymic activity of PEP carboxylase was greater in the early evening at a time when acid formation was greatest and that malate enzyme activity was highest in early morning when deacidification or acid decrease was greatest. QUEIROZ (54,66) also reported similar diurnal fluctuations of enzymic activities in *Kalanchoe*. Both workers suggested that the diurnal fluctuation of enzyme activity could account for the diurnal fluctuation of malate and other CO_2 fixation products. Further, both suggested that the fluctuation could be accounted for by the regulation of the enzyme by small substrate molecules.

Since PEP carboxylase is inhibited by l-malate and perhaps by ADP, the *in vivo* levels of these molecules could tend to regulate the flow of carbon. When l-malate accumulates, it would tend to reduce the activity of PEP carboxylase. QUEIROZ'S (54) comments about malate activation of malate enzyme could also be significant here.

As BRANDON (38) has shown, high temperatures as expected during the day light hours may tend to inhibit PEP carboxylase whereas low temperatures as expected at night would tend to favor PEP carboxylase. It is well known that high day temperatures and low night temperatures magnify the diurnal fluctuation of organic acids in succulents (67).

Changes in pool sizes as a function of light and dark were measured in *Bryophyllum* by KLUGE (68). Both 3-PGA and PEP decreased in the light. The decrease in PEP in the light might tend to limit the carboxylase reaction. An additional control which may be important in the diurnal fluctuation of organic acids is the inhibition of phosphofructokinase by PEP (69). Although the latter was shown to be true in an enzyme isolated from pea seeds, other work suggests that phosphofructokinase may be regulatory in plants (70).

As stated, WALKER and BROWN (57) reported that high CO_2 concentrations inhibited PEP carboxylase activity. Their experiments indicated that as CO_2 pressures increased in the range of 0 to 0.5%, the activity increases but was subsequently inhibited at higher concentrations. The CO_2 effect on the enzyme was similar to CO_2 effects on malate accumulation in *Kalanchoe* (THOMSON and RANSON'S experiments cited by WALKER AND BROWN (57)). Further, WALKER (61) showed that high CO_2 inhibited malate decarboxylation by the malate enzyme. Hence, CO_2 concentrations in the tissues could tend to regulate malate levels.

In general, it is clear that a variety of factors influence dark CO_2 fixation. In green plant tissues, light is important. In all tissues temperature is important. Endogenously, concentrations of substrates and products such as CO_2, PEP, ADP, and malate may be improtant in regulating the flow of carbon. High PEP may tend to inhibit the flow of carbon from glycolysis by inhibiting phosphofructokinase; low PEP may affect both PEP carboxylase and pyruvate kinase. High ADP may favor pyruvate kinase. High malate may tend to inhibit PEP carboxylase and favor pyruvate kinase.

EFFECT OF TEMPERATURE AND PHOTOPERIOD ON CO_2 FIXATION AND CRASSULACEAN ACID METABOLISM

It is a well known phenomenon that CAM activity, i.e., night CO_2 fixation and diurnal fluctuation of organic acids, is most pronounced when night temperatures are low and day temperatures are high (2,67,71,72). In part, the temperature effect is correlated with stomatal resistance to gas transfer. Under certain circumstances, stomates of succulent plants are open at night and closed during the day (63). The day and night opening of *Kalanchoe* stomates is a function of the thermoperiod (64). When grown under cool thermoperiods or with little diurnal fluctuation of temperature, stomatal opening and water loss were greater during the light period than the dark period. When grown under thermoperiods with large diurnal temperature fluctuations, more water loss and greater stomatal opening occurred during the dark period. CO_2 exchange, transpiration, and stomatal opening of pineapple follows a similar pattern (unpublished data of CONNELLY and BARTHOLOMEW).

The rate of CO_2 uptake and release is apparently not entirely a function of the stomatal resistance. KLUGE and FISCHER (73) showed that in *Bryophyllum* leaves, transpiration and CO_2 exchange were correlated when the leaf epidermis was intact. If, however, the epidermis was removed to eliminate stomatal effects, then transpiration was continuous but fluctuations in CO_2 exchange still occurred. They suggested that CO_2 concentration regulated stomatal opening (see 74).

BRANDON (38) reported that the temperature characteristics of CO_2 fixation enzymes and CO_2 decarboxylation enzymes was such that acid would accumulate at low temperatures and not at high temperature. Hence the thermal properties of the enzymic proteins also could account partly for the observed temperature effects.

The diurnal variation in CAM activity is also a function of photoperiod (54,75). There was more night net CO_2 fixation when plants were grown on short days than when grown on long days. The products of dark CO_2 fixation, however, are not significantly different when the photoperiod varies (76). No evidence was obtained to suggest that stomatal opening varied with photoperiod (64). Hence it seems that the capacity for net CO_2 fixation in the dark is a function of both temperature and photoperiod, but not entirely governed by leaf resistance to gas transfer. Depending on the photoperiod and thermoperiod, CO_2 assimilation may occur in light or dark. Significant CO_2 uptake occurs in the light under cool thermoperiods with little diurnal fluctuation on long days. Significant CO_2 uptake occurs in the dark under thermoperiods with large fluctuations (low night and high day temperatures) and under short days.

An endogenous rhythm of CO_2 evolution which is entrained by light is known in *Bryophyllum* (77).

RELATIONSHIP BETWEEN NONAUTOTROPHIC CO_2 FIXATION, CRASSULACEAN ACID METABOLISM, AND C_4 PHOTOSYNTHESIS

In C_4 photosynthesis, the first stable carboxylation product appears to be malate and aspartate. Unlike C_3 photosynthesis which also produces significant malate, the malate of C_4 photosynthesis seems to be an active intermediate (cf. 78,79,80). In nonautotrophic CO_2 fixation, malate is also the first stable product which tends to accumulate to any extent. It is visualized that the metabolic pathway mediating the synthesis of malate in all the types is basically similar. *Figure 2* shows the assumed metabolic relationship between nonautotrophic CO_2 fixation, CAM, and C_4 photosynthesis.

The source of PEP, the substrate for malate synthesis (through oxaloacetate), in the nonautotrophic CO_2 fixation pathway seems to be largely from preformed carbohydrates. The PEP may come directly from starch or other carbohydrate via the glycolytic pathway or perhaps, in the case of CAM, through pentose metabolism in which there is a precarboxylation step (see 3 for discussions of this). In C_4 photosynthesis, the source of PEP seems to be directly from pyruvate by action of pyruvate,Pi dikinase, a light dependent reaction (80).

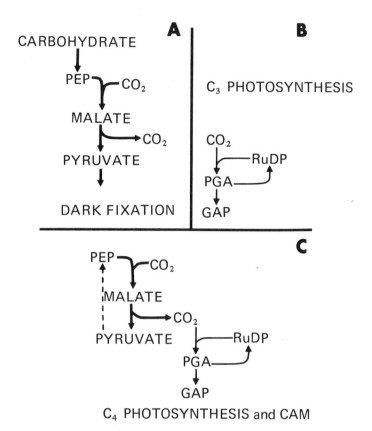

Figure 2 Diagram of the relationship between the dark or non-autotrophic CO_2 fixation pathway, Crassulacean acid metabolism (CAM) pathway, and the C_3 and C_4 photosynthetic pathways. A. Nonautotrophic CO_2 fixation pathway. B. C_3 photosynthetic pathway. C. C_4 photosynthesis and CAM (portions of the schemes in A and B are coupled).

Hence the carboxylation reaction of the latter (C_4 photosynthesis) is largely light dependent, whereas the carboxylation reactions of the former (i.e., nonautotrophic CO_2 fixation) are not directly light dependent. Probably for the same reasons, the subsequent turnover of malate in nonautotrophic CO_2 fixation proceeds in light and dark, whereas the subsequent utilization of malate in C_4 photosynthesis is light dependent (80).

It is probably important that in the light, the malate synthesized by *Sedum* (a CAM plant) is labeled in the C-4 position whereas in the dark the label is asymmetric with the ratio of C-4 to C-1 about 2 to 1 (AVADHANI, unpublished). Succulents may also have pyruvate,Pi dikinase (KLUGE, unpublished). Hence in succulents in the light the source of PEP could be from pyruvate or sources other than stored carbohydrates.

In CAM, providing the thermoperiod is right, malate accumulates at night by carboxylation. Decarboxylation presumably occurs in dark and light; however, in the light the CO_2 is refixed photosynthetically (81). In *Opuntia*, we measured the steady-state release of CO_2 by decarboxylation (of malate) in dark and light (18). The rate of CO_2 release in the dark was 8 times that in the light. By knowing the amount of carbohydrate synthesized from $^{14}CO_2$ in the light and the steady state decarboxylation rates in light and dark, it was possible to calculate that about 20% of the decarboxylation CO_2 in the light would appear in carbohydrates[1]. These data and interpretation suggest that the pathway for carbon flow from malate to carbohydrates is not substantially different in succulents and C_4 plants. In CAM plants, the pool of malate is formed largely at night whereas in C_4 plants the malate pool is synthesized in the light. The present interpretation of C_4 photosynthesis is that the malate is synthesized in mesophyll cells and carbohydrate synthesis from CO_2 derived from malate occurs in the bundle sheath cells (80). Hence, there is the suggestion that CAM and C_4 photosynthesis differ in that malate synthesis is separated from carbohydrate synthesis temporally in CAM plants and spatially in C_4 plants. Probably in both types some C_3 type photosynthesis occurs in the light and most certainly some malate is synthesized by β-carboxylation in the light in CAM plants.

A possible difference between β-carboxylation and decarboxylation in C_4 photosynthesis and nonautotrophic CO_2 fixation may be in compartmentation of the pathways. Much of the evidence suggests that nonautotrophic CO_2 fixation, including carboxylation and decarboxylation, is in the cytosol or nonparticulate fraction of the cell. One would assume that β-carboxylation in C_4 photosynthesis is chloroplastic; however, the available evidence does not suggest that PEP carboxylase is in the same compartment as NADP-triose phosphate dehydrogenase nor does it strongly suggest that it is a firm component of the chloroplast envelope. The second enzyme of the sequence, NADP malate dehydrogenase, does appear to be localized with the triose phosphate dehydrogenase in chloroplasts. If the PEP carboxylase is localized outside the chloroplasts, much of the malate could be synthesized by the soluble NAD malate dehydrogenase. The decarboxylation of malate in the nonautotrophic CO_2 pathway is assumed to take place in the cytosol and be mediated by a soluble-malate enzyme. In C_4 photosynthetic plants, one would predict that malate decarboxylation is within the chloroplasts, and in fact malic enzyme is located in the bundle sheath chloroplasts of maize (36). Decarboxylation of malate by a chloroplast-malic enzyme isoenzyme would effectively trap the free CO_2 in the chloroplasts for subsequent carboxylation by RuDP carboxylase. Since data suggest that succulents with CAM, *viz.*,

---

[1] *In the light about 50% of ^{14}C from $^{14}CO_2$ is in carbohydrates in a 2 hour period. About 50% of the ^{14}C is lost as CO_2 in a chase experiment. Of the 50% released as estimated from the dark rate, in the light about 35% is retained by refixation. Hence we predict that 50 x 35 or 17.5% of the ^{14}C would appear as carbohydrates. The actual measurement was 17.7%.*

Opuntia, have a chloroplast-malate enzyme isoenzyme (47), decarboxylation of malate and transfer of CO_2 to photosynthetic pathways may
also take place in the chloroplasts in CAM plants.

In summary, we visualize a major difference between nonautotrophic CO_2 fixation and C_4 photosynthesis to be the direct precursor
of the initial carboxylation substrate, PEP. In the former, the
source is stored carbohydrate and in the latter the source is directly from pyruvate or other three carbon compound generated by the
photosynthetic sequence. Secondly, the decarboxylation of malate is
presumably in the cytosol by a soluble-malate enzyme isoenzyme in
nonautotrophic CO_2 fixation and in the chloroplasts in C_4 photosynthesis by a chloroplast-malate enzyme isoenzyme. CAM is intermediate; perhaps the source of PEP differs in light and dark and perhaps
the decarboxylation differs in light and dark. In CAM, the effective
carboxylation with subsequent accumulation of malate and effective
decarboxylation and photosynthetic refixation are largely separated
in time; i.e., carboxylation in the dark and decarboxylation and refixation in the light.

REFERENCES

1. H.G. WOOD and C.H. WERKMAN, *J. Bacteriol.*, 30 (1935) 332.
2. H.B. VICKERY, *Plant Physiol.*, 29 (1954) 385.
3. S.L. RANSON and M. THOMAS, *A. Rev. Pl. Physiol.*, 11 (1960) 81.
4. J. WOLF, in Encyclopaedia of Plant Physiology, ed. W.H. RUHLAND,
      Springer, Berlin, 12 (1960) 809.
5. D.A. WALKER, *Biol. Rev.*, 37 (1962) 215.
6. H.G. WOOD and M.F. UTTER, Essays in Biochemistry, ed. P.N. CAMP-
      BELL and G.D. GREVILLE, Academic Press, London, 1 (1965) 1.
7. R.V.F. LACHICA, *Enzymologia*, 34 (1968) 281.
8. J. THURLOW and J. BONNER, *Archs. Biochem. Biophys.*, 25 (1948)
      280.
9. M. THOMAS and S.L. RANSON, *New Phytol.*, 53 (1954) 1.
10. P. SALTMAN, G. KUNITAKE, H. SPOLTER, and C. STITT , *Plant Phy-
      siol.*, 31 (1956) 464.
11. G. KUNITAKE, C. STITT, and P. SALTMAN, *Plant Physiol.*, 34 (1959)
      123.
12. M.L. STILLER, G.E. NEAL, and H. BEEVERS, *Plant Physiol.*, 33
      (1958) xxxiv.
13. K.L. WEBB and J.W.A. BURLEY, *Can. J. Bot.*, 43 (1965) 281.
14. G. JOSHI, T. DOLAN, R. GEE, and P. SALTMAN, *Plant Physiol.*, 37
      (1962) 446.
15. I.P. TING and W.M. DUGGER, *Science*, 150 (1965) 1727.
16. I.P. TING and W.M. DUGGER, *Plant Physiol.*, 41 (1966) 500.
17. I.P. TING and W.M. DUGGER, *Plant Physiol.*, 42 (1967) 712.
18. I.P. TING and W.M. DUGGER, *Bot. Gaz.*, 129 (1968) 9.
19. J.E. VARNER and R.C. BURRELL, *Archs. Biochem.*, 256 (1950) 280.
20. J.W. BRADBEER, S.L. RANSON, and M. STILLER, *Plant Physiol.*, 33
      (1958) 66.

21. P.N AVADHANI, Ph. D. Thesis, Durham University, Newcastle-upon-Tyne, (1957).
22. H.G. WAGER, *J. Exp. Bot.*, 18 (1967) 672.
23. W.E. SPLITTSTOESSER, *Proc. Am. Soc. Hort. Sci.*, 90 (1967) 235.
24. C.J. CLEGG and C.P. WHITTINGHAM, *Phytochemistry*, 9 (1970) 279.
25. M.D. HATCH and C.R. SLACK, *Biochem. J.*, 101 (1966) 103.
26. D.H. MacLENNAN, H. BEEVERS, and J.L. HARLEY, *Biochem. J.*, 89 (1963) 316.
27. S.H. LIPS and H. BEEVERS, *Plant Physiol.*, 41 (1966) 709.
28. S.H. LIPS and H. BEEVERS, *Plant Physiol.*, 41 (1966) 713.
29. C.B. OSMOND and G.G. LATIES, *Plant Physiol.*, 44 (1969) 7.
30. I.P. TING and W.C. ZSCHOCHE, *Plant Physiol.*, 45 (1970) 429.
31. R.S. BANDURSKI and C.M. GREINER, *J. biol. Chem.*, 204 (1953) 781.
32. M. MAZELIS and B. VENNESLAND, *Plant Physiol.*, 32 (1957) 591.
33. L.L. ROSENBERG, J.B. CAPINDALE, and F.R. WHATLEY, *Nature*, 181 (1958) 632.
34. E.J. JOHNSON and B.S. BRUFF, *Plant Physiol.*, 42 (1967) 1321.
35. J. GARNIER-DARDART, *Physiol. Veg.*, 3 (1965) 215.
36. C.R. SLACK, M.D. HATCH, and D.J. GOODCHILD, *Biochem. J.*, 114 (1969) 489.
37. S.K. MUKERJI and I.P. TING, *Phytochemistry*, 7 (1968) 903.
38. P.C. BRANDON, *Plant Physiol.*, 42 (1967) 977.
39. J. DANNER and I.P. TING, *Plant Physiol.*, 42 (1967) 719.
40. S.K. MUKERJI and I.P. TING, *Archs. Biochem. Biophys.*, in press.
41. R.K. YAMAZAKI and N.E. TOLBERT, *Biochim Biophys. Acta*, 178 (1969) 11.
42. V. ROCHA and I.P. TING, *Archs. Biochem. Biophys.*, 140 (1970) 398.
43. S.K. MUKERJI and I.P. TING, *Archs. Biochem. Biophys.*, 131 (1969) 336.
44. M.D. HATCH and C.R. SLACK, *Biochem. Biophys. Res. Commun.*, 34 (1969) 589.
45. H.S. JOHNSON and M.D. HATCH, *Biochem. J.*, 119 (1970) 273.
46. D.R. DILLEY, *Plant Physiol.*, 41 (1966) 214.
47. S.K. MUKERJI and I.P. TING, *Biochim. Biophys. Acta*, 167 (1968) 239.
48. P. FASELLA, F. BOSSA, C. TURANO, and A. FANELLI, *Enzymologia*, 30 (1966) 198.
49. R.S. BANDURSKI, *J. biol. Chem.*, 217 (1955) 137.
50. A. YOSHIDA, *J. biol. Chem.*, 240 (1965) 1118.
51. J.L. CANOVAS and H.L. KORNBERG, *Proc. R. Soc. (B.)*, 165 (1966) 189.
52. P. MAEBA and B.O. SANWAL, *Biochem. Biophys. Res. Commun.*, 21 (1965) 503.
53. I.P. TING, *Plant Physiol.*, 43 (1968) 1919.
54. O. QUEIROZ, *Physiol. Veg.*, 6 (1968) 117.
55. K. IZUI, A.I. WATANI, T. NISHIKIDO, H. KATSUKI, and S. TANAKA, *Biochim. Biophys. Acta*, 139 (1967) 188.
56. T.S. SMITH, *Archs. Biochem. Biophys.*, 125 (1968) 178.

57. D.A. WALKER and J.M.A. BROWN, *Biochem. J.*, 67 (1957) 79.
58. B.D. SANWAL, *J. biol. Chem.*, 244 (1969) 1831.
59. I.P. TING, *Archs. Biochem. Biophys.*, 126 (1968) 1.
60. I. HARARY, S.R. KAREY, and S. OCHOA, *J. biol. Chem.*, 203 (1953) 595.
61. D.A. WALKER, *Biochem. J.*, 74 (1960) 216.
62. M. KLUGE, *Planta*, 80 (1968) 359.
63. K. NISHIDA, *Physiologia Pl.*, 16 (1963) 281.
64. I.P. TING, M.L. THOMPSON, and W.M. DUGGER, *Am. J. Bot.*, 54 (1967) 245.
65. S.K. MUKERJI, *Indian J. Biochem.*, 5 (1968) 62.
66. O. QUEIROZ, *Physiol. Veg.*, 8 (1970) 75.
67. T.A. BENNET-CLARK, *New Phytol.*, 32 (1933) 37.
68. M. KLUGE, *Planta*, 85 (1969) 160.
69. G.J. KELLY and J.F. TURNER, *Biochem. J.*, 115 (1969) 481.
70. D.T. DENNIS and T.P. COULTATE, *Biochem. Biophys. Res. Commun.*, 25 (1966) 187.
71. O. QUEIROZ, *Physiol. Veg.*, 3 (1965) 203.
72. O. QUEIROZ, *Physiol. Veg.*, 4 (1966) 323.
73. M. KLUGE and K. FISCHER, *Planta*, 77 (1967) 212.
74. H. MEIDNER and T.A. MANSFIELD, *Biol. Rev.*, 40 (1965) 483.
75. F.G. GREGORY, I. SPEAR, and K.V. THIMANN, *Plant Physiol.*, 29 (1954) 220.
76. G.M. KUNITAKE, P. SALTMAN, and A. LANG, *Plant Physiol.*, 32 (1957) 201.
77. M.B. WILKINS, *Plant Physiol.*, 37 (1962) 735.
78. M.D. HATCH, C.R. SLACK, and H.S. JOHNSON, *Biochem. J.*, 74 (1967) 4477.
79. M. CALVIN and P. MASSINI, *Experientia*, 8 (1952) 445.
80. M.D. HATCH and C.R. SLACK, *A. Rev. Pl. Physiol.*, 21 (1970) 141.
81. S. KUNITAKE and P. SALTMAN, *Plant Physiol.*, 33 (1958) 400.

The "hard core" participants, from front-left to back-right :
C.B. Osmond, I.P. Ting, C.C. Black, E.R. Waygood, M. Gibbs, N.E.
Tolbert, R.O. Slatyer, J.W. Lyttleton, C.R. Slack, W.J.S. Downton,
H. Beevers, M.D. Hatch, K. Asada, U. Heber, M. Kluge, D.N. Moss, U.
Lüttge, P.N. Avadhani, Hillary S. Johnson, T.A. Bull, H.P. Kortschak,
R.M. Gifford, M.M. Ludlow, J.H. Troughton, I. Wardlaw, C. Marshall,
Jan. M. Anderson, J.T.O. Kirk, G. Roughan, T.F. Neales, O. Björkman,
W.A. Jackson, J.R. McWilliam, P.F. Brownell, Elza A. Chapman, R.G.
Everson, E.G. Brittain, N.K. Boardman, D.G. Bishop, W. Kowallik,
Margaret C. Anderson, D.J. Goodchild.

SECTION 2

Carbon Dioxide Assimilation

B. Research Papers and Assessments

THE CO_2 CARRIER BETWEEN MESOPHYLL AND BUNDLE SHEATH CHLOROPLASTS IN C_4 PATHWAY SPECIES

H.S. Johnson, C.R. Slack,[1] M.D. Hatch,[2] T.J. Andrews[3]

Botany Department, University of Queensland,
St Lucia, Qld., 4067, Australia

SUMMARY

The C_4 pathway, as presently envisaged, requires the transfer of CO_2 as the β-carboxyl of a dicarboxylic acid from the mesophyll to bundle sheath cells. Certain species contain sufficiently high levels of an NADP malate dehydrogenase and malic enzyme, appropriately located, for malate to act as the CO_2 carrier. Other species, however, contain only low levels of these enzymes, but very high levels of aspartate aminotransferase and alanine aminotransferase in both the mesophyll and bundle sheath. It is proposed that aspartate functions as the CO_2 carrier in these latter species.

INTRODUCTION

Investigations of the localization of enzymes implicated in the C_4 pathway of photosynthesis (1-4) have led to the conclusion that the two types of chloroplast present in species that photosynthesize via the pathway must function in a cooperative manner. Our initial work (1,2) was performed with leaves of maize (*Zea mays*) and we concluded from the results of nonaqueous density fractionation that enzymes operative in the incorporation of CO_2 into malate and aspartate were localized in the mesophyll chloroplasts whereas Calvin cycle enzymes and malic enzyme were located in the bundle sheath

---

[1] *Present address: Plant Physiology Division, DSIR, Palmerston North, New Zealand.*

[2] *Present address: Division of Plant Industry, CSIRO, Canberra City, 2601, Australia.*

[3] *Present address: Department of Biochemistry, Michigan State University, East Lansing, Michigan, 48823, U.S.A.*

chloroplasts. The demonstration by BJÖRKMAN and GAUHL (3) that ribu-lose-1,5-diphosphate (RuDP) carboxylase is present in respectable quantities in leaves of C_4 pathway species necessitates its inclu-sion in any proposed scheme and obviates the need for a transcarb-oxylation reaction, to move CO_2 from oxaloacetate to 3-phosphogly-cerate (3-PGA), that we postulated previously.

It is now believed (2,4, see also HATCH, this volume) that in maize the transfer of CO_2 from oxaloacetate to RuDP carboxylase is mediated by NADP malate dehydrogenase and malic enzyme which are lo-cated in the mesophyll and bundle sheath chloroplasts respectively. This view implies that malate acts as a CO_2 carrier between chloro-plasts in different cells. Leaves of certain C_4 pathway species, however, have been reported to contain levels of these malate-meta-bolizing enzymes that are inadequate to support the transfer of CO_2 at rates comparable with those of photosynthetic CO_2 fixation. In this paper we present evidence, reported in part elsewhere (5), which suggests that aspartate, rather than malate, functions as a CO_2 carrier in these species.

RESULTS AND DISCUSSION

Table 1 shows the activity of NADP malate dehydrogenase and malic enzyme in C_4 pathway species from seven genera. With the exception of *Gomphrena celosoides* the species fall into one of two groups. Leaf extracts of species in group 1 contain activities of these en-zymes that are adequate to support the observed rate of photosynthe-tic CO_2 fixation, but only a low activity of alanine aminotransfer-ase. Compared with group 1 species leaf extracts of group 2 species contain only a low activity of malic enzyme, less NADP malate dehyd-rogenase, more aspartate aminotransferase activity, and much more alanine aminotransferase activity. We consider it very unlikely that these differences in enzyme activity reflect an artefact of extrac-tion, firstly because one would have to postulate the inactivation of the malate-metabolizing enzymes in extracts of group 2 and the inactivation of the aminotransferases in group 1, and secondly be-cause the activity of several other enzymes is very similar in leaf extracts from these species (7,8). We suggest, therefore, that these activities reflect the enzyme levels *in vivo*.

The high levels of alanine and aspartate aminotransferase in group 2 species could be required for a CO_2 transfer system involv-ing aspartate in a manner analagous with that proposed for malate in the group 1 species. This system would require the presence of an aspartate decarboxylase in the bundle sheath chloroplasts to libe-rate CO_2 from aspartate exported by the mesophyll cells. We have been unable, however, to detect the presence of such an enzyme in leaf extracts of any group 2 species. As an alternative approach to the investigation of the mechanism by which aspartate could serve as a CO_2 carrier we decided to examine the inter- and intracellular lo-cation of the aspartate and alanine aminotransferases. We argued

TABLE I

Malic enzyme, NADP malate dehydrogenase and aminotransferases in leaf extracts of C_4 pathway species (μmole min^{-1} mg^{-1} chlorophyll). Extracts were prepared and enzymes assayed as described previously (2,6). Maximum rates of photosynthetic CO_2 fixation for these species range from 3-5 μmoles CO_2 min^{-1} mg^{-1} chlorophyll (7,8)

Species	Malic enzyme	NADP malate dehydro-genase	Aspartate aminotrans-ferase	Alanine aminotrans-ferase
Group I				
Zea mays	10	9	0.7	5.7
Saccharum (hybrid)	12	12	0.3	4.6
Sorghum (hybrid)	10	14	0.6	4.9
Group 2				
Eragrostis brownii	0.4	0.6	16.0	–
Chloris gayana	0.2	4.0	10.0	22.0
Amaranthus edulis	0.8	1.5	21.0	28.0
Amaranthus palmeri	1.3	2.4	32.0	38.0
Atriplex spongiosa	0.15	2.1	25.0	40.0
Gomphrena celosoides	5.5	2.3	16.0	15.0

that if the above system were operative then both transaminanases should be largely, or entirely, located in the mesophyll cells; the former to aminate oxaloacetate and the latter to catalyze the conversion of alanine to pyruvate.

The intracellular localization of these enzymes was examined in maize, a high malic enzyme species, and in *Amaranthus levitus*, a high aminotransferase species, by nonaqueous density fractionation (*Table 2*). The leaf from both species was destarched prior to lyophilization and hence the bundle sheath chloroplasts were isolated in the light fractions together with those of the mesophyll. Aspartate aminotransferase was distributed amongst the fractions from maize leaf in a manner similar to that of chlorophyll and the chloroplast enzyme NADP malate dehydrogenase. Hence, in maize this enzyme is localized entirely, or almost so, in the chloroplasts. In the fractions from *A. levitus*, however, there were two peaks of aspartate aminotransferase activity one corresponding to that of chlorophyll and the chloroplast enzyme, pyruvate,Pi dikinase, and the other in the heaviest fraction which contained nonchloroplast material. It would appear, therefore, that in this species a portion of the enzyme is associated with the chloroplasts and the remainder with some other cellular compartment. Examination of the distribution

TABLE 2

Percentage distribution of chlorophyll and enzymes in non-aqueously prepared fractions of leaves of maize and *Amaranthus levitus* (shown in brackets). Methods of nonaqueous density fractionation were as previously described (I).

Enzyme	<1.30	1.30-1.33	1.33-1.36	1.36-1.40	>1.40
Chlorophyll	48(65)	13(18)	13 (6)	17 (3)	7 (8)
NADP malate dehydrogenase	50	14	9	20	7
Pyruvate,Pi dikinase	(60)	(24)	(7)	(7)	(2)
Aspartate aminotransferase	51(32)	16(20)	12 (9)	14 (6)	7(33)
Alanine aminotransferase	16(17)	7(13)	13(10)	38 (8)	26(52)

of this enzyme in fractions from starched *A. levitus* leaves has indicated that the chloroplast fraction of the enzyme is associated with the mesophyll chloroplasts only. In fractions from both maize and *A. levitus* the alanine aminotransferase had a distribution pattern that was typical of an enzyme not associated with chloroplasts. Similar results have been obtained with *A. palmeri* (5).

Nonaqueous density fractionation can be used to determine whether an enzyme is a nonchloroplast or chloroplast enzyme, and to separate mesophyll from bundle sheath chloroplasts. However, it cannot be used to determine the intercellular distribution of enzymes not associated with chloroplasts. We have attempted to determine, therefore, the intercellular distribution of alanine aminotransferase and of the cytoplasmic component of aspartate aminotransferases by the method of differential maceration (3,4). Leaves of two species, *Amaranthus palmeri* and *Atriplex spongiosa*, which contain high levels of the two aminotransferases were used (*Table 3*). Extract 1 was obtained by a gentle maceration and the residue macerated more vigorously to give a second extract. Finally the remaining tissues were completely homogenized to provide extract 3 (5). RuDP carboxylase and phosphoribulokinase have been shown previously to be localized in the bundle sheath and pyruvate,Pi dikinase and phosphoenolpyruvate (PEP) carboxylase in the mesophyll (1-4). These enzymes were assayed in each extract to give an indication of the degree of breakage of the two cell types by the different macerations. With both species it was apparent that the proportions of the two aminotransferases released by each maceration were very similar, and furthermore, that they were intermediate between the values for enzymes associated

TABLE 3

Percentage distribution of enzymes in extracts prepared by
differential maceration (Ext. 1 and Ext. 3) and estimated distribution in mesophyll cells (M). Percentage of aminotransferase in mesophyll cells calculated as described; other enzyme
distributions assumed from previous studies (1,2).

Enzyme	Amaranthus			Atriplex		
	Ext. 1	Ext. 3	M	Ext. 1	Ext. 3	M
Ru-5-P kinase	15	37	0	32	39	0
RuDP carboxylase	17	34	0	32	40	0
PEP carboxylase	35	7	100	98	0	100
Pyruvate,Pi dikinase	32	9	100	97	0	100
Alanine aminotransferase	28	25	47	63	19	47
Aspartate aminotransferase	24	25	42	62	21	42

with only one of the two cell types. It is proposed, therefore, that
these aminotransferases are present in both the mesophyll and bundle
sheath cells. We estimated the distribution of each enzyme between
mesophyll and bundle sheath cells using the formula:

$$ax + b(1-x) = \text{per cent of enzyme in extract 1,}$$

where x and $(1-x)$ are the fractions of the total leaf complement of
the enzyme in mesophyll and bundle sheath cells, and a and b are the
percentages of the total contents of mesophyll and bundle sheath
cells in extract 1. The latter values were obtained from the percentage of the total leaf complement of the mesophyll enzymes and bundle sheath enzymes in this extract. The assumption that most of the
leaf aminotransferase is distributed between these cells is considered elsewhere (5).

The calculated values for the intercellular distribution of the
aminotransferases in both the *Amaranthus* and *Atriplex* species are
very similar; in each case there appears to be slightly less than
50% of the leaf complement of both enzymes in the mesophyll cells.

The above results indicate that C_4 pathway species containing only low levels of malic enzyme contain large amounts of aspartate and
alanine aminotransferases in both the mesophyll and bundle sheath

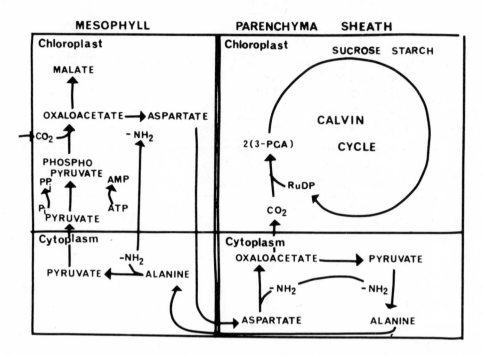

Figure 1 Scheme for operation of aspartate as a CO_2 carrier between mesophyll and bundle sheath in species possessing high levels of aspartate and alanine aminotransferases.

cells. The mesophyll cell complement of the aspartate aminotransferase is probably largely, or entirely, located in the mesophyll chloroplasts. However, aspartate aminotransferase of the bundle sheath and the alanine aminotransferase of both cell types are located in some nonchloroplast compartment. This inter- and intracellular distribution does not support the proposition discussed above, that aspartate could be formed in the mesophyll chloroplasts, decarboxylated in the bundle sheath chloroplasts and the alanine returned to the mesophyll.

A scheme that we think is consistent with data presently available is given in *Figure 1*. Oxaloacetate is aminated in the mesophyll chloroplasts by aspartate aminotransferase and the aspartate transferred to the bundle sheath. The presence of large amounts of aspartate aminotransferase in some nonchloroplast compartment of these cells suggests that aspartate is converted to oxaloacetate before being decarboxylated. Pyruvate released in the β-decarboxylation of oxaloacetate could be aminated by the bundle sheath alanine aminotransferase. It must be assumed that the bundle sheath cells export alanine to the mesophyll in order to maintain both carbon and amino group balance between the cells. The nonchloroplast alanine aminotransferase in the mesophyll could serve to reconvert the alanine to pyruvate. This scheme requires the presence of an oxaloacetate decarboxylase in the bundle sheath which, as yet, we have been unable to detect.

In conclusion we must emphasize that this distinction between C_4 pathway species in terms of the C_4 acid acting as a CO_2 carrier is a quantitative one, and should not be inferred to be absolute. In *Gomphrena* and probably other species both malate and aspartate may make a major contribution to CO_2 transport.

REFERENCES

1. C.R. SLACK, *Phytochemistry*, 8 (1969) 1387.
2. C.R. SLACK, M.D. HATCH, and D.J. GOODCHILD, *Biochem. J.*, 114 (1969) 489.
3. O. BJÖRKMAN and E. GAUHL, *Planta*, 88 (1969) 197.
4. J.A. BERRY, W.J.S. DOWNTON, and E.B. TREGUNNA, *Can. J. Bot.*, 48 (1970) 777.
5. T.J. ANDREWS, H.S. JOHNSON, C.R. SLACK, and M.D. HATCH, *Phytochemistry*, in press.
6. H.S. JOHNSON and M.D. HATCH, *Biochem. J.*, 119 (1970) 273.
7. C.R. SLACK and M.D. HATCH, *Biochem. J.*, 103 (1967) 660.
8. H.S. JOHNSON and M.D. HATCH, *Phytochemistry*, 7 (1968) 375.

TRANSIENT CHANGES IN LEVELS OF SOME COMPOUNDS IN SPINACH AND MAIZE LEAVES[1]

Erwin Latzko, Larry Laber[2], and Martin Gibbs

Department of Biology, Brandeis University, Waltham, Mass., 02154, U.S.A. and
Chemisches Institut, Technische Hochschule, Weihenstephan, Germany.

SUMMARY

Ribulose-1,5-diphosphate, 3-phosphoglycerate, phosphoenolpyruvate, and malate were assayed enzymically in spinach and maize leaves. Subsequent to illumination in the absence of CO_2, (preillumination) leaves were exposed either in light or in dark to a pulse of CO_2. In both leaves, ribulose-1,5-diphosphate increased linearly during pre-illumination and decreased rapidly following the introduction of CO_2 in the light or in the dark. 3-phosphoglycerate reacted the opposite to ribulose-1,5-diphosphate. In maize, phosphoenolpyruvate and malate decreased during preillumination but increased with the introduction of CO_2. The phosphoenolpyruvate level in spinach leaves remained constant.

The observed changes in the levels of these compounds are consistent with ribulose-1,5-diphosphate serving as a primary CO_2 acceptor in both plants.

INTRODUCTION

Evidence for the operation of a C_4 dicarboxylic pathway in a number of plants, including maize and sugarcane, has been obtained from kinetic labeling experiments during photosynthesis for short periods of time (1,2,3) and from enzyme investigations (4,5). It has been proposed that the C_4 dicarboxylic acid pathway operates exclusively in the chloroplasts of the mesophyll cells and interacts with the

[1] *This research was supported by a grant from the United States Atomic Energy Commission AT (30-1) 3447.*
[2] *Postdoctoral trainee of the United States Public Health Service, GM-1586-03.*

196

reductive pentose phosphate cycle which is located predominantly in the bundle sheath chloroplasts (6).

HATCH and SLACK (6) have suggested that plants utilizing the C_4 dicarboxylic acid pathway fix CO_2 primarily via phosphoenolpyruvate (PEP) carboxylase, and that the carbon flows by means of malic or aspartic acid to the reductive pentose phosphate cycle.

During recent work in this laboratory, it was not possible to demonstrate PEP carboxylase in maize chloroplasts capable of photosynthesizing CO_2 (7). These chloroplasts possessed an enzyme pattern characteristic of the reductive pentose phosphate cycle and did not metabolize malate or aspartate.

In the present study photosynthetic CO_2 fixation by leaves of maize and spinach was examined using a method based on that described by many investigators for algae (8,9,10,11). In this method plants were illuminated in absence of CO_2 (termed preillumination) and were given subsequently a pulse of CO_2 during illumination or after the lights were switched off. From this type of experiment the conclusion was made that 3-phosphoglycerate (3-PGA) was the first labeled compound formed during photosynthetic $^{14}CO_2$ incorporation in algae (8). Similarly WILSON and CALVIN (12) were able to demonstrate a marked accumulation of ribulose-1,5-diphosphate (RuDP) during illumination of *Chlorella* in the absence of CO_2. Addition of $^{14}CO_2$ led to a sharp decrease in the level of RuDP and an equivalent formation of carboxyl labeled 3-PGA. If PEP was the primary CO_2 acceptor in plants possessing a C_4 dicarboxylic acid pathway, illumination of intact leaves of these plants in the absence of CO_2, may be expected to cause an accumulation of PEP during illumination and after a CO_2 pulse, the subsequent formation of C_4 dicarboxylic acids such as malate and aspartate.

The purpose of this study was to examine this possibility. The levels of PEP, RuDP, 3-PGA, and malate were compared in spinach and in maize leaves.

METHODS

Segments of primary leaves of maize or of spinach leaves were preilluminated in 50 ml flasks at 30° and in the presence of 3800 ft-c furnished by incandescent lamps. Nitrogen containing 1.5% O_2 was bubbled into the flasks during the preillumination periods. After introduction of gaseous CO_2, the leaf segments, held together by thread, were placed into 80% (v/v) hot ethanol. The rate of CO_2 fixation in terms of chlorophyll or fresh weight was calculated from the radioactivity in the ethanol extract.

To extract the intermediates (RuDP, 3-PGA, PEP, and malate), leaves were submerged in liquid nitrogen and ground to a fine powder. The powder was extracted with boiling 80% (v/v) ethanol containing 50 mM MES buffer, pH 6.5. The clear supernatant fluid following centrifugation was used for assay.

RuDP was assayed as described previously (13). PEP and 3-PGA were

also determined using radiometric methods. Malate was determined by
the method of WILLIAMSON and CORKEY (14).

RESULTS

In preliminary experiments the optimum conditions for CO_2 incorporation in darkness following illumination in the absence of CO_2
were determined. A light intensity of 3800 ft-c and a temperature of
30° were found suitable and convenient. Under these conditions, preillumination for 30 minutes resulted in an 18 fold increase in CO_2
fixation during the subsequent dark period of 30 sec compared with
leaves kept throughout in darkness. This increase was approximately
25 fold when the preillumination time was increased to 60 minutes.
The enhanced dark fixation did not continue for more than 2 minutes
in maize or spinach leaves (*Figure 1*).

A linear increase in the level of RuDP in maize leaves was observed when the intact leaves were illuminated in the absence of CO_2
(*Figure 2*). After a pulse of CO_2 the level of RuDP fell sharply and
the level of 3-PGA and of PEP increased simultaneously. Similar
changes were noted when the CO_2 pulse was given after the lights
were switched off. When a CO_2 pulse was not given during the dark
period, the RuDP level decreased more gradually to a higher steady
level.

During the dark period following illumination the levels of 3-PGA
and of PEP first increased sharply. The addition of CO_2 had only
little effect on these changes.

Malate constantly decreased in maize leaves being illuminated in
the absence of CO_2. No response by this pool to the CO_2 pulse during
illumination could be observed (*Figure 3*). In contrast, an increase
in malate was observed after addition of CO_2 during the dark period.

An analogous experiment was carried out with spinach leaves in
order to follow the changes in the level of intermediates in a plant
possessing only the reductive pentose phosphate cycle (*Figure 4*).
The transitional changes in spinach leaves for both RuDP and the 3-
PGA were more pronounced. In spinach leaves the level of RuDP was
roughly 1/3 higher than under comparable conditions in maize leaves.
This level decreased sharply when a CO_2 pulse was given during the
preillumination period and after the lights were switched off. The
content of RuDP decreased only gradually to a higher steady state
level in darkness when CO_2 was not supplied.

The level of 3-PGA increased sharply after the CO_2 pulse during
illumination in a CO_2 free atmosphere. During the dark period following illumination 3-PGA first increased then decreased below the level initially observed in the light. The transient rise was more pronounced by the presence of CO_2.

The level of PEP was constantly less than 0.1 μmole mg^{-1} protein.
This level was unaffected by either CO_2 pulses or light dark transitions.

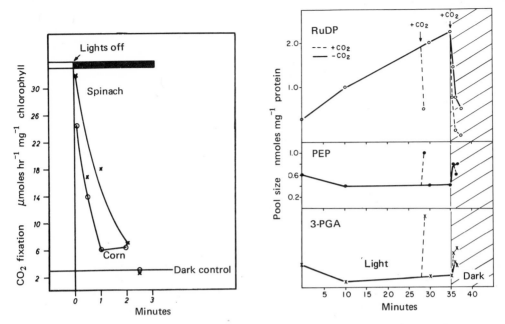

Figure 1 (left) Light enhanced dark fixation of $^{14}CO_2$ in spinach and maize leaves. Detached primary leaves were illuminated in flasks under N_2 containing 1.5% O_2, at 3800 ft-c and
30°. At the indicated times $^{14}CO_2$ was introduced into the
flasks and 1 minute later the leaves were killed by immersion
into 80% ethanol (v/v). The $^{14}CO_2$ fixation rates were determined as described in the Methods.

Figure 2 (right) Changes in the level of intermediates after
CO_2 pulses and during light-dark transitions in maize leaves.
Maize leaves at 30° were illuminated with 3800 ft-c in an atmosphere of N_2 containing 1.5% O_2. Each point on the graph represents one flask containing 5 primary leaves cut 8 days after planting. Metabolism was then stopped by immersion in liquid nitrogen. After 28 minutes of preillumination 10 μmoles
of CO_2 was introduced into one flask for 1 minute. All other
flasks were kept in light for a total of 35 minutes. The
lights were then switched off and 10 μmoles of CO_2 were supplied to some flasks while other flasks did not receive CO_2.
Samples were submerged in liquid nitrogen at the times indicated, ground to a fine powder and stored at -70°. The levels of
intermediates were determined in all samples as described under Methods. Dotted lines indicate addition of CO_2.

DISCUSSION

The spinach and maize leaves used in our experiment responded
during preillumination and subsequent CO_2 assimilation in a manner
analogous to that found with green algae (8,9,10,11,12) and with

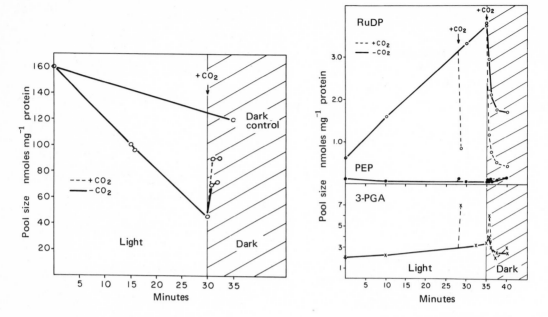

Figure 3 (left) Changes of the malate level in maize leaves during light-dark transition. Conditions as for Figure 2, except that the gaseous CO_2 pulse in the light was applied after 15 minutes of illumination and the light was removed after 30 minutes. Dotted lines indicate addition of CO_2.

Figure 4 (right) Changes in the levels of intermediates after CO_2 pulses and during light-dark transition in spinach leaves. Conditions as in Figure 2. Each point on the graph represents analysis of 1 sample. Dotted lines indicate addition of CO_2.

spinach chloroplasts (15). During preillumination, RuDP accumulated and when CO_2 was added after the light was removed, this compound decreased with a concomitant increase in 3-PGA.

In maize leaves, the concentrations of PEP showed a similar time course to that of 3-PGA, namely, a low level during preillumination without CO_2 and a simultaneous increase after a pulse of CO_2 during the dark phase. It would appear that PEP behaved as a product of CO_2 fixation rather than as a primary acceptor of CO_2.

On the basis of our results, we propose that RuDP can serve as a primary acceptor of CO_2 in maize leaves. A dicarboxylic acid does not appear to be required as a carrier of CO_2. Furthermore, it would appear that under the conditions of our experiments carbon can flow in maize leaves from RuDP to PEP via 3-PGA.

It is well to recall here the experiments of BRADBEER *et al.* (16) who dealt with carbon flow in Crassulacean leaves. Malate formed in the dark during exposure of leaves to $^{14}CO_2$ was labeled exclusively in the carboxyl groups with C-1 (α-COOH) containing approximately one third of the activity. To account for the asymmetry of labeling a mechanism involving RuDP and PEP carboxylases was proposed. Degradation of the malate formed in our experiments would be helpful in elucidating the flow of carbon in maize leaves following preillumination.

Finally, an important question involves the fate of malate in maize during the preillumination period. The level of malate exceeds by a factor of 20 the level of any other measured intermediates. During the 30 minutes of illumination in the absence of CO_2, roughly one-third of the malate pool was lost and cannot be accounted for as RuDP or PEP. The fact that PEP did not accumulate in the absence of CO_2 suggests that malate either was not converted in the light phase to PEP or if this was formed it was not metabolized by a pathway involving PEP carboxylase. There is no evidence of a light accelerated mitochondrial respiration of malate. Furthermore, the transient in RuDP during the dark phase cannot account for the far larger increase in malate. It is possible that the malate lost during preillumination is converted to carbohydrate and reformed from carbohydrate during the subsequent dark phase. Presumably this carbohydrate synthesis and breakdown involving malate is located exclusively in the cytoplasm.

REFERENCES

1. H.P. KORTSCHAK, C.E. HARTT, and G.O. BURR, *Plant Physiol.*, 40 (1965) 209.
2. M.D. HATCH and C.R. SLACK, *Biochem. J.*, 101 (1966) 103.
3. H.S. JOHNSON and M.D. HATCH, *Biochem. J.*, 114 (1969) 127.
4. C.R. SLACK, M.D. HATCH, and D.J. GOODCHILD, *Biochem. J.*, 114 (1969) 489.
5. M.D. HATCH, C.R. SLACK, and T.A. BULL, *Phytochemistry*, 8 (1969) 697.
6. M.D. HATCH and C.R. SLACK, *A. Rev. Pl. Physiol.*, 21 (1970) 141.
7. M. GIBBS, E. LATZKO, D. O'NEAL, and C.S. HEW, *Biochem. biophys. Res. Commun.*, 40 (1970) 1356.
8. E.W. FAGER, J.L. ROSENBERG, and H. GAFFRON, *Fed. Proc.*, 9 (1950) 535.
9. J.A. BASSHAM and M. KIRK, *Plant Cell Physiol.*, (spec. issue) (1963) 493.
10. R.K. TOGASKI and M. GIBBS, *Plant Physiol.*, 42 (1967) 991.
11. D. HOGETSU and S. MIYACHI, *Plant Physiol.*, 45 (1970) 178.
12. A.T. WILSON and M. CALVIN, *J. Am. chem. Soc.*, 77 (1955) 4948.
13. E. LATZKO and M. GIBBS, in Methoden der Enzymatischen Analyze, ed. H.U. BERGMEYER, Verlag Chemie, (1970) 1348.
14. J.R. WILLIAMSON and B.E. CORKEY, in Methods of Enzymology, ed. J.M. LOWENSTEIN, Academic Press, New York, XIII (1969) 435.
15. E. LATZKO and M. GIBBS, *Plant Physiol.*, 44 (1969) 396.
16. J.W. BRADBEER, S.L. RANSON, and M. STILLER, *Plant Physiol.*, 33 (1958) 66.

A COMPARATIVE STUDY OF THE ACTIVITIES OF PHOTOSYNTHETIC CARBOXYLATION IN A C_4 AND A CALVIN-TYPE PLANT (THE SITES OF CO_2 FIXATION IN C_4 PLANTS)

Jack Farineau[1]

Département de Biologie, Centre d'Etudes Nucléaires de Saclay,
B.P. n°2 -91-GIF-sur-Yvette, France.

SUMMARY

In corn leaves, the usual CO_2 substrate used for the operation of the Calvin cycle is an endogenous CO_2 reservoir fed by decarboxylation of C_4 acids and by respiration or photorespiration in an oxygenated atmosphere. Changes in the pool sizes of 3-phosphoglycerate and ribulose-1,5-diphosphate observed after removal of CO_2 in the light, demonstrate the operation of the Calvin cycle in corn leaf. Photosynthesis is proposed to occur with the participation of bundle sheath and mesophyll cells. The possibility of a translocation of phosphorylated intermediates between these cells is suggested. The changes in pool sizes of phosphoenolpyruvate and aspartate during the same transient conditions support the involvement of these compounds as substrate and product for the operation of a β-carboxylation pathway in mesophyll cells. The different pattern of labeling of C_4 acids and Calvin cycle compounds from $^{14}CO_2$ in pulse-chase experiments is explained by variations of the specific radioactivity of the CO_2 pool during pretreatments. This pool, fed by C_4 acids is the usual source of CO_2 for the operation of the Calvin cycle in corn.

INTRODUCTION

Studies of $^{14}CO_2$ incorporation by C_4 plants, in short term experiments, have led to the discovery of a β-carboxylation pathway operating in steady-state photosynthesis. Hence, it has seemed doubtful that the Calvin cycle pathway was taking place in these plants (1). However, recent results of pulse-chase type experiments performed with C_4 plants, showing a delayed labeling of the Calvin

---

[1] *Paper read to seminar by M.D. Hatch.*

cycle compounds (2), and the identification of high ribulose-1,5-di-
phosphate (RuDP) carboxylase activity in bundle sheath cells (3)
suggested that C_4 plants behave as Calvin type plants, except that
C_4 acids are the first labeled products of the $^{14}CO_2$ fixation in mesophyll cells. The problem of the entry of label into Calvin compounds can be resolved if it is assumed that C_4 acids are decarboxylated to give CO_2 for the Calvin cycle carboxylation reaction.

For our part, in the course of a study of photophosphorylation in
corn leaf, we have shown that the metabolism of phosphorylated intermediates of the pentose cycle seemed to be very similar to that
observed in Calvin type plants (4).

This paper reports data obtained from a study of substrates and
products of the two carboxylation processes [phosphoenolpyruvate
(PEP) carboxylase and RuDP carboxylase] in leaves of plants with
(corn), or without (*Phragmites*) the β-carboxylation pathway. Two
types of experiments were done; determination of the pool-sizes of
products and substrates of carboxylation reactions during various
transients and $^{14}CO_2$ incorporation *in vivo* by leaves placed in non-
physiological conditions: atmosphere of N_2 in the light or in the
dark. An interpretation relative to the localization of the sites of
carboxylation in leaves is given in conclusion.

EXPERIMENTAL

Corn, *Zea mays* var. INRA 260 (a C_4 plant), and *Phragmites commu-
nis* (a Calvin type plant) were grown on soil in a glasshouse. Experiments were performed in a thermostated reaction chamber with
small pieces (200 mg) of fully expanded leaves. The circulating gaseous medium in the chamber always lacked oxygen to suppress respiration and photorespiration. Photosynthesis was measured in a mixture
of N_2 with 0.5% CO_2 using monochromatic light of 650 nm ($\Delta\lambda$ = 10 nm;
50,000 ergs cm^{-2} sec^{-1}). The CO_2 incorporation in the reported data
is 80 μmoles mg^{-1} chlorophyll hr^{-1} for corn, 30-40 μmoles mg^{-1} chlorophyll hr^{-1} for *Phragmites*.

Pool sizes of phosphorylated intermediates in leaves were measured at isotopic equilibrium with ^{32}P according to a technique described earlier(4). Aspartate, malate, and 3-phosphoglycerate (3-PGA)
were separated by high voltage paper electrophoresis and the concentrated products determined with classical enzymic methods. Short
term $^{14}CO_2$ experiments were done using a technique similar to that
described by GALMICHE (5). After a period of photosynthesis in the
above conditions, CO_2 was removed and in some experiments the tissue
was held in the dark. At this stage a pulse of 80 μCi $^{14}CO_2$ was injected in the circulating gas and flushed out from the reaction
chamber with either CO_2+N_2 or with N_2. The labeled CO_2 remained 2
seconds in the reaction chamber, and the chase lasted 4 seconds. The
trap door was then opened and the leaf dropped into liquid N_2. It
was then ground, compounds being separated by paper electrophoresis
(5) and two dimensional separation by electrophoresis and chromato-

graphy was made using small aliquots to measure the radioactivity of individual compounds (4).

RESULTS AND DISCUSSION

Changes in pool sizes during transients

(a) Calvin cycle intermediates

When CO_2 was removed during illumination the changes in RuDP and 3-PGA pools were similar in leaves of corn and *Phragmites (Figure 1a,b)*. The RuDP pool (carbon acceptor) rose rapidly and that of

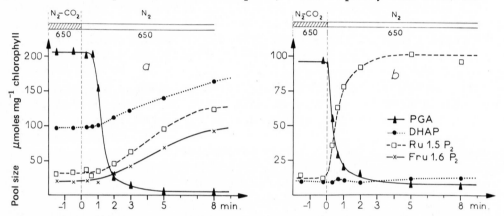

Figure 1 *Changes in pool sizes of some phosphorylated inter-mediates of the Calvin cycle in the light following removal of CO_2. (a) corn (b) Phragmites*

3-PGA declined sharply, this reciprocal relationship proving the involvement of a Calvin cycle carboxylation in both plants (6). When tissues were again exposed to CO_2, but in the dark, the RuDP pool declined and that of 3-PGA increased, in both species *(Figure 2a, 3a)*.

However there are also several important differences between species in the behavior of these phosphorylated Calvin cycle intermediates. Firstly, in *Phragmites* the changes are immediate, whereas in corn, changes in the pool size of 3-PGA, RuDP, dihydroxyacetone phosphate (DHAP), and fructose-1,6-diphosphate (FDP) are only seen 30-50 seconds after removal of CO_2. It seems clear that during this period corn leaves make use of an endogenous CO_2 source sufficient to allow quite normal photosynthesis (see further evidence below). Secondly, the pool sizes of DHAP and FDP are large in corn leaves, compared to *Phragmites (Figure 1)*, but still rise after removal of CO_2 and decrease again only after 15-30 min in CO_2 free N_2. Accumulation of these intermediates seems a logical consequence of the accumulation of RuDP. Finally, when corn leaves are transfered to the dark in the absence of CO_2, there is a sharp rise in 3-PGA level *(Figure 2b)*. In leaf of *Phragmites* the same treatment causes only a

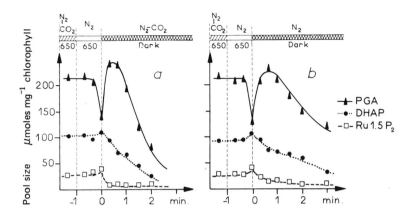

Figure 2 Changes in the pool sizes of some phosphorylated in-termediates of the Calvin cycle in corn leaves following re-moval of CO_2 in the light and subsequent transfer to the dark, (a) with or (b) without CO_2.

small increase in 3-PGA level, *(Figure 3b)* whereas if CO_2 is present in the dark phase, much 3-PGA is produced *(Figure 3a)*. These data suggest that the endogenous CO_2 supply in *Phragmites* is comparative-ly low and readily exhausted.

(b) β-carboxylation substrates and products

In corn, removal of CO_2 in the light, causes an immediate in-crease in the PEP pool and a sharp decrease in the aspartate pool *(Figure 4a)*. This result confirms that PEP and aspartate are related in a loose precursor-product relationship. Indeed PEP, oxalacetate, aspartate, and malate are related in the following sequence of re-actions (see also *Figure 5*).

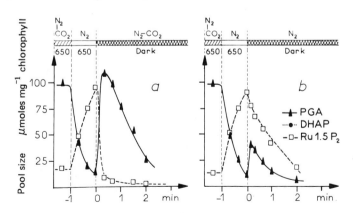

Figure 3 Changes in the pool sizes of some phosphorylated in-termediates of the Calvin cycle in Phragmites leaves. Condi-tions as in Figure 2.

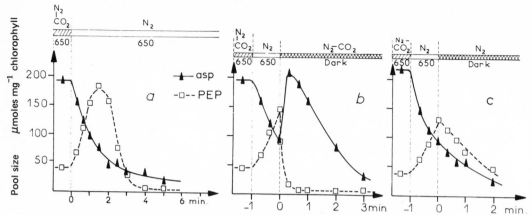

Figure 4 *Changes in the pool sizes of two compounds of the β-carboxylation pathway, following various transients.*

$$PEP + CO_2 \rightarrow oxaloacetate \rightarrow malate \rightarrow pyruvate + CO_2$$

$$\downarrow\uparrow \leftarrow NH_2$$

aspartate

As aspartate is in equilibrium with oxaloacetate, we suppose that the aspartate disappearance may be related to a synthesis of malate and then, decarboxylation of this last compound to give CO_2 and pyruvate.

A transitory increase of the aspartate level in the dark is observed only if CO_2 is present in the atmosphere *(Figure 4b,c)*. It is concluded that *atmospheric* CO_2 is the usual source of CO_2 for the β-carboxylation reaction. In the O_2 free conditions employed it is likely that Krebs cycle activity is reduced and does not contribute to malate or aspartate synthesis.

The changes in pool sizes of the two CO_2 acceptors, RuDP and PEP, during the zero CO_2 treatment in the light are quite different. RuDP level in corn rose slowly over a period of many minutes *(Figure 1a)* whereas that of PEP passes through a maximum within 2-3 min, then falls rapidly to a low level *(Figure 4a)*. This implies a block in pyruvate supply, possibly involving complex interactions between amino acids or keto acids.

No significant changes were observed in the malate pool of corn leaves during a 5 min transient of zero CO_2 in the light (not shown). Subsequently a small change of 100-200 nmoles mg^{-1} chlorophyll is observed after 10 min. This may correspond to the small fraction of the total malate pool (1-3 μmoles mg^{-1} chlorophyll) which turns over.

In *Phragmites* no changes in the aspartate or PEP pools are observed during transients, consistent with the absence of the β-carboxylation pathway in this species. Moreover, the PEP pool is very low (2-10 nmoles mg^{-1} chlorophyll).

Short term $^{14}CO_2$ fixation experiments in corn

In these experiments corn leaf tissue was exposed to $^{14}CO_2$ for 2 sec, followed by a 4 sec chase. Before exposure to $^{14}CO_2$, leaf tissue was allowed to photosynthesize under a range of conditions which, in the above experiments, have been shown to modify the pool sizes of photosynthetic intermediates. *Table 1* shows that in the

TABLE I

Effects of pretreatments which modify pool size of intermediates on the distribution of ^{14}C following 2 sec $^{14}CO_2$ fixation, 4 sec chase in the light. Total ^{14}C fixed shown as dpm $\times 10^{-6}$.

Treatment	Total fixed	C_4 acids % total	Calvin cycle % total	% Distribution within Calvin cycle	
#1 control $N_2 + CO_2$	1.80	79.0	21.0	3-PGA	92
#2 25 sec N_2 light	3.74	75.5	24.5	3-PGA sug DP*	88 6
#3 55 sec N_2 light	6.29	58.5	41.5	3-PGA sug DP DHAP	70 13 10
#4 180 sec N_2 light	4.57	51.5	48.5	3-PGA sug DP DHAP	48 29 15
#5 55 sec N_2 light 5 sec dark	6.42	98.0	2.0	3-PGA	96

* sug DP = total sugar diphosphates

control treatment in $N_2 + CO_2$ 20% of total label was transfered to 3-PGA in the standardized $^{14}CO_2$ pulse-chase. In #2, *Table 1*, 25 sec after removal of CO_2, there was no change in the distribution of ^{14}C amongst products of the pulse-chase. This is consistent with the data of *Figure 1a* which show no change in pool size of metabolites for about 50 sec and imply continuing photosynthetic fixation from an endogenous CO_2 source. In #3 and #4, 55 and 180 sec after removal of CO_2 there was a large increase in the proportion of ^{14}C moved to Calvin cycle intermediates in the pulse-chase, and a spread of label to compounds other than 3-PGA. Further, the total ^{14}C fixed in the pulse-chase declined after 55 sec pretreatment in N_2.

These data may be explained in terms of the following. The C_4

acids are labeled only during the 2 sec pulse but Calvin cycle products may be labeled during the whole 6 sec period for decarboxylation continues throughout. Kinetically speaking, there is a precursor-product relationship between C_4 acids and Calvin cycle products,
being separated by an intermediate CO_2 pool. *Figure 1a* shows that
the pools of Calvin cycle intermediates (RuDP, 3-PGA, DHAP, and FDP)
do not change significantly during the first 60 sec in a CO_2 free
atmosphere. This suggests that more rapid labeling of Calvin cycle
products in the above pulse-chase experiments is due to a higher
specific activity of the endogenous CO_2 pool. Two facts support this
conclusion. Firstly, carboxylation lowers the endogenous CO_2 pool
during the initial 60 sec in CO_2 free atmosphere. Secondly, the specific activity of the C_4 acids increases after transfer to a CO_2
free atmosphere *(Figure 4)* whereas the total ^{14}C fixed by these compounds increases *(Table 1 #2,3)*.

In #5, *Table 1*, 5 sec dark after 55 sec in CO_2 free atmosphere is
sufficient to block the transfer of ^{14}C to Calvin cycle intermediates. This block occurs without effecting the rate of [14]CO_2 fixation.
Figure 2b shows that 3-PGA synthesis continues when leaves are
transfered from light to dark in the absence of CO_2 i.e. RuDP carboxylase is active in the dark. The low labeling of 3-PGA in #5, *Table 1* means that the 3-PGA is synthesized in cells remote from atmospheric CO_2. Perhaps the labeling of the endogenous CO_2 pool is prevented in the dark, probably by a reduction in the rate of decarboxylation of C_4 acids.

CONCLUSIONS

The results of experiments with corn leaves demonstrate that an
endogenous CO_2 source, fed by carboxylation of C_4 acids plays a key
role in the operation of the Calvin cycle in leaves of C_4 plants.
Such a CO_2 source may be released in the atmosphere as a CO_2 burst,
at the onset of illumination (9). It can be used in a N_2 atmosphere
and light to give rise to an oxygen evolution (10). The results of
short term photosynthesis experiments with [14]CO_2 are explained by
the fact that atmospheric CO_2 is only passed to Calvin cycle chloroplasts after dilution in C_4 acid pools by synthesis and, perhaps, by
exchange reactions. (Unpublished experiments suggest the possibility
of exchange reactions between ^{14}C oxaloacetate and aspartate in
corn).

Figure 5 shows a model which states precisely the sites of operation of the Calvin cycle in corn leaf. We assume that most Calvin
cycle carboxylation occurs in bundle sheath cells which contain
75-80% of leaf RuDP carboxylase (3). Agranal chloroplasts of bundle
sheath cells of corn have a negligible reducing activity (11), so it
is necessary to suppose that the NADPH required for the operation of
the Calvin cycle is supplied by malate decarboxylation, catalyzed by
malic enzyme, which is abundant in bundle sheath cells (9). Nevertheless, reducing capacity is a limiting factor in those cells, as

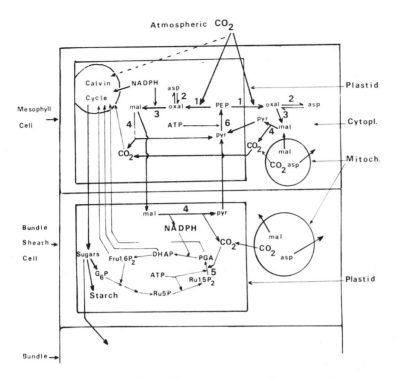

Figure 5 The sites of CO_2 fixation in the corn leaf. Dashed lines refer to pathways of secondary importance. The numbered reactions are catalyzed by, 1 - PEP carboxylase, 2 - aspartate aminotransferase, 3 - NADP specific malic dehydrogenase (16), 4 - malic enzyme, 5 - RuDP carboxylase, 6 - pyruvate,Pi diki-nase (7).

the carboxylation of one malate gives only one NADPH for two PGA resulting from the incorporation of one CO_2. A fraction of the PGA pool must escape by diffusion through the plastid membrane, and be translocated to mesophyll cells, where its reduction occurs. The PGA may be accompanied by DHAP and FDP, but transfer in opposite direction, i.e. from mesophyll to bundle sheath cells of these two compounds produced by the operation of the Calvin cycle in mesophyll cells, is also possible. The translocations of PGA, DHAP, and FDP between the two categories of leaf cells may account for the large pool of these compounds observed in corn and *Sorghum* (results not shown) during photosynthesis or after transfer to CO_2 free atmosphere. It is tempting to make an analogy that chloroplasts of bundle sheath and mesophyll cells would have the same behavior as spinach chloroplasts performing CO_2 fixation in *in vitro* experiments, with an important loss of the same diffusible Calvin intermediates (12). The decarboxylation of malate in bundle sheath cells yields CO_2 (stored as part of the endogenous CO_2 pool) and pyruvate which is translocated towards mesophyll cells. It is possible, that pyruvate is aminated to alanine, which is abundant in corn leaves (J.P. MICHEL, unpublished data), and which might be translocated in place of pyruvate (13).

The model supposes considerable movements of various compounds in opposite directions, between mesophyll and bundle sheath cells, but anatomical characteristics of those cells make the translocation possible (14). Most of our experiments are consistent with this model. Recent studies with isolated bundle sheath and mesophyll cells support these concepts (15).

REFERENCES

1. H.P. KORTSCHAK, C.E. HARTT, and G.O. BURR, *Plant Physiol.*, 40 (1965) 209.
2. H.S. JOHNSON and M.D. HATCH, *Biochem. J.*, 114 (1969) 127.
3. O. BJÖRKMAN and E. GAUHL, *Planta*, 88 (1969) 197.
4. J. FARINEAU, *Planta*, 85 (1969) 135.
5. J.M. GALMICHE, in La Photosynthèse, ed. du C.N.R.S., Coll. Int. du C.N.R.S. (1963) 589.
6. A.T. WILSON and M. CALVIN, *J. Am. chem. Soc.*, 77 (1955) 5948.
7. M.D. HATCH and C.R. SLACK, *A. Rev. Pl. Physiol.*, 21 (1970) 141.
8. J.M. GALMICHE, in Proceedings of International Symposium on Photosynthesis in Sugarcane, ed. J. COOMBS, Imperial College, University of London, and Tate and Lyle Ltd., (1968) 36.
9. J.A. BERRY, W.J.S. DOWNTON, and E.B. TREGUNNA, *Can. J. Bot.*, 48 (1970) 777.
10. P. GUERIN de MONTGAREUIL, M. ANDRÉ, and N. SEIMANDI, *C. R. Acad. Sci. (Paris)*, 265 (1967) 485.
11. K.C. WOO, Jan M. ANDERSON, N.K. BOARDMAN, W.J.S. DOWNTON, C.B. OSMOND, and S.W. THORNE, *Proc. natn. Acad. Sci. U.S.A.*, 67 (1970) 18.
12. J.A. BASSHAM, M. KIRK, and R.G. JENSEN, *Biochim. biophys. Acta*, 153 (1968) 211.
13. Y.S. KARPILOV, The Photosynthesis of Xerophytes. Academy of Sciences of the Moldavian S.S.R. Kishnev (1970).
14. W.M. LAETSCH, *Sci. Prog. Oxf.*, 57 (1969) 323.
15. G.E. EDWARDS, S.S. LEE, T.M. CHEN, and C.C. BLACK, *Biochem. biophys. Res. Commun.*, 39 (1970) 389.
16. M.D. HATCH and C.R. SLACK, *Biochem. biophys. Res. Commun.*, 34 (1969) 589.

LIGHT-ENHANCED DARK CARBON DIOXIDE FIXATION IN MAIZE LEAVES[1]

Muneaki Samejima and Shigetoh Miyachi

Faculty of Agriculture, and Institute of Applied Microbiology,
University of Tokyo, Tokyo, Japan

SUMMARY

Preillumination of maize leaves in the absence of CO_2 greatly enhanced the capacity for fixing $^{14}CO_2$ into malate and aspartate in the subsequent dark period. Practically no radioactivity was incorporated in phosphate esters and free sugars. The light-enhanced dark $^{14}CO_2$ fixation lasted for about 2 minutes and the levels of radioactivity in the C_4 dicarboxylic acids remained constant thereafter. The radioactivity in the C_4 acids decreased rapidly when $^{14}CO_2$ was removed from the atmosphere and the light was turned on again. Concomitant with the decrease in radioactivity in the C_4 acids, that in glycerate and phosphate esters rose rapidly, indicating that the transfer of the radioactivity from the C_4 acids to 3-phosphoglyceric acid is strictly light-dependent. From the studies on the effects of inhibitors on the light-dependent transfer, it was inferred that the transfer occurs by a transcarboxylation reaction.

INTRODUCTION

BENSON and CALVIN (1) discovered that the capacity of dark CO_2 fixation was markedly increased when *Chlorella* cells were illuminated before provision of CO_2 in the dark. The occurrence of such light-enhanced dark CO_2 fixation was found to be a property common to all classes of algae (2), as well as the leaves of higher plants (3). Chloroplasts isolated from spinach and soybean leaves have also been shown to manifest the same reaction (3).

In a previous study, HOGETSU and MIYACHI (4) analyzed products of light-enhanced dark $^{14}CO_2$ fixation in *Chlorella* cells. The initial

[1] This research was supported by a grant from the Ministry of Education (No. 84107 given to S.M.). We are grateful to Drs. E. Hase and A. Takamiya for suggestions made in the preparation of the manuscript.

fixation product was mainly 3-phosphoglyceric acid (3-PGA) and the percentage distribution of radioactivity incorporated in this initial product rapidly decreased in the following dark period. It was concluded that the initial ^{14}C incorporation into 3-PGA was due to carboxylation of ribulose-1,5-diphosphate (RuDP) which accumulated in *Chlorella* cells during preillumination.

It is known that in C_4 plants, CO_2 fixed appears initially in C_4 dicarboxylic acids such as malate, aspartate, and oxaloacetate (OAA), and the C-4 of these dicarboxylic acids is then transferred to C-1 of 3-PGA (5). There have been at least two proposals for the mechanism of the transfer reaction (6). One assumes that C-4 of the dicarboxylic acid, possibly OAA, is transferred by a transcarboxylation reaction to an acceptor to provide C-1 of 3-PGA. The other assumes that the dicarboxylic acids may move from the mesophyll cells to bundle sheath chloroplasts and there undergo decarboxylation by malic enzyme. The CO_2 so released would then be refixed by RuDP carboxylase.

In order to study the fate of the C-4 of the dicarboxylic acids, we extended our research on light-enhanced dark CO_2 fixation to preilluminated leaves of C_4 plants. The results obtained with maize leaves are described in this paper.

EXPERIMENTAL

Plant material

The expanding leaves of maize (*Zea mays*, Nagano No. 1) at the ages of 15 through 50 days after germination were used for experiments.

Labeling experiments

Rectangular pyrex tubes (volume, 90 ml) connected to a diaphragm pump with vinyl tubes were used as $^{14}CO_2$ fixation chamber. The light-enhanced dark $^{14}CO_2$ fixation was carried out under aerobic conditions according to the methods described elsewhere (3).

The effects of inhibitors on the transfer of radioactivity in C_4 dicarboxylic acids fixed in the light-enhanced dark fixation to other compounds were studied, using three leaf discs placed in a vessel of a Warburg manometer. After 5 minutes of the light-enhanced dark $^{14}CO_2$ fixation, $^{14}CO_2$ was removed from the vessel by evacuation and the leaf discs were infiltrated with inhibitor solution under continued darkness. The leaf discs were then illuminated again. After 5 minutes of illumination, boiling 80% ethanol (v/v) was poured into the vessel and the radioactive compounds were extracted (3). 3,4-dichlorophenyl-N,N-dimethyl urea (DCMU, 2.2 mM) was dissolved in 20% methanol (v/v). Iodoacetamide (IAM) and p-chloromercuribenzoate (PCMB) were prepared as described (7). Tricine buffer, pH 9.0 was added to the PCMB solution before the start of experiment. As a control to each inhibitor experiment, leaf discs were infiltrated with 30% methanol, tricine buffer, or water.

Unless otherwise mentioned, the light intensity applied was

30,000 lux at the leaf surface and the temperature was 23-25°. Time for preillumination was 20 minutes. Distribution of radioactivity was determined by 2-dimensional paper chromatography (8). Paper co-chromatography of serine and glycine, and of malic acid and glyceric acid were also carried out with acetone-diethylamine-water (16:1:3, v/v) and ethanol-19 N ammonia-water (8:1:1, v/v), respectively.

Enzyme assays

Enzyme activities were determined with leaf extracts treated on a column of Sephadex G-25. RuDP carboxylase in the extract was determined by the amount of $H^{14}CO_3^-$ fixed into 3-PGA. Phosphoenolpyruvate (PEP) carboxylase was determined by the procedures as described (9). Malic enzyme was determined in the direction of decarboxylation by following the absorbancy change at 340 nm (10).

RESULTS AND DISCUSSION

Figure 1 shows that dark incorporation of ^{14}C by preilluminated leaves proceeded rapidly at first, then slowed down following almost the same kinetics as previously observed in *Chlorella* cells and isolated spinach chloroplasts(3,4). Usually, the rapid incorporation lasted for 1-2 minutes after turning off the light.

Figure 2 shows that in maize leaves the radioactivity was mostly incorporated into malate and aspartate. This is in contrast to the case of *Chlorella* cells in which the main initial $^{14}CO_2$ fixation product was 3-PGA. It was assumed that the enhanced dark $^{14}CO_2$ fixation was due to carboxylation of PEP which had accumulated in the leaves during preillumination in the absence of CO_2. The OAA thus

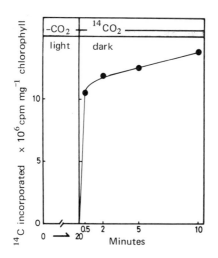 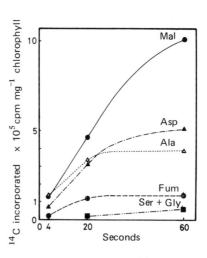

Figure 1 (left) Time course of light-enhanced dark $^{14}CO_2$ fixation subsequent to preillumination in CO_2 free air.

Figure 2 (right) Time courses of ^{14}C incorporation into products formed during the light-enhanced dark $^{14}CO_2$ fixation.

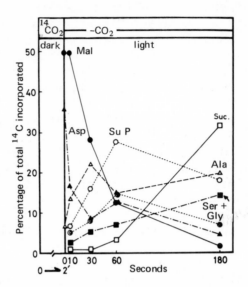

Figure 3 Effects of the second illumination on the fate of [14]*C products of the light-enhanced dark* [14]*CO₂ fixation.*

formed would be converted to malate or aspartate. Some radioactivity was found in alanine, fumaric acid, and also serine and glycine but practically no radioactivity was detected in phosphate esters and free sugars. It should be noted that neither glycine nor serine was detected as the product of light-enhanced dark CO_2 fixation in *Chlorella* cells (4) and the leaves of spinach and soybean (3), thereby indicating that the formation of glycine and serine is light-dependent in these C_3 plants. Probably the pathway for the formation of glycine and serine in C_4 plants is different from that in C_3 plants (11).

When maize leaf segments which had fixed $^{14}CO_2$ in the dark after preillumination were transferred to CO_2 free air and then the light was turned on again, radioactivity in aspartate and malate decreased rapidly as shown in *Figure 3*. In repeated experiments it was confirmed that the decrease of radioactivity in aspartate was always faster than that in malate. The radioactivity in glycerate and P-esters rose and then fell. On the other hand, the radioactivity in sucrose continued to increase after a lag period. The ^{14}C-glycerate would have been derived from labeled 3-PGA. These results indicate that, upon reillumination, radioactivity which had been incorporated into C_4 dicarboxylic acids was quickly transferred to sucrose via 3-PGA.

Although the data are not shown here, further incubation of the leaves in CO_2 free air in darkness did not affect the levels of radioactivity in the compounds which had been incorporated in the light-enhanced dark fixation. These results confirmed the earlier observation by HATCH and SLACK (5) that the transfer of C-4 of the dicarboxylic acid to 3-PGA is strictly light-dependent. Results

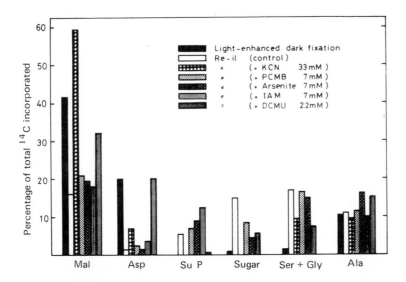

Figure 4 Effects of inhibitors on the light-dependent transfer of radioactivity. Duration of the preillumination, 15 minutes. The percentage distribution of radioactivity fixed was determined after 5 minutes of light-enhanced dark $^{14}CO_2$ fixation (black bars) and after 5 minutes of the second illumination in the presence of inhibitors (other bars).

which are not reported here showed that a significant transfer of radioactivity from C_4 dicarboxylic acids to P-esters occurred on illumination of the leaves at light intensities higher than 3,000 lux. When the light intensity for the second illumination was 500 lux, ^{14}C-malate decreased while ^{14}C-aspartate increased, but only a small amount of radioactivity appeared in P-esters.

The effect of inhibitors on the light-dependent transfer of the radioactivity in C_4 dicarboxylic acids to 3-PGA was studied. The black bars in *Figure 4* show the percentage distribution of radioactivity fixed into various products during light-enhanced dark $^{14}CO_2$ fixation. The other bars show the percentage distribution of radioactivity after the second illumination in the presence or absence of inhibitors. The figure shows that the light-dependent transfer of radioactivity was inhibited by KCN but not by SH-inhibitors such as PCMB. In the presence of KCN, the radioactivity in aspartate decreased and that in malate increased upon the second illumination, but practically no radioactivity in C_4 dicarboxylic acids was transferred to P-esters.

Table 1 shows that both KCN (5×10^{-3} M) and PCMB ($1-10 \times 10^{-4}$ M) almost completely inhibited RuDP carboxylase and malic enzyme extracted from maize leaves. The observation that PCMB (7×10^{-3} M) did not affect the light-dependent transfer of radioactivity, therefore, is not consistent with the view that these enzymes participate in the transfer reaction.

Figure 4 further shows that DCMU inhibited the light-dependent

TABLE I

Effect of inhibitors on enzyme activities extracted from maize leaves (percentage inhibition).

Inhibitors		RuDP carboxylase	PEP carboxylase	Malic enzyme
KCN	5×10^{-3} M	82.4	0.0	79.6
	5×10^{-4}	27.1	0.0	36.3
PCMB	1×10^{-3}	96.3	98.9	86.0
	1×10^{-4}	96.3	98.2	15.3
IAM	1×10^{-3}	27.6	5.5	
	1×10^{-4}	7.7	0.0	

transfer of ^{14}C. We therefore infer that the transfer requires light energy provided through a functional photosynthetic machinery.

According to the view that the transfer occurs through decarboxylation and the subsequent refixation of the released CO_2 (see Introduction), one may expect the incorporation of CO_2 into 3-PGA as the first step in C_4 photosynthesis if CO_2 could be carried directly into the bundle sheath chloroplasts. In order to fill the vascular bundle with $^{14}CO_2$, the leaf discs which had been kept in darkness were vacuum-infiltrated with a high concentration of ^{14}C-bicarbonate solution (70 mM) under continued darkness and then the light was turned on. It was found that after 10 seconds of illumination, 80% of radioactivity resided in aspartate and malate but practically no radioactivity was detected in 3-PGA.

These results seem to indicate that the transfer of the ^{14}C of the C_4 dicarboxylic acids to 3-PGA is mediated by a light-dependent transcarboxylation reaction rather than decarboxylation followed by refixation of the released CO_2. There is also a possibility that the movement of C_4 dicarboxylic acids from mesophyll to bundle sheath cells is light-dependent. In this connection, a study on the distribution of radioactivity between bundle sheath and mesophyll cells in maize leaves which fixed $^{14}CO_2$ in the dark after preillumination is being carried out in this laboratory.

REFERENCES

1. A.A. BENSON and M. CALVIN, *Science*, 105 (1947) 648.
2. R.K. TOGASAKI and M. GIBBS, *Plant Physiol.*, 42 (1967) 991.
3. S. MIYACHI and D. HOGETSU, *Plant Cell Physiol.*, 11 (1970) 927.
4. D. HOGETSU and S. MIYACHI, *Plant Physiol.*, 45 (1970) 178.
5. M.D. HATCH and C.R. SLACK, *Biochem. J.*, 101 (1966) 103.
6. M.D. HATCH and C.R. SLACK, *A. Rev. Pl. Physiol.*, 21 (1970) 141.

7. S. MIYACHI, *Plant Cell Physiol.*, 1 (1960) 117.
8. N. OGASAWARA and S. MIYACHI, *Plant Cell Physiol.*, 11 (1970) 1.
9. N. OGASAWARA and S. MIYACHI, *Plant Cell Physiol.*, 11 (1970) 411.
10. R.Y. HSU and H.A. LARDY, Methods in Enzymology, ed. J.M. LOWEN-STEIN, Academic Press, New York, 13 (1969) 230.
11. N.E. TOLBERT and R.K. YAMAZAKI, *Ann. N.Y. Acad. Sci.*, 168 (1969) 325.

TRANSFER OF CARBON, PHOSPHATE ENERGY, AND REDUCING EQUIVALENTS ACROSS THE CHLOROPLAST ENVELOPE

U. Heber and G.H. Krause

Botanisches Institut der Universität Düsseldorf,
4 Düsseldorf, Germany

SUMMARY

Penetration of oxaloacetate and malate across the envelope of intact spinach chloroplasts is demonstrated by photoreduction of added oxaloacetate. This reaction was found to be catalyzed by chloroplastic NAD-specific, not NADP-specific, malate dehydrogenase.

Pyruvate, phosphoenolpyruvate, and glycolate are able to penetrate the chloroplast envelope as shown by photo-induced chloroplast shrinkage reactions that depend on addition of those compounds.

Indirect transfer of phosphate energy via the triose phosphate oxidation system and the indirect transfer of reducing equivalents across the chloroplast envelope via a malate/oxaloacetate shuttle were studied in vitro. The latter links intra- and extraplastid NAD pools. During photosynthesis in green plant cells both systems may cooperate in the effective transfer of phosphate energy from the chloroplasts to the cytoplasm.

Direct measurements of chloroplastic NAD(P) reduction as well as the photoreduction of oxaloacetate suggest that not only NADP, but also NAD become reduced to a substantial extent upon illumination.

A proton gradient from the chloroplast stroma to the outside medium seems to exist in the light. In the presence of a functioning malate/oxaloacetate shuttle system this gradient would maintain a higher NADH/NAD ratio inside the chloroplasts than outside.

INTRODUCTION

In C_4 photosynthesis a cooperation of different cells containing two types of chloroplasts has been suggested by SLACK *et al.* (1). This requires the efficient transfer of various intermediary products and cofactors of photosynthesis. Although the transport phenomena reported in the present paper were studied with spinach chloroplasts they may prove to be relevant also for plants of the C_4

pathway. The rapid transfer of 3-phosphoglycerate (3-PGA), dihydroxy-acetone phosphate (DHAP), oxaloacetate (OAA), and malate, as well as of aspartate, glutamate, and α-ketoglutarate across the envelope of spinach chloroplasts is well established by both direct and indirect evidence. The addition of 3-PGA to intact, but not to broken chloroplasts in the light initiates immediate O_2 evolution corresponding to the reduction of up to 300 μmoles 3-PGA mg^{-1} hr^{-1} chlorophyll (2). The apparent K_m for 3-PGA (0.2 mM) during 3-PGA reduction by intact chloroplasts is similar to that of the phosphoglycerate kinase-glyceraldehyde phosphate dehydrogenase system, indicating that 3-PGA and the main product of its reduction, DHAP, do not encounter a significant barrier at the chloroplast envelope. However, other phosphorylated intermediates of photosynthesis such as ribulose-1,5-diphosphate (RuDP) and fructose-1,6-diphosphate (FDP), are effectively retained within the chloroplasts. Recent work of HELDT and RAPLEY (3) has shown that two highly effective transfer systems are involved in the transport of 3-PGA and DHAP, and OAA and malate, respectively. In the present study the OAA-malate transfer is also demonstrated by the ability of isolated intact chloroplasts to photoreduce added OAA. Furthermore, evidence is presented for the transfer of pyruvate and phosphoenolpyruvate (PEP) across the chloroplast envelope. Since no effective systems seem to exist for the direct transfer of phosphate energy (2,4) and reducing power (5) possible ways of indirect transport of those cofactors were investigated.

EXPERIMENTAL

Intact chloroplasts were isolated from freshly harvested leaves of *Spinacia oleracea* in a modification (6) of JENSEN and BASSHAM'S method (7). Contaminations by envelope-free chloroplasts were determined (2) and ranged in different experiments from 10 to 25%. CO_2 fixation (80-140 μmoles hr^{-1} mg^{-1} chlorophyll), 3-PGA reduction (130-290 μmoles hr^{-1} mg^{-1} chlorophyll), and OAA reduction were measured polarographically with a Clark type electrode.

Malate dehydrogenase activity of chloroplasts was measured in the presence of 0.5 mM OAA as the difference in NADPH oxidation before and after brief ultrasonication of the chloroplast suspension.

Shrinkage experiments were carried out by illuminating intact or osmotically shocked chloroplasts with a broad band of red light (620-760 nm, intensity 1.25 x 10^5 ergs cm^{-2} sec^{-1}) and measuring the light-induced absorbance change at 535 nm (8). Chloroplasts (65 μg chlorophyll ml^{-1}) were suspended in "solution C" of JENSEN and BASSHAM (generally without pyrophosphate and nitrate) and pyruvate, PEP, or glycolate were added after the first cycles of illumination and darkness. Light path was 0.5 cm.

Absorbance changes at 340 nm were recorded similarly with orange light of 4 x 10^4 ergs cm^{-2} sec^{-1}, as the exciting beam. To measure indirect transfer of reducing equivalents via a malate/OAA shuttle chloroplasts (65 µg chlorophyll ml^{-1}) were suspended in "solution C" containing, in addition, 50 mM malate, 0.7 mM NAD, and 20 µg ml^{-1} malate dehydrogenase. NAD/NADH ratios were varied by adding NADH or OAA. The samples were illuminated with red light of 6 x 10^4 ergs cm^{-2} sec^{-1} while the absorbance at 340 nm was recorded.

The proton uptake of osmotically shocked or intact chloroplasts was recorded by following the light-induced pH change in unbuffered "solution C" with a pH meter.

RESULTS AND DISCUSSION

Photoreduction of oxaloacetate

If 1 mM OAA is added to illuminated isolated chloroplasts, O_2 is evolved at rates of about 15 to 20 µmoles hr^{-1} mg^{-1} chlorophyll indicating reduction of 30 to 40 µmoles OAA hr^{-1} mg^{-1} chlorophyll. Uncoupling of the chloroplasts by NH_4^+ ions (20 mM) before illumination increased the OAA reduction up to 150 µmoles hr^{-1} mg^{-1} chlorophyll. The chloroplast preparations contained NAD dependent malate dehydrogenase at high activities (up to 600 µmoles reduced hr^{-1} mg^{-1} chlorophyll), but virtually no NADP specific activity could be detected. The reduction of OAA by intact chloroplasts at rates comparable with those of 3-PGA reduction therefore implies not only rapid penetration of the chloroplast envelope by OAA and malate, but also effective photoreduction of chloroplastic NAD.

Effect of pyruvate and PEP on photo-induced chloroplast shrinkage

The CO_2 acceptor of C_4 plants appears to be PEP which is assumed to be formed from pyruvate. There is the question as to whether these compounds can traverse intracellular membranes. Intact chloroplasts prepared from "starved" spinach leaves (leaves kept after harvesting for 24-48 hours at room temperature in the dark) showed photo-induced shrinkage only if pyruvate or PEP were added before illumination (Figure 1). At 20 mM, a concentration which caused response with intact chloroplasts, no significant photo-induced shrinkage could be seen with shocked, envelope-free chloroplasts. At higher concentrations there was greater shrinkage of intact chloroplasts and some shrinkage by broken chloroplasts was also apparent. As with pyruvate and PEP, shrinkage could also be induced by glycolate. Since shrinkage is considered to be the consequence of the protonation of anions available inside the thylakoid space, the shrinkage of intact chloroplasts induced by these compounds indicates that they not only reach the stroma region but actually penetrate the thylakoid membrane system.

Indirect transfer of phosphate energy and of reducing equivalents

The previously observed linkage between chloroplastic and cytoplasmic adenylate pools and the absence of sufficient direct transfer of ATP and ADP (2,4) gave rise to the question of indirect transfer of phosphate energy between stroma and cytoplasm. The chloroplastic and cytoplasmic triosephosphate oxidation systems, with 3-PGA and DHAP operating as transport metabolites, may be capable of

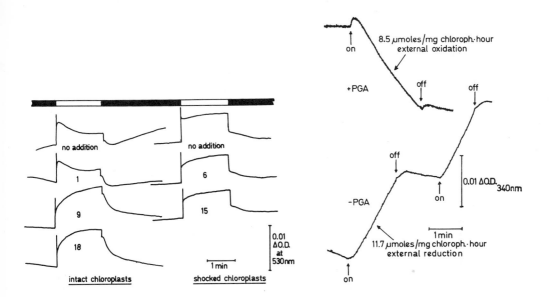

Figure 1 (left) Effect of pyruvate (20 mM) on photo-induced chloroplast shrinkage. Numbers show time (minutes) after pyruvate addition. Slow positive changes caused by light scattering denote chloroplast shrinkage, slow negative changes chloroplast swelling. Fast vertical changes are caused by light absorption.

Figure 2 (right) Transfer of reducing equivalents across the envelope of intact chloroplasts in the presence and absence of 3-PGA (2 mM). The reaction mixture contained in addition to chloroplasts, NAD, malate, and malate dehydrogenase.

effective ATP transfer *in vivo* (2,6). In this case ATP transfer would be coupled stoichiometrically with the transfer of reducing equivalents (9). Experiments with isolated chloroplasts have shown that indirect ATP transfer via the triose phosphate oxidation reaction is controlled by the external phosphate potential (ATP/ADP.Pi) and the redox state of external NAD. In the intact leaf the size of the NADH pool is only about 1% of that of ATP. Significant indirect ATP transfer in the light from chloroplasts to cytoplasm can, for equilibrium reasons, occur only if the NADH formed during cytoplasmic DHAP oxidation is rapidly oxidized. There is good reason to assume that rapid mitochondrial oxidation is not possible because of inhibition of the respiration in the light (10,11). Is there, then, a back transfer of excessive reducing equivalents into the chloroplasts where rapid oxidation could take place during CO_2 reduction?

  If chloroplast suspensions are illuminated in presence of externally added malate, NAD, and malate dehydrogenase the absorbance at 340 nm increases steadily indicating reduction of the added NAD (*Figure 2*). Since the chloroplast envelope has a very low permeability to pyridine nucleotides we interpret the light-induced reduction of

TABLE I

Reduction of external NAD (μmoles hr^{-1} mg^{-1} chlorophyll) by
isolated chloroplasts via the malate-oxaloacetate shuttle system in the absence of 3-PGA.

% external NAD reduced at start of experiment	External ratio NAD/NADH	Reduction rate
4	22	4.3
11	8	3.1
18	4.5	1.1
25	3	0
50	1	0

external NAD as being caused indirectly by a shuttle system involving malate and OAA. OAA is photo reduced inside the chloroplast to
malate which enters the extra-chloroplastic space disturbing the
equilibrium between pyridine nucleotide, malate, and OAA. It will
reduce external NAD thereby forming OAA which can again be reduced
in the chloroplasts to malate. Direction and rate of hydrogen transfer are controlled by the external ratio of NAD to NADH (*Table 1*).
3-PGA added to the system functions as an electron sink in the light
if the external NAD/NADH ratio is kept below a certain limit. Instead of reduction, an oxidation of external NADH now takes place
(*Figure 2*). At NAD/NADH ratios of about 20 to 50 (depending on the
individual chloroplast material) the oxidation stopped and was reversed to reduction at higher rates (*Table 2*).

These observations demonstrate that during photosynthesis, depending on the source-sink relationship, reducing equivalents are
transferred either from the chloroplasts to the outside medium or
vice versa. Under our experimental conditions the maximum transfer
rates in both directions (about 10 μmoles hr^{-1} mg^{-1}) seemed to be

TABLE 2

Oxidation of external NADH (μmoles hr^{-1} mg^{-1} chlorophyll) by
isolated chloroplasts via the malate-oxaloacetate shuttle system in the presence of 3-PGA.

% external NAD reduced at start of experiment	External ratio NAD/NADH	Oxidation rate
4.1	23	4.7
2.8	35	1
1.8	54	0
1.1	87	-0.4

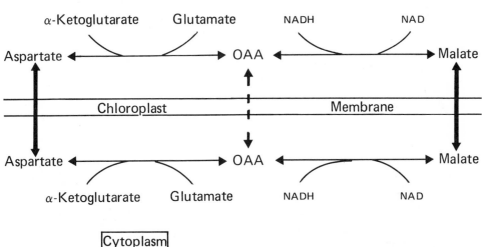

Figure 3 Pathways of indirect transfer of reducing equivalents between chloroplasts and cytoplasm.

limited by the low level of OAA. According to *Figure 3*, the participation of aspartate in the shuttle may considerably increase the capacity of the system *in vivo*. Both chloroplasts and cytoplasm contain glutamate-oxaloacetate transaminase. If during photosynthesis of leaf cells cytoplasmic NADH/NAD ratios are not kept low by cytoplasmic NADH oxidation, the generation of excessive reducing equivalents during indirect ATP transfer into the cytoplasm must necessarily result in a back flow of hydrogen into the chloroplasts via the malate-OAA shuttle system. In fact, *Table 2* shows that isolated chloroplasts actively reducing 3-PGA transfer hydrogen from the outside medium into the stroma, though only a few percent of the external NAD is in the reduced state. The following observation suggests that such a back flow might be possible against a gradient in the redox states of internal and external pyridine nucleotides.
Measurement of chloroplastic NAD(P) reduction
    Illumination of intact chloroplasts with red light causes an absorbance increase at 340 nm (*Figure 4*). We interpret this change as the reduction of internal pyridine nucleotides. DCMU completely inhibits the signal. It reappears on addition of ascorbate, dichlorophenolindophenol. Addition of malate decreases the signal (*Figure 4*), presumably because partial dark reduction of pyridine nucleotides, catalized by malate dehydrogenase, takes place. Conversely, OAA increases the absorbance change. It apparently oxidizes a substantial part of the endogenous pyridine nucleotides in the dark, thus producing a bigger pool for the subsequent light reduction. The size of the signal seen in the absence of malate and OAA corresponds to about 4×10^{-2} µmoles NAD(P) reduced mg^{-1} chlorophyll. This large

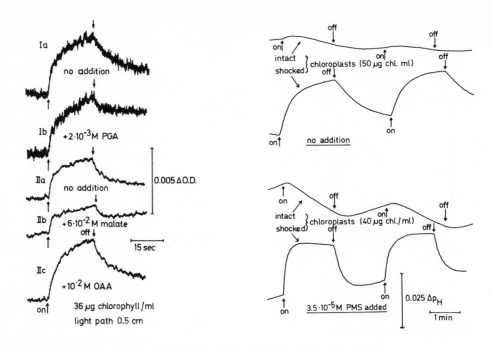

Figure 4 (left) Light-induced signals at 340 nm from intact chloroplasts indicating photoreduction of internal pyridine nucleotides. Signals are shown from two different experiments designated as I and II.

Figure 5 (right) Light-induced pH changes in suspensions of intact and osmotically shocked chloroplasts. Upper part: without PMS; lower part: with PMS added.

response together with the effects of malate and OAA on the signal suggest that not only most of the endogenous NADP, but also NAD, becomes reduced upon illumination. At higher light intensities the 340 nm change is not drastically diminished by the addition of 3-PGA. Despite its active photoreduction, 3-PGA apparently does not alter the internal redox state to a great extent. If indeed pyridine nucleotides are largely reduced inside the chloroplasts in the light, then oxidation of external NADH in the presence of 3-PGA and a functioning malate-OAA shuttle system indicates an uphill transfer of hydrogen into the chloroplasts. The question arises as to the driving force of such a transfer.

Proton gradient across the outer chloroplast membrane

With OAA and malate both capable of penetrating the chloroplast envelope, the malate-OAA shuttle should at first sight be expected to equalize the redox states of internal and external NAD systems. However, protons also participate in the reaction. A proton gradient from the chloroplast stroma to the outside medium could therefore provide the means to keep NAD more reduced inside than outside the chloroplasts.

Upon illumination, envelope-free chloroplasts transfer hydrogen ions into the intrathylakoid space, thereby increasing the pH of the suspending medium. The envelope of intact chloroplasts seems to be a barrier to this proton movement (*Figure 5*). Intact chloroplasts do not take protons up from the medium but even release protons to a small extent. This is seen most clearly in the absence of electron acceptors such as 3-PGA and CO_2. With acceptors present, a continuous large proton uptake overshadows the effects shown in *Figure 5*. That intact chloroplasts still transfer protons from the stroma region into the thylakoids can be shown by their photo-induced shrinkage reaction. Thus, in the light a hydrogen ion gradient seems to exist between stroma and the external medium, in addition to the known gradient between stroma and intrathylakoid space.

REFERENCES

1. C.R. SLACK, M.D. HATCH, and D.J. GOODCHILD, *Biochem. J.*, 114 (1969) 489.
2. U. HEBER and K.A. SANTARIUS, *Z. Naturforsch.*, 25b (1970) 718.
3. H.W. HELDT and L. RAPLEY, *FEBS Letters*, 10 (1970) 143.
4. H.W. HELDT, *FEBS Letters*, 5 (1969) 11.
5. U.W. HEBER and K.A. SANTARIUS, *Biochim. biophys. Acta*, 109 (1965) 390.
6. G.H. KRAUSE, *Z. PflPhysiol.*, (1971) in press.
7. R.G. JENSEN and J.A. BASSHAM, *Proc. natn. Acad. Sci. U.S.A.*, 56 (1966) 1095.
8. U. HEBER, *Biochim. biophys. Acta*, 180 (1969) 302.
9. R. STOCKING and S. LARSON, *Biochem. biophys. Res. Commun.*, 37 (1969) 278.
10. G. HOCH, O.v.H. OWENS, and B. KOK, *Archs Biochem. Biophys.*, 101 (1963) 171.
11. A. RIED, *Biochim. biophys. Acta*, 153 (1968) 653.

RELATIVE IMPERMEABILITY OF THE INTACT CHLOROPLAST ENVELOPE TO ATP[1]

D.M. Stokes[2] and D.A. Walker

Department of Botany, The University of Sheffield, S10 2TN, U.K.

SUMMARY

In a reconstituted broken chloroplast system 3-phosphoglycerate dependent oxygen evolution is inhibited by nigericin and ammonia, which are believed to discharge the proton gradient, and also by arsenate and phloridzin, which inhibit ATP formation without discharging the proton gradient. The inhibition of oxygen evolution by ammonia, nigericin, or arsenate is then completely restored by the addition of ATP. The system inhibited by phloridzin requires the addition of ATP and an uncoupling agent for complete restoration of oxygen evolution.

In mixtures containing spinach chloroplasts with intact envelopes and added 3-phosphoglycerate, oxygen evolution is also inhibited by nigericin, ammonia, arsenate, or phloridzin. However there is no reversal of the inhibition on the subsequent addition of ATP. We attribute the failure of ATP to reverse the effect of these inhibitors in whole chloroplasts to the inability of ATP to penetrate the intact chloroplast envelope at a rapid rate.

INTRODUCTION

Much of the current interest in what has become known at this meeting as the "C$_4$ Syndrome" relates to the degree of biochemical compartmentation reflected, or imposed, by Kranz anatomy. Indeed many of the current uncertainties revolve round the transport of organic acids and related phosphorylated compounds between cells and

[1] *This work was supported by grants from the U.K. Science Research Council and the Royal Society. We are grateful to Dr. David Hall and Dr. Robert Hill, F.R.S., for gifts of ferredoxin and phloridzin.*

[2] *Present address: Botany Department, University of Melbourne, Parkville, 3052, Australia.*

between chloroplasts and cytoplasm. Because the isolation of cells and chloroplasts is more readily achieved using spinach or peas than sugarcane or maize there is inevitably a greater body of relevant knowledge derived from work on the former, much of which will need to be independently re-established for C_4 plants. Central to the whole problem of intercellular transport is the question of whether or not ATP and ADP and inorganic phosphate move readily between the chloroplast and the cytoplasm. If photophosphorylation is to contribute to the energy economy of green cells then the chloroplast must either export ATP at appreciable rates (and import ADP or AMP at comparable rates) or alternatively export metabolites which can promote cytoplasmic or mitochondrial phosphorylation (for recent reviews see 1,2, and 3). Our own interest in this problem derives from an earlier conclusion (4) that intact chloroplasts are probably incapable of phosphorylating exogenous ADP at rapid rates and our continuing failure to affect the course of photosynthesis in intact isolated chloroplasts by ATP in a way which could not be attributed to external hydrolysis and penetration of orthophosphate (see e.g. ref. 5). In this contribution we wish to report further evidence that ATP does not appear to penetrate the intact chloroplast envelope at rapid rates.

EXPERIMENTAL

Whole spinach chloroplasts were isolated as described previously (6). For broken chloroplast preparations the chloroplast pellet was osmotically shocked in dilute assay medium and after a further centrifugation the supernatant retained as chloroplast extract. The pellet of broken chloroplasts was then resuspended in assay medium (6).

Spectrophotometric examination of the chloroplast extract was carried out as described previously (6). Oxygen was measured simultaneously in twin cells fitted with Clark-type electrodes (6).

RESULTS

In *Figure 1b* it will be seen that photosynthetic oxygen evolution is inhibited by ammonium chloride in mixtures containing intact chloroplasts and added 3-phosphoglycerate (3-PGA). No reversal of this inhibition could be detected upon the subsequent addition of ATP. *Figure 1d* shows very similar results obtained when nigericin was substituted for NH_4Cl. In this laboratory a large number of similar experiments have been carried out by Dr. L.J. LUDWIG using chloroplasts from spinach and peas, and he has shown that oxygen evolution associated with carbon assimilation is also inhibited by these uncouplers and that this inhibition is not reversed by added ATP.

In *Figure 2* curves a and b show oxygen evolution in a reconstituted chloroplast system (see Methods and ref. 6) in which ruptured chloroplasts, supplemented with chloroplast extract, ferredoxin and

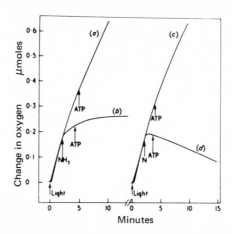

Figure 1 Oxygen evolution by whole chloroplasts. In addition to 2 ml of assay medium and 100 µg chlorophyll all reaction mixtures contained 3 µmoles 3-PGA. ATP (3 µmoles), NH₄Cl (120 µmoles) and Nigericin (N) (1.4 n moles) were added as indicated.

catalytic NADP, utilize added 3-PGA as a Hill oxidant. This system required the conversion of 3-PGA to 1,3-diphosphoglycerate (DPGA) (reaction (i)) at the expense of ATP generated by photophosphorylation (6) and is therefore susceptible to inhibition by uncoupling agents. Thus, the addition of nigericin (curve d) and ammonia (curve c) leads to a pronounced inhibition. However, unlike the system containing intact chloroplasts (*Figure 1*), it will be seen that the further addition of ATP led to restoration of oxygen evolution. Moreover, the evolution following the addition of the uncoupler and ATP was faster than that recorded in their absence. When sodium arsenate (6 µmoles) was used in place of nigericin or NH₄Cl the results were essentially the same, that is both whole and ruptured chloroplast systems were inhibited but ATP reversed the inhibition only in the reconstituted (ruptured chloroplasts) system.

The ATP-ase inhibitor, phloridzin (7), was also inhibitory in both systems (*Figure 3*). This inhibition could again be reversed only in the ruptured chloroplast system and this time both ATP *and* an uncoupler were required for reversal.

DISCUSSION

Chloroplasts with intact envelopes will assimilate carbon and evolve oxygen at rapid rates (8). Such intact chloroplasts will also evolve oxygen when provided with 3-PGA in a reaction for which there is initially no CO_2 requirement (9) and which is believed to involve the following reactions of the Calvin cycle in which 3-PGA is converted to glyceraldehyde-3-phosphate (GAP).

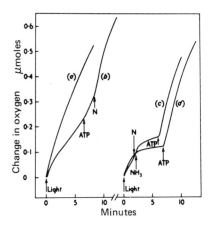 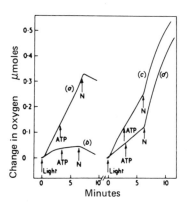

*Figure 2 (left) Oxygen evolution in a reconstituted chloro-
plast system containing washed envelope-free chloroplasts. Reaction mixtures (b) and (d) contained NADP (0.1 μmole), ferredoxin (7 n moles), 2 ml of assay medium, 100 μg chlorophyll,
and chloroplast extract containing 16.6 mg protein. Reaction
mixtures (a) and (c) were identical to (b) and (d) except that
the added chloroplast extract contained 20.3 mg of protein and
reaction mixture (a) contained ATP (3 μmoles) from the outset.
Otherwise Nigericin (N) (1.4 nmoles), ATP (3 μmoles) and NH₄Cl
(120 μmoles) were added as indicated.*

*Figure 3 (right) Oxygen evolution in whole chloroplasts and in
a reconstituted system containing chloroplasts osmotically
shocked in the electrode (see ref. 6). Reaction mixtures for
curves (a) and (b) contained whole chloroplasts (100 μg chlorophyll), 3-PGA (3 μmoles), and 2 ml of assay medium. In addition curve (b) contained phloridzin (2 μmoles). Reaction mixtures for curves (c) and (d) contained envelope-free chloroplasts (100 μg chlorophyll), ATP (0.1 μmole), NADP (0.1 μmole),
ferredoxin (3.2 nmoles), chloroplast extract containing 9.6 mg
protein and 2 ml of assay medium. Curve (d) contained in addition 2 μmoles of phloridzin. Nigericin (1.4 nmoles) and ATP (3
μmoles) were added as indicated.*

$$3\text{-PGA} + ATP \rightarrow DPGA + ADP \tag{1}$$

$$DPGA + NADPH \rightarrow GAP + NADP + Pi \tag{2}$$

The ability to evolve O_2 with CO_2 or 3-PGA as substrate is largely lost following rupture of the chloroplast envelope but in a reconstituted system containing added soluble chloroplast components
the ability to evolve O_2 with 3-PGA (but not with CO_2) is restored.
In the preceding experiments we have used these facts to compare the
behavior of intact and ruptured chloroplasts evolving oxygen with

3-PGA as the oxidant. In both systems it is assumed that 3-PGA is initially converted to DPGA (according to reactions (1) and (2), above) and this is supported by the pronounced inhibition which follows the addition of nigericin or NH_4Cl as uncoupling agents. At this pH, both of these agents produce a several fold acceleration in oxygen evolution if ferricyanide or substrate quantities of NADP are used as oxidants and indeed the accelerating effect of uncouplers on the Hill reaction is a very well established phenomenon. In the reconstituted (ruptured chloroplast) system therefore, it would be expected that the addition of ATP subsequent to an uncoupler would not only restore O_2 evolution, but might easily increase it beyond that initially observed. *Figure 2* shows that this prediction was borne out. With whole chloroplasts (*Figure 1*) the arguments are precisely the same and at least some measure of restoration would be predicted if ATP were free to enter the chloroplast at an appreciable rate. We suggest, therefore, that the simplest interpretation of these results is that the intact envelope is largely impermeable to exogenous ATP. This conclusion does not conflict with reports of adenine nucleotide movements which occur at rates of the order of 2 or 3 μmoles hr^{-1} mg^{-1} chlorophyll (10,11), because entry of ATP at this rate would not be fast enough to allow a significant restoration of oxygen evolution. Thus in *Figure 1* the uninhibited rate of oxygen evolution was 55 μmoles hr^{-1} mg^{-1} chlorophyll. Assuming that half a molecule of oxygen is evolved for each molecule of NADP reduced in photosynthesis this implies a rate of ATP consumption in reaction (1) of 110 μmoles hr^{-1} mg^{-1} chlorophyll and consequently ATP would need to penetrate the envelope at at least this rate in order to fully restore oxygen evolution to its uninhibited value.

The fact that 3-PGA-dependent oxygen evolution is markedly depressed by agents such as NH_4Cl and nigericin, which are believed to discharge the proton gradient, and also by arsenate and phloridzin which inhibit ATP formation without discharging the proton gradient (11,12), implies that these widely different inhibitors probably affect oxygen evolution only by virtue of their common ability to interfere with ATP formation. Indeed it is clear (*Figures 2,3*) that with added ATP the ruptured chloroplast system produces O_2 at least as well (or better) in the presence of these agents as in their absence. This strengthens our contention that the failure of ATP to reverse the inhibitions in the intact chloroplast system (*Figures 1, 3*) is attributable to its inability to penetrate the intact envelope at a sufficiently rapid rate.

REFERENCES

1. D.A. WALKER and A.R. CROFTS, *A. Rev. Biochem.*, 39 (1970) 389.
2. D.A. WALKER, *Nature*, 226 (1970) 1204.
3. U. HEBER, Proc. intern. Symp. Transport in Higher Plants (Reinhardsbrunn) in press.
4. D.A. WALKER, *Plant Physiol.*, 40 (1965) 1157.

5. D.A. WALKER, in Progress in Photosynthesis Research, ed. H. MET-
   ZNER, I.U.B.S., Tübingen, 1 (1969) 250.
6. D.M. STOKES and D.A. WALKER, *Plant Physiol.*, submitted for publication.
7. S. IZAWA, W.D. WIGET, and N.E. GOOD, *Biochem. biophys. Res. Commun.*, 22 (1966) 223.
8. D.A. WALKER and R. HILL, *Biochim. biophys. Acta*, 131 (1967) 330.
9. D.A. WALKER, W. COCKBURN, and C.W. BALDRY, *Nature*, 216 (1967)
   597.
10. W.H. HELDT, *FEBS Letters*, 5 (1969) 11.
11. U. HEBER and K.A. SANTARIUS, *Z. Naturforsch.*, 25 (1970) 718.
12. M. AVRON and J. NEUMANN, *A. Rev. Pl. Physiol.*, 19 (1968) 137.

STUDIES OF CHLOROPLASTS FROM Amaranthus AND MAIZE AFTER ISOLATION IN AQUEOUS SYSTEMS

J.W. Lyttleton

Applied Biochemistry Division, DSIR, Palmerston North, New Zealand

SUMMARY

Chloroplasts have been isolated from maize and Amaranthus by a technique which gave a modest proportion of apparently intact organelles. Such preparations yielded some enzymes associated with both mesophyll and bundle sheath cells; however neither adenylate kinase nor phosphoenolpyruvate carboxylase could be detected. Phosphoenolpyruvate carboxylase was present in once pelleted chloroplasts, but was lost almost completely after a single wash. It is suggested that this enzyme is cytoplasmic rather than associated with chloroplasts.

Chloroplast preparations isolated from Amaranthus contain a number of doublets in which two apparently mature chloroplasts are joined by a substantial linkage. Such paired chloroplasts have not been observed in preparations from other plants.

INTRODUCTION

The distribution between the mesophyll and bundle sheath cells of the enzymes involved in carbon fixation during photosynthesis by C_4 plants has been examined by selective grinding of leaves (1), by isolation of the two separate classes of cells (2), and by density gradient fractionation of nonaqueously isolated chloroplasts (3,4). General agreement has been reached that the enzymes concerned with the initial β-carboxylation and the regeneration of phosphoenolpyruvate (PEP) are located in the mesophyll, while the remaining enzymes, which operate a Calvin-type cycle on the CO_2 initially fixed in the mesophyll cells, are located in the bundle sheath.

Whereas most of these enzymes are located in the chloroplasts, the precise location of the PEP carboxylase is less clearly defined. Its association with chloroplasts during nonaqueous isolation has not been consistent (4), and this fact, along with some evidence from its sedimentation properties (5) has led to the suggestion that it might be bound to the membrane of the chloroplasts.

In this work, an attempt was made to isolate chloroplasts from *Amaranthus* and maize in an aqueous medium, to try to cast light on the association between PEP carboxylase and the mesophyll chloroplasts.

EXPERIMENTAL

Isolation of chloroplasts

The isolation medium used contained 3% (w/v) Ficoll, 0.4 M sorbitol, 2 mM dithiothreitol (DTT), 5 mM $MgCl_2$, 0.1% (w/v) bovine serum albumin (BSA), and 10 mM Na pyrophosphate, with pH adjusted to 7.5.

Leaves were taken from young plants of *Amaranthus lividus* or *Zea mays* after 18 hr dark followed by 30 min light. The chloroplasts were isolated either by grinding 5 g leaves by hand in a mortar and pestle into 10 ml extraction medium, or by slicing 3-4 g with a scalpel into 8-10 ml medium until the tissue was reduced to fragments less than 1 mm^2. In either case the extract was filtered through two layers of Miracloth, and then layered over 10 ml Ficoll/sorbitol medium containing 0.425 M rather than 0.4 M sorbitol. Centrifugation at 1000 g for 3 min pelleted most of the intact chloroplasts through the denser medium.

The pellet was resuspended in 5 ml extraction medium, and the chloroplasts sedimented at 700 g for 3 min. The supernatant was removed as completely as possible, and the pellet suspended in a measured volume of hypotonic buffer (25 mM Tris,SO_4 buffer pH 7.3, 10 mM EDTA, 2 mM DTT) equivalent to 8-10 times the volume of the pellet.

In order to compensate for possible contamination of the chloroplast pellet by soluble components of the original leaf extract, an amount of the second supernatant, equal in volume to the pellet itself was added to a volume of Tris,SO_4 buffer equal to that in which the pellet was suspended. After centrifuging at 10,000 g for 15 min, these two samples were tested for enzyme activity. Only when the activity of the chloroplast extract was significantly greater than that of the control blank was it considered that the enzyme arose from the chloroplast pellet.

Extraction and assay of enzymes

The activity of the enzymes in the leaves used for chloroplast isolation was tested on an extract made from 0.5 g leaf ground into 1 ml Tris,SO_4 buffer, using a conical glass homogenizer, centrifuged at 10,000 g for 10 min, and passed through a G-25 Sephadex column in Tris,SO_4 buffer.

Adenylate kinase, ribulose-1,5-diphosphate (RuDP) carboxylase, glyceraldehyde phosphate dehydrogenase, and pyruvate,Pi dikinase were assayed as in (4). PEP carboxylase was assayed at 340 nm when coupled to added malic dehydrogenase (6). Chlorophyll was estimated by the method of ARNON (7).

Electron microscopy

Isolated chloroplasts were examined by electron microscopy after fixation in glutaraldehyde, embedding, and poststaining with uranyl nitrate and lead citrate.

RESULTS

Enzyme studies

In *Table 1* are set out the range of enzymes found in *Amaranthus* and maize chloroplasts isolated by grinding or slicing techniques. The results given were consistently obtained in a series of three independent experiments with each plant. In all cases the enzymes were present at high level in the leaf extracts.

TABLE I

Qualitative survey of enzyme activities in chloroplast preparations.

Enzyme	Maize		*Amaranthus*	
	Sliced	Ground	Sliced	Ground
Glyceraldehyde-P dehydrogenase	++	++	++	++
PEP carboxylase	-	-	-	trace
Adenylate kinase	-	-	-	-
RuDP carboxylase	+	-	+	+
Pyruvate,Pi dikinase	+	+	+	+

In *Table 2* are set out comparative figures for the activity of the two carboxylases obtained from *Amaranthus* chloroplasts after the first and second pelleting step.

TABLE 2

Carboxylase activity and specific activity (chlorophyll base) in *Amaranthus* chloroplasts, expressed as percentage of activity in total leaf extract.

	Pellet I	Pellet 2
Activity		
PEP carboxylase	19.3	1.1
RuDP carboxylase	24.2	9.9
Specific activity		
PEP carboxylase	7.8	0.5
RuDP carboxylase	9.6	4.8

Plate 1 Class 1 chloroplasts isolated from Amaranthus leaves (1 micron bar)

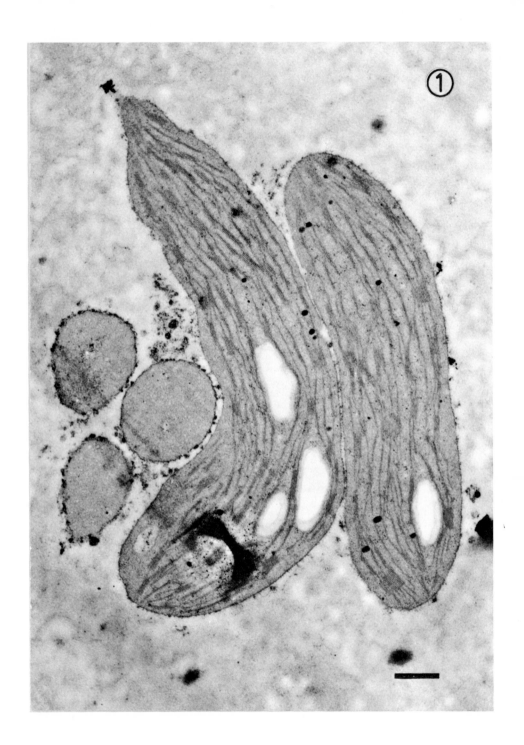

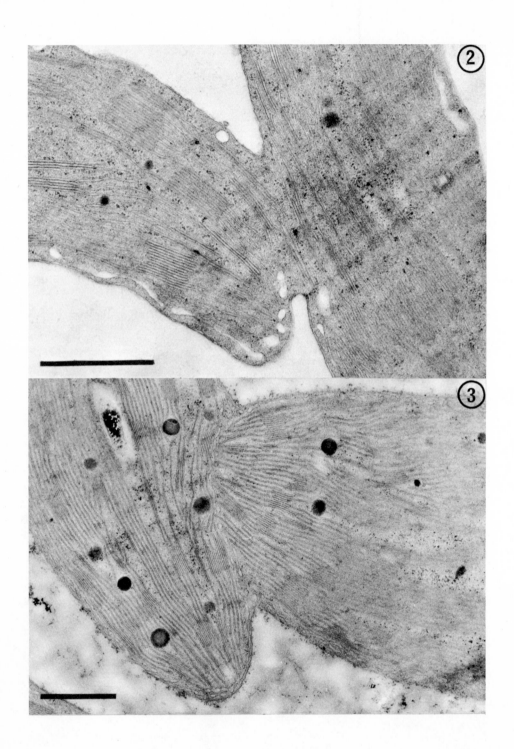

In this experiment the extract from the ground leaves was split into two equal fractions, and each banded over the more dense Ficoll/sorbitol medium to yield the first chloroplast pellet. One of these was treated as formerly to yield a second pellet, while the other was resuspended directly into Tris,SO₄ buffer. Contamination of this pellet by soluble enzymes was allowed for, using supernatant taken from just above the pellet.

Morphology of isolated chloroplasts

Although phase contrast observation is usually of great value in determining whether or not isolated chloroplasts have preserved their outer membrane intact (8), it has not been found equally helpful with these C₄ plants. Even resuspension of a pellet in distilled water, which brings about profound changes in isolated spinach chloroplasts, scarcely altered the appearance of those from *Amaranthus*.

Electron microscopy of the isolated chloroplasts was thus used to examine their quality, and indicated that at best about 15-20% of a preparation from *Amaranthus* appeared to be of Class 1. This type of chloroplast is shown in *Plate 1*, with good preservation of internal structure, relatively dense stromal region, and an apparently intact outer membrane.

It was noticed that when *Amaranthus* leaves were disrupted by slicing about 15% of the isolated chloroplasts appeared by light microscopy to be present as pairs. Such paired chloroplasts were not seen in preparations from *Amaranthus* made by grinding the leaves, nor in maize, pea, bean, or spinach chloroplasts prepared by either grinding or slicing.

Electron microscope examination of chloroplasts isolated from sliced *Amaranthus* leaves has shown pairs of apparently mature chloroplasts joined by a region in which the internal structure seems to connect one to the other without the interposition of any external membrane (*Plate 2,3*), although sometimes a layer of peripheral reticulum seems to mark the junction point (*Plate 3*). The appearance of these pairs did not resemble that normally associated with chloroplasts undergoing division.

DISCUSSION

The method of isolation used in this work is capable of giving good yields of Class 1 chloroplasts from spinach or pea leaves, but from the C₄ plants only a small proportion seemed to remain intact. None the less, active enzymes could be isolated from them, as indicated in *Table 1*.

With *Amaranthus*, the results for RuDP carboxylase and for

Plate 2 Paired chloroplasts from Amaranthus leaves without external membrane at junction.

Plate 3 Paired chloroplasts from Amaranthus leaves showing some endoplasmic reticulum at junction.

pyruvate,Pi dikinase suggested that chloroplasts from both mesophyll
and bundle sheath cells contributed to the extract. In the case of
maize, slicing as opposed to grinding was needed to obtain chloroplasts which would yield RuDP carboxylase, an observation which is
consistent with the mechanical difficulty of breaking bundle sheath
cells (1,9).

In no case was significant activity of adenylate kinase or PEP
carboxylase associated with chloroplasts even though the mesophyll
enzyme pyruvate,Pi dikinase was found in all cases, as was glyceraldehyde phosphate dehydrogenase. This latter enzyme is considered to
occur in both types of cell (4). The evidence is that while these
chloroplasts retain some enzymes on aqueous isolation, they do not
retain PEP carboxylase or adenylate kinase.

PEP carboxylase associated with the original chloroplast pellet
was almost totally removed by a single washing. By contrast RuDP
carboxylase was reduced much less. If the PEP carboxylase is in fact
membrane bound, it is lost by the very mild procedure of resuspension in the isolation medium. An equally plausible explanation might
be that fragments of adhering cytoplasm accompany the chloroplasts
during the first sedimentation, and are almost completely removed by
resuspension and pelleting.

It has been suggested (5) that because centrifugation of a chloroplast preparation from sugarcane at 140,000 g for 45 min sedimented 20% of the PEP carboxylase activity, the enzyme might be membrane
bound. This evidence should be considered in light of the fact that
calculation (10) predicts that a molecule with sedimentation coefficient of about 13 S, typical of plant PEP carboxylase (11), will
completely sediment in a high speed rotor in about 300 min when exposed to fields of 140,000 g.

The significance of the observed pairing of isolated *Amaranthus*
chloroplasts is unknown. Although in the bundle sheath cells of this
plant highly convoluted chloroplasts are seen, this type of association has not been described. It has, however, been observed in the
bundle sheath cells of *Portulaca oleracea* (W.M. LAETSCH, personal
communication). It is doubtful whether the phenomenon is of importance in understanding the C_4 pathway of photosynthesis.

ACKNOWLEDGMENTS

Thanks are due to Miss E. Todd for technical assistance, and to
Mr. D. Hopcroft for carrying out the electron microscopy.

REFERENCES

1. J.A. BERRY, W.J.S. DOWNTON, and E.B. TREGUNNA, *Can. J. Bot.*, 48
   (1970) 777.
2. G.E. EDWARDS, S.S. LEE, T.M. CHEN, and C.C. BLACK, *Biochem. biophys. Res. Commun.*, 39 (1970) 389.
3. C.R. SLACK, *Phytochemistry*, 8 (1969) 1387.

4. C.R. SLACK, M.D. HATCH, and D.J. GOODCHILD, *Biochem. J.*, 114
   (1969) 489.
5. C.W. BALDRY, C. BUCKS, and J. COOMBS, *Biochem. biophys. Res.
   Commun.*, 37 (1969) 828.
6. D.A. WALKER, *Biochem. J.*, 67 (1957) 73.
7. D.I. ARNON, *Plant Physiol.*, 24 (1949) 1.
8. J.W. LYTTLETON, *Expl Cell Res.*, 26 (1962) 312.
9. O. BJÖRKMAN and E. GAUHL, *Planta*, 88 (1969) 197.
10. Model L Instruction Manual, Beckman Instruments Inc. Palo Alto,
    (1963) 49.
11. H. MARUYAMA, R. EASTERDAY, H. CHANG, and M.D. LANE, *J. biol.
    Chem.*, 241 (1966) 2405.

PHOTOSYNTHETIC CARBON METABOLISM OF ISOLATED CORN CHLOROPLASTS[1]

Denny O'Neal, Martin Gibbs, and Dwight Peavey

Department of Biology, Brandeis University,
Waltham, Massachusetts, 02154, U.S.A.

SUMMARY

Chloroplasts have been obtained from primary leaves of corn seed-lings which can assimilate $^{14}CO_2$ at rates as high as 45 μmoles CO_2 hr^{-1} mg^{-1} chlorophyll, with usual rates of 8-16. The products are similar to those seen in spinach and pea chloroplasts, i.e. 3-phos-phoglycerate (40-66% of total label), triose phosphates (3-25%), hexose monophosphates (6-29%), with the remainder of the label dist-ributed in pentose phosphates, starch, and occasionally glycolate, malate, aspartate, and maltose. A primer is not required for CO_2 as-similation. The rate of fixation is stimulated by dithiothreitol, low concentrations of inorganic phosphate, and by several phosphate esters, especially triose phosphates, ribose 5-phosphate, and fruct-ose 1,6-diphosphate. The distribution of ^{14}C among the products is influenced by inorganic phosphate, dithiothreitol, some phosphate esters, and nitrogen and oxygen. Label in glycolate is increased by low bicarbonate and high oxygen. Fixation of CO_2 is sharply reduced by inhibitors and uncouplers. That a complete Calvin cycle is oper-ating is seen by the distribution of label in each carbon atom of 3-phosphoglycerate.

INTRODUCTION

In the last five years research from several laboratories has ap-peared which suggests that in certain groups of plants, such as sug-arcane and maize, there seems to be a unique pathway for photosyn-thetic CO_2 fixation. This whole subject has recently been thoroughly reviewed (1). This pathway is characterized by the rapid appearance of ^{14}C-labeled malate and aspartate following introduction of $^{14}CO_2$ to leaves in the light, followed by the appearance of and increase of radioactivity in 3-phosphoglycerate (3-PGA) coupled with a

---

[1] *This research was supported by the National Science Foundation.*

decline in the levels of malate and aspartate. The enzyme mediating the initial fixation of CO_2 in these plants is phosphoenolpyruvate (PEP) carboxylase. This pathway, referred to as the "C_4 dicarboxylic acid pathway" envisions malate or aspartate as the donors of CO_2 to ribulose-1,5-diphosphate (RuDP) carboxylase, via the decarboxylation of the β-carboxyl of malate and the refixation of the CO_2 by RuDP carboxylase.

One approach to studying the photosynthetic CO_2 fixation of C_4 plants is to isolate from these plants chloroplasts which can actively fix CO_2. GIBBS *et al*. (2) have recently obtained chloroplasts from maize which can assimilate CO_2 at rates as high as 45 µmoles of CO_2 hr^{-1} mg^{-1} chlorophyll. Some of the properties of these chloroplasts will be the subject of this report.

EXPERIMENTAL

The growth conditions for *Zea mays* seedlings have appeared previously (2). Nearly all of the studies described here involved var. Early Fortune although var. Golden Beauty was used in some experiments with similar results. Primary leaves were taken from plants 4-5 days after planting of seeds imbibed one day in tap water. The methods for preparation of chloroplasts and assay of $^{14}CO_2$ fixation have been described previously (2). The labeled products were separated by one dimensional paper chromatography in Wood "GW-3" solvent (3), or by two dimensional chromatography in phenol-H_2O and 1-butanol/propionic acid/water. The activity of each compound was determined either on the paper by an end window gas flow counter or by cutting out the spots and counting in scintillation vials containing dioxane-PPO-POPOP. For electron microscopy the chloroplasts were fixed in 2% glutaraldehyde and stained with osmium tetroxide.

RESULTS AND DISCUSSION

The chloroplasts were usually at least 25-50% intact as judged from their appearance under a phase contrast microscope and in electron micrographs. An electron micrograph of a chloroplast preparation appears in *Plate 1*. Some chloroplast preparations have fixed CO_2 at rates up to 45 µmoles CO_2 hr^{-1} mg^{-1} chlorophyll although the usual rates are in the order of 8-16.

Following onset of illumination, there is usually a lag in CO_2 fixation for the first 2-4 min. The degree of lag varies widely between individual preparations. The lag is sometimes reduced slightly by dihydroxyacetone phosphate (DHAP), glyceraldehyde-3-phosphate (GAP), 3-PGA, sedoheptulose-7-phosphate (S7P), ribose-5-phosphate (R5P), and fructose-1,6-diphosphate (FDP), but not by glucose-6-phosphate (G6P), or fructose-6-phosphate (F6P). The maximum rate of CO_2 fixation usually occurs between 4-9 min, decreasing thereafter.

The rate of fixation is not significantly affected by the presence of Mg^{2+} or Mn^{2+} during incubation. Sodium or potassium phosphate, when present at 0.1-0.2 mM. stimulate the rate up to 80%, but higher concentrations inhibit (up to 60% inhibition at 2.5 mM Pi). Alarge part of this Pi inhibition (at 2 mM Pi) is reversed by the addition of 3-PGA, FDP, and R5P. Sodium pyrophosphate at 0.1-10 mM stimulated the rate 20-60% (always less than monophosphate) but did not inhibit even at 10 mM. Dithiothreitol was the most effective of several thiols tested, maximum stimulation occurring at 2-8 mM. Where dithiothreitol was added at a concentration of 4-6 mM, CO_2 fixation increased between 3 to 10-fold, results varying widely between experiments. Sodium ascorbate (up to 10 mM) usually had little effect on CO_2 fixation rate, unlike the case in spinach (4), nor did it affect label distribution.

The pH optimum for CO_2 fixation was between 7.7-7.9, falling off sharply below 7.5 and above 8.0, similar to spinach chloroplasts (5). The concentration of $NaH^{14}CO_3$ giving 50% maximum rates was between 0.4-0.8 mM, also similar to spinach chloroplasts (6).

CO_2 fixation was sensitive to several inhibitors of photosynthesis. When chloroplasts were incubated without dithiothreitol or phosphate esters, trifluoro carbonylcyanide phenylhydrazine (F_3 CCCP) at 10^{-5} M caused 70% inhibition, 2×10^{-6} M dichlorodiphenyl-dimethyl urea (DCMU) caused over 90% inhibition, and 5×10^{-6} M sodium arsenite or iodoacetamide resulted in 55-60% inhibition. Chlorophyll concentration in these experiments was 7-26 µg in 1.5 ml.

The distribution of ^{14}C (*Plate 2*) ranged over the following values compiled from a number of experiments performed over a period of several months: 3-PGA, 40-66%; DHAP, 3-25%; sugar diphosphate, 1-4%; hexose monophosphates, 6-29%; pentose monophosphates and S7P, 4-9%; glycolate, 0-10%; starch, 0-15%; other compounds (malate, aspartate, maltose), 0-6% and unknowns, 0-5%. In many experiments only phosphate esters were present.

The inclusion of one of several phosphate esters in the incubation medium increases the rate of CO_2 fixation, as is the case in spinach chloroplasts (7). Greatest stimulation (up to 100%) resulted from the addition of GAP and DHAP, closely followed by FDP and R5P. S7P stimulated 20-40%, while PEP, 3-PGA, G6P, and F6P caused only slight or no stimulation. Triose phosphates, R5P, and FDP shift the distribution of label in favor of 3-PGA (up to 88% of the total) at the expense of hexose monophosphate, pentose phosphates, and sometimes triose phosphates. These compounds also increase the label in starch (up to 3-fold) and decrease label in glycolate.

Reduced or oxidized pyridine nucleotides and ATP had no pronounced or reproducible effect on product distribution, but occasionally stimulate fixation up to 40%.

Plate 1 Electron micrograph of isolated maize chloroplasts. x 9900. Courtesy of Dr. Shimon Klein.

Plate 2 Products of 8 min $^{14}CO_2$ photosynthesis by chloroplasts isolated from primary leaves of maize.

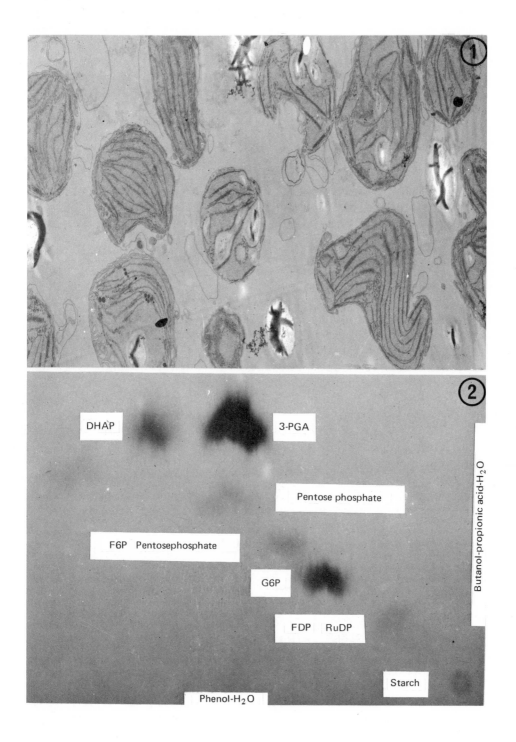

1

2

DHAP   3-PGA

Pentose phosphate

F6P Pentosephosphate

G6P

FDP RuDP

Starch

Phenol-H_2O

Butanol-propionic acid-H_2O

Dithiothreitol sometimes affected label distribution, decreasing label in 3-PGA up to 40%, and increasing label in starch and hexose monophosphates up to 330%. Inorganic phosphate also usually shifted product distribution. As Pi was increased from 0 to 2.5 mM the label in insolubles decreased sharply, decreased label in G6P and F6P by 40%, and increased DHAP by up to 5-6 fold. Pyrophosphate had only a small influence on label distribution.

Under low $NaHCO_3$ concentrations (below 1 mM), 100% oxygen caused significant inhibition of CO_2 fixation, and led to increased label in glycolate, but even so glycolate seldom accounted for over 10% of the total label.

The chloroplasts were unable to metabolize exogenously supplied $L(^{14}C)$-malate (uniformly labeled) or $L(^{14}C)$-aspartate (β-carboxyl labeled), even when the preparation was supplemented with supernatant, ATP, or α-ketoglutarate.

Additional evidence supporting the hypothesis that these chloroplasts have a complete carbon reduction cycle is the observation that label is found in all three carbon atoms of 3-PGA after 8 min of photosynthesis.

Furthermore, these plastid preparations have been assayed for many of the Calvin cycle enzymes (2) and were found to contain sufficient activities to catalyze the observed rates of CO_2 fixation. Finally, behavior of total pool sizes of sugar phosphates in the chloroplasts when illumination is interrupted is similar to spinach chloroplasts (8) i.e., RuDP, triose phosphates, and FDP fall sharply, and 3-PGA rises.

In conclusion, these maize chloroplasts are similar to spinach and pea chloroplasts with respect to $^{14}CO_2$ fixation products, kinetics, response to O_2, phosphate esters, and inhibitors (5-10). The enzyme complements and total pool sizes are also comparable. It is apparent that these maize chloroplasts are capable of performing a normal reductive pentose phosphate cycle at reasonable rates.

REFERENCES

1. M.D. HATCH and C.R. SLACK, *A. Rev. Pl. Physiol.*, 21 (1970) 141.
2. M. GIBBS, E. LATZKO, D. O'NEAL, and Choy-Sin HEW, *Biochem. biophys. Res. Commun.*, 40 (1970) 1356.
3. T. WOOD, *J. Chromat.*, 6 (1961) 142.
4. Z. PLAUT and M. GIBBS, *Plant Physiol.*, 45 (1970) 470.
5. M. GIBBS and N. CALO, *Plant Physiol.*, 34 (1950) 318.
6. R.G. JENSEN and J.A. BASSHAM, *Proc. natn. Acad. Sci. U.S.A.*, 56 (1966) 1095.
7. E.S. BAMBERGER and M. GIBBS, *Plant Physiol.*, 40 (1965) 919.
8. J.A. BASSHAM and R. JENSEN, in Harvesting the Sun, eds. A. SAN PIETRO, F.A. GREEN, and T. ARMY, Academic Press, New York, (1967) 79.
9. M. GIBBS, E. LATZKO, R.G. EVERSON, and William COCBURN, in Harvesting the Sun, eds. A. SAN PIETRO, F.A. GREEN, and T. ARMY,

Academic Press, New York, (1967) 111.
10. D.A. WALKER and A.R. CROFTS, *A. Rev. Biochem.*, 39 (1970) 389.

CARBON DIOXIDE FIXATION BY MAIZE CHLOROPLASTS ISOLATED BY THE 'LACERATION TECHNIQUE'[1]

E. Roy Waygood, Sudha K. Arya, and Regis Mache[2]

Department of Botany, University of Manitoba,
Winnipeg, Manitoba, Canada

SUMMARY

Maize (Zea mays L.) chloroplasts were isolated from leaves by the 'laceration technique' of MACHE and WAYGOOD using the mannitol containing media of SHEPHARD et al. A maximum rate of 113 μmoles CO_2 hr^{-1} mg^{-1} chlorophyll was recorded. This rate was considerably enhanced by the supernatant obtained after centrifugation of the chloroplasts for 50 seconds at 2000 x g.

It was shown that the products of $^{14}CO_2$ fixation included organic and amino acids, sugar phosphates, and sucrose. The supernatant also showed considerable dark $^{14}CO_2$ fixation yielding a similar array of radioactive products. ATP was present in the supernatant and it has been considered that the source of reducing power is an NADP-mannitol dehydrogenase active in the supernatant.

INTRODUCTION

While it has been firmly established that illuminated spinach chloroplasts can fix CO_2 at physiological rates (1), there have been only two reports of CO_2 fixation by chloroplasts from monocotyledonous plants at rates of 2.9 and 19.9 μmoles CO_2 hr^{-1} mg^{-1} chlorophyll (2,3).

Using the 'laceration technique' (4) for the isolation of chloroplasts and the media of SHEPHARD et al. (5) we have obtained considerably higher rates of CO_2 fixation by maize chloroplasts fortified with 16 mM phosphoenolpyruvate (PEP) and under other conditions as

[1] *Acknowledgment is made to the National Research Council of Canada for their award of a grant-in-aid (A2698) to one of us (ERW).*

[2] *Present address: Faculte des Sciences de Grenoble, Laboratoire de Physiologie Vegetale, Grenoble, France.*

reported herein.

EXPERIMENTAL

Plant material

Plants of *Zea mays* L. var. Early King were grown in a 13 hr photoperiod in the greenhouse. Leaves were harvested 5 hr after the
commencement of the photoperiod, washed in water and their cut ends
placed in distilled water in a beaker and allowed to stand for 2 hr
at 4°.

Chloroplast isolation

About 10 g of leaves were placed in a glass chromatographic solvent trough (34 cm in length) containing 25 ml of solution H (5) pH
7.8. The leaves were gently lacerated (stroking motion) parallel to
the veins with three to five small closely spaced (1 mm apart) scal-
pels (Cat. No. 411, Irex German Surgicals, 788 Adelaide Street, Toronto) with their ends embedded in a lucite block to form a handle.
After laceration the leaves were removed and the suspension was filtered through Kleenex tissues or Miracloth and centrifuged at 0° for
50 sec at 2000 x g. The pellet, a mixture of bundle sheath and mesophyll chloroplasts was then suspended in 0.5 ml of solution W (5) pH
7.2. For some experiments the 2000 x g supernatant (supernatant-a)
was centrifuged for 10 min at 29,000 x g (supernatant-b) or 100,000
x g (supernatant-c). All procedures were conducted in the cold.

Photosynthetic $^{14}CO_2$ fixation

Photosynthesis was carried out in 6 ml volume standard single
side arm Warburg flasks illuminated from above by a quartz 750 w
flood lamp giving an intensity of 4000 ft-c at the base of the
flask. Each flask contained 50 µl of chloroplast suspension (ca.
0.06 mg chlorophyll); 375 µl solution A (5), pH 7.2; 50 µl of 160 mM
Na PEP, and 20 µl of 375 mM NaHCO$_3$ with 5 µl of NaH^{14}CO$_3$ (5 µc) in
the side arm. The final volume was 0.5 ml and the control flask contained 0.5 ml H$_2$O. The flasks were flushed with N$_2$ for 3 min with
stopcocks closed and then illuminated for 3 min without shaking. At
zero time the NaH^{14}CO$_3$ was tipped into the main compartment and the
flasks were shaken. After 6 min the reaction was stopped by the addition of 2 ml of boiling methanol. The mixture was placed in a vial
with 0.5 ml 2N HCl and flushed with air for 5 min. The solution was
extracted with CHCl$_3$ several times and the water soluble fraction
was dried in a stream of air at 25° and dissolved in 0.2 ml water.
Radioactivity was determined by adding 20 µl of water soluble fraction to 10 ml of standard scintillation fluid and measured in a
Picker Nuclear Liquimat 220. Chlorophyll was determined by the method of ARNON (6).

RESULTS AND DISCUSSION

PEP concentration

Preliminary experiments (*Table 1*) showed that the presence of PEP was essential for $^{14}CO_2$ fixation by maize chloroplasts in the media of SHEPHARD *et al.* (5). The optimal concentration of PEP was 16 mM which was used in all subsequent experiments.

TABLE I

Effect of PEP concentration (mM) and supernatant-a on photo-synthetic CO_2 fixation by maize chloroplasts (μmoles hr^{-1} mg^{-1} chlorophyll). Conditions as under Experimental.

PEP concentration	Fixation rate
0	1.4
8	28
8 + supernatant-a	138
16	37
16 + supernatant-a	217
32	29
32 + supernatant-a	157

Table 1 shows that the addition of the 2000 x g supernatant-a considerably augmented CO_2 fixation on a chlorophyll basis. There was, however, no light activating factor in supernatant-a as that described by WILDNER and CRIDDLE (7). *Table 2* shows that there was little fixation of CO_2 by chloroplasts in darkness even in the presence of PEP, but addition of the 2000 x g supernatant-a increased CO_2 fixation markedly.

TABLE 2

Fixation of CO_2 by maize chloroplasts in light and darkness (μmoles hr^{-1} mg^{-1} chlorophyll). Conditions as under Experimental.

Reaction conditions		Fixation rate
Chloroplasts (no PEP)	Light	1.4
Chloroplasts (16 mM PEP)	Light	68.3
Chloroplasts (16 mM PEP)	Dark	3.0
Chloroplasts (+ 0.05 ml supernatant-a + 16 mM PEP)	Light	291.7

Chlorophyll and bicarbonate

The optimum chlorophyll concentration was 0.06 mg per 0.5 ml (*Table 3*). This concentration was used in all subsequent experiments except those determining the optimum $NaHCO_3$ concentration in which it was 0.02 mg per 0.5 ml. Bicarbonate was optimum at 15 mM or 7.5

μmoles per 0.5 ml. *Table 3* shows the maximum rate of 113 μmoles CO_2 hr^{-1} mg^{-1} chlorophyll recorded in our experiments by chloroplasts in the absence of supernatant-a.

TABLE 3

Effect of chlorophyll concentration (mg in 0.5 ml) on CO_2 fixation by maize chloroplasts (μmoles hr^{-1} mg^{-1} chlorophyll). Conditions as under Experimental.

Chlorophyll concentration	Fixation rate
0.02	104
0.04	107
0.06	113
0.08	89
0.10	74

Initial product

Oxaloacetate has been shown to be the initial product of photosynthetic CO_2 fixation in sugarcane leaves (8). Experiments were undertaken to determine whether radioactive oxaloacetate 2,4 dinitrophenylhydrazone could be identified as a product of $NaH^{14}CO_3$ fixation in 10 sec of photosynthesis by maize chloroplasts. Reaction systems, either illuminated or dark and, with or without supernatant-a, were reacted with 2,4 dinitrophenylhydrazine at various temperatures from either -80° to +80°. The 2,4 dinitrophenylhydrazones were extracted with $CHCl_3$ and the radioactivity determined in both the $CHCl_3$ soluble and the water soluble fractions. The results indicated that both light and the presence of supernatant-a enhanced $^{14}CO_2$ incorporation into both the $CHCl_3$ and water soluble fractions. The $CHCl_3$ soluble fractions were cochromatographed with authentic oxaloacetate 2,4 dinitrophenylhydrazone using TLC in the solvent described by KRUPKA and TOWERS (9) and all showed radioactivity at the position of the oxaloacetate 2,4 dinitrophenylhydrazone (R_f 0.19). Another unknown radioactive 2,4 dinitrophenylhydrazone gave the same brown color with NaOH (9), but it moved to an R_f of 0.23.

Products of $^{14}CO_2$ fixation by chloroplasts and the supernatant

The products of $^{14}CO_2$ fixation by chloroplasts and supernatant-a and supernatant-b in both light and darkness were chromatographed initially in one dimension on TLC as shown in *Plate 1*. All treatments gave radioactive products which included organic and amino acids, sugar phosphates, acid phosphates, and sucrose. In the absence of PEP there were no hot methanol stable radioactive products.

A two dimensional radioautogram (10) on Whatman No. 1 paper is shown in *Plate 2* listing the products that could be identified by elution and subsequent cochromatography.

(a) Organic acids

Organic acids were separated from amino acids, acid phosphates, and sugar phosphates, etc. by high voltage thin layer electrophoresis according to the procedure of COOK and BIELESKI (11). *Plate 3* shows that malate and glycolate were the two major acids produced in 6 min photosynthesis.

(b) Sugars and sugar phosphates

The sugar phosphates and sucrose area was eluted from a typical one dimensional chromatogram as shown in *Plate 1*. This was developed two dimensionally (12) as shown in *Plate 4*. The products were identified by cochromatography and by treatment with acid phosphatase under toluene followed by cochromatography of the free sugars with authentic compounds.

$^{14}CO_2$ fixation by the supernatants in darkness

One of the more puzzling features of this study was the considerable dark $^{14}CO_2$ fixation by the supernatants-a and -b. Chloroplasts did not fix significant amounts of $^{14}CO_2$ in the dark (*Table 2*), but the supernatants on a chlorophyll basis gave exaggerated readings of over 800 µmoles CO_2 hr^{-1} mg^{-1} chlorophyll. The absolute amount of $^{14}CO_2$ fixation into organic acids, amino acid, sugar phosphates, and sucrose was similar to that accomplished by chloroplasts centrifuged down for 50 sec at 2000 x g. *Plate 1* shows the qualitative results in a one-dimensional chromatogram. Supernatant-a did contain smaller chloroplasts (3.0 x 6.0 µ to 6.5 x 7.0 µ) as compared to the larger size of the chloroplasts in the 2000 x g pellet (5.2 x 7.5 to 7.5 x 10.2 µ). However supernatant-a and -b were practically devoid of chloroplasts yet they fixed $^{14}CO_2$ in both light and darkness, but only in the presence of PEP (*Plate 1*).

Plate 5 shows a two-dimensional TLC developed from supernatant-b in the dark. The fingers represent the same array of radioactive

---

Plate 1 Radioautogram of products of $^{14}CO_2$ fixation by chloroplasts with PEP and with either supernatant-a or supernatant-b. A. Chloroplasts, light; B. Chloroplasts and supernatant-a, light; C. Chloroplasts and supernatant-b, light; D. Supernatant-a, dark; E. Supernatant-a, light; F. Supernatant-b, dark; G. Supernatant-b, light (4000 ft-c, 6 min). One dimensional in butanol:propionic acid:H_2O (10:5:7). TLC, M and N Cellulose 300 without binder, 250 µ thickness.

Plate 2 Paper radioautogram of products of $^{14}CO_2$ fixation by chloroplasts with PEP and supernatant-a at 4000 ft-c for 6 min. 1st dimension, 48 hr, in 88% phenol:H_2O:CH_3COOH:1M EDTA (840:160:10:1), 2nd dimension, 24 hr, in butanol:propionic acid:H_2O. Equal parts of butanol:H_2O (370:25) and propionic acid:H_2O (18:22). Whatman No. 1. Numbers refer to: 1. ribulose 1,5-P; 2. fructose 1.6-P; 3. glucose 6-P; 4. fructose 6-P; 5. ribose 5-P; 6. 3-phosphoglycerate; 7. unknown; 8. ribulose 5-P; 9. PEP; 10, 11. unknown; 12. aspartate; 13. serine; 14. glutamate; 15. sucrose; 16. glycine; 17. malate.

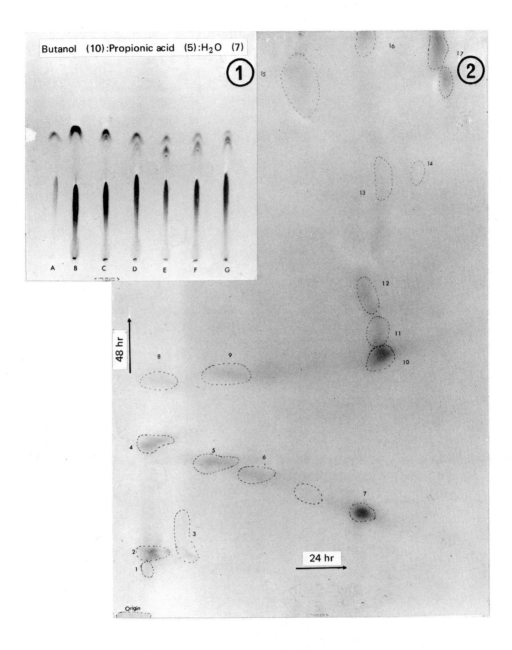

Butanol (10):Propionic acid (5):H₂O (7)

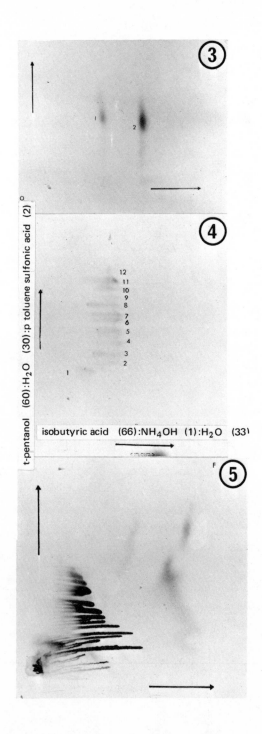

t-pentanol (60):H$_2$O (30):p toluene sulfonic acid (2)

isobutyric acid (66):NH$_4$OH (1):H$_2$O (33)

compounds as shown in *Plate 2*. This pattern of radioactive products was typical of either supernatant-a or -b in darkness or in light.

Assays for ATP in these supernatants indicated its presence at concentrations of 0.2 µM which may or may not be sufficient energy. However, the origin of reducing power was suggested to us by Dr. R. ROHRINGER of the Canada Department of Agriculture, Winnipeg (personal communication) as perhaps an NADP-mannitol dehydrogenase. The media of SHEPHARD *et al.* (5) contains 0.6 M mannitol and we have demonstrated that the supernatant rather than the chloroplast fraction contained this enzyme.

However, considerably more investigations will have to be made on these systems to elucidate their mechanism of CO_2 fixation especially in the light of the recent report by PAN and WAYGOOD (13) of a thermostable cyanide sensitive PEP carboxylase found in both bundle sheath and mesophyll chloroplasts of maize and other nonphotosynthetic organisms.

REFERENCES

1. M. GIBBS, E. LATZKO, R.G. EVERSON, and W. COCKBURN, in Harvesting the Sun, eds. A. SAN PIETRO, F.A. GREEN, and T. ARMY, Academic Press, New York, (1967) 111.
2. C.W. BALDRY, C. BUCKE, and J. COOMBS, *Biochem. biophys. Res. Commun.*, 37 (1969) 828.
3. M. GIBBS, E. LATZKO, D. O'NEAL, and Choy-Sin HEW, *Biochem. biophys. Res. Commun.*, 40 (1970) 1356.
4. R. MACHE and E.R. WAYGOOD, *FEBS Letters*, 3 (1969) 89.
5. D.C. SHEPHARD, W.B. LEVIN, and R.G.S. BIDWELL, *Biochem. biophys. Res. Commun.*, 32 (1968) 413.
6. D.I. ARNON, *Plant Physiol.*, 24 (1949) 1.
7. G.F. WILDNER and R.S. CRIDDLE, *Biochem. biophys. Res. Commun.*, 37 (1969) 952.
8. M.D. HATCH and C.R. SLACK, *Biochem. J.*, 101 (1966) 103.

Plate 3 Chromatogram showing the presence of (1) malate and (2) glycolate as the two major organic acids produced after 6 min photosynthesis.

Plate 4 Radioautogram of products of $^{14}CO_2$ fixation by chloroplasts (2-dimensional) with PEP and supernatant-a at 4000 ft-c. 1. glucose 6-P; 2. sucrose; 3. glucose 1-P; 4. fructose 1,6-P; 5. ribulose 1,5-P; 6. fructose 6-P; 7. ribose 5-P; 8. 3-phosphoglycerate; 9. ribulose 5-P; 10. 3-phosphoglyceraldehyde; 11. unknown; 12. dihydroxyacetone phosphate. (Sugar phosphate and sucrose area was eluted from (B) in Plate 1).

Plate 5 Radioautogram of products of $^{14}CO_2$ fixation by supernatant-b at 4000 ft-c, for 6 min in dark. TLC, M and N Cellulose, 2-dimensional with solvents as is Plate 2.

9. R.M. KRUPA and G.H.N. TOWERS, *Can. J. Bot.*, 36 (1958) 165.
10. T.A. PEDERSEN, M. KIRK, and J.A. BASSHAM, *Biochim. biophys. Acta*, 112 (1966) 189.
11. A.R. COOK and R.L. BIELESKI, *Anal. Biochem.*, 28 (1969) 428.
12. P.P. WARING and Z.Z. ZIPORIN, *J. Chromatog.*, 15 (1964) 168.
13. D. PAN and E.R. WAYGOOD, *Can. J. Bot.*, (1971) in press.

SUGAR SYNTHESIS IN MESOPHYLL CHLOROPLASTS OF SUGARCANE[1]

Hugo P. Kortschak

Physiology and Biochemistry Department, Experiment Station,
Hawaiian Sugar Planters' Association, Honolulu, Hawaii 96822,
U.S.A.

SUMMARY

In sugarcane leaves fructose was labeled more rapidly from $(3-^{14}C)$ labeled aspartate and malate than from $(4-^{14}C)$ labeled acids. This contrasted with the results obtained for glucose and presented some inconsistencies in relation to the present interpretation of the C_4 pathway. It is proposed that fructose may be formed from aspartate via pyruvate and 3-PGA in the mesophyll cells.

Both the HATCH and SLACK (1) and the KARPILOV (2) cycles proposed for the C_4 pathway of photosynthesis utilize only carbon 4 of the C_4 acids in the synthesis of sugars. The synthesis of sugars from the other carbons presents no problem in itself, since pyruvate and 3-phosphoglyceric acid (PGA) are readily interconvertible. However, when synthesis from $3-^{14}C$ aspartate is compared with synthesis from $4-^{14}C$ aspartate, the former invariably results in a higher activity in free fructose *(Table 1)*. Tracing the label through the intricacies of the Calvin cycle provides no explanation. Since in both cases all sugars must be formed from 3-phosphoglycerate (3-PGA) the only explanation appears to be that synthesis from the two substrates must be spatially separated in the leaf. As both types of synthesis are photosynthetic, the obvious first assumption is that the separation involves the two kinds of chloroplasts (2).

Bundle sheath chloroplasts form considerable quantities of starch (3). Greened parenchyma tissue gave free glucose only (4). EDWARDS *et al.* (5) found bundle sheath chloroplasts of *Digitaria* were unable to utilize pyruvate. We may be certain that synthesis of sugars from pyruvate, with a surplus of free fructose produced, is not taking place in the bundle sheath chloroplasts.

Mesophyll chloroplasts do not lack the enzymes required for starch synthesis, as this synthesis takes place when glucose is infiltrated

---

[1] *Published with the approval of the Director as Paper No. 266 in the Journal Series of the Experiment Station, Hawaiian Sugar Planters' Association.*

in the dark or when the bundle sheath cells are overloaded (6). The lack of starch synthesis under normal conditions must be due to a relative lack of glucose.

TABLE I

Effect of position of label on the distribution of radioactivity in fructose and glucose formed in sugarcane leaves [Radioactivity in sugars from $(3-^{14}C)/(4-^{14}C)$ aspartate expressed as a percentage]

Time (minutes)	Fructose	Glucose
4	4.3	0.0
8	7.5	1.0
60	2.0	0.3
60	6.5	0.7
60	0.6	0.4
60*	4.0	1.0
60**	2.0	0.2

* using $(3-^{14}C)$ pyruvate
** using $(3-^{14}C)$ malate

Although HATCH and SLACK (7) conclude that fructose diphosphate aldolase and fructose diphosphatase are absent from mesophyll cells, their data are compatible with as much as 20% of the total of these enzymes, as well as others of the Calvin cycle, being present here. Since mesophyll cells are smaller than the bundle sheath cells, concentrations actually may be not much lower. If the results of HATCH and SLACK are recalculated in the units used by EDWARDS *et al.* (5), units of enzyme activity per mg chlorophyll, the enzymes of the Calvin cycle appear to be much more concentrated in the bundle sheath chloroplasts, with the typical C_4 cycle enzymes evenly distributed between both types of chloroplasts; nearly all the chlorophyll was found by HATCH and SLACK in the mesophyll cell fraction. This question of the distribution of enzymes cannot be considered completely settled.

It may be that synthesis of sugars from carbon-4 of the C_4 acids is predominantly in the bundle sheath chloroplasts where production of glucose is favored, whereas direct conversion of pyruvate to 3-PGA and then to sugars takes place in the mesophyll cells. If this suggestion of separate syntheses is true, it implies of course that at least a substantial part of the PGA found in the mesophyll cells does not enter the bundle sheath chloroplasts.

The proportion of sugar synthesis that takes place in the mesophyll can be large. Accumulation of label in sugars and sugar phosphates from aspartic 3- or $4-^{14}C$ was equally rapid *(Table 2)*.

TABLE 2

Incorporation of $3-^{14}C$ and $4-^{14}C$ aspartate into sugars and sugar phosphates in sugarcane leaves (Percentage of radioactivity)

Sugars	From $3-^{14}C$ (minutes)			From $4-^{14}C$ (minutes)		
	2	4	8	2	4	8
Sucrose	4	5	25	7	17	51
Glucose			1		1	1
Fructose		13	15		3	2
Glucose-P	15	10	17	17	16	13
Fructose-P	2	4	4	4	6	1
Total Sugars	21	32	62	27	43	68

In the previously proposed cycles, pyruvate is simply carboxylated and decarboxylated with no significant losses. With a significant fraction lost by conversion to 3-PGA and sugars, the question arises as to a source to maintain the supply required. The pyruvate pools are so small that they are usually not listed in tables of radioactive products of short-time photosynthesis. In some of our (unpublished) experiments, total pyruvate reached 1.5% of total counts, compared to 30% in 3-PGA. If sugar synthesis from pyruvate represents only exchange, the results of equilibrium with 3-PGA, the pyruvate in tests with $^{14}CO_2$ should be labeled like 3-PGA with practically no detectable delay. Since HATCH and SLACK found labeling of carbons 2 and 3 of malic acid, which come from pyruvate, to be very much slower than labeling of 2 and 3 of 3-PGA, this can hardly be a sufficient source of supply. The fact that these internal carbons do accumulate radioactivity, being one-fourth as nearly saturated as carbon 4 at 150 seconds, shows that a part may come from PGA.

If it is true that there is a source of unlabeled pyruvate continually diluting the pool used for capturing CO_2, then the specific activity of carbons 2 and 3 of malic acid can never equal the saturated value for carbon 4. Unfortunately, the available data do not reach the times required to test this. Extrapolation of the data of HATCH and SLACK, plotted on a log basis, indicates that this could not occur until 10^4 or 10^5 seconds, a time scale that introduces too many complications in the way of secondarily labeled compounds to be considered.

The only apparent available source appears to be deamination of alanine. Alanine is supplied from the stalk at a concentration of around 10-20 mM and could thus furnish a considerable amount of pyruvate. We have shown that synthesis of sucrose from exogenous alanine does occur; however, supplying alanine to detached blades did not increase the photosynthesis rates.

At low concentrations of CO_2 it would be expected that photosynthesis would be largely limited to the mesophyll chloroplasts, assuming that these have at least a small complement of all the required enzymes. This should again lead to an excess of fructose over glucose, compared to higher CO_2 concentrations, and this, in fact, was found *(Table 3)*.

TABLE 3

Radioactivity in free glucose and fructose formed in $^{14}CO_2$ of different concentrations [Radioactivity in sugars formed in 0.01% $^{14}CO_2$/0.05% $^{14}CO_2$, expressed as a percentage]

Time (minutes)	Fructose	Glucose
0.25	1.1	0.0
20	14.9	14.3
20	1.8	1.4

Sucrose synthesis should proceed even in the absence of CO_2, and this has also been found. Using ^{14}C-alanine, CO_2 was kept low by carrying out photosynthesis in the presence of 10% KOH. Of the alanine metabolized in 2 hours, 43% was in sucrose as compared to 55% in the control.

It would be difficult indeed to prove this hypothesis as long as we are unable to separate the two types of chloroplasts while retaining full activity, including synthesis of sugars, but it would appear to explain some otherwise puzzling facts.

REFERENCES

1. M.D. HATCH and C.R. SLACK, *Biochem. J.*, 101 (1966) 103.
2. Y.S. KARPILOV, *Proc. Moldavian Sci. Res. Inst.*, 11 (1970) 33.
3. F.A.F.C. WENT, *Archief voor de Java-Suikerendustrie*, IV (1896) 525.
4. H.P. KORTSCHAK and L.G. NICKELL, *Plant Physiol.*, 45 (1970) 515.
5. G.E. EDWARDS, S.S. LEE, T.M. CHEN, and C.C. BLACK, *Biochem. biophys. Res. Commun.*, 39 (1970) 389.
6. W.M. COELINGH and V.J. KONINGSBERGER, *Archief voor de Suiker-Industrie in Nederlandsch-Indie*, XL (3) (1932) 1325.
7. C.R. SLACK, M.D. HATCH, and D.J. GOODCHILD, *Biochem. J.*, 114 (1969) 489.

TWO PHOSPHATASES ASSOCIATED WITH PHOTOSYNTHESIS AND THE GLYCOLATE PATHWAY[1]

D.D. Randall[2] and N.E. Tolbert

Department of Biochemistry, Michigan State University, East Lansing, Michigan, 48823, U.S.A.

SUMMARY

The isolation and properties of a P-glycolate phosphatase and a 3-phosphoglycerate phosphatase from leaves is described. From the activity and location, possible functions for these enzymes in photorespiration and photosynthesis are suggested.

RESULTS AND DISCUSSION

A specific P-glycolate phosphatase and another phosphatase, which preferentially catalyzes the hydrolysis of 3-phosphoglycerate (3-PGA), are very active in 38 plants and algae surveyed (1). Results from some plants are shown in *Table 1*. In C_3 plants the ratio of activities for P-glycolate phosphatase:3-PGA phosphatase range from 2:1 to 10:1 with an average of about 4:1. In Waring blender homogenates of C_4 plants, that represent mainly the mesophyll cells, this ratio is reversed being about 1:2 to 1:4. This change in activities between mesophyll cells of C_4 plants relative to C_3 plants reflects an increase in 3-PGA phosphatase and a decrease in P-glycolate phosphatase. However in bundle sheath cells of C_4 plants the ratio of activity for these two phosphatases is similar (4:1) to that in all C_3 plants. Activities of both phosphatases vary widely among the plants ranging from 1 to 20 μmoles min^{-1} mg^{-1} chlorophyll. Maximum activities of both phosphatases were found in bean leaves (20 surveyed). Activities of both phosphatases seem adequate to account for

[1] *This work was supported in part by National Science Foundation Grant GB-17543 and published as Journal Article No. 5357 of the Michigan Agricultural Experiment Station.*

[2] *Present address: Clayton Foundation Biochemical Institute, University of Texas, Austin, Texas, 78712.*

TABLE I

Specific activity (μg P min^{-1} mg^{-1} chlorophyll) of phosphatases in leaf extracts

Species	P-glycolate phosphatase	3-PGA phosphatase	Ratio P-glycolate to 3-PGA
C$_3$ plants			
Tobacco	160	24	7
Spinach	296	58	5
Cotton	390	101	4
Sugar beet	714	130	5
Soybean	293	157	2
Elodea	142	46	3
Chlamydomonas	114	14	8
Scenedesmus	207	12	17
Sedum spectabile	180	98	2
White oak	130	13	10
Elm	87	32	3
C$_4$ plants*			
Sorghum mesophyll	48	176	0.25
bundle sheath	8	28	4
Corn mesophyll	85	38	0.5
bundle sheath	185	29	6
Atriplex rosea mesophyll	56	132	0.5
bundle sheath	42	21	2
Crabgrass mesophyll	10	75	0.13
bundle sheath	55	94	2

* Activities in the two cell types was ascertained after differential grinding, first by a Waring blender for 2 minutes for mesophyll cells and then by a roller mill for the bundle sheath cells (1,3).

the rate of formation of glycolate and glycerate during photosynthesis.

Phosphatase isolation and properties

Both phosphatases are absent in etiolated leaves and are formed during greening in the light (1,2). P-glycolate phosphatase is located in the chloroplasts of C$_3$ plants and bundle sheath cells of C$_4$ plants. Only a few percent of this phosphatase activity remains with chloroplasts isolated in a NaCl medium (2), but 20% remains with the chloroplasts isolated in sorbitol media (1). All of the phosphatase is found in the chloroplasts when isolated by the freeze-dry nonaqueous procedure (1).

By these same procedures 3-PGA phosphatase activity has been found

TABLE 2

Properties of phosphatases

	P-glycolate phosphatase	P-glycerate phosphatase
Specificity	Only P-glycolate	Most active with 3-PGA
Location	Chloroplast	Chloroplast and cytosol, mesophyll of C_4 plants, starch grains
pH optimum	6.3	5.9-6.3
Mol. wt.	Approx. 15,000	Approx. 160,000
K_m	2.5×10^{-3} M	8×10^{-4} M
Dialysis or EDTA	Inactivates	No effect
Cofactor	Mg^{++} or Zn^{++}	None
Stabilizer	Citrate	None
Inhibitors	None	Zn^{++} or Cu^{++}
Lability	Heat stable 55°	Heat labile 55°

at several sites. It is in the soluble or cytosol fraction of de-starched corn leaves (1), and by differential grinding procedures was found mainly in the mesophyll cells of several C_4 plants (1,3). In C_3 plants, the activity is found in both chloroplasts and the soluble cytoplasm, but the exact distribution has not yet been determined. In addition, 5 to 20% of the 3-PGA phosphatase is tightly bound to starch grains in spinach leaves (4).

P-glycolate phosphatase has been purified 110-fold and character-ized from spinach and wheat leaves (5,6,7). 3-PGA phosphatase has been purified 2530-fold and characterized from sugarcane leaves (8) and from starch grains of spinach leaves (4). Some of the properties of these phosphatases are listed in *Table 2*. The two enzyme activi-ties are due to different proteins. P-glycolate phosphatase is speci-fic, while 3-PGA phosphatase hydrolyzes P-glycerate faster than any other ester. It hydrolyzes P-glycolate at 1/10 the rate of 3-PGA. Similarities between the two phosphatases are in the chemical struc-ture of the substrates, the pH optimum of 5.9 to 6.3, the same loca-tion on chloroplasts, and their postulated function. Their pH opti-mum and substrate specificity is atypical of acid phosphatases. The two phosphatases are very different in other aspects. P-glycolate phosphatase is of small molecular weight but with typical phosphatase properties, such as a Mg^{++} or Zn^{++} requirement and inhibition by EDTA, NaF, and cysteine. 3-PGA phosphatase is of large molecular

weight, and it tends to further aggregate. It is a very stable enzyme, has no cation requirement and in fact is stable to EDTA treatment and nearly unaffected by phosphatase inhibitors. It is however
inhibited by 10^{-4} M Zn^{++} or Cu^{++} in contrast to a Zn^{++} or Mg^{++} requirement by P-glycolate phosphatase.

The stability properties of P-glycolate phosphatase are unusual.
As isolated from leaf extracts it has bound to it in an ionic manner
many molecules of tricarboxylic acid whose presence are necessary for
its stability. The bound acid must be a tricarboxylic acid, and P-
glycolate phosphatase from tobacco leaves contains isocitrate and
citrate, while the phosphatase from wheat contains mainly cis-aconi-
tate. In extracts of leaves or algae in the presence of these tricarboxylic acid, the enzyme is stable and very active in boiling ethanol
or methanol for several minutes (7,9). As a result, P-glycolate in
extracts is all hydrolyzed when killing tissue by such procedures,
although a pool of P-glycolate is a known product of short-time $^{14}CO_2$
fixation experiments (10). On the other hand, 3-PGA phosphatase is
heat labile, and its action is stopped quickly by the usual killing
procedures.

Speculation on function of the phosphatases

As discussed elsewhere at these meetings glycolate is a major product of photosynthesis and is formed only in the light and in chloroplasts or algae. P-glycolate is also formed during short periods of
$^{14}CO_2$ fixation (10). A specific phosphatase for hydrolysis of P-glycolate in the chloroplast should be involved in the biosynthetic and
metabolic pathways for glycolate. Glycolate is excreted by algae (11)
and chloroplasts (12) and in higher plants the glycolate is then
metabolized in the leaf peroxisomes (13). Participation of P-glycolate phosphatase in this excretion has been postulated (7,12) on the
comparative basis that substrates and enzymes involved in membrane
transport may be phosphate esters and phosphatases. In the case of
these C_2 compounds, this hypothesis has been supported with data on
the formation of P-glycolate by isolated chloroplasts, by excretion
of glycolate, and by rapid reabsorption of glyoxylate-^{14}C but not
glycolate-^{14}C by chloroplasts (12). Consequently the beginning of the
glycolate pathway of metabolism may start with P-glycolate hydrolysis
in the chloroplasts. This hypothesis is tenuous until more is known
about glycolate biosynthesis during photosynthesis. Data from GIBBS'
group have supported the biosynthesis of glycolate from fructose-6-
phosphate by a peroxidation of a thiamine pyrophosphate-C_2 complex
formed from it by transketolase (14, also see this symposium). This
mechanism does not produce P-glycolate, and, if it is correct, then
the function for a specific P-glycolate phosphatase in the chloroplast is unknown. We have not been able to detect a glycolate kinase.

Since 3-PGA is an intermediate in both the photosynthetic carbon
cycle and hexose metabolism, regulation of its pool size by a phosphatase has many implications. Certainly any hydrolysis of 3-PGA as
it is formed during CO_2 fixation should reduce the rate of photosynthesis. The labeling pattern of glycerate formed during

photosynthesis with $^{14}CO_2$ is similar to that of 3-PGA (15,16), but the rate of labeling of the glycerate pool is much slower or its specific activity lower than that of 3-PGA. However, the ultimate pool size of glycerate is relatively large particularly in such plants as bean and tobacco leaves, which have high levels of 3-PGA phosphatase. We have also observed considerable free glycerate-^{14}C in experiments with corn and sugarcane leaves. Earlier experiments on ^{14}C-labeled products of photosynthesis had noted the presence of glycerate-^{14}C, but considered it to be due to general phosphatase action during killing procedures. HESS and TOLBERT (16) argued that glycerate came from hydrolysis of 3-PGA *in vivo* because the labeled pool of glycerate in tobacco leaves increased with time relative to the 3-PGA pool. After 4 seconds of CO_2 fixation there was 1/5 as much labeled glycerate as 3-PGA, but by 60 seconds there was more glycerate-^{14}C than 3-PGA-^{14}C, and in all cases both products were labeled in the carboxyl groups to about the same degree. These data indicate that most of the glycerate is formed initially from 3-PGA and that the rate of labeling and turnover of glycerate is relatively fast, though slower than the labeling of the first products of the photosynthetic carbon cycle.

Thus it appears as if chloroplasts rapidly synthesize P-glycolate and P-glycerate and excrete free glycolate and glycerate perhaps by the action of the two different phosphatases. Physiological data bearing on a function or proof for this transport is lacking. We speculate that both phosphatases may be involved in some carbon and/or ionic transport system between the chloroplasts and the cytosol. Perhaps related is the rapid appearance of 3-PGA in the cytoplasm during photosynthesis (17,18).

A striking analogy concerning the function of the two phosphatases is evident in the subsequent metabolic sequence between glycolate and glycerate. This pathway has been called the glycolate pathway, and it is located in leaf peroxisomes (13), except for the interconversion between glycine and serine. In this pathway two glycolates are converted to one serine and then the carbon flows on to glycerate (see TOLBERT, this volume). In the reverse direction glycerate is converted to serine and glycine plus a C_2 complex, but since the reactions from glycolate to glycine are physiologically irreversible, carbon can flow from glycerate only to glycine but not on to glycolate. Thus both phosphatases initiate carbon metabolism that leads to glycine and serine from two alternative sources and by two pathways, both of which are in the peroxisomes. Labeling patterns in glycine and serine formed during $^{14}CO_2$ fixation by C_3 plants are uniform and have proven that most of the serine arises from P-glycolate and glycolate (19). On the other hand, serine formed by the C_4 plant, corn, is initially more carboxyl labeled (19,20), as would occur during formation from 3-PGA and glycerate. However even in the C_4 plants the serine becomes uniformly labeled rapidly as glycolate conversion to serine is also occurring.

Massive flow of carbon from P-glycolate and glycolate represents

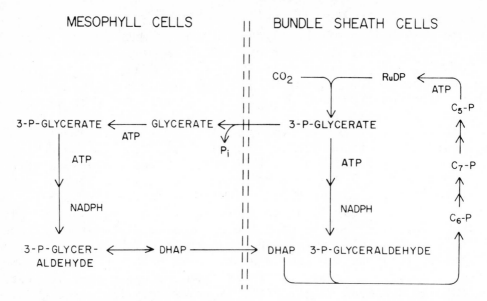

*Figure 1 Proposed scheme for 3-PGA transport from bundle sheath
to mesophyll cells in C₄ plants*

photorespiration and appears to be a wasteful respiratory process.
Flow of carbon from 3-PGA to serine and glycine via 3-PGA phosphatase
would be more economical and efficient, since it would bypass CO_2 and
energy losses of photorespiration. However, an unfavorable equilibrium and enzymatic pH optimum between glycerate and hydroxypyruvate
may prevent extremely rapid oxidation of glycerate in the physiological pH range (21). A diurnal variation for 3-PGA phosphatase in sugarcane leaves (1) activity was highest in the latter part of the day
when less active photosynthesis and lower light intensity would have
curtailed glycolate formation. It would appear that then carbon could
flow from 3-PGA to glycine, serine, and C_1 compounds to continue the
photosynthetic supply of these structural components.

The presence of lowered amounts of P-glycolate phosphatase in the
mesophyll cells of C_4 plants is consistent with the reduced number of
peroxisomes there for glycolate metabolism. However, the mesophyll
cells of the C_4 plants have a relatively high activity for 3-PGA
phosphatase yet low peroxisomal activity for glycerate metabolism,
which suggest an additional function for this phosphatase. The presence of a C_4 dicarboxylic acid cycle of CO_2 fixation in these mesophyll cells, has resulted in a search for a carbon transport process
between the mesophyll and bundle sheath cells. In the schemes proposed by HATCH and SLACK (22,23,24), it is possible that 3-PGA and
glycerate may be involved in this transport. *Figure 1* shows how 3-PGA
phosphatase may operate to achieve this. In this hypothesis 3-PGA is
formed in the bundle sheath cell, where there is inadequate reducing
power. About half of the 3-PGA pool is found in the mesophyll cell
(24) and presumably it was rapidly transported there during photosynthesis for reduction. 3-PGA phosphatase of the mesophyll cytosol,

would hydrolyze the ester and create a concentration gradient for flow of this component from the bundle sheath cells to the mesophyll cells. Rephosphorylation of the glycerate in the mesophyll chloroplasts by the glycerate kinase (24) with photosynthetic ATP would complete the transport of 3-PGA from the site of carboxylation to the site of adequate reducing power. This transport would have occurred at the expense of the energy of one phosphate bond. The concentration of 3-PGA, and inhibitor of ribulose-1,5-diphosphate (RuDP) carboxylase in the bundle sheath cells would be kept low. In the mesophyll chloroplasts the 3-PGA would be reduced to triose phosphate, as proposed by HATCH and SLACK, and transported back to the bundle sheath cells as dihydroxyacetone phosphate. This seems like a lot of movement between cells, but it may result in increased efficiency of CO_2 fixation. If movement of photosynthetic products between mesophyll and bundle sheath cells does occur, it is preferable to envisage facilitated diffusion by an enzyme such as 3-PGA phosphatase rather than passive diffusion through the thick bundle sheath cell walls or limited plasmodesmata. As presently envisaged (23), CO_2 fixation in C_4 plants requires more ATP than in C_3 plants with only the photosynthetic carbon cycle. The activity of the phosphatase to facilitate 3-PGA transport would increase by one ATP the energy balance for each molecule of 3-PGA transported. Since only half of the 3-PGA is believed to be transported back to the mesophyll cells, while the other half is reduced in the bundle sheath cells, this transport process would require a net addition of 1/2 ATP for each CO_2 reduced to the level of a carbohydrate.

REFERENCES

1. D.D. RANDALL, N.E. TOLBERT, and D. GREMEL, *Plant Physiol.*, submitted.
2. Y.L. YU, N.E. TOLBERT, and G.M. ORTH, *Plant Physiol.*, 39 (1964) 643.
3. D.W. REHFELD, D.D. RANDALL, and N.E. TOLBERT, *Can. J. Bot.*, 48 (1970) 1219.
4. D.D. RANDALL and N.E. TOLBERT, *Plant Physiol.*, submitted.
5. K.E. RICHARDSON and N.E. TOLBERT, *J. biol. Chem.*, 236 (1961) 1285.
6. D.E. ANDERSON and N.E. TOLBERT, *Meth. Enzym.*, IX (1966) 646.
7. D.E. ANDERSON, Ph.D. thesis, Michigan State University, East Lansing, (1969).
8. D.D. RANDALL and N.E. TOLBERT, *J. biol. Chem.*, submitted.
9. J. ULLRICH, *Biochim. biophys. Acta*, 71 (1963) 589.
10. A.A. BENSON, J.A. BASSHAM, M. CALVIN, A.G. HALL, H.E. HIRSCH, S. KAWAGUCHI, V. LYNCH, and N.E. TOLBERT, *J. biol. Chem.*, 196 (1952) 703.
11. N.E. TOLBERT and L.P. ZILL, *J. biol. Chem.*, 222 (1956) 895.
12. P.C. KEARNEY and N.E. TOLBERT, *Archs Biochem. Biophys.*, 98 (1962) 164.

13. N.E. TOLBERT, *A. Rev. Pl. Physiol.*, 22 (1971) in press.
14. M. GIBBS, *Ann. N.Y. Acad. Sci.*, 168 (1969) 356.
15. W.J. BRUIN, E.B. NELSON, and N.E. TOLBERT, *Plant Physiol.*, 46 (1970) 386.
16. J.L. HESS and N.E. TOLBERT, *J. biol. Chem.*, 241 (1966) 5705.
17. J.A. BASSHAM and R.G. JENSEN, in Harvesting the Sun, eds. A. SAN PIETRO, F.A. GREEN, and T.J. ARMY, Academic Press, New York, (1967) 79.
18. U.W. HEBER, in Biochemistry of Chloroplasts, II, ed. T.W. GOODWIN, Academic Press, New York, (1967) 71.
19. R. RABSON, N.E. TOLBERT, and P.C. KEARNEY, *Archs Biochem. Biophys.*, 98 (1962) 154.
20. H.S. JOHNSON and M.D. HATCH, *Biochem. J.*, 114 (1969) 127.
21. N.E. TOLBERT, R.K. YAMAZAKI, and A. OESER, *J. biol. Chem.*, 245 (1970) 5129.
22. T.J. ANDREWS, H.S. JOHNSON, C.R. SLACK, and M.D. HATCH, *Phytochemistry*, (1971) in press.
23. M.D. HATCH and C.R. SLACK, *A. Rev. Pl. Physiol.*, 21 (1970) 141.
24. C.R. SLACK, M.D. HATCH, and D.J. GOODCHILD, *Biochem. J.*, 114 (1969) 489.

CARBONIC ANHYDRASE, PHOTOSYNTHESIS, AND LIGHT-INDUCED pH CHANGES

D. Graham, C.A. Atkins,[1] M.L. Reed,[2]
B.D. Patterson, and Robert M. Smillie

Plant Physiology Unit, CSIRO, Division of Food Research, and
School of Biological Sciences, Macquarie University,
North Ryde, N.S.W., 2113, Australia

SUMMARY

When *Chlorella pyrenoidosa* or *Chlamydomonas reinhardi* are grown in
high CO_2 (5% in air) they lack carbonic anhydrase activity. After
transfer to low CO_2 (0.03% in air) the high CO_2-grown *Chlorella* is
unable to photosynthesize until its level of carbonic anhydrase
rises. Within 90 min the rate of O_2 evolution increases about 8-fold
and the carbonic anhydrase activity about 100-fold. However, enzymes
of the Calvin cycle and of β-carboxylation do not change in activity.
In the higher plants studied, carbonic anhydrase appears to be con-
stitutive, and is little affected by CO_2 concentrations up to 10% in
air.

In *Chlorella* cells which contain carbonic anhydrase, photosynthesis is inhibited by a specific inhibitor of the enzyme, Diamox (90%
at 25 mM). This inhibition, is overcome by high CO_2 concentration.

A number of soluble and particulate isoenzymes of carbonic anhydrase have been separated in leaf extracts by gel electrophoresis.
These isoenzymes may reflect diverse functions or sites of action of
the enzyme.

Although carbonic anhydrase appears essential for the utilization
of low levels of CO_2, no evidence was found either for a molecular
interaction with ribulose diphosphate carboxylase, or for a direct
role of the enzyme in providing H^+ ions for a pH gradient in chloroplasts.

---

[1] *Rothmans postdoctoral Fellow at the University of Sydney.*

[2] *Present address: Chancellor College, Limbe, Malawi.*

INTRODUCTION

Carbonic anhydrase catalyzes the reaction:

$$CO_2 + H_2O \rightleftarrows H^+ + HCO_3^- \qquad (1)$$

The uncatalyzed reaction is relatively slow, especially in the forward direction (1). Recent evidence indicates that in plants which
lack the C_4 pathway of photosynthesis the enzyme is localized in the
chloroplasts (2) and that it is required for photosynthesis by spinach chloroplasts (3) and *Chlorella* cells (4,5).

TABLE I

Summary of effects of CO_2 concentration during growth or subsequent photosynthesis upon the photosynthetic induction period
and the path of carbon in *Chlorella pyrenoidosa* (6).

CO₂ concentration		Induction period	Principal path of carbon fixation
During growth*	During photo-synthetic ex-periments**		
0.03	∿40	short	Calvin cycle
0.03	∿1	short	Calvin cycle
5	∿40	short	Calvin cycle
5	∿1	long	β-carboxylation

  * Percentage by volume
** μMolar

Table 1 summarizes the results of previous work (6) which showed
the effect of growth in high (5%) or low (0.03%) CO_2 on the photosynthetic induction period and the path of fixed carbon in *Chlorella*.
Cells grown in high CO_2 and then transferred to 0.03% CO_2 are unable
to photosynthesize at a significant rate (*Figure 1*) but after about
30 min the rate of O_2 evolution begins to rise, and reaches a steady
rate after about 90 min. The principal pathway of carbon fixation
during the first 30 min of the induction phase is through β-carboxylation with malate and aspartate as major products (6). During the
whole induction phase there was no marked variation in the activity
of the enzymes of the Calvin cycle, or of β-carboxylation (4,7).
  After growth in high CO_2 the cells contain negligible carbonic anhydrase activity, but after transferring the cells to low CO_2, the
increase in the activity of the enzyme parallels the increase in photosynthetic rates (*Figure 1*). A similar effect occurs in *Chlamydomonas reinhardi* under these conditions, with an increase in carbonic
anhydrase activity of about 1,000-fold.
  Three possible functions for carbonic anhydrase in relation to

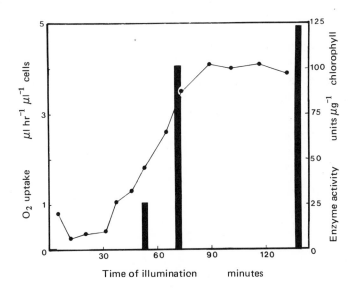

Figure 1 Increase in rate of photosynthesis (closed circles) and activity of carbonic anhydrase (histogram) during the induction period of Chlorella pyrenoidosa grown in 5% CO_2 and illuminated in alkaline buffer with 0.48 μM free CO_2.

photosynthesis have been suggested (5):-

(i)   It may be a permease which facilitates the transport of CO_2 to the sites of carboxylation in the chloroplast.

(ii)   It may physically associate with ribulose-1,5-diphosphate (RuDP) carboxylase, either to increase the local availability of CO_2, or to increase the affinity of the enzyme for CO_2.

(iii) It may be essential for the production of a light-induced pH gradient in the chloroplasts, in which H^+ ions are derived from the hydration of CO_2 according to reaction I.

In this paper the last two possibilities are considered.

RESULTS AND DISCUSSION

Table 2 shows the difference in carbonic anhydrase activity between algal cells grown in 0.03% or 5% CO_2. *Euglena gracilis*, which has very little carbonic anhydrase, requires a higher level of CO_2 than that in air for significant autotrophic growth.

In *Chlorella* cells which contained carbonic anhydrase photosynthesis in low CO_2 was inhibited more than 90% by 25 mM Diamox, a specific inhibitor of the enzyme (*Table 3*). In high CO_2, Diamox had no effect. These results show that carbonic anhydrase is required for photosynthesis at low concentrations of CO_2 such as are found in air, but not at high concentrations.

Carbonic anhydrase in leaves of higher plants

In *Chlorella* and *Chlamydomonas* high levels of CO_2 appear to repress the formation of carbonic anhydrase, although whether actual

TABLE 2

Carbonic anhydrase activity (enzyme units mg^{-1} chlorophyll) of extracts from algae grown in air or 5% CO_2. (Enzyme assayed and expressed according to EVERSON and SLACK (2), showing mean ±S.E. and number of experiments in parenthesis).

Chlorella pyrenoidosa		
Air grown	188 ± 32	(6)
5% CO_2 grown	0	(5)
Chlamydomonas reinhardi		
Air grown	2041 ± 35	(5)
5% CO_2 grown	9 ± 4	(4)
Euglena gracilis		
Air grown	56 ± 35	(5)
5% CO_2 grown	0	(2)

synthesis of the protein is repressed is uncertain at present. Various higher plants were grown in different concentrations of CO_2 in an attempt to repress the enzyme in leaves. While there was some evidence of a decrease in activity at 10% CO_2 (*Table 4*), the effect was not marked when compared with the results obtained with *Chlorella* and *Chlamydomonas*. High CO_2 concentrations cause stomatal closure in leaves (9); thus it is possible that the internal CO_2 level was relatively low due to depletion by photosynthesis. This experiment then does not conclusively show that carbonic anhydrase is not repressed by high CO_2, in higher plants.

TABLE 3

Effect of Diamox on apparent photosynthesis (APS, µl O_2 hr^{-1} µl^{-1} packed cells) in *Chlorella pyrenoidosa* (measured in alkaline buffer after 90 min illumination).

CO_2 concentration		APS
During growth*	During photosynthesis measurements**	
5	78.8	12.5
5	78.8 + 25 mM Diamox	13.2
5	0.48	5.7
5	0.48 + 25 mM Diamox	0.15
0.03	0.48	3.3
0.03	0.48 + 25 mM Diamox	0.2

* Percentage by volume
** µMolar

TABLE 4

Effect of CO_2 concentration (% by volume) during growth on carbonic anhydrase activity in leaf extracts (enzyme units mg^{-1} chlorophyll).

| | CO_2 concentration | | | |
	0.03	1	5	10
C$_3$ plants				
Pisum sativum	7880	-	6250	5600
Typha sp.	9040	-	-	-
C$_4$ plants				
Zea mays	508	490	545	353
Sorghum bicolor	741	-	-	-

EVERSON and SLACK (2) showed that some plants with the C$_4$ pathway of photosynthesis have relatively low activities of carbonic anhydrase when compared to plants having the Calvin cycle only. The results in *Table 4* confirm these findings, although *Chlorella*, a Calvin-type species, has very low activities (*Table 2*). More recently we have obtained somewhat higher activities from *Zea mays* and *Amaranthus hybridus* (up to 1200 units mg^{-1} chlorophyll), by extracting the tissue under nitrogen in the presence of 20 mM mercaptoethanol. It seems possible that some of the variation between species results from different degrees of oxidation during extraction. The carbonic anhydrase from maize leaves was essentially confined to the mesophyll cells (*Table 5*), and the trace of activity in the bundle sheath can be attributed to contamination with other fractions. Similar results were obtained with the C$_4$ Dicotyledon, *Amaranthus hybridus* L. (sens lat.).
Isozymes of carbonic anhydrase
Carbonic anhydrase from a variety of species (*Figure 2*) has been separated using gradient gel electrophoresis (10). These gels

TABLE 5

Distribution of carbonic anhydrase activity (enzyme units mg^{-1} chlorophyll) in leaves of maize. (The mesophyll and bundle sheath cells were isolated by a differential grinding technique (8).)

Whole leaf homogenate	1149	690
Mesophyll cell homogenate	853	509
Bundle sheath cell homogenate	0	23

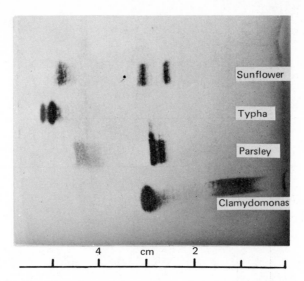

Figure 2 *Gel electrophoresis of carbonic anhydrase isozymes*
(10) derived from Chlamydomonas *and leaves of higher plants.*

separate largely on the basis of molecular weight (11). The enzyme
from parsley, which has been purified by TOBIN (12), was separated on
these gels into two major isozymes (*Figure 2*), whose molecular
weights were greater than 100,000. *Helianthus annuus* (sunflower) also
has a third isozyme whose molecular weight is less than 100,000. In
some monocotyledons, such as *Tradescantia albiflora* and *Zantedeschia
aethipica* (arum lily) isozymes of molecular weight less than 100,000
predominate. In *Typha orientalis* (reedmace) all the isozymes have
molecular weights less than 100,000. In *Amaranthus hybridus* and *Atriplex spongiosa* some of the carbonic anhydrase activity does not enter
gels, and appears to be associated with a particulate fraction. While
the picture is fragmentary at present, a common feature of the plants
studied (fourteen species) was the existence of more than one isozyme
of carbonic anhydrase in each plant. Carbonic anhydrase from maize
and *Sorghum* appears to be unstable to gel electrophoresis, and consequently we have been unable to resolve isozymes from these sources.
It seems possible that the different isozymes which we have found in
leaves may perform different functions, or have different locations
in the cell.

Function of carbonic anhydrase

Using gel electrophoresis, we found that carbonic anhydrase activity was always separated from Fraction I protein. While this does
not eliminate the possibility of an interaction *in vivo*, it does not
support the idea of an association of carbonic anhydrase with RuDP
carboxylase.

When algal cells are suspended in an unbuffered medium and illuminated, the pH of the medium rises. It has been suggested that the
light-induced pH changes shown by whole cells of *Chlamydomonas* (13)
and *Dunaliella* (14) are related in some way to the light-induced pH
gradients found with isolated chloroplasts (15) and so to

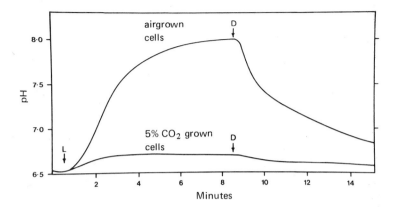

Figure 3 Light-induced pH changes in air-grown and 5% CO_2-grown Chlamydomonas reinhardi. Cells were suspended in distilled water and pH measured with a glass electrode at 25°. (L = light on; D = light off).

photophosphorylation (16). Light-induced pH changes were measured in suspensions of *Chlamydomonas reinhardi* grown either in air, when carbonic anhydrase was present, or in 5% CO_2, when carbonic anhydrase was repressed. The results are shown in *Figure 3*. Light induced a much larger pH change in cells which contained carbonic anhydrase, i.e. those which were grown in air. However, when photosynthesis was measured, these pH changes could be accounted for by the movement of CO_2 into the cells (17). When photosynthesis was limited in the cells grown in 5% CO_2 by lack of carbonic anhydrase, CO_2 uptake was also limited. This resulted in a proportionately smaller change in the pH of the medium.

Our experiments with isolated chloroplasts have so far given no clear indication that carbonic anhydrase is involved directly in the light-induced pH gradient. For example, chloroplasts isolated from *Chlamydomonas* or pea leaves which showed light-induced pH changes did not respond to CO_2, which in the presence of carbonic anhydrase might be expected to produce H$^+$. Diamox, an inhibitor of carbonic anhydrase, had no effect on the pH change. Added carbonic anhydrase (purified from bovine erythrocytes) also had no effect. However, these negative results may reflect the condition of the chloroplasts, and some recent experiments reported elsewhere in this publication (18) suggest that carbonic anhydrase may function in the production of pH gradients.

While there is ample evidence that carbonic anhydrase participates in photosynthesis, its precise function remains obscure.

REFERENCES

1. T.H. MAREN, *Physiol. Rev.*, 47 (1967) 595.
2. R.G. EVERSON and C.R. SLACK, *Phytochemistry*, 7 (1968) 581.
3. R.G. EVERSON, *Phytochemistry*, 9 (1970) 25.

4. M.L. REED and D. GRAHAM, *Plant Physiol.*, 43 (1968) S29.
5. D. GRAHAM and M.L. REED, *Nature*, (1971) in press.
6. D. GRAHAM and C.P. WHITTINGHAM, *Z. PflPhysiol.*, 58 (1968) 418.
7. M.L. REED, Ph.D. Thesis, Sydney University, Sydney, (1970).
8. D.G. BISHOP, K.S. ANDERSEN, and R.M. SMILLIE, This volume, (1971).
9. H. MEIDNER and T.A. MANSFIELD, *Biol. Rev.*, 40 (1965) 483.
10. B.D. PATTERSON, C.A. ATKINS, D. GRAHAM, and R.B.H. WILLS, (submitted for publication).
11. J. MARGOLIS and K.G. KENRICH, *Anal. Biochem.*, 25 (1968) 347.
12. A.J. TOBIN, *J. biol. Chem.*, 245 (1970) 2656.
13. S. SCHULDINER and I. OHAD, *Biochim. biophys. Acta*, 180 (1969) 165.
14. A. BEN-AMOTZ and B.Z. GINZBURG, *Biochim. biophys. Acta*, 183 (1969) 144.
15. A.T. JAGENDORF and E. URIBE, *Proc. natn Acad. Sci. U.S.A.*, 55 (1966) 170.
16. P. MITCHELL, *Biol. Rev.*, 41 (1966) 445.
17. C.A. ATKINS and D. GRAHAM, (submitted for publication).
18. R.G. EVERSON, This volume, (1971).

CARBONIC ANHYDRASE IN PHOTOSYNTHESIS

R.G. Everson

Department of Biology and Environmental Science,
Queensland Institute of Technology, Brisbane, Qld., 4000, Australia

SUMMARY

Diamox (5-acetamido-1,3,4-thiadiazole-2-sulphonamide) inhibits photosynthesis in leaves of plants with and without the C_4 pathway. Diurnal fluctuation of carbonic anhydrase activity and a relationship between carbonic anhydrase activity and light intensity is such as to suggest some degree of feedback control related to photosynthesis. On the other hand, light activation of carbonic anhydrase has not been established.

The question arises of whether the biological function of carbonic anhydrase is in the catalysis of CO_2 hydration and mobilization, particularly in connection with light-dependent pH changes. Absence of carbonic anhydrase from the chloroplast fraction of the tropical grasses makes it clear that if the enzyme were to act in response to light-dependent cation movements there may be related differences in the photosynthetic organization of chloroplasts of C_4 and C_3 pathway plants.

INTRODUCTION

Carbonic anhydrase is widespread in leaf tissue and has recently been purified from parsley (1). This enzyme has a high molecular weight and yields six identical sub units, each of which has a similar molecular weight to the mammalian enzyme.

Partial dependence of photosynthetic carbon dioxide fixation upon carbonic anhydrase has been claimed for spinach chloroplast preparations *in vitro* (2,3) and in algae (4,5,6). Although reported to be absent from leaves of Panicoid grasses, fractionation of leaves in nonaqueous solvents showed carbonic anhydrase to be present in all species assayed but at a distinctly lower level in plants with the C_4 pathway of photosynthesis (7). Furthermore, in the C_4 plants carbonic anhydrase did not appear to be specifically associated with the chloroplast as in other leaves analyzed.

TABLE I

Inhibition of spinach carbonic anhydrase by sulfonamides. Assays were performed at 0° by the Wilbur-Anderson veronal-indicator method as described previously (3). 400-600 units of carbonic anhydrase were used in each assay. The source of enzyme was Sigma "Ribulose 1,5 diphosphate carboxylase" - a partially purified Fraction I protein preparation from spinach leaves. Five mg of each inhibitor was dissolved in 0.5 ml methanol and 0.5 ml water or dilute alkali was added to make up volume. Blank assays were run on inhibitor solutions at the appropriate concentrations. Compounds (i)-(v) were supplied gratis by the Cyanamid Corp., (Lederle Division) and compound (vi) by Upjohn.

	Compound	Concentration (M)	% inhibition
(i)	5-acetamido-1,3,4-thiadiazole-2-sulfonamide (Diamox)	1×10^{-4}	76
	"	2×10^{-5}	50
(ii)	5-n butyrylamino-1,3,4-thiadiazole-2-sulfonamide	1×10^{-4}	80
(iii)	5-phthaloylamino-1,3,4-thiadiazole-2-sulfonamide	1×10^{-4}	96
	"	1×10^{-5}	65
(iv)	5-benzenesulfonamido-1,3,4-thiadiazole-2-sulfonamide	1×10^{-4}	57
(v)	5-p toluenesulfonamido-1,3,4-thiadiazole-2-sulfonamide	1×10^{-4}	67
(vi)	6 ethoxybenzothiazole-2-sulfonamide (ethoxyzolamide)	1×10^{-5}	94
	"	1×10^{-6}	68
	"	4×10^{-7}	50
	"	2×10^{-7}	30

In view of the differences between the C_4 and the orthodox Calvin cycle plants with respect to the quantity of and distribution of carbonic anhydrase further comparative studies have been carried out.

RESULTS

Inhibition of spinach leaf carbonic anhydrase by thiadiazole-2-sulfonamides

Several thiadiazole-2-sulfonamides show similar potency to Diamox (K_1 = 2×10^{-5} M as inhibitors of spinach leaf carbonic anhydrase (3). However, 6-ethoxybenzothiazole sulfonamide (ethoxyzolamide) is a much more effective inhibitor than any of the thiadiazole-2-sulfonamides tested. Ethoxyzolamide inhibits spinach leaf carbonic anhydrase activity by 50% at a concentration of 4×10^{-7} M (*Table 1*). This is about five hundred-fold the concentration required for equivalent inhibition of the enzyme from mammalian tissue (8). In view of its specificity, lack of systemic toxicity when tested on intact plants, and low K_i, ethoxyzolamide would appear to have potential as a useful inhibitor for studies on plant carbonic anhydrase and its physiological function.

Comparative inhibition of carbonic anhydrase from maize, Amaranthus, and spinach

For all experiments crude extracts were prepared from leaf-punch samples extracted in a chilled mortar and pestle using 0.05 M veronal buffer pH 8.0 containing 1 mM EDTA. After centrifuging 2 minutes at 2000 g, the supernatant was assayed as previously (3).

TABLE 2

Comparative percentage inhibition of carbonic anhydrase from plant sources

Plant	PCMB* 10^{-3}M	PCMB* 10^{-3}M	Arsenite 10^{-3}M	BAL* 5×10^{-4}M	Diamox 5×10^{-5}M	NaCl 5×10^{-2}M	Air 24 hrs
Zea mays	99	100	92	–	85	53	100
Amaranthus viridis	86	100	54	53	82	–	100
Spinacia oleracea	0	8	18	63	60	70	25

* PCMB, *p*-chloromercuribenzoate; BAL, 2,3-dimercaptopropanol.

Approximately 200 units of enzyme from extracts of maize and *Amaranthus* (C_4 plants) and spinach (a C_3 plant) were assayed for carbonic anhydrase activity in the presence of a number of inhibitors of the enzyme. Inhibition was essentially similar in all cases (*Table 2*), except for the resistance of spinach leaf enzyme to sulfhydryl inhibitors. This characteristic was not found in other C_3 plants.

Inhibition by Diamox of photosynthesis in intact leaves

In intact leaves of both spinach (*Figure 1a*) and sugarcane, var. Pindar, (*Figure 1b*) Diamox applied from the leaf base or petiole progressively inhibits photosynthesis. These experiments, using the infra-red analyzer were performed with the co-operation of Dr. T.A. BULL, and showed spinach to be more susceptible to inhibition by Diamox than sugarcane.

Thus, at ambient CO_2 levels, carbonic anhydrase activity is essential for maintaining maximal rates of photosynthesis of two species with differing pathways of initial CO_2 fixation. The lack of effect of sulfanilimide in these experiments is consistent with the absence of inhibition of CO_2 fixation in isolated spinach chloroplasts by this compound (3).

Generally similar results were obtained in experiments using an oxygen electrode with immersed leaf segments (epidermis detached) from spinach and *Amaranthus*. Diamox-inhibition of photosynthesis in immersed leaf segments could be partly overcome by increasing the bicarbonate concentration.

Diurnal and light-dependent fluctuation in carbonic anhydrase activity

TABLE 3

Carbonic anhydrase activity (enzyme units mg^{-1} chlorophyll) in leaves from plants placed in high and low light (100 and 8 × 10^3 lux respectively) at 18° for four days.

Species	Light intensity	
	High	Low
Pisum sativum	3125	1410
Spinacia oleracea	6950	1275
Amaranthus viridis	2675	198
Zea mays	1135	590

In leaves of young spinach plants in daylight, enzyme activity per unit leaf area rose to double the value in darkened controls during the morning and fell to a common base level during the afternoon. Such diurnal fluctuations were not found in *Amaranthus* although shaded leaves consistently had less activity than ones in full daylight. WAYGOOD and CLENDENNING (9) have shown that the level of the enzyme in *Tropaeolum* and *Petroselinum* plants has a long term dependence on light; the carbonic anhydrase level fell on darkening the plant for 4 days. This observation has been confirmed by placing several other plants including both C_4 and C_3 species in high and low light intensities in growth cabinets (*Table 3*).

Short term effects of darkening were studied on plants grown under ambient illumination in the greenhouse. Whereas a component of the

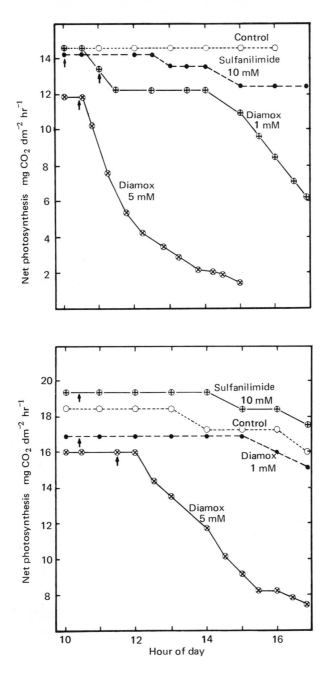

Figure 1 Inhibition of photosynthesis in detached leaves by Diamox, (a) (top) spinach, (b) (bottom) sugarcane.

short term fluctuation in carbonic anhydrase level appeared to be light-dependent (*Table 4*), a major component of activity was not affected by light in either spinach or the two C₄ plants assayed. For example, in *Table 4*, the last sample from corn, an already low value in the light is not further depressed by darkness. Activity in such extracts was not increased by incubation with 5 mM dithiothreitol

(DTT).

In view of the report (4) that carbonic anhydrase was active in *Chlorella* only at low CO_2 concentrations evidence of a similar control in leaves was sought, but as anticipated, the complicating effect of stomatal response to both light and CO_2 concentrations hindered these efforts.

TABLE 4

Changes in carbonic anhydrase level (enzyme units cm^{-2} leaf) on darkening.

Species	Time in darkness (minutes)			
	0	15	30	60
Zea mays	3300	–	2300	–
"	2200	–	1100	1100
"	1400	–	1400	1400
Amaranthus edulis	11000	–	3300	4200
Spinacia oleracea	12900	6300	4900	3200

DISCUSSION

From the foregoing results it would appear that carbonic anhydrase plays a part in maintaining maximum rates of photosynthesis both in plants with the C_4 pathway of photosynthesis, in which CO_2 is trapped via a carboxylation reaction prior to its participation in the Calvin cycle, and in plants in which the Calvin cycle operates with the ambient carbon dioxide reacting directly with ribulose-1,5-diphosphate (RuDP) carboxylase.

Requirements for maintaining a CO_2 trap in the latter group of plants (via a pH gradient) have been suggested in connection with photosynthesis by spinach chloroplasts (3). As carbonic anhydrase cannot influence equilibrium concentrations of CO_2 and HCO_3 , it might be supposed that the enzyme catalyzes a permease reaction (7) or has a catalytic role in making CO_2 available for RuDP carboxylase within an alkaline environment which would favor bicarbonate accumulation. Theoretically either role could be effectively carried out by carbonic anhydrase (10).

However, an additional physiological role can be suggested for this enzyme in buffering rapidly from ambient CO_2 the fast pH changes that follow illuminating or darkening the thylakoids of chloroplasts (11,12). This well-documented response of chloroplasts to light could in theory be utilized in concentrating CO_2 for priming the Calvin cycle.

A positive partial pressure of CO_2 is maintained within leaves during photosynthesis via the generation of CO_2 from photorespiration and could provide a source of carbonic acid to buffer sudden cation fluxes. In C_4 plants the whole bundle sheath cell may act as a CO_2 concentrating organ, and carbonic anhydrase may not then be needed to carry out the mobilization of CO_2.

In preliminary experiments by the author at $0°$ light-induced pH changes were much smaller with whole spinach chloroplasts than with chloroplast fragments, and 1 mM Diamox usually reduced the rate and magnitude of the change in whole chloroplasts but not in chloroplast fragments. Experiments so far have been complicated by the presence of dilute phosphate buffer and by the weakly acid inhibitor and further work is in progress.

REFERENCES

1. A.J. TOBIN, *J. biol. Chem.*, 245 (1970) 2656.
2. R.G. EVERSON, *Nature*, 222 (1969) 876.
3. R.G. EVERSON, *Phytochemistry*, 9 (1970) 25.
4. M.L. REED and D. GRAHAM, *Plant Physiol.*, 43 (1968) 329.
5. M. IKEMORI and K. NISHIDA, *Physiologia Pl.*, 21 (1968) 292.
6. G.W. BOWES, *Plant Physiol.*, 44 (1969) 726.
7. R.G. EVERSON and C.R. SLACK, *Phytochemistry*, 7 (1968) 581.
8. T.H. MAREN, *Physiol. Rev.*, 47 (1967) 597.
9. E.R. WAYGOOD and K.A. CLENDENNING, *Can. J. Res. C.*, 28 (1950) 673.
10. T. ENNS, *Science*, 155 (1967) 44.
11. J. NEUMANN and A.T. JAGENDORF, *Archs Biochem. Biophys.*, 107 (1964) 109.
12. P.S. NOBEL, *Biochim. biophys. Acta*, 172 (1969) 134.

ADDENDUM: EFFECTS OF AN INHIBITOR OF CARBONIC ANHYDRASE ON LIGHT-INDUCED pH CHANGES IN PEA CHLOROPLASTS

R.G. Everson and D. Graham

Department of Biology and Environmental Science, Queensland Institute of Technology, Brisbane, 4000, and Plant Physiology Unit, CSIRO, Division of Food Research and School of Biological Sciences, Macquarie University, North Ryde, 2113, Australia

Preparations containing predominantly Class I chloroplasts were obtained from pea leaves by modifications of the methods of EVERSON

(1) and WALKER *et al.* (2). These preparations were able to carry out a light-induced pH increase of about 0.1 pH unit at 8° with an initial pH about 6.5.

Addition of 6×10^{-5} M ethoxyzolamide (Upjohn) a specific inhibitor of carbonic anhydrase in the dark resulted in a substantially faster initial rate and about 50% greater extent of pH increase on subsequent illumination.

The effect of ethoxyzolamide was overcome and the pH change was depressed by the subsequent addition of 2×10^{-3} M NaHCO$_3$. Chloroplast preparations broken by washing in hypotonic medium showed a higher rate and a greater extent of light-induced pH increase (approx. 0.4 pH unit) which was not significantly altered by ethoxyzolamide. The broken chloroplast preparations contained carbonic anhydrase which was a substantial proportion of that found in the whole chloroplast (Class I) preparations.

These results, which are opposite in sign from those referred to above by one of us (R.G.E.), may be interpreted to indicate that carbonic anhydrase mediates in a buffer action at the unknown, but presumably low, CO_2 concentrations of the chloroplast preparations. The effect of carbonic anhydrase is to counteract the rapid light-induced pH rise. When the enzyme is inhibited an excess of CO_2 provided by the addition of bicarbonate presumably may serve the same function. The presence of the outer limiting membrane in the Class I chloroplasts appears essential for carbonic anhydrase to affect the pH change since in broken chloroplast suspensions still containing the enzyme the pH change was maximal and unaffected by the inhibitor, ethoxyzolamide. The enzyme thus appears to play a significant role in the light-induced pH change by whole chloroplasts. However, the condition of the chloroplasts is apparently critical to the response obtainable.

1. R.G. EVERSON, *Phytochemistry*, 9 (1970) 28.
2. D.A. WALKER, C.W. BALDRY, and W. COCKBURN, *Plant Physiol.*, 43 (1968) 1419, and D.A. WALKER, personal communication, (1970).

STUDIES ON CO_2 FIXATION BY SUCCULENT PLANTS IN THE LIGHT

Manfred Kluge[1]

Botanisches Institut, Technische Hochschule,
6100 Darmstadt, Germany.

SUMMARY

The distribution of radioactivity in products of $^{14}CO_2$ fixation in the light in Bryophyllum tubiflorum varied at different times of the light-dark cycle. When tissue malate level was low, $^{14}CO_2$ fixation in the light into malate exceeded that in sugars. When tissue malate level was high, sugars were the predominant products of $^{14}CO_2$ fixation in the light. The data are discussed in terms of a feedback control of phosphoenolpyruvate carboxylase activity. The relationship between Crassulacean acid metabolism and C_4 photosynthesis is discussed.

Many species of succulent plants are well known to accumulate large amounts of malate during the night period. The malate synthesis is due to a dark fixation of CO_2. These succulents fix not only the respiratory CO_2 produced in the dark, but also consume much of the CO_2 from the atmosphere surrounding the plants. Net CO_2 uptake during the night can be detected easily when the gas exchange of succulents is measured by infra red gas analysers (1).

In contrast, a striking depression of net CO_2 uptake during most of the light period is shown by these plants. Normally there is net CO_2 fixation for only about one hour in the beginning of the light phase and again during some hours at the end of the light phase. Preferential photosynthetic consumption of CO_2 derived from the internal malate pool, rather than CO_2 from the atmosphere, seems to be responsible for the depression of CO_2 uptake during the light. The extent in time of this depressed CO_2 uptake from the air is closely correlated with the size of the malate pool at the beginning of the light period and with the velocity of malate consumption by photosynthesis(2).

Thus CO_2 from two sources, the malate pool and the surrounding

---

[1] *Address during 1970-71, Research School of Biological Sciences, Australian National University, supported by Deutsche Forschungsgemeinschaft.*

atmosphere, seems to be available for the CO_2 fixation by succulents during the light period. However, the problem of how CO_2 is fixed in the light by succulents is still open. This problem became evident when $^{14}CO_2$ was fed to phyllodia of illuminated *Bryophyllum* at different times of the day-night cycle. The standard conditions of these $^{14}CO_2$ feeding experiments were 10 min preillumination (8000 lux) followed by $^{14}CO_2$ fixation for further 15 min (8000 lux).

When $^{14}CO_2$ was fed in the light to phyllodia sampled early in the day period, sucrose was found to be the main labeled product. In subsequent feeding experiments done during the light period, the proportion of radioactivity transfered to sucrose decreased, with more label being transfered to malate. Finally, malate was found as the most heavily labeled product of $^{14}CO_2$ fixation in the light during the late day period (3). Malate seemed to be the predominant labeled product during those phases of the day-night cycle, in which net CO_2 uptake was observed. Furthermore, with some approximation radioactive malate was found to be the dominant product when the phyllodia were in the de-acidified state and sucrose to be the most heavily labeled product of $^{14}CO_2$ fixation in the light in the acidified state during the natural diurnal cycle.

Considering these findings, a direct causal connection was proposed between the malic acid content in the tissue and the readiness with which radioactive carbon can be transfered into malate during the light fixation of $^{14}CO_2$. If in fact the extent of malate labeling is controled directly by the malate content of the tissue, it should be possible to influence the labeling patterns produced during $^{14}CO_2$ light fixation by manipulating the malate level of the tissue experimently.

This manipulation was done in two types of experiment. In some experiments the de-acidification of part of a phyllodia population was accelerated by application of higher light intensities than to the controls. Those phyllodia with lower malate content always fixed more radioactivity into malate than those with higher malic acid content. Sucrose labeling was the inverse of malate *(Figure 1)*. There was no distinction in the labeling patterns between the two groups of phyllodia in the early light period, when the difference in the malate content was not yet apparent, and the labeling patterns became most distinct when the difference in the malic acid level between the two phyllodia populations reached its maximum during the afternoon. In further experiments phyllodia with different malate content at the beginning of the light period were obtained by inhibition of malate accumulation during the previous night in one part of the phyllodia population and by promoting malate synthesis in the other part (application of different temperatures during the night). When $^{14}CO_2$ was fed to the well acidified phyllodia, most of the radioactive carbon was accumulated in sucrose *(Figure 2)*. On the other hand, those phyllodia in which malate accumulation was inhibited fixed most of the radioactivity into malate, as did all the phyllodia the previous evening. However, the difference in the

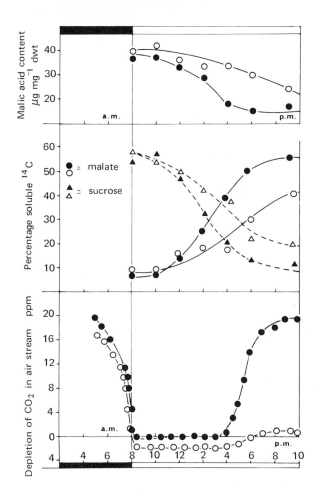

Figure 1 Labeling of malate and sucrose during 15 min $^{14}CO_2$ photosynthesis in phyllodia of Bryophyllum tubiflorum with slow (open symbols 8000 lux) and accelerated (filled symbols 17000 lux) de-acidification.

labeling patterns between the phyllodia with high and those with low malate content was observed only after a preillumination of about one hour. When the two groups of phyllodia were fed with $^{14}CO_2$ just after the transfer to light at the end of the dark period, no differences were observed in labeling patterns between acidified and non-acidified tissue, although the total malic acid content was no different than that one hour later, when great differences in the labeling patterns were apparent.

At this state of investigation the following *hypothesis* would fit the findings to some extent. In the light, two CO_2 fixing systems are operating and competing for CO_2 (or perhaps also for CO_2 acceptors). One of these two systems should be identical with the pathway of β-carboxylation of phosphoenolpyruvate (PEP), involving PEP carboxylase and the subsequent reduction of the product (oxaloacetate) to malate via malate dehydrogenase. This system is also responsible

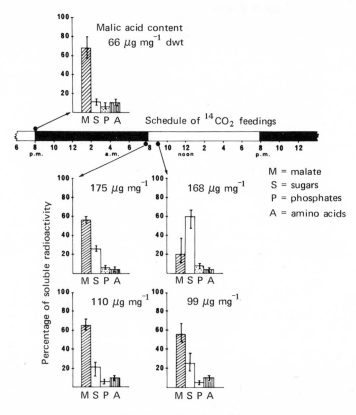

Figure 2 Labeling patterns after 15 min $^{14}CO_2$ fixation in the light in phyllodia of Bryophyllum tubiflorum differing in malate content (controled by regulation of temperature in the dark phase).

for the powerful CO_2 fixation and malate synthesis by succulents during the night and would be identical with the pathway of the primary CO_2 fixation during the light in C_4 plants. The other CO_2 fixing system in succulents in the light should be identical with that of the Calvin cycle ribulose-1,5-diphosphate (RuDP) carboxylase. If PEP carboxylase is not inhibited (de-acidified tissue), most of the $^{14}CO_2$ applied to the succulent tissue in the light would be fixed by this system and would therefore flow into malate, for PEP carboxylase has a much higher affinity to CO_2 than RuDP carboxylase. However, PEP carboxylase seems to be inhibited by malate (4) and so a high malic acid content in the cells would induce a feedback inhibition of this enzyme. In highly acidified tissue the malate producing PEP carboxylase system therefore can not compete for CO_2. Most of applied $^{14}CO_2$ could now be fixed by the Calvin cycle and flow into sucrose.

The changed labeling patterns described involve not only fluctuation in malate acid content of the tissue, but also the compartmentation of the metabolic pathways. Feedback inhibition of PEP carboxylase can only be expected if the malate content rises up to a critical concentration in the cytoplasm (plasm organelles are here included) where the enzymes are localized. Light treatment seems to enhance the permeability of the malate storing vacuole, increasing the outflow of malate from this store, flooding the cytoplasm with malate and finally resulting in feedback inhibition of PEP carboxylase. At the moment this seems to be the simplest explanation of the finding that the labeling patterns of acidified and non-acidified phyllodia differed only after a preillumination.

Considering the striking importance of the CO_2 fixation via PEP carboxylase in succulents as well as in C_4 plants, the assumption that the path of photosynthetic CO_2 fixation may be fundamentally similar in the two types of plants, seems relevant. Indeed succulents fix most, and C_4 plants probably fix all exogenous CO_2 via PEP carboxylase first into C_4 dicarboxylic acids (mainly malate). From there the carbon is transfered into carbohydrates. The pathway which transfers carbon from malate to sugars should be the same in succulents and in C_4 plants. Malic enzyme is involved in the decarboxylation of malate, producing CO_2, and this carbon dioxide probably is refixed in Calvin cycle photosynthesis. However CO_2 refixation by PEP carboxylase must be prevented if the carbon is to finally flow into carbohydrates.

In succulent plants this is probably attained by feedback inhibition of the PEP carboxylase pathway as long as malate remains in the vacuoles to be transformed into carbohydrates. In C_4 plants the CO_2 fixing pathways are spatially separated, and here refixation of CO_2 from malate by PEP carboxylase is prevented by transporting the product of β-carboxylation pathway (malate) from the tissue where it was produced into another tissue before decarboxylation and the refixation of CO_2 into carbohydrates and compounds other than C_4 acids.

It is not yet clear if there are differences between C_4 plants and succulents in the fate of the C_3 residue (pyruvate) produced by decarboxylation of malate in the light. C_4 plants are known to contain a powerful pyruvate,Pi dikinase, recycling pyruvate into the CO_2 acceptor PEP. Our most recent experiments give some evidence that this enzyme might also be active in succulents. The pyruvate to PEP converting enzyme activity was comparable to that of RuDP carboxylase. NADP dependent malate dehydrogenase, another key enzyme of C_4 metabolism, could also be demonstrated in succulents. At the moment it is difficult to decide if pyruvate,Pi dikinase and NADP-malate dehydrogenase are of as fundamental importance for CO_2 fixation during the light in succulents as they are in C_4 plants.

REFERENCES

1. E.L. NUERNBERGK, *Planta*, 56 (1961) 28.
2. M. KLUGE, *Planta*, 80 (1968) 359.
3. M. KLUGE, *Planta*, 88 (1969) 113.
4. O. QUEIROZ, *C. R. Acad. Sci. (Paris)*, 265 (1967) 1928.

CRASSULACEAN ACID METABOLISM AND THE C$_4$ PATHWAY OF PHOTOSYNTHESIS IN SUCCULENT PLANTS

P.N. Avadhani, C.B. Osmond,[1] and K.K. Tan[2]

Department of Botany, University of Singapore, Singapore 10

SUMMARY

Leaves of Sedum prealtum *and* Bryophyllum calycinum *which exhibit marked Crassulacean acid metabolism, showed malate as the principal product of 5 sec* $^{14}CO_2$ *fixation in the light. This malate was largely labeled in the C-4 position in contrast to the "Crassulacean malate", and malate formation in the light was prominent only in tissues with low acid content. When acid content was high, leaves of* Sedum *in the light fixed* $^{14}CO_2$ *into sugars and phosphorylated compounds while in* Bryophyllum *the fixation of external* CO_2 *was drastically reduced. Bisulfite compounds inhibited fixation both in light and in the dark. In* Bryophyllum, *the fixation of* CO_2 *in light also decreased with age and maturity of leaves, concomitant with an increase in the magnitude of Crassulacean acid metabolism.*

INTRODUCTION

The diurnal fluctuation of acidity characteristic of the leaves of "Crassulacean" plants is principally due to the changes in malate concentration (1) produced via phosphoenolpyruvate (PEP) carboxylase and malate dehydrogenase (2,3). It is, therefore, of considerable interest that the photosynthetic CO_2 fixation in the C$_4$ plants is also mediated by PEP carboxylase (4). In the following experiments comparative fixation of $^{14}CO_2$ in light and dark by leaves of *Sedum prealtum* and *Bryophyllum calycinum* was followed. The effect of α-hydroxypyridine methane sulfonate (HMS) on the two processes was investigated. The nature and magnitude of fixation by leaves containing different

---

[1] *Research School of Biological Sciences, Australian National University, Canberra City, 2601, Australia.*

[2] *Present address: Department of Botany, Manchester University, Manchester, U.K.*

levels of acid was also studied. In the case of *Bryophyllum*, the effect of age on photosynthetic CO_2 fixation and Crassulacean acid metabolism (CAM) is compared.

EXPERIMENTAL

Leaf samples were taken at appropriate times in the day-night cycle and acid content measured by titration of extracts. Samples were exposed to $^{14}CO_2$ in the light (25 mW cm^{-2}) after 10 min pretreatment or in the dark and the distribution of radioactivity in extracts determined as described elsewhere (5). Inhibitor solutions were injected into leaf tissue as described previously (6).

RESULTS AND DISCUSSION

$^{14}CO_2$ *fixation in Sedum prealtum in the light or dark*

The products of dark fixation in *Sedum* leaves exposed to $^{14}CO_2$ after deacidification in the light are shown in *Table 1*. The principal product of $^{14}CO_2$ fixation was malate which on degradation, showed close to the characteristic 66% labeling of the C-4 of malate (7).

TABLE I

Percentage distribution of radioactivity among products of fixation of $^{14}CO_2$ in the dark and light by leaf slices of *Sedum prealtum*.

Compound	Dark fixation		Light fixation		
	30 min	60 min	5 sec	30 sec	3 min
malate	80	84	59	49	39
amino acids	5	4	13	15	15
phosphorylated compounds	0	0	23	28	22
sugars	0	0	8	6	24
C-4 of malate	62	60	100	89	53

A similar $^{14}CO_2$ fixation experiment was done in the light (25 mW cm^{-2}) using tissue that had been deacidified overnight in the light at 28°. *Table 1* shows that malate was initially the principal radioactive product but with prolonged treatment in $^{14}CO_2$, the label in sugars increased. Unlike the dark fixation malate, malate formed in the light in short term exposures was almost 100% C-4 labeled. The proportion of ^{14}C in this carbon of malate declined with time. These results are reminiscent of C_4 photosynthesis and implicate PEP carboxylase in the initial carboxylation in the light. They further suggest parallel or subsequent incorporation of label into Calvin cycle

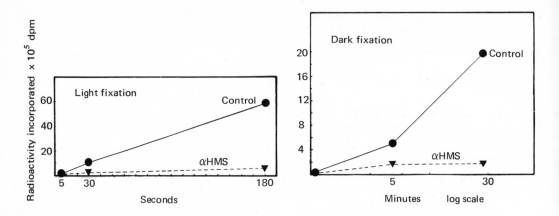

Figure 1 Effect of HMS on $^{14}CO_2$ fixation in the light and dark in leaves of Sedum prealtum.

intermediates.

Effect of bisulfite compounds

Treatment of leaves with 10 mM HMS inhibited both light and dark CO_2 fixation in Sedum leaves (Figure 1), suggesting the involvement of PEP carboxylase in both processes (6). However, recent experiments (unpublished) have shown that the effects of HMS on PEP carboxylase may be masked by nonspecific effects on photosynthesis and plant cell membranes.

CO_2 fixation in the light and tissue acid content

When Sedum leaves were harvested at different periods of the day-night cycle and exposed to $^{14}CO_2$ in the light for 5 sec, following 5 min preillumination, it was found that malate was rapidly labeled on-ly in tissues of low acid content. Figure 2 shows that malate label declined and sugar phosphate label increased as tissue acid content increased in the night period. Throughout these experiments the C-4 carbon of malate contained 81-95% of the total malate radioactivity.

These data suggest that CO_2 fixation in the light may proceed via PEP carboxylase or via ribulose-1,5-diphosphate (RuDP) carboxylase and that endogenous malate concentration appears to be one of the principal regulating factors.

In Bryophyllum similar results were obtained during deacidification in the light following the night period. Leaves were exposed to $^{14}CO_2$ for 5 sec and the fixation of ^{14}C as a function of tissue mal-ate content is shown in Table 2. As the malate content of the tissue declined during deacidification the fixation of exogenous CO_2 in-creased. Malate and aspartate were the main products of $^{14}CO_2$ fixa-tion in low-acid tissue.

The effect of age on $^{14}CO_2$ fixation in the light

The effect of malate content on $^{14}CO_2$ fixation, shown in Figure 2, prompted a study of $^{14}CO_2$ fixation in the light as a function of leaf age in Bryophyllum. CAM increases with age (1) and Table 3 shows the magnitude of the diurnal cycle in terms of total acidity and malate

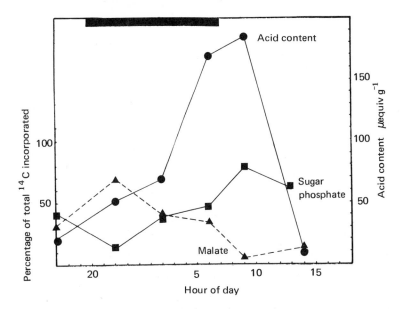

Figure 2 Distribution of radioactivity amongst products of 5 sec $^{14}CO_2$ fixation in the light in relation to the acid content of Sedum prealtum leaves throughout the day-night cycle.

TABLE 2

Relationship between malate content (μequiv g^{-1}) and $^{14}CO_2$ fixation (dpm g^{-1}) in mature *Bryophyllum* leaves.

Malate content	$^{14}CO_2$ fixation
65	47
62	317
50	1733
38	5715

content. When partly deacidified leaves were exposed for 5 sec to $^{14}CO_2$ in the light, it was found that the ability to fix CO_2 in the light declined markedly with age, concomitant with the increase in diurnal fluctuation of malate content. In all experiments malate and aspartate were the main products of $^{14}CO_2$ fixation in the light. These data suggest that with the increase in CAM, dependence of *Bryophyllum* leaves on the photosynthetic fixation of CO_2 declines.

CONCLUSIONS

These experiments suggest that succulent plants with CAM may fix exogenous CO_2 in the light via PEP carboxylase or RuDP carboxylase. The significance of these processes in the carbon balance of CAM is

TABLE 3

Acid fluctuation (μequiv g^{-1}) in *Bryophyllum* leaves from plants of different ages and its effect on $^{14}CO_2$ fixation (cpm g^{-1}). (Malate content in parenthesis).

Age (weeks)	0 hr	16 hr (dark)	24 hr (light)	$^{14}CO_2$ fixation (5 sec)
2	26	34	21	854
8	21 (33)	114 (95)	76 (43)	169
12	21 (25)	158 (128)	27 (18)	90
Mature plant	39 (11)	223 (131)	65 (18)	4

unknown but they indicate the nature of reactions involved in refixation in the light of CO_2 evolved from dark acidification malate. The experiments show that tissue malate content is an important factor, controlling first the extent of CO_2 fixation in the light via PEP carboxylase or RuDP carboxylase, and second the magnitude of CO_2 fixation in the light. The mechanism of this interaction is not clear at this time.

Studies on the distribution of label within malate formed during light and dark $^{14}CO_2$ fixation suggest that the source of PEP may differ in the two conditions. In the dark PEP is probably already labeled and is derived from a double carboxylation reaction as suggested earlier (6). In the light the PEP is unlabeled and may be derived from pyruvate via a pyruvate,Pi dikinase of the type reported in C_4 leaves (8, see KLUGE, this volume).

At present it is difficult to see how malate synthesis in the light can contribute to net carbon gain in CAM plants. In these plants both carboxylase enzymes appear to be in the same cell. Labeling of malate in the light may simply reflect the greater affinity of PEP carboxylase for CO_2. Presumably, both carboxylases are active in the chloroplasts of CAM leaves (9). If malate formed in the light is decarboxylated by malic enzyme, the $^{14}CO_2$ released presumably mixes with the free CO_2 pool in the cell and some may be fixed by RuDP carboxylase. If on the other hand, this malate is rapidly transported to the vacuole, in the manner of dark fixation malate, it may contribute to net carbon gain during subsequent deacidification. Malate accumulates during net CO_2 fixation in prolonged light periods (10) suggesting some light fixation malate is transported to the vacuole. The decrease in malate label with time during $^{14}CO_2$ fixation in the light and the randomization of malate radioactivity (*Table 1*), argue against this interpretation.

It is known that the ability of succulent leaves to exhibit CAM increases with the maturation of the leaf and the environmental conditions (1). Thus the data presented here indicate the possibility that the role of photosynthetic CO_2 fixation could also exhibit corresponding variations. If this is so, these plants may represent a situation intermediate between the C_3 and the C_4 types of higher plant photosynthesis.

ACKNOWLEDGMENTS

This work was initiated during an ANU Visiting Fellowship awarded to P.N.A. We wish to further acknowledge the generous financial assistance provided by the Commonwealth Foundation to one of us (P.N.A.) to attend this Conference.

REFERENCES

1. S.L. RANSON and M. THOMAS, *A. Rev. Pl. Physiol.*, 11 (1960) 81.
2. C.H. COLES and E.R. WAYGOOD, *Can. J. Bot.*, 35 (1957) 25.
3. D.A. WALKER, *Biol. Rev.*, 37 (1962) 215.
4. M.D. HATCH and C.R. SLACK, *A. Rev. Pl. Physiol.*, 21 (1970) 141.
5. C.B. OSMOND and P.N. AVADHANI, *Aust. J. biol. Sci.*, 21 (1968) 917.
6. C.B. OSMOND and P.N. AVADHANI, *Plant Physiol.*, 45 (1970) 228.
7. J.W. BRADBEER, S.L. RANSON, and M.L. STILLER, *Plant Physiol.*, 33 (1958) 66.
8. M.D. HATCH and C.R. SLACK, *Biochem. J.*, 112 (1969) 549.
9. S.K. MUKERJI and I.P. TING, *Phytochemistry*, 7 (1968) 903.
10. M. KLUGE, *Planta*, 88 (1969) 113.

CO_2 FIXATION: ASSESSMENT

D.A Walker

Botany Department, University of Sheffield,
Sheffield S10, 2TN, England.

As now formulated, all photosynthetic carbon traffic in C_4 plants passes through the reaction catalyzed by ribulose-1,5-diphosphate (RuDP) carboxylase, just as it does in C_3 plants. This statement recognises a fact, sometimes overlooked, that the Benson-Calvin cycle remains as the only known carboxylating system which can lead to a simultaneous accumulation of product and an increase in the concentration of its CO_2-acceptor. In addition it would appear that C_3, C_4, and plants with Crassulacean acid metabolism (CAM) also operate a second carboxylation, catalyzed by phosphoenolpyruvate (PEP) carboxylase, leading to the appearance of labeled oxaloacetate, malate, and aspartate.

In terms of overall carbon traffic two factors then distinguish C_4 from C_3 plants.

(i) In C_3 plants, photosynthetically produced malate is probably an end product of photosynthesis (in the same sense as sucrose is an end product) and its further metabolism probably involves reactions such as those of the Krebs cycle. In contrast, malate formed in the light in C_4 plants is clearly an intermediate of photosynthesis and where $^{14}CO_2$ feeding is followed by $^{12}CO_2$ feeding there is an almost quantitative transfer of label from malate to sugar phosphates.

(ii) There is present in C_4 plants, and possibly in CAM plants, an active pyruvate-phosphate dikinase. This together with PEP carboxylase, malic dehydrogenase, and malic enzyme allows the operation of a carboxylation-decarboxylation cycle which cannot in itself lead to net incorporation of carbon but which may do so if linked to the Benson-Calvin cycle or to catabolic generation of phosphoenolpyruvate. It is suggested, largely on the basis of enzyme distribution studies, that this system operates in mesophyll chloroplasts but there seems to be no compeling reason why some of the reactions (e.g. the carboxylation of PEP) should not occur, at least in part, in the mesophyll cytoplasm. Direct inferential evidence for the synthesis of PEP by the dikinase was presented by C.C. BLACK who showed that pyruvate (but not sugar phosphates) stimulated light assimilation of CO_2 by mesophyll cells isolated from *Digitaria*.

COMPARTMENTATION

It seems clear from the work of HATCH and SLACK, BJÖRKMAN, and others that the operation of C_4 assimilation calls for some degree of compartmentation in addition to the presence of the necessary enzymes. This compartmentation is physically manifested by the presence of "Kranz-type" anatomy which essentially comprises 3 concentric cylinders of tissue - the vascular bundle, bundle sheath, and mesophyll cells. Apparently in some species of C_4 plants the distribution of enzymes within these compartments is such that β-carboxylation occurs within the mesophyll and malate (derived from the carboxylation product) passes into the bundle sheath where it is decarboxylated by malic enzyme so that there is, in effect, an inward transport of CO_2 and reduced NADP. The released CO_2 is believed to be fixed by RuDP carboxylase. Pyruvate and some 3-phosphoglycerate (3-PGA) is then probably exported from the bundle sheath to the mesophyll. The extent to which CO_2 enters the bundle sheath directly and the extent to which PEP is derived from pyruvate or from 3-PGA is still uncertain and may well vary from plant to plant or be governed by other metabolic considerations. In some C_4 plants, aspartate is apparently a more important short term fixation product than malate and may provide an alternative pool. This is said to be correlated with lower levels of malic enzyme and higher levels of amino transferases. Present investigation into the metabolic processes of "aspartate formers" will be complicated by the fact that actual net changes in amino acids will always be smaller than those suggested by ^{14}C studies because of rapid, enzyme-catalyzed "exchange transamination". Thus a small, rapidly-fluxing, pool of (^{14}C)oxaloacetate will inevitably pass label to (^{12}C)aspartate in the presence of the appropriate amino transferase, even if the aspartate pool is otherwise entirely static. The recently reported enzyme levels in "aspartate formers" is suggestive but unsuccessful first attempts to demonstrate high levels must continue to be regarded with extreme caution.

In examining compartmentation it is possibly instructive to compare C_4 and CAM plants. The suggestion by LAETSCH that C_4 metabolism is "CAM mit Kranz" is worth serious consideration. In crude terms there seems little doubt that C_4 plants operate C_3 photosynthesis and CAM simultaneously in the light. This, according to HATCH and SLACK, would be possible (a) because diphosphoglyceric acid produced in the light would maintain a sufficiently high level of oxidized NADP to allow rapid oxidative decarboxylation of malate, (b) because the compartmentation of enzymes between mesophyll and bundle sheath would prevent back reactions such as immediate refixation of released CO_2 by PEP carboxylase - a reaction which BJÖRKMAN believes may occur in some of his C_3-C_4 crosses.

Although it is not the purpose of this particular assessment to review the whole C_4 syndrome in depth it seems necessary to make one extrapolation at this point. That is, that in accord with LAETSCH'S suggestion, compartmentation allows a spacial separation of biochemical events which is achieved by a diurnal separation of Crassulacean plants. One ecological advantage of the C_4 syndrome appears to be that it gives a high ratio of CO_2 fixed/water lost by transpiration. This is postulated to be a consequence of the PEP carboxylase mediated fixation of CO_2 in the mesophyll and the consequent transport of malate or aspartate to the bundle sheath. Subsequent decarboxylations would then allow a high steady state level of CO_2 within the bundle sheath existing side by side with a low CO_2 concentration in the mesophyll. This system (incorporating as it does both the high affinity for CO_2 characteristic of PEP carboxylase and well developed compartmentation) may operate efficiently even in the presence of a diffusive resistance imposed by partly closed stomata. As NEALES has shown, this may be taken to extremes in Crassulacean plants where lack of compartmentation necessitates temporal rather than spacial separation of the relevant biochemical processes and the entire CO_2 uptake may occur in the dark with stomata remaining closed during the day.

This assessment must also include three unambiguous observations which are difficult to equate with the proposed carbon traffic as presently detailed

(a)   GIBBS has shown that when maize leaves are pre-illuminated in CO_2-free air the subsequent dark assimilation of $^{14}CO_2$ leads to malate with the characteristic 1/3:2/3 labeling pattern associated with CAM. This implies that under *these* conditions RuDP carboxylation *precedes* PEP carboxylation and that there are no pools of 3-PGA and PEP to which the carboxylating mechanisms have free access.

(b)   Chloroplasts, prepared by GIBBS from young maize leaves, are able to carry out photosynthetic carbon assimilation with CO_2 but are unable to utilize malate.

(c)   Bundle sheath cells prepared by BLACK from *Digitaria* can assimilate CO_2 but not malate.

It is just conceivable that the observations of BLACK (i.e. that mesophyll cells respond to pyruvate whereas bundle sheath cells respond to ribose-S-phosphate but not malate) may be all explained in terms of different cell structure and closed plasmodesmata. However the added observation of lack of utilization of malate by chloroplasts must also raise, once again, the question of the degree of cytoplasmic involvement.

THE C_4 PATHWAY: ASSESSMENT

C.R. Slack

Plant Physiology Division, DSIR, Palmerston North, New Zealand.

There is now a large measure of agreement regarding the overall mechanism of CO_2 fixation by the C_4 pathway. However, several points of detail remain to be elucidated and these are currently the subject of considerable discussion. In essence the pathway consists of two interlinked cycles, one containing phosphoenolpyruvate (PEP) carboxylase, the other ribulose-1,5-diphosphate (RuDP) carboxylase, which are joined by a decarboxylation reaction (1,2). It is recognised that the successful operation of this pathway is largely dependent upon the spatial separation of these two carboxylases and on a Kranz-type leaf anatomy that is a characteristic of C_4 pathway species (3).

Investigations of the distribution of enzymes between the dimorphic chloroplasts of the mesophyll and bundle sheath of maize have largely been responsible for formulating the present schemes of the C_4 pathway (4,5). It was found that enzymes implicated in the formation of PEP and in the reduction and amination of oxaloacetate were localized in the mesophyll chloroplasts, whereas the leaf's complement of Calvin cycle enzymes and malic enzyme were restricted to the bundle sheath chloroplasts. Consequently it was proposed (1,5) that CO_2 is incorporated into dicarboxylic acids in the mesophyll and transferred, as the β-carboxyl of malate, to the bundle sheath chloroplasts where it is liberated by malic enzyme and incorporated into sugars via the Calvin cycle. The pyruvate, liberated in the oxidative decarboxylation of malate, being returned to mesophyll chloroplasts, and there phosphorylated by pyruvate,Pi dikinase. These proposals have been largely substantiated by investigations of the CO_2-fixing capabilities of mesophyll and bundle sheath cells *in vitro* (2). However, the inability of isolated bundle sheath cells to utilize malate as a donor of CO_2 (2), despite the fact that they contain malic enzyme and Calvin cycle enzymes, raises the question (see below) as to the mechanism by which malate could be transferred from the mesophyll to bundle sheath cells.

A topical point of discussion, in view of the demonstration that the bundle sheath chloroplasts contain a full complement of Calvin

cycle enzymes, is whether or not this cycle can function in an autonomous manner to fix externally derived CO_2. It has generally been
considered that the flux of CO_2 from substomatal cavities directly
to the bundle sheath may be severely limited by a scarcity of connecting air channels, and by the large amount of PEP carboxylase in
the mesophyll cells, which has been envisaged as an efficient trap
for CO_2 entering the leaf, and consequently that the Calvin cycle of
the bundle sheath can only function independently of the reactions
of the mesophyll cells to a very limited degree. This belief has
been supported by the demonstration, with maize, that essentially
all carbon which enters 3-PGA is derived from the β-carboxyls of the
dicarboxylic acids, malate and aspartate (6). The reported absence
of carbonic anhydrase from the bundle sheath of maize (7) may be an
added restriction to the fixation of CO_2 directly by the Calvin cycle, since carbonic anhydrase appears to be essential for high rates
of Calvin cycle photosynthesis. Additional information is required,
however, about the location of carbonic anhydrase in other C_4 pathway species in view of the report that this enzyme is present in
high levels in *Atriplex rosea* (3). The possibility that a fraction
of the leaf's Calvin cycle enzymes may, in some species, be located
in the mesophyll cells is related to the question of whether some
CO_2 is fixed directly by this cycle. Available evidence suggests
that the mesophyll chloroplasts of maize contain, at the most, only
a very small fraction of the leaf's complement of Calvin cycle enzymes (4). In *Digitaria sanguinalis*, however, this fraction could be
somewhat larger (2). The potential of mesophyll chloroplasts to carry out Calvin cycle photosynthesis may vary, therefore, in different
C_4 pathway species. However, it is conceivable that the availability
of CO_2 to small amounts of RuDP carboxylase in mesophyll cells could
be severely restricted by the large amount of PEP carboxylase present in these cells.

   The localization of PEP carboxylase is only partially understood.
There is good evidence that the enzyme is localized in the mesophyll
(2,5,8), but the evidence for its association with chloroplasts is
conflicting. Chloroplasts isolated from C_4 pathway species by mild
procedures have contained only a small proportion of the enzyme present within the leaf (9,10). On the other hand, mesophyll chloroplasts prepared in non-aqueous media have in certain experiments
contained most of the leaf's complement of the enzyme (4). The current view, therefore, is that PEP carboxylase has, at the most, only
a tenuous association with the mesophyll chloroplasts, and could
perhaps be linked to the chloroplast envelope. Other evidence, however, such as the rate of PEP carboxylase synthesis following the
illumination of etiolated leaves is consistent with the view that
the synthesis of the enzyme is closely linked with that of other enzymes present in mesophyll chloroplasts (11).

   The demonstration that illuminated mesophyll cells can incorporate CO_2 into dicarboxylic acids when supplied with pyruvate (2)
supports the view that pyruvate,Pi dikinase, located in the chloro-

plasts of these cells, can function to provide PEP from pyruvate. However, the fact that these cells can also utilize 3-PGA as a source of PEP (2) has raised the question as to which of the two C_3 acids is the more important precursor of PEP *in vivo*. The results of $^{14}CO_2$ labeling studies with maize, in high intensity illumination and under steady state conditions, have indicated that only a small fraction of the PEP that is carboxylated is derived from 3-PGA (12). However, other experiments in which maize leaves were illuminated in CO_2 free air then exposed to $^{14}CO_2$ in the dark have indicated that under these conditions 3-PGA is the principle source of PEP (13). It must be assumed therefore, that the relative importance of pyruvate and 3-PGA as precursors of PEP is to some degree dependent upon environmental conditions.

A considerable number of C_4 pathway species have now been classified as either "malate formers" or "aspartate formers" on the basis of the relative amount of ^{14}C present in these dicarboxylic acids after exposing leaves for brief periods to $^{14}CO_2$ (14). It has been stressed that this difference in the amount of label in the two acids need not in itself represent a biochemical difference because (a) the difference could be merely due to a variation in the pool sizes of aspartate and malate and (b) because aspartate aminotransferase can catalyze the exchange of ^{14}C between oxaloacetate and aspartate without a net movement of carbon. However, an examination of enzyme activities in the leaves of a limited range of C_4 pathway species does suggest that the above classification may have a biochemical basis (15). Leaf extracts of the "malate formers" examined to date, have contained high activities of NADP malate dehydrogenase and malic enzyme whereas those of the "aspartate formers" contain much less of these enzymes but high activities of alanine aminotransferase and aspartate aminotransferase. It has been tentatively suggested, therefore, that there may be two variations of the C_4 pathway, one, operative in the Panicoid grasses, in which malate acts as a CO_2 carrier between the mesophyll and bundle sheath, and one in which aspartate functions as a CO_2 carrier (1,14,15).

A major requisite for the successful operation of the C_4 pathway, as presently envisaged, is that certain intermediates of the pathway can move between the mesophyll and bundle sheath chloroplasts at rates sufficient to maintain the observed maximum photosynthesis rates. Information as to how this transport might be achieved is not available. Calculations of the maximum rates of diffusive transport based upon the rates of ion diffusion in cells indicates that the major limitation to movement by this process is the cross sectional area of plasmodesmata connecting the mesophyll and bundle sheath cells. No precise estimates of this area have been made, as yet, but certain calculations do indicate that there may be insufficient plasmodesmata to support the required transport rate by diffusion. One explanation suggested for the inability of malate to serve as a CO_2 donor in isolated bundle sheath cells (see above) is that access via plasmodesmata is blocked by closure of these channels during isolation. However, pyruvate has been shown to readily enter isolated mesophyll cells (2), in which presumably the plasmodesmata are also closed. This difference suggests either that the mesophyll cell tonoplast is permeable to pyruvate whereas the tonoplast of the bundle sheath is impermeable to malate, or that some form of activated transport of malate is operative *in vivo*. The numerous observations that the movement of carbon from malate to 3-PGA is light dependent could be indicative of a light-activated transport mechanism. There are, however, several other possible explanations for the light-dependent movement of carbon from malate to 3-PGA glycerate.

It might be argued that the acceptance of the C_4 pathway as an effective route for CO_2 fixation must await an understanding of the mechanism of intercellular transport. This seems to me to be an unwarranted objection to the proposed schemes since high rates of movement of sucrose must also occur in Calvin cycle species from chloroplast-containing cells to the phloem. In these species the distance over which photosynthate is transported would in most instances be considerably greater than the average distance between mesophyll and bundle sheath chloroplasts.

It is now apparent that the C_4 pathway and Crassulacean acid metabolism (CAM) represent rather similar modifications of the Calvin cycle (16). In the latter, incorporation of CO_2 into dicarboxylic acids is temporally separated from CO_2 fixation by the Calvin cycle, whereas in the former the two processes are spatially separated. Both processes utilize the same enzymes with perhaps the exception of NADP malate dehydrogenase and pyruvate,Pi dikinase, which may be specific to the C_4 pathway. There has been, however, a preliminary report that pyruvate,Pi dikinase is present in a CAM species (17). If this finding is confirmed it will explain the old observation that pyruvate can be incorporated into sugars by tissue of CAM species without being first metabolized to CO_2 (18). In these species the enzyme would serve a somewhat different role to that in C_4 pathway species, producing PEP for conversion to 3-PGA, rather than for carboxylation by PEP carboxylase.

REFERENCES

1. M.D. HATCH, this volume (1971).
2. G.E. EDWARDS and C.C. BLACK, this volume (1971).
3. O. BJÖRKMAN and M. NOBS, this volume (1971).
4. C.R. SLACK, M.D. HATCH, and D.J. GOODCHILD, *Biochem. J.,* 114 (1969) 489.
5. J.A. BERRY, W.J.S. DOWNTON, and E.B. TREGUNNA, *Can. J. Bot.,* 48 (1970) 777.
6. H.S. JOHNSON and M.D. HATCH, *Biochem. J.,* 114 (1969) 127.
7. D. GRAHAM, C.K. ATKINS, M.L. REED, B.D. PATERSON, and R.M. SMILLIE, this volume (1971).
8. O. BJÖRKMAN and E. GAUHL, *Planta,* 88 (1969) 197.
9. J.W. LYTTLETON, this volume (1971).

10. E.R. WAYGOOD, S. ARYA, and R. MACHE, this volume (1971).
11. M.D. HATCH, C.R. SLACK, and T.A. BULL, *Phytochemistry*, 8 (1969) 697.
12. M.D. HATCH and C.R. SLACK, *Biochem. J.*, 101 (1966) 103.
13. M. GIBBS, E. LATSKO, and L. LABER, this volume (1971).
14. W.J.S. DOWNTON, this volume (1971).
15. H.S. JOHNSON, C.R. SLACK, M.D. HATCH, and T.J. ANDREWS, this volume (1971).
16. I.P. TING, this volume (1971).
17. M. KLUGE, unpublished communication, this Seminar.
18. H. BEEVERS, M.L. STILLER, and V.S. BUTT, in Plant Physiology, ed. F.C. STEWARD, Academic Press, New York, IV B (1966) 198.

TRANSPORT OF SOLUTES DURING C_4 PHOTOSYNTHESIS: ASSESSMENT

F.A. Smith

Department of Botany, University of Adelaide, Adelaide, 5001, Australia.

Throughout the conference there was much discussion about the pathways and mechanisms involved in the transport of organic metabolites within leaves of C_4 plants. The aim of this review is to summarize these aspects of C_4 photosynthesis and discuss their importance.

TRANSPORT OF ORGANIC COMPOUNDS BETWEEN LEAF CELLS

The C_4 pathway involves two-way transport of organic metabolites between chloroplasts of mesophyll cells and chloroplasts of bundle sheath cells. Distances involved may be of the order of 10-100 μ in different plants. For any one metabolite three distinct transport processes are involved: movement out of one chloroplast type, movement between cells through cytoplasm (including plasmodesmata), and uptake into the second chloroplast type. The first and third processes involve movement across membranes.

The compounds thought to be transported include malate and pyruvate (in "malate formers"), or aspartate and alanine (in "aspartate formers"). In addition, transport of 3-phosphoglycerate (3-PGA) and triose phosphate was postulated [1]. HEBER [2] has shown that the first five of these compounds can move across envelopes of chloroplasts isolated from C_3 plants, and he suggested that the proposed triose phosphate movement might be accounted for by transport of dihydroxyacetone phosphate. The fact that the chloroplast envelope is impermeable to other photosynthetic products suggests that specific transport mechanisms of unknown nature are involved rather than simple diffusion. It is not clear whether such transport requires a direct supply of metabolic energy. Nevertheless, there is evidence that the chloroplast envelope is not a major barrier to the transport of the metabolites involved in the C_4 pathways.

In considering intercellular transport, the speed with which ^{14}C-labeled compounds are transfered from the mesophyll suggests that pool sizes must be small. OSMOND [3] found label in the bundle sheath within 2 sec; in discussion HATCH suggested that the time for half-exchange between the two cell types is about 10 sec. Thus there

can be no question of equilibration with the whole of the cell; in
other words the tonoplast must be an effective barrier preventing
transport to the vacuole. This may be an important specialization.
At present, little is known about the mechanisms controling the
movement of organic acids within plant cells. LIPS and BEEVERS (4,5)
and OSMOND and LATIES (6) have studied the compartmentation of malate in root cells, and an extension of this type of approach to
leaf cells would be very valuable. It would be interesting to compare compartmentation in leaves from C_3 plants, C_4 plants, and
plants having Crassulacean acid metabolism. In the latter plants, at
least, there must be high concentrations of organic acids in the
vacuole, but little is known of turn-over rates (7).

Some of those present at the conference wondered why the organic
acids moving out of the chloroplasts in C_4 plants were not metabolized by mitochondria; but there was also the suggestion that respiration (e.g. of malate) by mitochondria in the bundle sheath might
supply ATP and also help to "concentrate" CO_2 to the levels apparently required for the efficient operation of ribulose-1,5-diphos-
phate (RuDP) carboxylase.

OSMOND (3) has analyzed rates of malate formation and movement to
the bundle sheath, allowing for the probability that movement must
be through plasmodesmata. His calculated diffusion constants were
similar to those for efflux of ions from the free space of root tissues (8). A further useful comparison was with "symplastic" (intercellular) transport of inorganic ions in roots or leaves. TYREE (9)
has made a detailed study of symplastic transport and states that
"for most solutes self diffusion through the bulk cytoplasm ought to
be more rapid than cyclosis at the rate of 5 cm hr^{-1} for distances
up to roughly 50 µ". The rates of malate movement calculated from
(3) are roughly 3-10 cm hr^{-1}, over distances of 10µ-100µ. However,
TYREE points out that if the pathway is tortuous (i.e. between organelles, through thin layers of cytoplasm pressed against cell walls)
the diffusion pathway can be longer, and cyclosis may be implicated.
At present there is no evidence for cyclosis in mesophyll or bundle
sheath cells, nor is there any evidence for mass flow of solution
between cells, which would also speed up transfer of solutes.

LÜTTGE (10) mentioned possible similarities with salt-secreting
mechanisms in leaf glands or bladders. The metabolic control of salt
secretion is complex. It must involve transfer of inorganic ions
across plasmalemma and tonoplast membranes, and through cytoplasm,
and is directly dependent on metabolic energy at least for the membrane transport processes (11,12). However, the scheme suggested by
OSMOND (3) avoids transport across membranes (other than the chloroplast envelope).

Whatever the nature of the intercellular transport processes they
will be complicated under normal conditions by flow of water in the
transpiration stream. Frictional interactions between solutes and
water would be most significant in the plasmodesmata.

Other points relevant to the present discussion include the

demonstration by LAETSCH (13) that in Cyperaceae there is an endo-
dermis-like layer of cells between mesophyll and bundle sheath. This
would greatly increase the diffusion pathway. ESAU (14) has mentioned that the bundle sheath has features comparable to those of the
endodermis in roots. Casparian strips have been detected in young
leaves from Gramineae and Cyperaceae (15). This again raises the
possibility of similarities between movement of solutes in leaves of
C_4 plants, and in roots.

In C_4 plants with centripetal chloroplasts in the bundle sheath,
the length of the cytoplasmic pathway would be at a maximum. The
physiological advantage of this is obscure. However, such chloroplasts would be advantageous in providing efficient transport of
carbohydrates to the phloem. The large mitochondria found in many
bundle sheath cells could also be involved here.

ACCUMULATION OF CO_2 IN THE BUNDLE SHEATH CELLS

It was suggested (1) that the transport of malate and its decarboxylation in the bundle sheath acts as a "CO_2 pump", providing ample
substrate for RuDP carboxylase (K_m for the isolated enzyme = 0.45mM).
This raised the problem of why such high levels of CO_2 could not
diffuse out of the bundle sheath, whereas organic acids can apparently move in out readily. Membrane biophysicists consider that
membranes are very permeable to CO_2 and would not approve of the
idea of a "CO_2 pump". However, they would be happy with the idea of
bicarbonate accumulation within chloroplasts. Bicarbonate accumulation would result in (and be assisted by) a high pH in the stroma,
if HCO_3^- exchanged with OH^- across the chloroplast envelope.

The alternatives of CO_2 and HCO_3^- as substrates for RuDP carboxylase were mentioned only briefly. Present evidence mainly favors CO_2
(16). A possible function of carbonic anhydrase would be to "trap"
CO_2 and act in physical association with RuDP carboxylase as mentioned by GRAHAM *et al.* (17) and EVERSON (18).

The question of how C_3 plants might accumulate and maintain high
CO_2 levels within the chloroplast remains unresolved. Many aspects
of these problems have recently been discussed by RAVEN (16); see
also the review by WALKER and CROFTS (19).

INTRACELLULAR TRANSPORT

Several aspects of transport of metabolites within cells have
been mentioned above, including the transfer of metabolites across
the chloroplast envelope. Especially relevant are the various shuttles which allow linkage between chloroplastic and cytoplasmic pools
of ATP or reducing power. These include the shuttles for 3-PGA/tri-
ose phosphate (or dihydroxyacetone phosphate) and malate/oxaloace-
tate (2), and the glycolate/glyoxylate shuttle (20). In C_4 plants,
these shuttles allow transfer of both organic carbon and reducing
power between mesophyll and bundle sheath, via the cytoplasmic

diffusion pathway.

WALKER (in discussion; see also ref. 19) supported the view that the chloroplast membrane is impermeable to ATP and ADP. If this is correct, then it is difficult to see how cyclic photophosphorylation could provide the energy for metabolic processes outside the chloroplast. These questions concerning transfer of photosynthetic energy between chloroplast and cytoplasm have wide implications which go far beyond the scope of this review. They are of particular importance with respect to the possible energy sources for light-dependent solute transport into intact green cells of all types (10: see also ref. 21 for a general review).

Transport processes between organelles are of key importance in TOLBERT'S scheme for peroxisomal respiration (22). This involves shuttling of organic compounds between chloroplasts and peroxisomes and from peroxisomes to mitochondria, followed by a return shuttle to the chloroplasts. The distances here are of course very small, but many (specific) transmembrane processes must be involved. The scheme illustrates very neatly the interactions between different organelles and shows that it is very unrealistic to regard such organelles as autonomous within the cell.

GENERAL CONCLUSIONS

It has been suggested that C_4 plants contain a highly evolved and efficient photosynthetic system. Certainly the effect of the factors discussed above is to make the overall control of C_4 photosynthesis much more complicated than the situation in C_3 plants, where the reactions are carried out in one organelle. At the present time it seems best to assume that there is rapid transfer of solutes through the chloroplast envelope *in vivo* and to accept OSMOND'S suggestion that transfer of solutes between cells can occur efficiently by simple diffusion. Nevertheless, it should be stressed that such diffusion could be greatly affected by extraneous environmental factors leading to changes in water flow through the leaf. In other words it may be necessary to include "transport resistance" for organic compounds as a factor determining the overall efficiency of C_4 photosynthesis.

REFERENCES

1. M.D. HATCH, this volume (1971).
2. U. HEBER, this volume (1971).
3. C.B. OSMOND, *Aust. J. biol. Sci.*, (1971) in press.
4. S.H. LIPS and H. BEEVERS, *Plant Physiol.*, 41 (1966) 709.
5. S.H. LIPS and H. BEEVERS, *Plant Physiol.*, 41 (1966) 713.
6. C.B. OSMOND and G.G. LATIES, *Plant Physiol.*, 44 (1969) 7.
7. D.H. MACLENNAN, H. BEEVERS, and J.L. HARLEY, *Biochem. J.*, 89 (1963) 316.
8. M.G. PITMAN, *Aust. J. biol. Sci.*, 18 (1965) 541.

9. M.T. TYREE, *J. Theoret. Biol.*, 26 (1970) 181.
10. U. LÜTTGE, this volume (1971).
11. M.R. ATKINSON, G.P. FINDLAY, A.B. HOPE, M.G. PITMAN, H.D.W. SAD-
    DLER, and K.R. WEST, *Aust. J. biol. Sci.*, 20 (1967) 589.
12. C.B. OSMOND, U. LÜTTGE, K.R. WEST, C.K. PALLAGHY, and B. SHACHER-
    HILL, *Aust. J. biol. Sci.*, 22 (1969) 797.
13. W.M. LAETSCH, this volume (1971).
14. K. ESAU, Plant Anatomy, John Wiley, N.Y., (1953) 431.
15. D.S. VAN FLEET, *Bull. Torrey bot. Club*, 77 (1950) 340.
16. J.A. RAVEN, *Biol. Rev.*, 45 (1970) 167.
17. D. GRAHAM, C.K. ATKINS, M.L. REED, B.D. PATERSON, and R.M. SMIL-
    LIE, this volume (1971).
18. R.G. EVERSON, this volume (1971).
19. D.A. WALKER and A.R. CROFTS, *A. Rev. Pl. Physiol.*, 21 (1970) 389.
20. T.J. ANDREWS, R.K. YAMAZAKI, S.L. VANDOR, and N.E. TOLBERT, this
    volume (1971).
21. E.A.C. MACROBBIE, *Q. Rev. Biophysics*, 3 (1970) 251.
22. E.M. NELSON, M.J. BRUIN, and N.E. TOLBERT, this volume (1971).

SECTION 3

Chloroplast Structure and Function

A. Review Papers

THE PHOTOCHEMICAL SYSTEMS IN C_3 AND C_4 PLANTS

N.K. Boardman

Division of Plant Industry, CSIRO, Canberra City, 2601, Australia

SUMMARY

The current status of photosynthetic electron transport and its relationship to photophosphorylation is briefly reviewed. At present, there is no compelling evidence to abandon the generally accepted Z scheme for noncyclic electron flow.

The photochemical systems in the mesophyll chloroplasts of C_4 plants or in the grana-containing bundle sheath chloroplasts do not appear to differ from those in C_3 plants. The agranal bundle sheath chloroplasts of Sorghum bicolor are highly deficient in photosystem II; the role of these chloroplasts in the bundle sheath cells may be to provide ATP by cyclic electron flow. Photosystem I of the agranal bundle sheath chloroplasts is similar to photosystem I of spinach chloroplasts in size and composition.

Chlorophyll-deficient mutants of C_3 plants resemble C_4 plants in their light saturation characteristics. The photochemical activities of the mutant plants are higher than those of normal plants when expressed on a chlorophyll basis. It is concluded that the photosynthetic units of the mutant plants are smaller than those of normal plants.

A possible relationship between grana and photosystem II activity is discussed.

INTRODUCTION

Current concepts of the energy conversion process of photosynthesis in which light energy is absorbed by the pigments of the chloroplast and transformed into chemical energy have come mainly from studies on algae and green plants, which assimilate carbon dioxide via the Calvin cycle. An added complexity occurs in plants in which carbon dioxide is introduced into the C_4 dicarboxylic acids, aspartic acid, and malic acid, as well as into 3-phosphoglyceric acid (3-PGA). As discussed elsewhere in this volume (1) plants with the C_4 dicarboxylic acid pathway have a different anatomy from purely Calvin

cycle or C_3 plants. Chloroplasts occur in the cells of the bundle
sheath surrounding the vascular bundles as well as in the mesophyll
cells. In many C_4 plants the morphology of the bundle sheath chloroplasts differs from that of the mesophyll chloroplasts. Moreover, the
stoichiometry of CO_2 fixation in C_4 plants appears to be different
from that in C_3 plants. To operate the Calvin cycle, 3 moles of ATP
and 2 moles of NADPH are required for each mole of CO_2 fixed, but C_4
plants seem to need two extra moles of ATP i.e. 5 moles of ATP and 2
moles of NADPH per mole of CO_2 (2,3).

In this paper, the current status of photosynthetic electron
transport, and its relationship to photophosphorylation is briefly
reviewed. A comparison is made between the photochemical systems in
C_3 and C_4 plants. It is seen that certain mutants of C_3 plants, which
are deficient in chlorophyll show similarities in their light-saturation behavior to C_4 plants. Finally, there is some discussion on the
possible relationship between chloroplast morphology and photochemical activity.

PHOTOSYNTHETIC ELECTRON TRANSPORT AND PHOTOPHOSPHORYLATION

It is currently believed that the reduction of NADP is driven by
light absorbed by two pigment assemblies which act sequentially to
transfer electrons from water to NADP by the well known Z scheme (4,
5,6) (*Figure 1*). The size of the photosynthetic unit is shown as 400
chlorophyll molecules since spinach chloroplasts contain one molecule
of cytochrome f and one molecule of P-700 per 400 chlorophyll molecules. The chlorophyll appears to be divided about equally between
the two photosystems (4).

Quanta of light absorbed by the pigments of photosystem II (PS-II)
are transferred to the trap pigment, shown as a form of chlorophyll
absorbing at 682 nm and termed *chl a*-682. Experimental support for
chl a-682 comes from the studies of DORING *et al.* (7) who observed a
rapid light-induced absorbance change at 682 nm in digitonin-prepared
subchloroplast fragments enriched in PS-II. Excitation of *chl a*-682
catalizes the transfer of an electron from Y to E, giving a strong
oxidant Y^+ and a reductant E^-. Electron flow from water to Y requires manganese, possibly in the form of a manganese-protein complex
(4) and chloride ion (5). The numbers in front of an electron transport component in *Figure 1* indicate the number of molecules of that
component per photosynthetic unit of 400 chlorophyll molecules. E, Q,
and P are hypothetical electron carriers, which are proposed to explain the fluorescence kinetics of isolated chloroplasts. Recent
fluorescence work in our laboratory with spinach chloroplasts (8) indicates 3Q and 18P per photosynthetic unit. P may be identical to
plastoquinone A (4).

KNAFF and ARNON (9) recently reported a light-induced absorbance
change at 550 nm in isolated chloroplasts at room temperature and at
77°K, which they suggested may be due to the reduction of a component
in PS-II, termed C-550. It is possible, therefore, that E may be

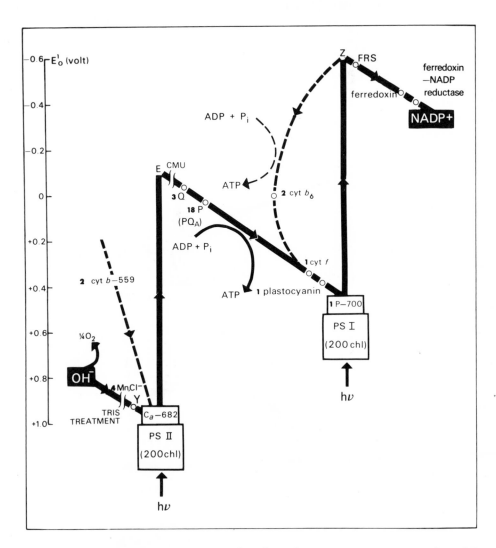

Figure 1 Scheme for photosynthetic electron transport in chloroplasts.

identical to C-550. We have confirmed the spectral change at 550 nm both in leaves and isolated chloroplasts (10).

The scheme presented previously (4) showed cytochrome b-559 on the electron transport chain after P, but our more recent studies (10,11) suggest that cytochrome b-559 is not a component of the coupled electron transport pathway between the light reactions. It may be on a side pathway from PS-II. Cytochrome f is now placed before plastocyanin. Experimental support for this order comes from the electron transport studies of LEVINE (6) in mutant strains of *Chlamydomonas reinhardi* which lack either plastocyanin or cytochrome f, and from the photochemical properties of subchloroplast fragments enriched in photosystem I (PS-I) (5). Previously, it was considered that cytochrome f was in close proximity to the trap chlorophyll of PS-I, because of earlier reports that cytochrome f was photooxidized at liquid nitrogen temperatures. Our recent studies (10) both with leaves and chloroplasts failed to provide any evidence for the photo-oxidation of cytochrome f at liquid nitrogen temperature. Instead, we confirmed the finding of KNAFF and ARNON (12) that cytochrome b-559 is photooxidized at liquid nitrogen temperature by light absorbed by PS-II.

Quanta absorbed by the light harvesting assemblies of PS-I are transferred to the trap pigment of PS-I, which is a form of chloro-phyll absorbing at 700 nm and termed P-700. On excitation of P-700, an electron is transferred from plastocyanin to Z, the primary elec-tron acceptor of PS-I (as yet not identified). The electron flows from Z^- apparently to FRS, a substance recently isolated by YOCUM and SAN PIETRO (13,14) and TREBST (15) and termed ferredoxin reducing substance. FRS transfers the electron to ferredoxin, then to ferre-doxin-NADP reductase and finally to NADP.

The well known inhibitor of oxygen evolution, 3(3,4-dichlorophe-nyl)-1,1-dimethylurea (DCMU) is considered to inhibit electron trans-port between E and Q, (4,5). Treatment of chloroplasts with 0.8 M Tris buffer pH 8.0 inhibits electron flow at a site between water and PS-II (16). Cytochrome b-559 is in a reduced state in isolated chlo-roplasts (17) and its photooxidation is observed when electron flow between water and PS-II is inhibited e.g. by tris treatment (12,18), by carbonyl cyanide chlorophenylhydrazone (CCCP) (19,11), or low tem-perature (12,10).

PS-I is known as the far-red system because its absorption spect-rum extends further to the red than that of PS-II, and at wavelengths greater than 700 nm a high fraction of the quanta absorbed by chloro-plasts are absorbed by PS-I.

What is the evidence for the series formulation of the photosys-tems?
(i)   The antagonistic effects of far-red and red light on the oxida-tion and reduction of the electron carriers, plastoquinone cytochrome f and P-700.
(ii)   The antagonistic effect of far-red and red light on the fluore-scence yield of chloroplasts. Fluorescence yield is determined by the redox state of E. In far-red light, E is oxidized and the fluore-scence is quenched; in red light, E is reduced and fluorescence is enhanced.
(iii) Enhancement studies on CO_2 fixation or O_2 evolution. The photo-synthetic rate in red and far-red light given together is greater than the sum of the rates of the individual beams (20). Enhancement has also been observed for the Hill reaction of isolated chloroplasts with NADP as oxidant by GOVINDGEE and coworkers (21), by JOLIOT, JOL-IOT, and KOK (22), by AVRON and BEN HAYYIM (23), and for fragments of red algae by GORDON (24). In contrast to these reports McSWAIN and ARNON (25) failed to observe enhancement for NADP reduction by spin-ach chloroplasts.

(iv) Photosynthetic studies with mutants of *Chlamydomonas reinhardi* also support the series formulation (6).

I now turn to the question of ATP formation. In noncyclic phosphorylation, ATP formation is coupled to electron flow from water to NADP, but there is still much uncertainty as to whether there are one or two sites of phosphorylation on the noncyclic pathway. Earlier measurements gave a P/e_2 ratio of 1, suggesting one site of phosphorylation (4). However, several investigators have reported P/e_2 ratios in excess of one (between 1.2 and 1.6) for phosphorylation coupled to the reduction of ferricyanide or NADP. These ratios have been interpreted in terms of two coupling sites in noncyclic electron flow (26). The interpretation, however, is complicated by the high rates of electron flow observed with chloroplasts under nonphosphorylating conditions, so that P/e_2 ratios in excess of one cannot be definitively interpreted in terms of 2 coupling sites. If there is only one coupling site on the noncyclic pathway, and even if phosphorylation is a highly efficient process, only 2 moles of ATP are formed per 2 moles of NADP reduced, and this is insufficient ATP even for photosynthesis in C_3 plants.

In isolated chloroplasts, another type of phosphorylation has been described. Termed cyclic phosphorylation, it requires the addition of an exogenous electron carrier such as phenazine methasulfate (PMS), pyocyanin, or ferredoxin. There is no net change in oxidation or reduction, it is insensitive to DCMU and it is driven by light absorbed by PS-I. It is not known with certainty whether cyclic phosphorylation operates *in vivo*, but it seems likely in view of the finding of TANNER and KANDLER (27) that a light dependent uptake of glucose or acetate in algae will proceed in the presence of DCMU. ARNON (28) considers that ferredoxin is the natural cofactor for cyclic phosphorylation. FORTI (29) believes that cyclic flow goes through NADP reductase. Cytochrome b_6 is shown as a component of the cyclic pathway; it is known that cytochrome b_6 is localized in PS-I, and the light-induced absorbance changes that have been observed for cytochrome b_6 are compatible with its postulated role in cyclic electron transport, but definitive evidence is lacking.

It seems likely, although by no means established that the sites for cyclic phosphorylation and noncyclic phosphorylation are different. Recently, LABER and BLACK (30) found that cyclic phosphorylation was more resistant than noncyclic phosphorylation to heptane extraction of the chloroplasts. A feature of cyclic phosphorylation at least with PMS as cofactor is the high light intensities required to saturate it, compared with noncyclic phosphorylation.

KNAFF and ARNON (18) have recently proposed a scheme of three light reactions in green plant photosynthesis. In their scheme, PS-II consists of two "short wavelength" light reactions (denoted IIb and IIa) operating in series and joined by an electron transport chain, containing C-550, plastoquinone, cytochrome b-559 and plastocyanin. Reduction of NADP and the associated noncyclic phosphorylation is considered to be driven by light absorbed solely by PS-II. In untreated chloroplasts, cytochrome b-559 is thought to be kept reduced by electron flow from water via light reaction IIb. Inhibition of electron flow from water by tris treatment permits photooxidation of cytochrome b-559 by light reaction IIa.

KNAFF and ARNON (18) consider that PS-I, the "long-wavelength" system is in parallel with PS-IIa and IIb, and its role is confined to cyclic electron flow and cyclic phosphorylation. Cytochromes f and b_6 are components of cyclic electron transport, and according to KNAFF and ARNON (18) are not involved in noncyclic electron flow from water to NADP.

The high quantum efficiencies reported by SAUER and PARK (31) for the reduction of dichlorophenolindophenol in the Hill reaction (1.9-2.8 quanta equivalent^{-1} in the wavelength range 635-680 nm) are difficult to explain by the scheme of KNAFF and ARNON (18), unless the fraction of chlorophyll associated with PS-I is small. But it is not possible to account for the high quantum efficiencies of reactions activated by PS-I (4), if only a small fraction of the chlorophyll is in this photosystem. For example, KELLY and SAUER (32) observed a requirement of 2 quanta equivalent^{-1} in the wavelength range 620-680 nm for the reduction of cytochrome c with trimethyl-p-benzoquinone as electron donor. Beyond 700 nm, the quantum requirement declined to 1 quantum equivalent^{-1}.

COMPOSITION OF THE PHOTOSYSTEMS

A partial fractionation of the photochemical systems is obtained by incubating chloroplasts with the neutral detergent, digitonin, and separating the subchloroplast fragments by differential centrifugation (33,34). *Table 1* summarizes the composition of the subchloroplast fragments. The smaller fragments (D-144) which sediment at forces of 144,000 g have a *chl a/chl b* ratio of 5.3, compared with 2.8 for chloroplasts, a xanthophyll/β-carotene ratio of 1.7 compared with 2.6 for chloroplasts, a *chl*/P-700 ratio of 205 compared with 440, a *chl/cyt f* ratio of 363 compared with 430 and a low manganese content. Cytochrome b-559 is absent from the small fragments and they have a *cyt b_6/cyt f* ratio of 1.9. Cytochrome f and cytochrome b_6 are partly lost from the D-144 fraction by the further action of digitonin, whereas P-700 is tightly bound. The small fragments show PS-I activity but little or no PS-II activity (33). The larger fragments, (D-10) which sediment at 10,000 g are enriched in cytochrome b-559 and in manganese, and they have less P-700 and cytochrome f on a chlorophyll basis than do chloroplasts. The *chl a/b* ratio is lower than for chloroplasts and the xanthophyll/β-carotene is higher. D-10 fragments are active in the Hill reaction with either ferricyanide or trichlorophenolindophenol (TCIP) as oxidant, but rates of reduction with NADP as oxidant are much smaller than those with ferricyanide and TCIP (33). Thus, the small fragments have the photochemical properties of PS-I, and the large fragments are enriched in PS-II.

TABLE I

Composition of subchloroplast fragments

Ratio of components	Chloroplasts	D-10	D-144
chl a/b	2.8	2.3	5.3
xanthophyll/β-carotene	2.6	3.8	1.7
chl/P-700	440	690	205
chl/cyt f	430	730	363
chl/Mn	73	52	250
chl/cyt (b_6 + b-559)	118	120	187
cyt b-559	+	+ (enriched)	–

PHOTOCHEMICAL SYSTEMS IN C_4 PLANTS

A comparison of the photochemical systems in C_4 plants with those in C_3 plants, such as spinach is complicated by the presence of two types of chloroplasts in C_4 plants, located respectively in the mesophyll and bundle sheath cells. A further complication is introduced by morphological differences of bundle sheath chloroplasts in different species of C_4 plants; for example, the bundle sheath chloroplasts of *Sorghum bicolor* lack grana, while those of *Atriplex spongiosa* and *Panicum miliaceum* show good development of grana. Mesophyll chloroplasts of the different species of C_4 plants contain grana.

DOWNTON, BERRY, and TREGUNNA (35) concluded that the agranal bundle sheath chloroplasts of *Sorghum sudanense* lack noncyclic flow from water. They treated leaf sections with the dye, tetranitro blue tetrazolium chloride (TNBT), which acts as a Hill oxidant. Photoreduction of the dye was observed in the mesophyll chloroplasts, but not in the bundle sheath chloroplasts. By contrast, in leaf sections of *Panicum miliaceum*, both mesophyll and bundle sheath chloroplasts reduced the dye. BLACK and MAYNE (36) reported that leaf extracts and isolated chloroplasts of a number of C_4 plants had a higher concentration of P-700, relative to chlorophyll and a higher ratio of *chl a/chl b* than several species of C_3 plants. They suggested that C_4 plants either have a more active PS-I or a smaller photosynthetic unit size.

The development of a method for separating mesophyll and bundle sheath chloroplasts was important for studies on the photochemical systems of bundle sheath chloroplasts, and for a comparison of the photochemical systems in the mesophyll chloroplasts of C_3 and C_4 plants. Recently, a method was described for obtaining mesophyll chloroplasts and bundle sheath chloroplast fragments from the leaves of C_4 plants (37,38).

A summary of the photochemical activities, pigment compositions and fluorescence properties of the isolated mesophyll and bundle

sheath chloroplasts of *S. bicolor* and *A. spongiosa* are presented
elsewhere in this volume (38). The mesophyll chloroplasts of both
species and the bundle sheath chloroplasts of *A. spongiosa* photoreduced NADP in the Hill reaction, but the agranal bundle sheath chloroplasts of *S. bicolor* were inactive. The pigment composition and
fluorescence properties of the bundle sheath chloroplasts of *S. bicolor* show a striking similarity to those of the PS-I subchloroplast
fragments of spinach. The mesophyll chloroplasts of *S. bicolor* closely resemble spinach chloroplasts (38). A comparison of the *chl*/P-700
ratios (Table 4, ref. 38) with those of spinach chloroplasts and PS-I
subchloroplast fragments (*Table 1*) suggests that the photosynthetic
unit of *S. bicolor* bundle sheath chloroplasts is similar in size to
that of PS-I subchloroplast fragments. The mesophyll chloroplasts of
S. bicolor and both mesophyll and bundle sheath chloroplasts of *A.
spongiosa* have a photosynthetic unit size similar to spinach chloroplasts.

We conclude that the photosynthetic unit size of PS-I and PS-II in
mesophyll chloroplasts of C_4 plants are comparable to the sizes of
the spinach photosystems. The agranal bundle sheath chloroplasts of
S. bicolor are highly deficient in PS-II, but the photosynthetic unit
size of its PS-I is about the same as for spinach PS-I.

PHOTOCHEMICAL SYSTEMS IN MUTANT C_3 PLANTS

Chlorophyll-deficient mutants of C_3 plants such as the tobacco mutants, studied by HOMANN and SCHMID (39) and our own pea mutant (40)
resemble C_4 plants in some respects. High light intensities are required to saturate photosynthesis, and the saturating rates are higher than for normal C_3 plants, if expressed on a chlorophyll basis.
Saturation curves for CO_2 uptake for normal and mutant pea leaves are
shown in *Figure 2*. Normal pea leaves saturate at a light intensity of
about 60,000 lux, whereas the mutant leaves were still showing a
slight increase at 113,000 lux (40). Similar differences in saturating light intensities were observed for the Hill reaction of chloroplasts isolated from normal and mutant leaves (40).

Table 2 summarizes the photochemical activities of normal and mutant pea, expressed on a chlorophyll basis. In all cases the photochemical activities of the mutant pea are more than double the corresponding activities of normal pea. When TCIP is used as oxidant in
the Hill reaction, the discrepancy is greater.

Compositional data for the normal and mutant pea are shown in *Table 3*. The *chl a/chl b* ratio for the mutant is 15 compared with 2.9
normal pea. The total chlorophyll content of the normal pea leaf is
2.6 times that of the mutant on an area basis and 2.1 on a fresh
weight basis. Examination of the cytochromes indicated that the mutant pea chloroplasts contained all 3 cytochromes; cytochrome *f*, cy-
tochrome b_6, and cytochrome *b*-559 in approximately the same proportion as normal pea. Thus the cytochrome b_6/cytochrome *b*-559 ratios
are the same for normal and mutant chloroplasts. This suggests that

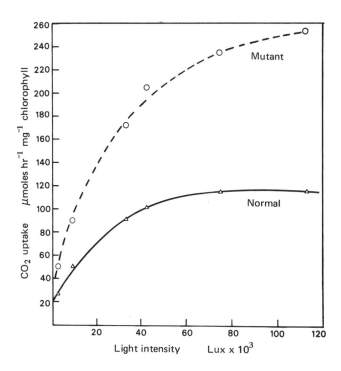

Figure 2 Light saturation curves for CO_2 uptake by normal and mutant pea plants (cf 40).

the mutant is not deficient in PS-II, and a similar conclusion is reached from the photochemical activities (*Table 2*). However, the chlorophyll/cytochrome ratios of the normal chloroplasts are approximately double those of the mutant. This indicates that the mutant leaf has approximately the same amount of cytochromes as the normal leaf on an area or fresh weight basis. On an area basis the CO_2 uptake for normal and mutant pea leaves at 113,000 lux were approximately the same, 19.3 mg dm^{-2} hr^{-1} for the mutant and 21.2 mg dm^{-2} hr^{-1}

TABLE 2

Photochemical activities of normal and mutant pea with saturating light (μmoles hr^{-1} mg^{-1} chlorophyll).

	Normal	Mutant	Mutant / Normal
CO_2 uptake	115	252	2.2
Hill reaction			
NADP	80	190	2.4
Ferricyanide	280	850	3.0
TCIP	220	950	4.3

TABLE 3

Composition of normal and mutant pea.

	Normal	Mutant	Normal / Mutant
chl a/b	2.9	15	
chl a + b (μg cm^{-2})	45	17.4	2.6
chl a + b (μg g^{-1} fresh wt)	2150	1020	2.1
chl/cyt b	106	58	1.8
chl/cyt f	409	279	1.8
cyt b$_6$/cyt b-559	1.1	1.0	

for the normal.

We conclude, therefore, that the mutant pea has the normal complement of electron transfer components, but the light harvesting cen-
ters contain about half the amount of the total chlorophyll i.e. the
photosynthetic unit size of the mutant is about one half that of the
normal.

HIGHKIN'S barley mutant (41) which completely lacks chlorophyll *b*
shows some similarities to the pea mutant, although the barley mutant
is not as deficient in total chlorophyll as the pea mutant. The normal barley contains 1.3 times the chlorophyll of the mutant barley.
The photochemical activities (*Table 4*) either of leaf sections or of
isolated chloroplasts, when expressed on a chlorophyll basis bear an
inverse relationship to the chlorophyll content (41,42).

Thus, in the mutant plants the photosynthetic units are smaller

TABLE 4

Chlorophyll content (mg g^{-1} dry wt) and photochemical properties of normal and mutant barley with saturating light (μmoles
hr^{-1} mg^{-1} chlorophyll).

	Normal	Mutant	Normal / Mutant	Mutant / Normal
chl a + b	11.3	7.9	1.3	
Photosynthetic rate of leaf sections	7.2	10.2		1.4
Hill reaction				
NADP	76	121		1.6
Ferricyanide	355	495		1.4
TCIP	244	350		1.4

than in the normal plants and this accounts for the higher intensities required to saturate the mutant plants and for the higher photosynthetic rates when expressed on a chlorophyll basis. This is in contrast to the C_4 plants where the photosynthetic units are about the same size as in normal C_3 plants, and therefore, we cannot account for the higher saturation rates for CO_2 fixation in C_4 plants on the basis of a smaller photosynthetic unit. The studies of CHEN, BROWN, and BLACK (2) and our own work with the mesophyll chloroplasts of *S. bicolor* have shown that the Hill reaction of C_4 plant chloroplasts saturates at approximately the same intensity as for C_3 plant chloroplasts.

Finally, I will conclude with some comments on the possible relationship between structure and PS-II activity. HOMANN and SCHMID (39) studied the photochemical activity of chloroplasts prepared from a number of chlorophyll deficient tobacco mutants. The chloroplast from the yellow sections of a variegated mutant were virtually inactive in the Hill reaction, but they photoreduced NADP if provided with an artificial electron donor, and they were capable of cyclic phosphorylation. Electron micrographs indicated that the chloroplasts of the yellow sections were devoid of grana, and HOMANN and SCHMID (39) suggested that grana were necessary for PS-II activity. Our studies with the agranal bundle sheath chloroplasts of *S. bicolor* support this suggestion. However, the yellow sections of the tobacco mutant had a *chl a/b* ratio of about 3, even though they were very deficient in total chlorophyll. Therefore it is not established whether the chloroplasts of the yellow sections were deficient in PS-II or whether they had an inactive PS-II.

Further support for a correlation between the presence of grana and an active PS-II comes from the fractionation studies of SANE, GOODCHILD, and PARK (43), described elsewhere in this volume. In our work on the structure and photochemical activities of developing pea plastids (44,45) there was a good correlation between grana formation and the development of Hill reaction activity.

However, studies with the barley mutant apparently do not support the view that appressed lamellae are essential for PS-II activity. The barley mutant chloroplasts which showed high PS-II activities (*Table 4*) had much fewer grana than normal barley chloroplasts (46). A mutant of *Chlamydomonas reinhardi* (ac-31) has also been described (47), which has normal PS-II activity, but essentially no grana.

Concerning the photochemical systems, I believe that there is no compelling evidence at this time to abandon the generally accepted Z scheme for noncyclic electron flow. We have seen that in bundle sheath chloroplasts of *S. bicolor* PS-I may exist without PS-II. It is not established whether cyclic phosphorylation is an integral part of the noncyclic scheme, as shown in *Figure 1*, or whether cyclic phosphorylation is performed on PS-I units which exist independently of PS-II units.

REFERENCES

1. W.M. LAETSCH, this volume (1971).
2. T.M. CHEN, R.H. BROWN, and C.C. BLACK, *Plant Physiol.*, 44 (1969) 649.
3. M.D. HATCH and C.R. SLACK, in Progress in Phytochemistry, eds. L. REINHOLD and Y. LIWSCHITZ, Interscience, London, 2 (1970) 35.
4. N.K. BOARDMAN, *Adv. Enzymol.*, 30 (1968) 1.
5. G. HIND and J.M. OLSON, *A. Rev. Pl. Physiol.*, 19 (1968) 249.
6. R.P. LEVINE, *A. Rev. Pl. Physiol.*, 20 (1969) 523.
7. G. DORING, J.L. BAILEY, W. KREUTZ, W. WEIKARD, and H.T. WITT, *Naturwissenschaften*, 55 (1968) 219.
8. S.W. THORNE and N.K. BOARDMAN, *Biochim. biophys. Acta*, (1971) in press.
9. D.B. KNAFF and D.I. ARNON, *Proc. natn Acad. Sci. U.S.A.*, 63 (1969) 963.
10. N.K. BOARDMAN, J.M. ANDERSON, and R.G. HILLER, *Biochim. biophys. Acta*, (1971) in press.
11. R.G. HILLER, J.M. ANDERSON, and N.K. BOARDMAN, submitted for publication.
12. D.B. KNAFF and D.I. ARNON, *Proc. natn. Acad. Sci. U.S.A.*, 63 (1969) 956.
13. C.F. YOCUM and A. SAN PIETRO, *Biochem. biophys. Res. Commun.*, 36 (1969) 614.
14. C.F. YOCUM and A. SAN PIETRO, *Archs Biochem. Biophys.*, 140 (1970) 152.
15. C. REGITZ, R. BERZBORN, and A. TREBST, *Planta*, 91 (1970) 8.
16. T. YAMASHITA and W.L. BUTLER, *Plant Physiol.*, 43 (1968) 1978.
17. N.K. BOARDMAN and J.M. ANDERSON, *Biochim. biophys. Acta*, 143 (1967) 187.
18. D.B. KNAFF and D.I. ARNON, *Proc. natn Acad. Sci. U.S.A.*, 64 (1969) 715.
19. G. HIND, *Photochem. Photobiol.*, 7 (1968) 369.
20. R. EMERSON, R. CHALMERS, and C. CEDERSTRAND, *Proc. natn Acad. Sci. U.S.A.*, 43 (1957) 133.
21. R. GOVINDGEE, GOVINDGEE, and G. HOCK, *Plant Physiol.*, 39 (1964) 10.
22. P. JOLIOT, A. JOLIOT, and B. KOK, *Biochim. biophys. Acta*, 153 (1968) 635.
23. M. AVRON and G. BEN HAYYIM, in Progress in Photosynthesis Research, ed. H. METZNER, IUBS, Tübingen, III (1969) 1185.
24. S.A. GORDON, *Plant Physiol.*, 38 (1963) 153.
25. B.D. McSWAIN and D.I. ARNON, *Proc. natn Acad. Sci. U.S.A.*, 61 (1968) 991.
26. S. IZAWA, G.D. WINGET, and N.E. GOOD, *Biochim. biophys. Res. Commun.*, 22 (1966) 223.
27. W. TANNER and O. KANDLER, in Progress in Photosynthesis Research, ed. H. METZNER, IUBS, Tübingen, III (1969) 1217.

28. D.I. ARNON, *Physiol. Rev.*, 47 (1967) 317.
29. G. FORTI and G. ZANETTI, in Progress in Photosynthesis Research, ed. H. METZNER, IUBS, Tübingen, III (1969) 1213.
30. L.J. LABER and C.C. BLACK, *J. biol. Chem.*, 244 (1969) 3463.
31. K. SAUER and R.B. PARK, *Biochemistry*, 4 (1965) 2791.
32. J. KELLY and K. SAUER, *Biochemistry*, 4 (1965) 2798.
33. J.M. ANDERSON and N.K. BOARDMAN, *Biochim. biophys. Acta*, 112 (1966) 403.
34. N.K. BOARDMAN, *A. Rev. Pl. Physiol.*, 21 (1970) 115.
35. W.J.S. DOWNTON, J.A. BERRY, and E.B. TREGUNNA, *Z. PflPhysiol.*, 63 (1970) 194.
36. C.C. BLACK and B.C. MAYNE, *Plant Physiol.*, 45 (1970) 738.
37. K.C. WOO, Jan M. ANDERSON, N.K. BOARDMAN, W.J.S. DOWNTON, C.B. OSMOND, and S.W. THORNE, *Proc. natn Acad. Sci. U.S.A.*, 67 (1970) 18.
38. J.M. ANDERSON, K.C. WOO, and N.K. BOARDMAN, this volume, (1971).
39. P.H. HOMANN and G.H. SCHMID, *Plant Physiol.*, 42 (1967) 1619.
40. H.R. HIGHKIN, N.K. BOARDMAN, and D.J. GOODCHILD, *Plant Physiol.*, 44 (1969) 1310.
41. H.R. HIGHKIN and A.W. FRENKEL, *Plant Physiol.*, 37 (1962) 814.
42. N.K. BOARDMAN and H.R. HIGHKIN, *Biochim. biophys. Acta*, 126 (1966) 189.
43. P.V. SANE, D.J. GOODCHILD, and R.B. PARK, *Biochim. biophys. Acta*, 216 (1970) 162.
44. S.W. THORNE and N.K. BOARDMAN, *Plant Physiol.*, 47 (1971)·
45. N.K. BOARDMAN, J.M. ANDERSON, A. KAHN, S.W. THORNE, and T.E. TREFFRY, in Autonomy and Biogenesis of Mitochondria and Chloroplasts, eds. N.K. BOARDMAN, A.W. LINNANE, and R.M. SMILLIE, North-Holland, Amsterdam, (1971) in press.
46. D.J. GOODCHILD, H.R. HIGHKIN, and N.K. BOARDMAN, *Expl Cell Res.*, 43 (1966) 684.
47. U.W. GOODENOUGH, J.J. ARMSTRONG, and R.P. LEVINE, *Plant Physiol.*, 44 (1969) 1001.

ADDENDUM

N.K. Boardman

Division of Plant Industry, CSIRO, Canberra City, 2601, Australia

In 1969, HUZISIGE *et al.* (1) described a method for the preparation of chloroplast fragments with photosystem II activity only. The fragments were active in the Hill reaction with dichlorophenolindophenol (DCIP) as oxidant, but they were unable to photoreduce NADP either with water or ascorbate-DCIP as electron donor. Their content of cytochrome f was negligible.

Recently, ARNON et $al.$ (2) reported that photosystem II fragments, prepared by HUZISIGE'S method showed some activity in the Hill reaction with either DCIP or NADP as oxidant. The fragments, which were unable to photoreduce NADP with ascorbate-DCIP as electron donor lacked P-700 and a functional cytochrome f.

The photochemical activities reported by ARNON et $al.$ (2) are consistent with the scheme of KNAFF and ARNON (3). On the other hand, the results obtained by HUZISIGE et $al.$ support the series formulation of photosystems I and II. If the findings of ARNON et $al.$ are confirmed, and if the photosystem II fragments are representative of photosystem II in the chloroplast, some radical changes will be required in current thinking of photosynthetic electron transport.

1. H. HUZISIGE, H. USIYAMA, T. KIKUTI, and T. AZI, $Pl.$ $Cell$ $Physiol.$, 10 (1969) 441.
2. D.I. ARNON, R.K. CHAIN, B.D. McSWAIN, H.Y. TSUJIMOTO, and D.B. KNAFF, $Proc.$ $natn$ $Acad.$ $Sci.$ $U.S.A.$, 67 (1970) 1404.
3. D.B. KNAFF and D.I. ARNON, $Proc.$ $natn$ $Acad.$ $Sci.$ $U.S.A.$, 64 (1969) 715.

CHLOROPLAST STRUCTURAL RELATIONSHIPS IN LEAVES OF C₄ PLANTS[1]

W.M. Laetsch

Department of Botany, University of California,
Berkeley, 94720, California, U.S.A.

SUMMARY

All C_4 plants possess "Kranz" type leaf anatomy, characterized by a layer of bundle sheath cells surrounding the vascular tissue. This layer contains starch accumulating chloroplasts that may differ in ultrastructure from mesophyll chloroplasts. This structural dimorphism is extreme in sugarcane, the bundle sheath chloroplasts of which lack grana, but in other C_4 grasses grana may be well developed. Where structural dimorphism is lacking there is considerable size dimorphism and numerous and large mitochondria are present in bundle sheath cells. Bundle sheath cells contain more microbodies than mesophyll cells and implications of this are discussed. Similar structural variations are found in C_4 members of the Cyperaceae and in C_4 dicotyledons.

A unique structural feature of C_4 plant chloroplasts is the peripheral reticulum which is seen as a series of anastomosing tubules, contiguous with the inner limiting chloroplast membrane. It is present in etioplasts and chloroplasts and is generally better developed in mesophyll chloroplasts.

The architecture of C_4 plant leaves permits gaseous diffusion between mesophyll cells up to the wall of the bundle sheath cells. The access of CO_2 to bundle sheath chloroplasts is complicated by the presence of a suberin layer in the bundle sheath cell wall.

It is suggested that the "light C_4 plants" carry out carbon fixation and photosynthesis in a manner similar to Crassulacean acid metabolism plants, the mesophyll cells carrying out the dark steps and the bundle sheath cells, the light steps of Crassulacean acid metabolism.

---

[1] The author's research reported in this paper was supported by a grant from Tate & Lyle, Ltd., and by NSF Grant GB-12964.

INTRODUCTION

One of the most striking features of C_4 plants is the correlation of a distinctive leaf anatomy and frequent specialization of chloroplasts with a specific CO_2 assimilation system. One of the first to call attention to the possible functional significance of leaves with "Kranz" leaf anatomy was HABERLANDT (1) who correlated such plants with xerophytic conditions. He also commented on the possibility of chloroplast specialization in the concentric mesophyll and in the starch-accumulating bundle sheath. The latter distinction was noted by a number of other workers in subsequent years and RHOADES and CARVALHO (2) provided evidence from their work with variegated leaf mutants of Zea mays that carbon assimilation took place in the mesophyll cells and that carbon was transferred to the bundle sheath. The first published electron micrographs of dimorphic chloroplasts in a C_4 plant were those of HODGE et al. (3). They considered the bundle sheath chloroplasts of Zea mays to be agranal, but it has since been shown on numerous occasions that these chloroplasts do have reduced grana (4). The first extensive work on the ultrastructure of chloroplasts of C_4 plants was that of JOHNSON (5), who in 1964, examined general features of chloroplast structure in a variety of grasses and observed certain aspects of the development of chloroplasts in Z. mays. Studies on sugarcane chloroplasts began at about the same time (6) and considerable work on C_4 dicots as well as additional monocots has been conducted since that time (7-12).

VARIETIES OF CHLOROPLAST DIMORPHISM

Extreme structural dimorphism such as that found in sugarcane chloroplasts has come to be associated with C_4-type photosynthesis, but the full spectrum of structural dimorphism can be observed in the tribes of tropical grasses possessing this pathway. *Muhlenbergia racmosa* in the Sporoboleae has fully-developed grana in the bundle sheath chloroplasts and structural dimorphism is absent (*Plates 1,2*).

Plate 1 Bundle sheath cell chloroplast of Muhlenbergia racemosa. Grana (G) are well developed and many starch grains (S) are present. x 8,320.

Plate 2 Mesophyll cell of M. racemosa. Grana (G) are similar to those in bundle sheath chloroplasts. Starch (S) is present, and the peripheral reticulum (R) is well developed. x 11,960.

Plate 3 Bundle sheath chloroplast of Spartina foliosa. Grana (G) and starch grains (S) are present. x 7,280.

Plate 4 Mesophyll cell chloroplast of S. foliosa. Grana (G) are similar to those in bundle sheath cell chloroplasts. Starch is not present and the peripheral reticulum (R) is well developed. x 12,600.

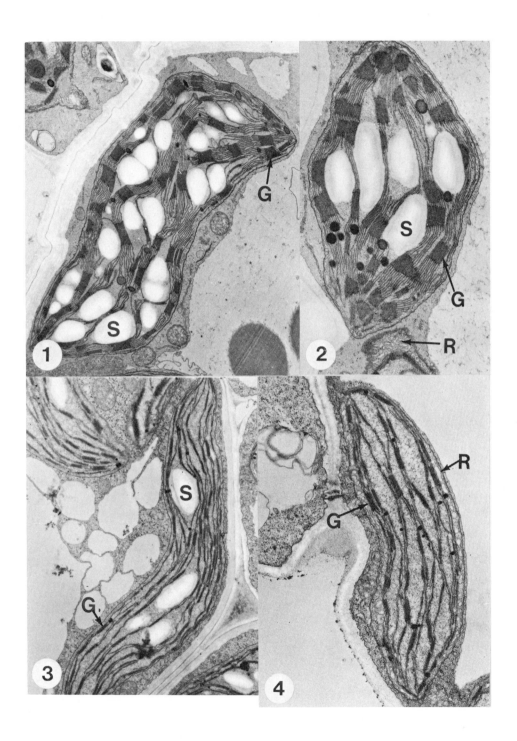

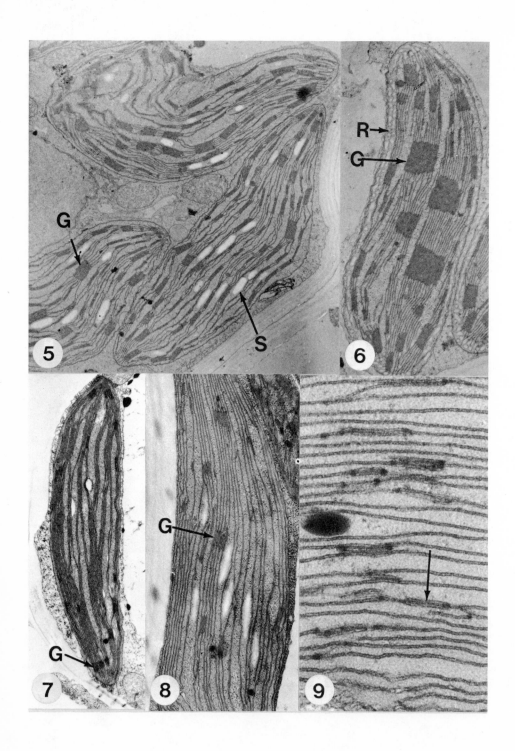

Chlorideae also have bundle sheath chloroplasts with well-developed grana. Those of *Spartina foliosa* have grana equal in size to those of the mesophyll cell chloroplasts (*Plates 3,4*), but *Bouteloua curtipendula* has better grana development in the mesophyll cell chloroplasts (*Plates 5,6*). These figures demonstrate the relatively better development of the peripheral reticulum in the mesophyll cell chloroplasts. They also illustrate the generally greater accumulation of starch in the bundle sheath chloroplasts. These two features are of general occurrence in C_4 plants.

Extensive reduction of grana in bundle sheath cell chloroplasts is characteristic of the Paniceae. These chloroplasts in *Echinochloa colona* (*Plate 7*) illustrate the frequent occurrence of grana with reduced diameters and smaller numbers of thylakoids relative to grana in mesophyll cell chloroplasts. The grana are frequently more numerous in the peripheral regions of the bundle sheath chloroplasts while the lamellae traversing the central stroma are unappressed. The grana in bundle sheath chloroplasts of the Maydeae are perhaps even more reduced than grana in the Paniceae. The most extreme structural dimorphism in chloroplasts in the tropical grasses is found in the Andropogoneae to which sugarcane and *Sorghum* belong. Grana development in this group is variable. The rudimentary grana of *Miscanthus sacchariflorus* (*Plate 8*) are reminiscent of those in the Maydeae. The bundle sheath chloroplasts of *Saccharum*, however, are normally agranal (9). This is true of *S. officinarum*, *S. robustum*, *S. sinensis*, and *S. spontenaum*, so it is probably a constant feature of the genus. This condition does not appear to be influenced by light intensity, since we have observed chloroplasts in leaves grown and fixed in full tropical sunlight as well as those grown in growth chambers with moderate light intensities and in glasshouses with low intensities. When sugarcane is grown at 20° and below, however, appressed thylakoids are observed in the bundle sheath chloroplasts (*Plate 9*).

Plate 5 Bundle sheath cell chloroplast of Bouteloua curtipendula. Grana (G) are well developed and starch (S) is present. x 9,360.

Plate 6 Mesophyll cell chloroplast of B. curtipendula. Grana are larger than those in the bundle sheath chloroplasts. Starch is absent and the peripheral reticulum (R) is well developed. x 12,600.

Plate 7 Bundle sheath cell chloroplast of Echinochloa colona. Grana (G) are rudimentary and are localized in the peripheral regions of the chloroplast. x 9,800.

Plate 8 Bundle sheath cell chloroplast of Miscanthus sacchariflorus. Grana (G) are rudimentary. x 22,080.

Plate 9 Bundle sheath cell chloroplast of sugarcane grown at 20°. Regions where two thylakoids are appressed are frequently seen (arrow). x 66,500.

The grasses demonstrate that C_4 photosynthesis is strictly correlated with Kranz leaf anatomy, but not with any degree of structural
dimorphism of chloroplasts. In those cases where structural dimorphism is absent or not strongly marked, there is a very noticeable size
dimorphism with the bundle sheath plastids being considerably larger
than the mesophyll plastids. Bundle sheath cells with chloroplasts
possessing well-developed grana also have more numerous and larger
mitochondria with more cristae than the mesophyll cells (*Plates 10,
11*). The latter situation is reminiscent of the C_4 dicots.

Tropical species of the Cyperaceae have been shown to possess C_4
photosynthesis (13), but little attention has been paid to their
structure. The leaves of *Cyperus esculentus* have the basic Kranz anatomy, but a layer of small cells without chloroplasts is interposed
between the bundle sheath cells with their agranal chloroplasts and
the mesophyll cells with granal chloroplasts (*Plates 12,13*). Unlike
the grasses, the latter cells have a greater cross-sectional area
than the bundle sheath cells. Of particular interest is the electron-
opaque layer in the primary wall of the cells between the bundle
sheath and mesophyll cells. This layer is analogous with that found
in the bundle sheath cell walls of the grasses. The agranal chloroplasts in the bundle sheath cells are illustrated in *Plate 13*. The
"swirly" appearance of the stroma lamellae is not a rare occurrence
in agranal chloroplasts. There has been concern in previous papers in
this meeting about the problem of transport of photosynthetic products between the bundle sheath and mesophyll cells, and the presence
of this additional cell layer in *Cyperus* raises some interesting
questions about transport.

The full spectrum of chloroplast structural variation observed in
the C_4 monocots is parallelled in the C_4 dicots. Genera such as *Amaranthus, Atriplex,* and *Portulaca* have little, if any, structural dimorphism although size dimorphism in the chloroplasts of the two cell
layers is frequently observed. *Froelichia gracilis* (Amaranthaceae),
however, has greatly reduced grana development in the bundle sheath
chloroplasts (*Plate 14*), relative to the mesophyll chloroplasts
(*Plate 15*). Regions where two thylakoids are appressed are frequently

---

*Plate 10 Bundle sheath cell mitochondrion of Muhlenbergia racemosa.
Note the numerous cristae. x 64,600.*

*Plate 11 Mesophyll cell mitochondrion of M. racemosa. Note difference
in size and number of cristae as compared with Plate 10. x 64,600.*

*Plate 12 Bundle sheath cell (BS) and mesophyll cell (M) of Cyperus
esculentus. Note difference in size of the two cells and electron-opaque layer(arrow) in the wall of cells between the BS and M. x 4100.*

*Plate 13 Bundle sheath cell chloroplast of C. esculentus. This chlo-
roplast is agranal with the exception of a few regions where two thylakoids are appressed. x 16,900.*

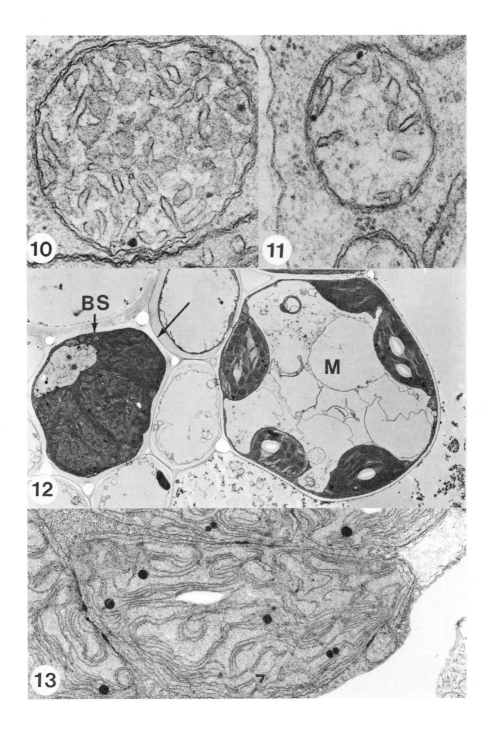

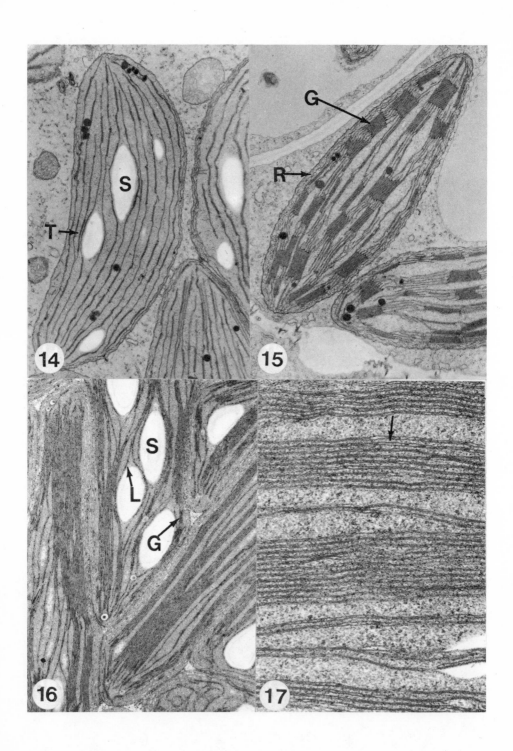

observed, but single thylakoids are extremely numerous. An even more advanced case of grana reduction is provided by *Euphorbia maculata* (*Plate 16*). The stroma lamellae closely parallel each other but are only occasionally appressed. This species as well as close relatives, have low CO_2 compensation points (14) and their leaves possess Kranz anatomy and structurally dimorphic chloroplasts.

The degree of grana development in the bundle sheath cells of C_4 plants appears on the basis of present information, to be relatively constant at the generic level, but the great variation found in higher taxonomic categories makes it imperative that statements about the degree of dimorphism be based on observation rather than assumption.

The primarily agranal bundle sheath chloroplasts in C_4 dicots and monocots have been shown to fluoresce primarily in the region of 730 nm by photographing thin sections of live leaves with infra-red film (Lynne ELKIN, personal communication), and isolated bundle sheath cell chloroplasts from *Sorghum bicolor* have been shown to possess only the long wave-length fluorescence and to lack photosystem II activity (15). *In vivo* demonstrations of the ability of sugarcane bundle sheath chloroplasts to perform photosystem II (PS-II) and photosystem I (PS-I) reactions have been made by using the reduction of the dyes, nitro blue tetrazolium (NBT), tetranitro blue tetrazolium (TNBT), and iodo nitro tetrazolium (INT). The dye-reduction was light and DCMU-dependent. The best dye-reduction occurred when the bundle sheath cells were damaged; indicating the dyes have difficulty penetrating the intact cell wall. The most satisfactory dye for this experiment is NBT and the least satisfactory are TNBT and INT (William NOLAN, personal communication).

Agranal chloroplasts lack short wave-length fluorescence when examined both *in vivo* and *in vitro*, but sugarcane bundle sheath cell chloroplasts appear capable of PS-II activity in the former state. *Plate 17* might shed some light on this paradox. It illustrates the alignment of stroma lamellae of a bundle sheath chloroplast in *Echinochloa colona*. This pattern of tight clustered lamellae is typical of primarily agranal plastids in C_4 plants. The lamellae tend to be

---

Plate 14 Bundle sheath cell chloroplasts of Froelichia gracilis. Note starch grains (S) and appressed thylakoids (T). x 12,600.

Plate 15 Mesophyll cell chloroplasts of F. gracilis. They have normal grana (G) and a well-developed peripheral reticulum (R) x 14,000.

Plate 16 Bundle sheath cell chloroplasts of Euphorbia maculata. They store large amounts of starch (S) and are largely agranal with closely approximated lamellae (L). Rudimentary grana (G) are present; particularly at the ends of the chloroplasts. x 11,960.

Plate 17 Stroma lamellae in bundle sheath cell chloroplast of Echinochloa colona. The interlamellar spaces (arrow) are approximately the dimensions of a single lamella. x 53,200.

separated by a space of 125-150 A°, which is the thickness of a single lamella. This pattern is reminiscent of that in agranal mutants
of *Chlamydomonas* capable of oxygen evolution (1). It is possible that
this alignment permits PS-II activity, but upon isolation and breaking of the chloroplasts, both this alignment and PS-II activity is
lost. If this is true, then the assumption that long wave-length
fluorescence is correlated with PS-I ahould be re-evaluated.

An extremely interesting, if complex example of chloroplast size
dimorphism is found in the genus *Mollugo* (Aizoaceae). Typical "Kranz"
leaf anatomy is found in *M. cerviana*, a species native to the old
world tropics, and its bundle sheath cells have extremely large granal chloroplasts relative to those in the mesophyll and large mitochondria with numerous cristae (*Plate 18*). The chloroplasts also have
the peripheral reticulum. In contrast, *M. verticillata* (*Plate 19*), a
species of the new world tropics, has a prominent bundle sheath with
numerous chloroplasts but lacks a concentric mesophyll. The bundle
sheath chloroplasts have a uniform distribution in the cell rather
than having a centripetal location as in *M. cerviana*, but it is interesting that the chloroplasts next to the vascular bundle are larger than those in the rest of the cell. The chloroplasts in this species do not have the peripheral reticulum. The large mitochondria associated with the C_4 dicots also have a centripetal position. This is
a case of chloroplast and mitochondria size dimorphism in the same
cell, so *M. verticillata* might prove to be a very complex intermediate between C_4 and C_3 plants. It will be surprising if more such intermediates are not discovered.

There has been a great deal of emphasis on structural aspects of
chloroplasts of C_4 plants, but relatively little attention has been
directed to the mitochondria. Dimorphism of size and frequently of
structure can be observed between bundle sheath and mesophyll cell
mitochondria in C_4 dicots and some C_4 grasses (*Plates 10,11*), and
more attention should be paid to their role in the metabolism of C_4
plants. It should be emphasized that metabolites are being transported both into the bundle sheath cells and out from them into the adjacent phloem. The bundle sheath cells can be viewed as "pumping sta-
tions" and the large and probably very active mitochondria might be
supplying the energy.

DEVELOPMENT OF DIMORPHIC CHLOROPLASTS

On the basis of both structure and function, agranal chloroplasts
of bundle sheath cells are essentially amyloplasts. It is known that
these chloroplasts in both *Zea mays* and sugarcane possess grana while
immature and lose their grana when mature (5,9). They descend from

*Plate 18 Large bundle sheath cells and portion of mesophyll cell (M)
of Mollugo cerviana. Note difference in size of chloroplasts in the
two cell types. The large chloroplasts and mitochondria in the bundle
sheath cells have a centripetal orientation. x 4,500.*

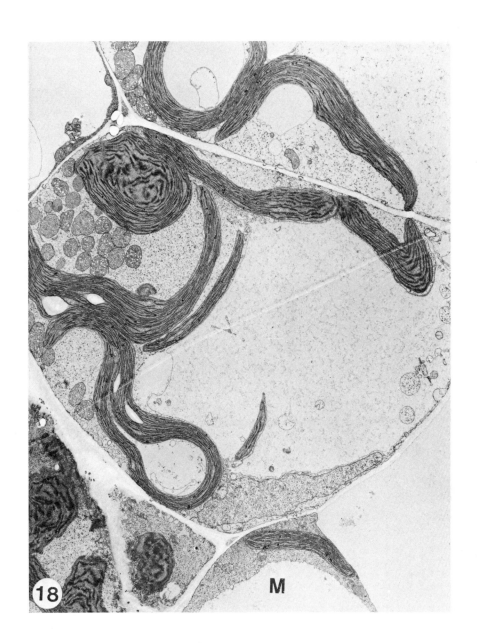

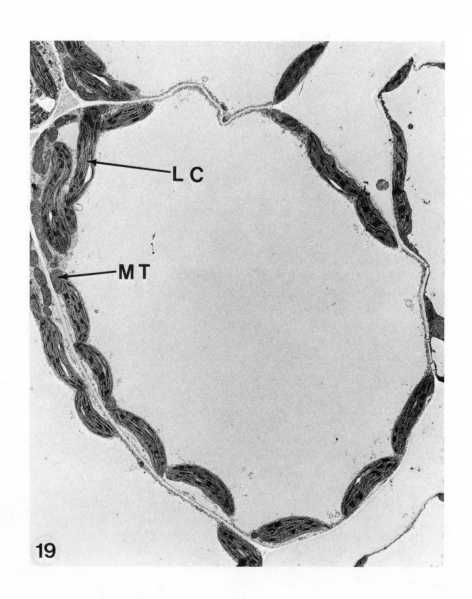

the same proplastid population in the leaf meristem as the mesophyll plastids, so they are phenotypic variants of the same genotype. Further confirmation of this viewpoint is obtained from studies of sugarcane callus proplastids, and the differentiation of chloroplasts in leaves of buds induced from this callus (author's unpublished data).

It is not surprising that dimorphic chloroplasts have the same origin, but the nature of the trigger inducing the dedifferentiation of bundle sheath chloroplasts presents an intriguing question. This event is correlated with the increasingly differentiated state of bundle sheath cells. As it initiates its role of a "starch sheath" and the adjacent phloem differentiates, it is reasonable to assume there is a fairly high sugar concentration in these cells. We have found that sucrose inhibits chlorophyll synthesis during the etioplast to chloroplast transformation in tobacco leaves (17) and that mature tobacco chloroplasts lose most of their grana when leaf tissue is cultured in the dark on sucrose and mineral salts (18). There are other reports of inhibition of chlorophyll synthesis in higher plants by sucrose (19). The loss of internal membranes which occurs in tobacco when the tissue is cultured on sucrose is always correlated with a great increase in the amount of starch in the chloroplasts. This is very much like the situation in sugarcane where the presence of large amounts of starch is correlated with the reduction of grana.

THE PERIPHERAL RETICULUM

The only unique structural feature of C_4 plant chloroplasts is the peripheral reticulum (PR). This is a series of anastomosing tubules contiguous with the inner limiting chloroplast membrane which is found in the peripheral stroma (*Plate 20*). Connections between this membrane system and the thylakoid membrane system have been reported (7), but they are difficult to observe in the material we have examined (4,8-10). This membrane system develops as the chloroplasts mature. In young chloroplasts, such as those in leaf primordia of *Portulaca oleracea* (*Plate 21*), the PR cannot be distinguished from the characteristic invaginations from the inner plastid membrane. A variety of C_4 dicots and monocots have been examined and their proplastids do not differ from proplastids in C_3 plants. The PR manifests itself in chloroplasts at intermediate stages of maturity and reaches its greatest development in leaves which have just reached full expansion. It is well developed in sugarcane etioplasts (9), but is poorly developed in etioplasts of C_4 dicots. The PR is generally better developed in mesophyll cell chloroplasts. *Plate 22* illustrates a profile of a peripheral section of a *P. oleracea* bundle sheath chloroplast which demonstrates the system of anastomosing tubules of the

---

Plate 19 Bundle sheath cell of M. verticillata. The chloroplasts have a uniform distribution in the cell. The large chloroplasts (LC) and mitochondria (MT) are located in a centripetal position adjacent to the vascular tissue.

PR. This image is in contrast to those in the adjacent mitochondrial
profiles with their tubular cristae. Profiles of chloroplast sections
rich in PR and those of mitochondria bear a superficial resemblance,
but the basic differences of the morphological patterns can be clearly distinguished.

   The PR disappears very early during leaf senescence. The chloroplast profile illustrated in *Plate 23* is from a *P. oleacea* leaf which
has lost about 40% (w/w) of its chlorophyll. The fragmentation of
stroma lamellae, condensation of grana, and accumulation of osmiophilic globules are morphological patterns typical of senescing chloroplasts. The PR is absent, although the peripheral vesicles might be
remnants. It is clear that the response of the PR to senescence differs from the thylakoid membrane system. The disappearance of this

---

*Plate 20 Mesophyll cell chloroplast of Portulaca oleracea. Note the
anastomosing tubules of the peripheral reticulum (R). It is contiguous with the chloroplast envelope (arrow). x 43,700.*

*Plate 21 Young chloroplasts in leaf primordia of P. oleracea. Note
blebby invaginations (B) from inner membrane of chloroplast envelope.
x 14,560.*

*Plate 22 Profile of section through peripheral region of P. oleracea
chloroplast. Note the difference in the peripheral reticulum (R) and
the cristae (C) in the adjacent mitochondria. x 47,500.*

*Plate 23 Chloroplast in senescent leaf of P. oleracea. Note grana (G)
and osmiophilic globules (O). The vesicles (V) are probably remnants
of the peripheral reticulum. x 12,600.*

*Plate 24 Tobacco chloroplast fixed in glutaraldehyde - OsO₄ at room
temperature. Note homogeneous stroma (S). x 22,880.*

*Plate 25 Mesophyll cell chloroplast of Amaranthus edulis fixed with
glutaraldehyde - OsO₄ at room temperature. Note well-developed peripheral reticulum (R). x 15,400.*

*Plate 26 Mesophyll cell chloroplast of A. edulis fixed with glutaral-
dehyde - KMnO₄. Note vesicular remnants of peripheral reticulum (R).
x 15,080.*

*Plate 27 Mesophyll cell chloroplast of A. edulis fixed in KMnO₄. Note
absence of peripheral reticulum. x 18,720.*

*Plate 28 Chloroplast in bundle sheath cell of A. edulis cotyledon.
Peripheral reticulum is absent. x 14,000.*

*Plate 29 Chloroplast of Opuntia sp. Peripheral reticulum-like tubules
(T) occur in the peripheral stroma. x 17,160.*

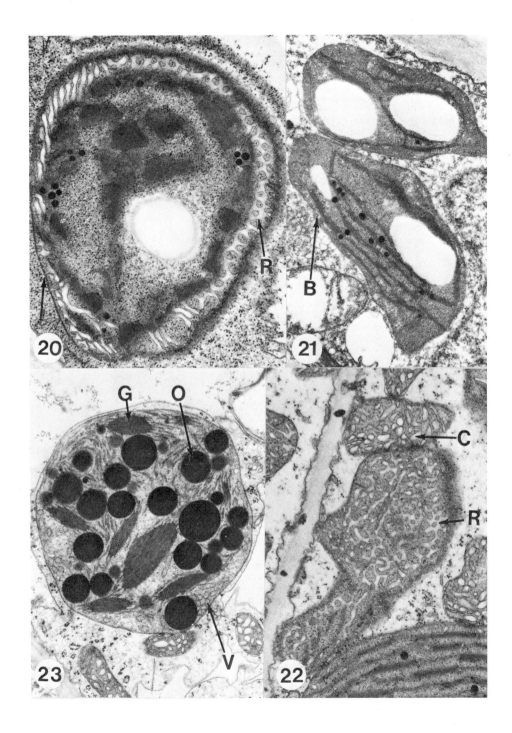

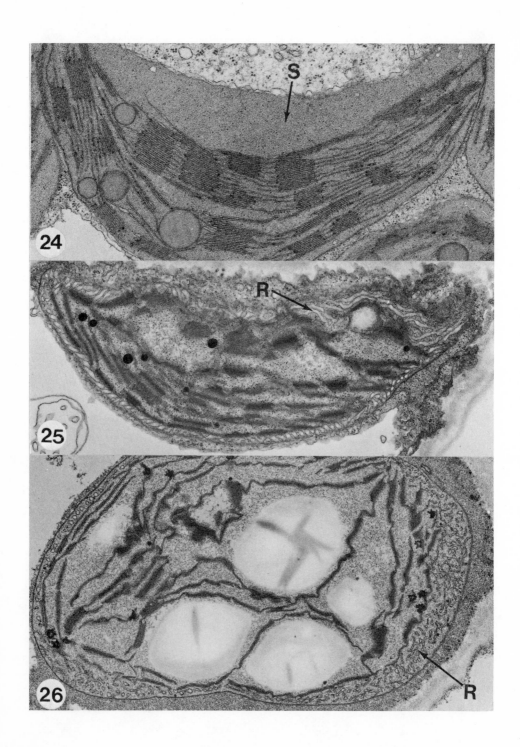

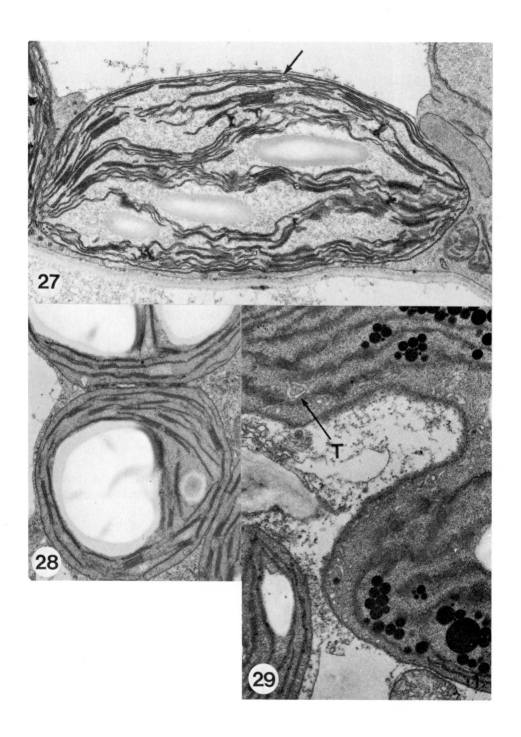

membrane system during senescence presents an opportunity to assay
possible functional correlates.

A nagging question about the PR has concerned its reality. It was
not observed in KMnO$_4$-fixed leaves of C$_4$ plants (3,5,6), but has been
widely observed in glutaraldehyde-osmium tetraoxide (G-OT) fixed
leaves of these plants, so it is at least a persistent artifact. A
variety of C$_4$ plants (*Amaranthus edulis, Portulaca oleracea, Euphorbia maculata, Saccharum officinarum*) and a C$_3$ plant (*Nicotiana tabacum*) were subjected to different fixation procedures to learn more
about the behavior of the PR. Chloroplast profiles of *A. edulis* are
illustrated, but the behavior of the PR in the other C$_4$ plants was
identical. The peripheral stroma of tobacco chloroplasts fixed in
G-OT is homogeneous and lacks PR (*Plate 24*). It was not present in
tobacco in any of the following fixation procedures. This is typical
of C$_3$ leaves including C$_3$ grasses. Fixation with G-OT at room temperature provides the image observed in the *A. edulis* chloroplast profile in *Plate 25*. The same fixation at 4° gives the same image. The
PR is observed in chloroplasts when leaves are fixed only in glutaraldehyde or OsO$_4$. Glutaraldehyde-KMnO$_4$-fixation destroys the integri-
ty of the PR, but its remnants can be observed as vesicles and lamellar fragments in the peripheral stroma (*Plate 26*). Chloroplasts subjected to KMnO$_4$-fixation do not have the PR (*Plate 27*). A lamellar
sheet close to the chloroplast envelope (arrow), and sometimes connected with it, might represent a transformed PR, but the typical image produced by the other fixation procedures is lost. The evidence
in *Plates 26,27* suggests that this membrane system responds to KMnO$_4$-
fixation in a fundamentally different fashion from the thylakoid and
outer envelope membrane systems, and this behavior indicates a different chemical composition. The PR greatly extends the surface area of
the inner membrane of the chloroplast envelope, so it is possibly involved in the transport of materials between the chloroplasts and the
cytoplasm. Its reaction to KMnO$_4$-fixation suggests a protein-enriched
membrane, so it is also tempting to speculate about its association
with a carboxylation enzyme system.

The determination of the function of the PR probably must await
the isolation of intact and functional C$_4$ plant chloroplasts, but a
certain amount of information can be obtained from comparative

---

*Plate 30 Selected portion of x-section of sugarcane leaf. Cell of up-
per epidermis (E), mesophyll cells (M), bundle sheath cell (BS), lower epidermis (E$_1$).*

*Plate 31 Contact between mesophyll cell (M) and bundle sheath cell
(BS) of sugarcane.*

Plate 32 End view of mesophyll cell of sugarcane.

*Plate 33 Two adjacent mesophyll cells of sugarcane. Note air spaces
(AS).*

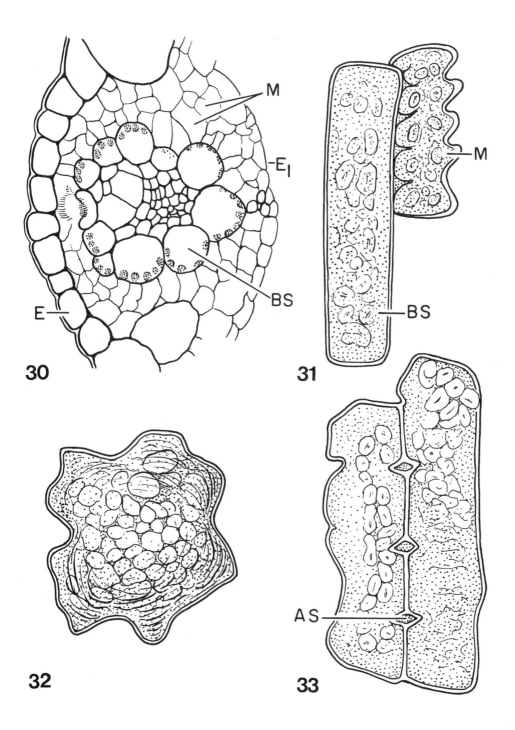

30 **31**

32 **33**

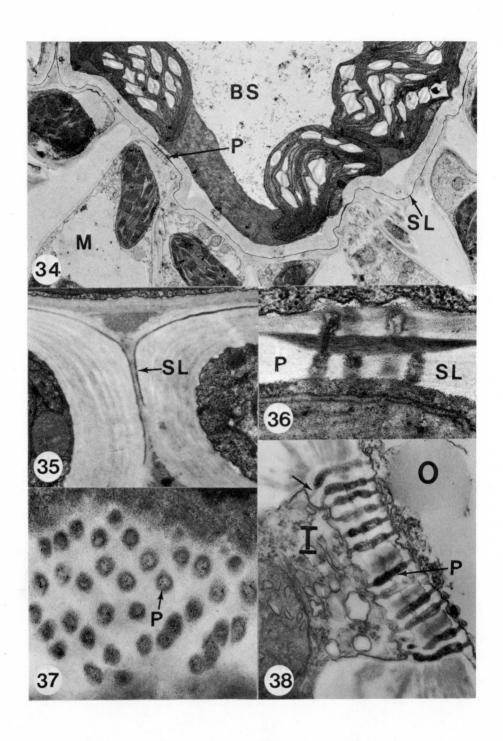

studies. Chloroplasts in the bundle sheath cells of *A. edulis* cotyledons do not have the PR (*Plate 28*). The cotyledons have a prominent bundle sheath with chloroplasts and a mesophyll with dorsal-ventral symmetry. Its anatomy is intermediate, and the lack of PR suggests the cotyledons of this well known C_4 species have C_3 photosynthesis. We have recently examined labeling patterns in *A. edulis* cotyledons and leaves and in tobacco leaves following short exposures to $^{14}CO_2$, and the cotyledons appear to have both C_3 and C_4 labeling patterns.

The PR is found in chloroplasts of all plants whose primary photosynthetic products are C_4 dicarboxylic acids ("light C_4 plants"), and it is of interest to know if it is found in chloroplasts of plants with Crassulacean acid metabolism (CAM) which use the same carboxylation enzyme system to produce C_4 dicarboxylic acids at night ("dark C_4 plants"). *Plate 29* shows a profile of an *Opuntia* chloroplast, and the peripheral stroma has a system of tubules reminiscent of PR. Similar images are observed in species of *Ferocactus*. Other CAM plants, such as pineapple, lack any suggestion of PR, so well-developed PR is only associated so far with "light C_4 plants". This correlation is evidence against the hypothesis that the PR is associated with the carboxylation enzyme of C_4 plants.

CELL RELATIONSHIPS AND ORGANELLE DISTRIBUTION

We have a great deal of information about the structural features of organelles in C_4 plants, but there has been little previous information on the architecture of C_4 plant leaves and about the connections between mesophyll and bundle sheath cells. A cross-section of a sugarcane leaf is represented in *Plate 30*, and it is fairly

---

Plate 34 Bundle sheath (BS) and mesophyll cells (M) of sugarcane. Note suberized layer (SL) in the primary wall of bundle sheath cell and pit field with plasmodesmata (P) traversing this layer. x 3,960.

Plate 35 Walls of adjacent bundle sheath cells showing termination of suberized layer (SL). x 17,500.

Plate 36 Plasmodesma (P) traversing the thickened suberized layer (SL) in pit field. Note constricted plasmodesma in the region where it traverses suberized layer. The constricted region is filled by an electron-opaque core. x 47,500.

Plate 37 Pit field with x-sections of plasmodesmata (P). Electron-opaque core is not observed in most profiles of plasmodesmata. x 47,500.

Plate 38 Pit field with plasmodesmata (P) in wall between cells of inner (I) and outer (O) bundle sheath in Aristida ascensionis. A sub-erized layer is not visible in these walls. Note constricted ends of plasmodesmata (arrow) and thin wall. x 34,200.

TABLE I

Cell and chloroplast relationships in C$_3$ and C$_4$ leaves (M, mesophyll; BS, bundle sheath).

Species	Cell size (μ)	Cell ratio M/BS	Plastids per cell	Plastids per cm^{-2} leaf	ratio M/BS	Chlorophyll* μg mg^{-1}	ratio a/b
Nicotiana tabacum	62×20	-	93.3	4.67	-	2.67	3.06
Euphorbia maculata							
M	29×10	1.7	12.9	4.42	1.19	3.03	2.45
BS	28×26		18.7				
Amaranthus edulis							
M	38× 8	11.3	10.5	3.08	2.29	1.56	3.90
BS	40×24		38.7				
Saccharum officinarum							
M	58×16	5.1	30.7	1.63	3.70	2.73	3.56
BS	113×18		41.9				

typical of the C_4 grasses. More than one layer of mesophyll cells is often found; an important point when considering gas diffusion or transport of metabolites. *Table 1* shows that the bundle sheath cells are longer than the mesophyll cells but are about the same diameter, and there are about five mesophyyl cells/bundle sheath cell. The bundle sheath cells have a regular outline, whereas, a series of peg-like projections constitute the surface of mesophyll cells. This means that the bundle sheath cells form a completely closed ring around the vascular tissue, while there are air spaces between mesophyll and bundle sheath cells (*Plate 31*). An end-on view of a mesophyll cell is illustrated in *Plate 32*, and the numerous projections result in articulations between mesophyll cells separated by air spaces (*Plate 33*). Gases would not have to pass through mesophyll cells in the diffusion path between epidermal cells and bundle sheath cells, but it is likely that they would, since the air spaces have very small dimensions.

An interesting aspect of the relationship between bundle sheath and mesophyll cells in grasses is the electron-opaque layer in the primary wall of the bundle sheath cells (*Plate 34*). This has been shown to be a suberized layer in *Zea mays* and *Avena* (20). It is most frequently found only in the wall adjacent to mesophyll cells and terminates in side walls of adjacent bundle sheath cells (*Plate 35*). Instances where this layer has completely surrounded the bundle sheath cell have been observed in sugarcane, so the extent of its development might be a function of age and growth conditions. The layer is thicker in pit fields where it is traversed by prominent plasmodesmata (*Plate 36*). The latter are constricted from about 650 A° to about 325 A° at the point where they traverse this layer. The electron-opaque core of a plasmodesma can be observed at this point. All the plasmodesma in a pit field are seen in *Plate 37*. The numbers range between 30 and 40. An even more complex arrangement of plasmodesmata in a pit field between cells of the inner and outer bundle sheath in *Aristida ascensionis* is seen in *Plate 38*. The constricted portions of the plasmodesma have the same dimensions in *Plate 36*, but the constrictions are at either end of the plasmodesma rather than in the middle. In this species the plasmodesma traverse a segment of cell wall much thinner than the adjacent wall.

The suberized layer appears to inhibit the passage of dyes outward from the bundle sheath (20), but it is not known if it serves as an "inner cuticle" or a barrier to water and small molecules. If this was the case, the plasmodesmata would be the sole transport route between the two cell layers. It would also be interesting to know if this suberized layer was a barrier to gas diffusion. The suberized layer is not found in the C_4 dicots. The transport of materials between mesophyll and bundle sheath cells is of crucial importance in the scheme for C_4 photosynthesis, and connections between cell layers should receive increasing attention.

The variation in the mesophyll/bundle sheath cell ratio in three different C_4 plants is seen in *Table 1*. The large ratio in *A. edulis*

is due to the long narrow shape of the mesophyll cells and the radial
alignment of their longitudinal axes. The number of plastids in each
cell type in the different species varies as does the total plastid
number cm^{-2} leaf. This figure is larger in tobacco, which is a C_3
plant. There are more mesophyll cell chloroplasts than bundle sheath
cell chloroplasts in *Amaranthus* and sugarcane, but the much larger
size of the bundle sheath chloroplasts in the former results in an
equal or even larger total volume of the bundle sheath chloroplasts.
The mesophyll chloroplasts would, of course, have a much greater relative surface area. The total chlorophyll is given in the same table
and it varies considerably, as do chlorophyll *a/b* ratios. There is no
correlation between agranal chloroplasts and the relative amount of
chlorophyll *a*.

TABLE 2

Distribution of organelles between mesophyll (M) and bundle
sheath cells (BS) of C_4 plants.

Species	Microbodies BS/M	Mitochondria BS/M
Amaranthus edulis	2.6	4.7
Saccharum officinarum	2.8	1.9

The distribution of microbodies and mitochondria in *A. edulis* and
sugarcane is seen in *Table 2*. It is clear that there are more of
these organelles in the bundle sheath cells. Both of these organelles
are presumably involved in photorespiration, and their concentration
in the bundle sheath suggests this cell layer is the major site of
photorespiration in C_4 plants. This supports the contention that the
difficulty in detecting photorespiration in C_4 plants is because most
of the CO_2 is evolved in the bundle sheath and re-assimilated in the
mesophyll cells (4). The concentration of microbodies in the bundle
sheath cells also explains the low yields when they are isolated from
C_4 plants, since bundle sheath cell walls are only broken with extensive grinding, and the major population of microbodies is probably
destroyed in the process.

"KRANZ" AND CAM PLANTS

The "light" and "dark" C_4 plants can be considered species of the
same biochemical genus, so it is profitable to make a comparative
study of the two groups. An examination of the order, Caryophyllales,
shows a striking taxonomic relationship. The Amaranthaceae, Chenopodiaceae, and Portulacaceae form a closely related cluster of families
in the order, and they have the largest number of species in the dicots with C_4 photosynthesis. The Cactaceae and Aizoceae also are in

this order and are considered to be more closely related to each other than to the above families (21). A large number of CAM plants are found in these two families. The primitive family in the order is the Phytolaccaceae and as a result of divergent evolution, the order has a group of "light C_4 plants" and a group of "dark C_4 plants". The same basic biochemical machinery has been used for different ends. The "dark C_4 plants" evolved in deserts with constant high day time temperatures, and they developed succulence. The "light C_4 plants" are generally weedy and probably evolved in environments where aridity was intermittent. This was also the problem for the grasses, which presumably evolved on tropical savannahs. Thus, the "light C_4 plants" had to combat both aridity and mesophytic competitors.

TABLE 3

Comparison of space and time functions in C_4 plants.

	Mesophyll cells	Bundle sheath cells
Light C_4 plants (Spatial specialization of chloroplasts)*	1. Carboxylation yields C_4 dicarboxylic acids 2. No CO_2 fixation during H_2O stress	1. Decarboxylation of C_4 acids 2. Refixation of CO_2 yielding PGA 3. Succulence
	CAM cells, dark	CAM cells, light
CAM plants (Temporal specialization of chloroplasts)	1. Carboxylation yields C_4 dicarboxylic acids	1. Decarboxylation of C_4 acids 2. Refixation of CO_2 yielding PGA 3. No primary CO_2 fixation during H_2O stress

* "Escape" from photorespiration and photosaturation results in high rate of photosynthesis.

The "light C_4 plants" really carry out carbon fixation and photosynthesis in a manner similar to the CAM plants, but they do it in different places in the leaf rather than at different times. *Table 3* presents a comparison of events in the spatially specialized chloroplasts of "light C_4 plants" and the temporally specialized chloroplasts of "dark C_4 plants". Events in the mesophyll cells in the former group are similar to dark events in the latter. The same relationship holds for bundle sheath cells and CAM plant cells in the light. The spatial specialization of chloroplasts has perhaps helped the "light C_4 plants" to cope with the problems of photosaturation and photorespiration. In combination with an efficient carboxylation enzyme system, this has resulted in the high CO_2 assimilation/water uptake ratio desirable in an arid environment. In conditions where water is not limiting, the minimizing of photosaturation and photorespiration results in the competitive advantage of high growth rates. The spatial specialization of chloroplasts might be considered as an adaptation to provide the flexibility to adapt to high temperatures and high light intensity in both arid and mesic environments.

The biochemical similarity between "light and dark C_4 plants" is obvious. The most obvious difference is the compartmentalization of the photosynthetic tissues in "light C_4 plants" provided by the "Kranz" anatomy. What we are really dealing with is plants whose unique photosynthetic functional and structural features can be summarized as "CAM mit Kranz". It is very likely that the biochemical peculiarities of "light C_4 plants" are a function of spatial patterns, so the basic problem revolves around the circuitry resulting from the compartmentalization.

ACKNOWLEDGMENTS

Expert technical assistance in electron microscopy was provided by Mr. Ian Price. I am also grateful for the assistance of Miss Barbara Brown and Mr. Jose Bonner.

REFERENCES

1. G. HABERLANDT, Physiologische Pflanzenanatomie, (1904).
2. M.M. RHOADES and A. CARVALHO, *Bull. Torr. bot. Club*, 71 (1944) 335.
3. A.J. HODGE, J.D. McLEAN, and F.V. MERCER, *Biophys. biochem. Cytol.*, 1 (1955) 605.
4. W.M. LAETSCH, *Sci. Prog. Oxf.*, 57 (1969) 323.
5. Sr. M.C. JOHNSON, Ph.D. Thesis, University of Texas, Austin, (1964).
6. W.M. LAETSCH, D.A. STETLER, and A.J. VLITOS, *Z. PflPhysiol.*, 54 (1965) 472.
7. J. ROSADO-ALBERIO, T.E. WEIER, and C.R. STOCKING, *Plant Physiol.*, 43 (1968) 1325.
8. W.M. LAETSCH, *Am. J. Bot.*, 55 (1968) 875.
9. W.M. LAETSCH and I. PRICE, *Am. J. Bot.*, 56 (1969) 77.
10. W.M. LAETSCH, in Progress in Photosynthesis Research, ed. H. METZNER, IUBS, Tübingen, I (1969) 36.
11. T. BISALPUTRA, W.J.S. DOWNTON, and E.B. TREGUNNA, *Can. J. Bot.*, 47 (1969) 15.
12. C.B. OSMOND, J.H. TROUGHTON, and D.J. GOODCHILD, *Z. PflPhysiol.*, 61 (1969) 218.
13. M.D. HATCH, C.R. SLACK, and H.S. JOHNSON, *Biochem. J.*, 102 (1967) 417.

14. D.N. MOSS, E.G. KRENZER, and W.A. BRUN, *Science*, 164 (1969) 187.
15. K.C. WOO, Jan M. ANDERSON, N.K. BOARDMAN, W.J.S. DOWNTON, C.B. OSMOND, and S.W. THORNE, *Proc. natn Acad. Sci. U.S.A.*, 67 (1970) 18.
16. U.W. GOODENOUGH, J.J. ARMSTRONG, and P.R. LEVINE, *Plant Physiol.*, 44 (1969) 1001.
17. W.M. LAETSCH, Proceedings of an International Symposium on Plant Stimulation, Sofia, (1966) 691.
18. W.M. LAETSCH, *Am. J. Bot.*, 54 (1967) 639.
19. A.K. KHUDAIRI, *Physiologia Pl.*, 23 (1970) 613.
20. T.P. O'BRIEN and D.J. CARR, *Aust. J. biol. Sci.*, 23 (1970) 275.
21. A. CRONQUIST, The Evolution and Classification of Flowering Plants, Houghton Mifflin, Boston, (1968).

Workshop circle left to right : N.E. Tolbert, H. Beevers, Hillary S. Johnson, T.A. Bull and C.C. Black.

SECTION 3

Chloroplast Structure and Function

B. Research Papers and Assessments

PHOTOCHEMICAL PROPERTIES OF MESOPHYLL AND BUNDLE SHEATH CHLOROPLASTS FROM C$_4$ PLANTS

Jan M. Anderson, K.C. Woo,[1] and N.K. Boardman

Division of Plant Industry, CSIRO, Canberra City, 2601, Australia

SUMMARY

Mesophyll and bundle sheath chloroplasts were isolated from Sorghum bicolor, Zea mays, and Atriplex spongiosa. The agranal bundle sheath chloroplast fragments of Sorghum bicolor were inactive in the Hill reaction, those of Zea mays showed traces of activity, and the grana-containing bundle sheath chloroplasts of Atriplex spongiosa showed good Hill reaction activity. Mesophyll chloroplasts from all species exhibited good photosystem II activity. All chloroplasts, whether mesophyll or bundle sheath, had photosystem I activity.

The bundle sheath chloroplast fragments of S. bicolor had a chlorophyll a/chlorophyll b ratio of 5.7, and in their pigment composition and fluorescence properties they closely resembled the photosystem I subchloroplast fragments obtained from spinach chloroplasts by fragmentation with digitonin. Thus, the bundle sheath chloroplasts of S. bicolor were enriched in β carotene, contained twice as much P-700 as spinach chloroplasts per mole of chlorophyll, and cytochrome b-559 was barely detectable. The grana-containing bundle sheath chloroplasts of A. spongiosa resembled mesophyll chloroplasts in their properties.

It is concluded that the agranal bundle sheath chloroplasts of S. bicolor and Z. mays are deficient in photosystem II.

INTRODUCTION

Leaves of plants with the C$_4$ pathway contain two layers of chloroplast-bearing cells; an inner bundle sheath layer surrounding the vascular bundles and a surrounding mesophyll layer (1). Mesophyll chloroplasts of different species of C$_4$ plants contain grana, but the bundle sheath chloroplasts show different degrees of grana

---

[1] *Present address: Federal Experimental Station, Serdang, Selangor, West Malaysia.*

development, depending on the species. DOWNTON, BERRY, and TREGUNNA
(2) treated leaf sections of some C_4 plants with the Hill oxidant,
tetranitro blue tetrazolium chloride (TNBT) and observed that the
bundle sheath chloroplasts of *Sorghum sudanense* which lack grana
(agranal) were incapable of photoreducing TNBT unless an electron donor was provided to photosystem I (PS-I).

Recently, a method was described for separating mesophyll and bundle sheath chloroplasts from the leaves of several C_4 plants (3). We
have examined the photochemical properties, pigment composition, and
fluorescence spectra of the isolated mesophyll and bundle sheath
chloroplasts (3). The results show that the agranal bundle sheath
chloroplasts of *Sorghum bicolor* are deficient in photosystem II (PS-
II). The grana-containing mesophyll chloroplasts of *S. bicolor* and
the granal bundle sheath chloroplasts of *Atriplex spongiosa* have a
functional PS-II.

RESULTS AND DISCUSSION

Seedlings of *Sorghum bicolor* L. (var. Texas 610), *Zea mays* L.
(var. NES 1002) and *Atriplex spongiosa* were grown in a glasshouse for
2-3 weeks. Mesophyll chloroplasts and fragments of bundle sheath
chloroplasts were isolated by a differential grind procedure, as described previously (3). It is essential to examine the material by
phase-contrast microscopy during preparation to ensure that the bundle sheath cells are not contaminated with mesophyll cells. Bundle
sheath chloroplast fragments were then sedimented by centrifugation
for 20 min at 10,000 g. A large proportion (*ca* 95%) of the chlorophyll of the preparation at the bundle sheath cell stage is recovered
in the 10,000 g pellet.

TABLE I

Hill reaction with NADP as oxidant (PS-I + PS-II, μ atoms [O]
hr^{-1} mg^{-1} chlorophyll) and NADP reduction (μmoles hr^{-1} mg^{-1}
chlorophyll) in chloroplasts from leaves of C_3 plants (3).

Chloroplast type	Hill reaction	NADP reduction
S. bicolor		
mesophyll	159	38
bundle sheath	0	43
Z. mays		
mesophyll	168	41
bundle sheath	trace	30
A. spongiosa		
mesophyll	150	–
bundle sheath	137	–

Hill reaction activities of the isolated chloroplasts are shown in *Table 1*. High rates of oxygen evolution were obtained with the mesophyll chloroplasts from the three species, and from the bundle sheath chloroplast fragments of *A. spongiosa*. In contrast, the bundle sheath chloroplast fragments of *S. bicolor* were inactive in the Hill reaction and oxygen evolution was not detectable. The bundle sheath chloroplast fragments from *Z. mays* gave traces of oxygen. An active PS-I, however, was present in the bundle sheath chloroplasts as evidenced by the rates of NADP reduction with ascorbate-dichlorophenolindophenol (DCIP) as an electron donor (*Table 1*). These results suggest that the bundle sheath chloroplast fragments of *S. bicolor* and *Z. mays* either are deficient in PS-II, or they contain an inactive PS-II. A similar conclusion was reached from observations of light-induced absorbancy changes of cytochrome *f* (3). With *S. bicolor* bundle sheath chloroplast fragments, the extent of the photooxidation of cytochrome *f* was independent of the wave length of actinic light over the range 650-732 nm. In contrast, the photooxidation of cytochrome *f* in the mesophyll chloroplasts of *S. bicolor* (and in both mesophyll and bundle sheath chloroplasts of *A. spongiosa*) was driven by 703, 714, and 732 nm light (PS-I) but not by 650, 663, or 675 nm light (PS-I + PS-II). In the presence of DCMU, which inhibits electron flow between the photosystems, cytochrome *f* oxidation in mesophyll chloroplasts becomes independent of wave length of actinic light. Thus, the DCMU-inhibited mesophyll chloroplasts resemble the bundle sheath chloroplasts of *S. bicolor*.

TABLE 2

Comparison of the composition of mesophyll (M) and bundle sheath (BS) chloroplasts from *Sorghum bicolor* with spinach chloroplasts and chloroplast fragments.

Components	Spinach chloroplast	*Sorghum* chloroplasts M	*Sorghum* chloroplasts BS	Spinach D-144 (PS-I)
chl a/chl b	2.8	3.1	5.7	5.3
$\frac{chl\ a + chl\ b}{carotenoid}$	6.3	6.3	6.3	6.8
$\frac{xanthophyll}{\beta\text{-carotene}}$	2.57	2.85	1.44	1.94
Carotenoid composition (%)				
β-carotene	28	26	41	34
lutein	45	38	26	30
violaxanthin	17	24	23	23
neoxanthin	10	12	10	10

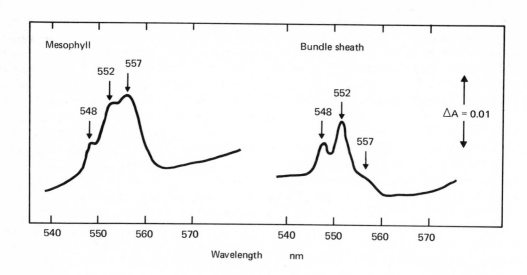

*Figure 1 Ascorbate-reduced minus ferricyanide-oxidized differ-
ence spectra of mesophyll and bundle sheath chloroplasts from
Sorghum bicolor at 77°K (3).*

The results presented above do not permit us to distinguish whether bundle sheath chloroplast fragments isolated from *S. bicolor* are
deficient in PS-II or whether they possess an inactive PS-II. However, a comparison of the pigment composition and fluorescence properties of *S. bicolor* bundle sheath chloroplast with those of the PS-
I subchloroplast fragments from spinach shows that the bundle sheath
chloroplast fragments are highly deficient in PS-II.

Mesophyll chloroplasts of *S. bicolor* have a *chl a/chl b* ratio of
3.1, which is comparable to that of spinach chloroplasts, but the
bundle sheath chloroplast fragments with a *chl a/chl b* of 5.7 (*Table*
2) more closely resemble the PS-I subchloroplast fragments (4). The
carotenoid data reported in *Table 2* also indicate that the bundle
sheath chloroplasts of *S. bicolor* are similar to the PS-I fragments,
whereas the mesophyll chloroplasts resemble spinach chloroplasts. For
example, the bundle sheath chloroplasts are rich in β-carotene and
they contain less of the xanthophylls compared with the mesophyll
chloroplasts of either *S. bicolor* or spinach.

A cytochrome difference spectrum recorded at liquid nitrogen temperature (*Figure 1*) indicates that the bundle sheath chloroplasts of
S. bicolor are highly deficient in cytochrome *b*-559 (the α-band of
cytochrome *b*-559 shifts to 557 nm at 77°K). The difference spectrum
of the mesophyll chloroplasts of *S. bicolor* is similar to that reported previously for spinach chloroplasts (5). Digitonin fragmentation of spinach chloroplasts indicates that cytochrome *b*-559 is localized in PS-II since it is absent from the PS-I subchloroplast fragments (5).

Molar ratios of chlorophyll/cytochrome are shown in *Table 3*. Comparable ratios are obtained for the mesophyll chloroplasts of *S.*

TABLE 3

Molar ratios of chlorophyll/cytochrome

Chloroplast type	$\dfrac{chl^*}{cyt\ f}$	$\dfrac{chl}{cyt\ b^{**}}$	$\dfrac{cyt\ b}{cyt\ f}$	$\dfrac{cyt\ b_6}{cyt\ b\text{-}559}$
S. bicolor mesophyll	437	119	3.7	1.1
S. bicolor bundle sheath	320	161	1.9	trace
spinach	430	118	3.6	1.0

* $chl = chl\ a + chl\ b$
** $cyt\ b = cyt\ b_6 + cyt\ b$

bicolor and spinach, but the bundle sheath chloroplasts of *S. bicolor* have only a trace of cytochrome b-559 and consequently their $cyt\ b/cyt\ f$ ratio is about half that of the mesophyll chloroplasts. The ratios reported in *Table 3* for *S. bicolor* bundle sheath chloroplasts are similar to the corresponding ratios for the PS-I fragments from spinach (5,6).

P-700 was determined by two methods. First, chloroplasts were treated with 1% Triton and the light-induced decrease in absorbance at 698 nm measured in a Chance-Aminco dual wavelength spectrophotometer. The reference wave length was 739 nm, and the wave length of the actinic light was 633 nm. Second, chloroplasts were treated with potassium ferricyanide in the presence of 1% Triton to chemically oxidize P-700, and the absorbance decrease at 698 nm was obtained from a ferricyanide-oxidized *minus* ascorbate-reduced difference spectrum recorded in a Cary Model 14R spectrophotometer. *Chl*/P-700 ratios obtained by the two methods were in reasonable agreement.

TABLE 4

Chlorophyll/P-700 ratio for chloroplasts from C_4 leaves as determined by two methods.

Chloroplast type		Chlorophyll/P-700 (a)*	(b)**
S. bicolor	Mesophyll	496	486
	Bundle sheath	254	276
A. spongiosa	Mesophyll	387	389
	Bundle sheath	414	456

* P-700 determined from oxidized *minus* reduced difference spectrum
** P-700 determined from light-induced absorbancy change

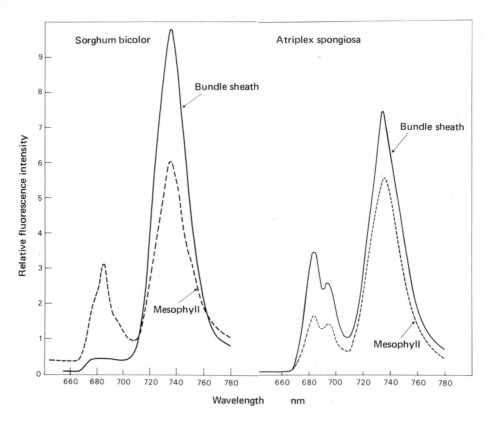

Figure 2 *Fluorescence emission spectra of mesophyll and bundle
sheath of* Sorghum bicolor *and* Atriplex spongiosa *at 77°K (3).*

The significant finding is that the agranal bundle sheath chloroplasts of *S. bicolor* contain about twice as much P-700 per mole of
chlorophyll compared with the mesophyll chloroplasts (*Table 4*). This
supports our view that the bundle sheath chloroplasts of *S. bicolor*
are very deficient in PS-II. In contrast, the bundle sheath chloroplasts of *A. spongiosa* resemble the mesophyll chloroplasts in their
P-700 content. The *chl*/P-700 ratios obtained for mesophyll chloroplasts from *S. bicolor*, and bundle sheath or mesophyll chloroplasts
from *A. spongiosa* are in reasonable agreement with the ratio obtained
with spinach chloroplasts (6). We conclude that the photosynthetic
unit size in mesophyll chloroplasts, whether from *S. bicolor*, *A.
spongiosa*, or spinach is 400-500 chlorophylls. Bundle sheath chloroplasts of *A. spongiosa*, which contain PS-II, have a similar photosynthetic unit, but the bundle sheath chloroplasts of *S. bicolor* contain
a photosynthetic unit with half the amount of chlorophyll.
      The data presented in this paper shows that the composition of the
bundle sheath chloroplasts of *S. bicolor* is in remarkable agreement
with that of the PS-I fragments obtained from spinach chloroplasts by
digitonin treatment. We have also incubated mesophyll chloroplasts
from *S. bicolor* with digitonin, and obtained PS-I fragments which are
indistinguishable from the PS-I subchloroplast fragments from

spinach.

Fluorescence spectroscopy at the temperature of liquid nitrogen supports our view that the bundle sheath chloroplasts are deficient in PS-II. Spinach chloroplasts, on cooling to 77°K, show a 3-banded fluorescence spectrum with peaks at 683, 695, and 735 nm. Fluorescence spectra of the subchloroplast fragments prepared by digitonin incubation indicated that the bands at 683 and 695 originate from PS-II and the 735 nm band comes primarily from PS-I (7).

Fluorescence emission spectra at 77°K of chloroplasts from *S. bicolor* and *A. spongiosa* are shown in *Figure 2*. The mesophyll chloroplasts from both plants and the bundle sheath chloroplasts from *A. spongiosa* show 3-banded spectra which resemble the spectrum of spinach chloroplasts. The quantum yields of fluorescence and the percentage of the fluorescence emitted at 735 nm (*Table 5*) are in reasonable agreement with the spinach values. In contrast, the fluorescence spectrum of *S. bicolor* bundle sheath chloroplasts shows only a small emission in the 680-690 nm region, and it resembles the fluorescence spectrum of spinach PS-I fragments (7). The 735 nm band accounts for 95% of the total fluorescence emission (*Table 5*).

TABLE 5

Quantum yields of fluorescence at 77°K in chloroplasts from leaves of C_4 plants (3) and in spinach chloroplasts.

Chloroplast type	ϕ_{total}*	$\dfrac{\phi_{735}}{\phi_{total}}$ (%)**
S. bicolor Mesophyll	0.13	79
Bundle sheath	0.19	95
A. spongiosa Mesophyll	0.16	76
Bundle sheath	0.14	74
Spinach	0.20	75
Spinach subchloroplast fragments (PS-I)	0.13	97

* ϕ_{total} is the fraction of adsorbed quanta emitted as fluorescence.

** $\dfrac{\phi_{735}}{\phi_{total}}$ (%) is the percentage of the fluorescence energy emitted at the 735 nm band.

We also measured the fluorescence spectrum at 77°K of intact bundle sheath cells attached to lengths of vascular tissue, before grinding in the "Janke Kunkle" mill. The spectrum was very similar to that of the chloroplast fragments. Thus, it is extremely unlikely that the grinding procedure has specifically released PS-I fragments from the bundle sheath chloroplasts.

From the studies reported here, we conclude that the agranal bundle sheath chloroplasts of *S. bicolor* are deficient in the pigment assemblies of PS-II, while the bundle sheath chloroplasts of *A. spongiosa* which contain grana have an active PS-II.

The deficiency in the production of reducing power (NADPH) by the bundle sheath chloroplasts of *S. bicolor* raises problems for the reduction of 3-phosphoglycerate. As indicated elsewhere in this volume (8) ribulose-1,5-diphosphate carboxylase appears to be confined to bundle sheath chloroplasts. It has been proposed, however, that NADPH in the agranal bundle sheath chloroplasts of *S. bicolor* is produced by the oxidative decarboxylation of malate transported from the mesophyll chloroplasts.

REFERENCES

1. W.M. LAETSCH, this volume, (1971).
2. W.J.S. DOWNTON, J.A. BERRY, and E.B. TREGUNNA, *Z. PflPhysiol.*, 63 (1970) 194.
3. K.C. WOO, Jan M. ANDERSON, N.K. BOARDMAN, W.J.S. DOWNTON, C.B. OSMOND, and S.W. THORNE, *Proc. natn. Acad. Sci. U.S.A.*, 67 (1970) 18.
4. J.M. ANDERSON and N.K. BOARDMAN, *Biochim. biophys. Acta*, 112 (1966) 403.
5. N.K. BOARDMAN and J.M. ANDERSON, *Biochim. biophys. Acta*, 143 (1967) 187.
6. N.K. BOARDMAN, this volume, (1971).
7. N.K. BOARDMAN, S.W. THORNE, and J.M. ANDERSON, *Proc. natn. Acad. Sci. U.S.A.*, 56 (1966) 586.
8. M.D. HATCH, this volume, (1971).

LIGHT REACTIONS IN C₄ PHOTOSYNTHESIS[1]

B.C. Mayne, G.E. Edwards, and C.C. Black

Charles F. Kettering Research Laboratory, Yellow Springs, Ohio,45387,
and Department of Biochemistry, University of Georgia,
Athens, Georgia, 30601, U.S.A.

SUMMARY

*Plants containing the C₄ dicarboxylic acid type photosynthetic
carbon metabolism have been utilized to study their light absorbing
pigments and various photochemically induced electron transport reactions in an effort to understand their high rates of net photosynthesis. Isolated mesophyll cells and bundle sheath cells from the C₄
plant Digitaria sanguinalis have been extensively utilized in these
studies. The characteristics studied in each cell type include: the
chlorophyll absorption and a to b ratios; fluorescence emission spectra; delayed light emission; variable fluorescence yield; Hill reaction activity; and distribution of the enzymes ferredoxin NADP reductase, glyceraldehyde 3-P dehydrogenase, and malic enzyme. We conclude
that both cell types contain a complete electron transport pathway
from oxygen evolution to the reduction of pyridine nucleotides. However, there is a quantitative difference in the distribution of Photosystem I and II activity in the cell types with the bundle sheath
cells having a 2 or 3 fold higher Photosystem I activity than the mesophyll cells.*

INTRODUCTION

The rate of net photosynthetic carbon dioxide fixation with intact
leaves of C₄ plants responds to increasing light intensity in a hy-
perbolic fashion with saturation often occurring at intensities near
full sunlight (see EDWARDS and BLACK, this volume, Figure 1), and the
maximum rate can be 2 or 3 times greater in C₄ plants than in C₃
plants in natural environments. Presently adequate photochemical and

---

[1] *This research was supported in part by NSF Grants GB 7772 (C.C.B.)
and GB 17156 (B.C.M.). C.F. Kettering Research Laboratory Contribution No. 414.*

361

biochemical explanations for higher rate of C_4 photosynthesis which saturates at such high light intensities are not available.

Previously we proposed that cyclic photophosphorylation, hence Photosystem I (PS-I), should be quite active to support the high photosynthetic capacity of C_4 plants and presented data to support the hypothesis (1). The proposal that C_4 plants may have a higher ATP requirement for net CO_2 fixation than C_3 plants was based on theoretical calculations of energy requirements for CO_2 fixation (*Table 1*).

TABLE I

Energy requirements for net carbon dioxide fixation in plants with various types of carbon metabolism.

Major type of metabolism	Theoretical stoichiometry*
C_3 plants (reductive pentose cycle)	I CO_2:3 ATP:2 NADPH
C_4 plants (C_4 cycle + reductive pentose cycle)	I CO_2:5 ATP:2 NADPH
Succulents with Crassulacean acid metabolism	I CO_2:5.5 ATP:2 NADPH

* The stoichiometrics were calculated for cycles including the complete regeneration of all components and the reduction of carbon to the level of a hexose.

Although the energy requirements differ it should be noted that all of these plants apparently ultimately utilize the reductive pentose phosphate cycle for net synthesis of hexoses (1,2). Since PS-I catalyzed reactions are the primary activities in chloroplasts which do not saturate near 2,000 to 3,000 ft-c, as do these reactions involving Photosystem II (PS-II) (1,3), we began a study of the light reactions in C_4 plants.

RESULTS AND DISCUSSION

Studies with leaf extracts and crude chloroplast preparations

With isolated bermudagrass chloroplasts the PMS dependent cyclic photophosphorylation (PS-I activity) is quite active and the apparent K_m for Pi and ADP is about 10 to 15 fold lower than the K_m with spinach chloroplasts (1). The ratio of chlorophyll $a:b$ is consistently higher in C_4 plants than C_3 plants. In addition it was noted in crude leaf extracts and chloroplasts that on a chlorophyll basis C_4 plants have more light-induced reversible bleaching at 700 nm (P-700) than C_3 plants. The reversible absorption in the region of 700 nm essentially is identical in C_4 and C_3 leaf extracts and chloroplasts (3,4). In other studies we isolated ferredoxin from corn, bermudagrass, nutsedge, pigweed, and sugarcane to characterize this important electron transport protein in C_4 plants and found that, on a total chlorophyll basis, C_4 plants contain about twice as much ferredoxin as C_3 plants (5,6).

A general summary of the comparative data on crude leaf extracts and chloroplasts from C_4 and C_3 plants is given in *Table 2*. Generally we interpret the data in *Table 2* as supporting the hypothesis that PS-I is more concentrated in C_4 plants than in C_3 plants, or, as an alternative explanation, the photosynthetic unit is smaller in C_4 plants (3,5).

TABLE 2

Summary of data on electron transport components and associated reactions in leaf extracts and crude chloroplasts (1,3,5,6) from a range of C_3 and C_4 species.

	C_4 plants	C_3 plants
Cyclic photophosphorylation apparent K_m*		
Pi	50 μM	0.4-2 mM
ADP	25 μM	0.5-0.6 mM
Ratio chlorophyll a/b	3.89 ± 0.6	2.78 ± 0.35
Ratio P-700 to total chlorophyll	1.83 ± 0.21	1.2 ± 0.11
Ratio extractable ferredoxin to total chlorophyll	5.04	2.38

* Bermudagrass chloroplasts (C_4) and spinach chloroplasts.

Studies with isolated cells

Because of the striking anatomical differences in leaves of C_4 species, both within C_4 genera and between C_4 and C_3 plants (7,8,9), it seemed necessary for an understanding of leaf photosynthesis that the individual cells and chloroplasts be examined. Thus when we separated the mesophyll cells and bundle sheath cells from the C_4 plant *Digitaria sanguinalis* (crabgrass) we undertook an investigation of the photochemical properties of these distinct cell types (8,10,11, 12). Fully differentiated crabgrass mesophyll cell chloroplasts contain well developed grana and readily store starch while the bundle sheath chloroplasts also synthesize starch but contain only rudimentary grana (8).

(a) Chlorophyll content

Absorption spectra of each cell type show that not only are there differences in chlorophyll $a:b$ ratios, as observed in leaf extracts from C_4 plants (*Table 2*), but each cell type has a characteristic ratio. In the low temperature absorption spectra the chlorophyll b peak

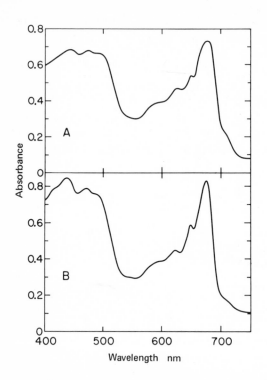
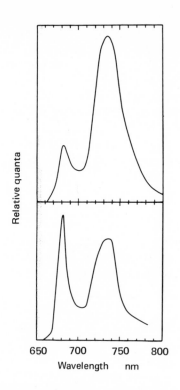

Figure 1 (left) Absorption spectra of cell extracts at liquid
nitrogen temperature. (A) Bundle sheath cells. (B) Mesophyll
cells. (From reference 12 courtesy of Plant Physiol.).

Figure 2 (right) Fluorescence emission spectra of cell extracts
at liquid nitrogen temperature. Upper trace bundle sheath
cells. Lower trace mesophyll cells. (From reference 12 courtesy
of Plant Physiol.).

is more prominent in the mesophyll cell extract than in the bundle
sheath cell extract (Figure 1). The 75% peak height band width of the
bundle sheath cell extracts was approximately twice as wide as that
of the mesophyll cell extracts and the 700 nm absorption shoulder
twice as large. The increase in the 75% peak height band width of
bundle sheath cell extracts appears to be due to a relatively greater
amount of a component absorbing at approximately 670 nm in the bundle
sheath cells.

   In the blue absorption band, the spectra differ in the wavelength
of the "chlorophyll a peak", 438 nm in mesophyll cell extracts and
444 nm in bundle sheath cell extracts. The chlorophyll a blue peak,
relative to the other peaks, also was higher in mesophyll cell extracts. In day to day experiments there is some variation in the
chlorophyll a:b ratios but the average ratio for bundle sheath cells
and mesophyll cells is 4.51 ± 0.59 and 2.99 ± 0.12 respectively (12).

   (b) Fluorescence emission spectra
   The room temperature fluorescence spectra of the two cell types

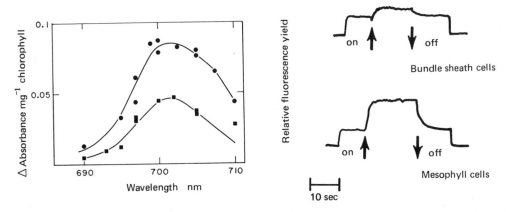

Figure 3 (left) Difference spectra of the light-induced P-700 absorption in cell extracts: bundle sheath cells o; mesophyll cells ▪. (From reference 12 courtesy of Plant Physiol.).

Figure 4 (right) The relative fluorescence yield of isolated bundle sheath and mesophyll cells. The cells were suspended in a solution containing the following: 0.05 M K MES, pH 6.05, 0.005 M MgCl₂, and 0.3 g ml⁻¹ Ficoll (2 x 10⁴ molecular wt). The chlorophyll concentration was 6 µg ml⁻¹. The measuring light beam was approximately 100 erg sec⁻¹ cm⁻² wavelength 435 nm. The actinic light approximately 1.5 x 10⁴ erg sec⁻¹ cm⁻² isolated from the output of tungsten-iodine lamp with 3 cm saturated CuSO₄ solutions and a Corning filter No. 4-94. Actinic light: on ↑; off ↓. Method of measurement was similar to that of YAMASHITA and BUTLER (24).

are similar but the fluorescence of the mesophyll cell extract is always larger than the fluorescence of the bundle sheath cell extract. On the other hand, the low temperature fluorescence spectra of the cell extracts show a striking difference between the two cell types (*Figure 2*). The bundle sheath cell extracts showed a much greater fluorescence in the long wavelength band at 730 nm than the mesophyll cells. Although there was some variation in the ratio of the peaks between experiments, in all cases the bundle sheath cells have a much greater ratio f730 to f684 than the mesophyll cells, the respective values in *Figure 2* being 2.84 and 0.82.

(c) P-700 content

It was previously reported that plants which fix carbon via the C_4 cycle had a larger light-induced reversible P-700 absorbance change on a total chlorophyll basis than plants which utilize the C_3 cycle (3). In view of this, we particularly were interested in measuring the reversible 700 nm absorption change in extracts from the two cell types. *Figure 3* shows that although both cells have essentially the

same reversible absorption spectrum in the region of P-700 absorption, on a chlorophyll basis the bundle sheath cells clearly have a
greater ΔOD at 700 nm.

(d) Delayed light emission

We have presented data on the delayed light emission of isolated
cells of crabgrass (12). In these experiments the amount of delayed
light emission is 2-fold greater in mesophyll cells but there is no
difference in the kinetics of the decay (12).

(e) Variable fluorescence yield

The variable fluorescence yield of isolated mesophyll and bundle
sheath cells is shown in *Figure 4*. The variable portion of fluorescence usually is considered to be that which is modulated by Duysens'
Q (13) and is indicative of PS-II.

(f) Ferredoxin NADP reductase

In the process of photochemically reducing NADP with spinach chloroplasts, ferredoxin NADP reductase is the terminal enzyme (4). Current formulations would place this enzyme in PS-I. The data in *Table
3* show that the enzyme is in both mesophyll and bundle sheath cells.

TABLE 3

Ferredoxin pyridine nucleotide reductase activities (μmoles
hr^{-1} mg^{-1} chlorophyll) in extracts from leaves, mesophyll
cells, and bundle sheath cells of *Digitaria sanguinalis* (12).

Source	Activity measured	Electron donor NADPH	NADH
Mesophyll	Diaphorase (Ferricyanide)	774	208
Bundle sheath		224	380
Whole leaf		586	398
Mesophyll	Diaphorase (DCIP)	46	12
Bundle sheath		12	16
Whole leaf		19	16
Mesophyll	Menadione reductase	122	0
Bundle sheath		28	6
Whole leaf		34	5
Mesophyll	Transhydrogenase (Deamino-NADPH)	40	–
Bundle sheath		9	–
Whole leaf		6	–

The enzyme activity by several assays appears to be 2 to 4 fold higher in the mesophyll cells than in the bundle sheath cells.

(g) Oxygen evolution

Table 4 presents the results of an oxygen evolution study with the
isolated cell types in the presence of exogenous electron acceptors.

It is assumed that benzoquinone is a typical noncyclic Hill oxidant and that diquat also is a typical Hill oxidant which catalyzes a Mehler reaction. Some variation in absolute rates of oxygen production is observed between experiments but these data are presented to demonstrate that both cells do evolve oxygen and that, on a chlorophyll basis, the mesophyll cells are about twice as active as bundle sheath cells in these typical noncyclic electron transport reactions. All of these reactions were sensitive to DCMU with 90 to 100% inhibition being observed at 1 μm DCMU.

TABLE 4

Photochemical oxygen production (μmoles hr^{-1} mg^{-1} chlorophyll) by isolated mesophyll cells and bundle sheath cells of *Digitaria sanguinalis* (12). Cells were isolated in 0.4 M sorbitol and 0.02 M MOPS buffer, pH 6.5 and assayed in the same medium at 30°.

Electron acceptor added			Ratio of mesophyll: bundle sheath
	Mesophyll	Bundle sheath	
None	0-2	3-5	
Benzoquinone 0.25 mM (Air)	85	43	1.97
Benzoquinone 0.25 mM (N$_2$)	145	74	1.96
Diquat 0.2 mM plus 2 mM KCN (Air)	58*	31	1.87

* With diquat alone as an electron acceptor oxygen uptake was measured.

DISCUSSION

The present study with photosynthetic cell types from crabgrass leaves allows a more detailed characterization and understanding of photosynthesis in C₄ plants. In this volume (see EDWARDS and BLACK) we have presented our interpretation of photosynthesis in mesophyll and bundle sheath cells in C₄ plants, integrating both carbon flow and electron flow. The following is a more detailed discussion of the light reactions in C₄ plants. A general summary of the observed spectral and electron transport activities is presented in *Table 5*. We

TABLE 5

Summary of spectral and electron transport characteristics of
mesophyll cells and bundle sheath cells from *Digitaria sanguinalis* (12).

Characteristic	Mesophyll:bundle sheath
Absorption at 700 nm	1:2
P-700 change	1:2
Ratio of f730:f685	1:3
Chlorophyll *a:b* ratio	3:4.5
Delayed light emission	2:1
Variable fluorescence yield	2 or 3:1
Ferredoxin NADP-reductase	3 or 4:1
Hill reaction activity	2:1
Glyceraldehyde 3-P dehydrogenase	1:1
Malic enzyme	1:20

will analyze the results in relation to the generally accepted two
light reaction scheme for photosynthesis (4).

That both cell types contain PS-II activity is shown by the following: oxygen evolution in the presence of typical Hill oxidants
(*Table 4*); the oxygen evolution activity is sensitive to DCMU; the
low temperature fluorescence emission spectra (*Figure 2*) in chloroplast extracts from both cells peak around 685 nm (14); the msec delayed light emission and the variable fluorescence yield are further
indication of PS-II activity (13,14,15) in both cell types.

PS-I activity is also present in both mesophyll and bundle sheath
cell types as judged by P-700(*Figure 3*); chlorophyll *a* content (*Table
5, Figure 1*], and fluorescence emission spectra (*Figure 2*). Diquat
probably accepts electrons from PS-I and is active in oxygen metabolism with both cells (*Table 4*). It is debatable whether or not ferredoxin NADP reductase is a component of cyclic electron flow but the
data in *Table 3* indicate that this enzyme, which reduces pyridine
nucleotides photochemically, is active in both cell types. Thus we
interpret the present data as evidence that both types of photosynthetic cells in crabgrass leaves contain a complete electron transport system from oxygen evolution to the reduction of pyridine nucleotides.

There is a quantitative difference in the distribution of the photosynthetic electron transport components in the two cell types. The
data of *Table 5* are consistent with the mesophyll cells containing
more noncyclic electron flow than the bundle sheath cells: the oxygen
evolution data show about a 2:1 ratio of mesophyll to bundle sheath
activity; the ferredoxin NADP-reductase activity is 3 to 4 times
higher in mesophyll cells; the intensity of low temperature fluorescence emission at 684 nm indicates a high ratio of PS-II to PS-I

activity in the mesophyll cells; and approximately twice as much delayed light emission and variable fluorescence are observed in mesophyll cells. On the other hand the bundle sheath cells have much
higher PS-I activity than the mesophyll cells (*Table 5*) as indicated
by: the higher P-700 in bundle sheath cells; the higher chlorophyll
a:b ratio in bundle sheath cells; the increased absorption in the region of 700-705 nm in bundle sheath extracts (*Figure 1*), shown by
BUTLER and BISHOP (16) to be related to PS-I activity (*Figure 1*); and
the very high fluorescence emission at 730 nm relative to 685 nm (14)
(*Figure 2*). It is possible that part of the increased amount of P-700
of the bundle sheath cells is due to the form of P-700 recently reported to be linked to stroma lamellae (17,18). Alternatively the electron transport data can be explained within the framework of KNAFF
and ARNON'S (19) electron transport scheme or similar proposals (20)
rather than two photosystems (4). We have refrained from utilizing
these more recent electron transport schemes but see no compelling
reason for only considering the data in relation to the traditional Z
scheme.

We conclude that the mesophyll cells are about 2 to 4 times more
active than bundle sheath cells in noncyclic electron flow, which results in the production of reducing power, NADPH, and the associated
ATP. Although the bundle sheath cells also generate reducing capacity
and ATP via noncyclic electron flow, they have a 2 to 3 times greater
capacity for cyclic electron flow than mesophyll cells.

We have proposed a scheme to accomodate the CO_2 fixation and subsequent carbon metabolism reactions of the mesophyll cells which fix
CO_2 via the carboxylation of PEP and bundle sheath cells which fix
CO_2 via the carboxylation of RuDP (see EDWARDS and BLACK, this volume). In the mesophyll cells CO_2 is fixed and one NADPH is utilized
to produce malate. The malate moves to the bundle sheath cells where
malic enzyme is localized (*Table 5*) and is decarboxylated, producing
pyruvate, CO_2, and NADPH. The CO_2 released in the bundle sheath cell
is fixed via the Calvin cycle and the NADPH produced by malic enzyme
is utilized by glyceraldehyde 3-P dehydrogenase to reduce the 3-PGA
to glyceraldehyde 3-P. If the bundle sheath cell operates by the
classical Calvin cycle (*Table 1*), then noncyclic electron flow in the
bundle sheath cells would need to provide only one NADPH per CO_2 ori-
ginating from malate decarboxylation. So to operate the Calvin cycle
in conjunction with the C_4 pathway the energy production by light reactions in the bundle sheath chloroplasts would be 3 ATP:1 NADPH per
CO_2 fixed. The theoretical energy requirements of the C_4 pathway in
the mesophyll cells would be 2 ATP:1 NADPH per CO_2 fixed if pyruvate
is the precursor for PEP. The present data which indicates more cyclic electron flow in the bundle sheath cells relative to mesophyll
cells and more noncyclic electron flow in mesophyll cells than in
bundle sheath cells are consistent with the above theoretical energy
requirements. The theoretical energy requirements of the two cell
types would be altered to the extent which 3-PGA or 2-PGA may be
transported from bundle sheath cells to mesophyll cells and reduced

there to form starch and to the extent which pyruvate is metabolized
in other reactions.

Although this hypothesis for C₄ photosynthesis may be correct for
crabgrass, it appears that other C₄ plants such as bermudagrass may
lack sufficient malic enzyme to operate this exact cycle and an alternative pathway involving aspartic acid has been proposed (21) (also see EDWARDS and BLACK, this volume).

The results of another investigation with isolated chloroplasts
from C₄ plants by WOO *et al.* (22) reports the absence of PS-II activity in isolated bundle sheath chloroplasts of *Sorghum bicolor* and *Zea
mays*. These results with isolated chloroplasts from plants which are
structurally similar to crabgrass are contrary to those reported here
with whole cells. The reasons for their failure to detect PS-II activity in their isolated bundle sheath chloroplast preparations from
these two plants are unknown. KARPILOV *et al.* (23) recently presented
fluorescence and Hill activity data on C₄ plants which qualitatively
agrees with the data in this manuscript.

REFERENCES

1. T.M. CHEN, R.H. BROWN, and C.C. BLACK, *Plant Physiol.*, 44 (1969)
   649.
2. M.D. HATCH and C.R. SLACK, *A. Rev. Pl. Physiol.*, 21 (1970) 141.
3. C.C. BLACK and B.C. MAYNE, *Plant Physiol.*, 45 (1970) 738.
4. B. KOK, in Plant Biochemistry, eds. J. BONNER and J.E. VARNER,
   Academic Press, New York, (1965) 903.
5. S.S. LEE, J. TRAVIS, and C.C. BLACK, *Archs Biochem. Biophys.*, 141
   (1970) 676.
6. S.S. LEE and C.C. BLACK, *Fed. Proc.*, 29 (1970) 535.
7. W.M. LAETSCH, *Sci. Prog. Oxf.*, 57 (1969) 323.
8. C.C. BLACK and H.H. MOLLENHAUER, *Plant Physiol.*, 46 (1970) 828.
9. G. HABERLANDT, Physiological Plant Anatomy, Macmillan and Co.
   Ltd., London.
10. G.E. EDWARDS, S.S. LEE, T.M. CHEN, and C.C. BLACK, *Biochem. bio-
    phys. Res. Commun.*, 39 (1970) 389.
11. G.E. EDWARDS and C.C. BLACK, *Plant Physiol.*, 47 (1971) 18.
12. B.C. MAYNE, G.E. EDWARDS, and C.C. BLACK, *Plant Physiol.*, 47
    (1971) in press.
13. L.N.M. DUYSENS and H.E. SWEERS, in Studies on Microalgae and Pho-
    tosynthetic Bacteria, ed. S. MIYACHI, special issue of *Plant
    Cell Physiology*, Tokyo, (1963) 353.
14. W.L. BUTLER, *Archs Biochem. Biophys.*, 93 (1961) 413.
15. W.F. BERTSCH, J.R. AZZIAND, and J.B. DAVIDSON, *Biochim. biophys.
    Acta*, 143 (1967) 129.
16. W.L. BUTLER and N.I. BISHOP, in Photosynthetic Mechanisms of
    Green Plants, eds. B. KOK and A.T. JAGENDORF, NAS-NRC, Washing-
    ton, (1963) 91.
17. P.V. SANE and R.B. PARK, *Biochem. biophys. Res. Commun.*, 41
    (1970) 206.

18. P.V. SANE, D.J. GOODCHILD, and R.B. PARK, *Biochim. biophys. Acta,* 216 (1970) 162.
19. D.B. KNAFF and D.I. ARNON, *Proc. natn. Acad. Sci. U.S.A.,* 64 (1969) 715.
20. H.J. RURAINSKI, J. RANDLES, and G.E. HOCH, *Fed. Eup. Soc. exp. Biol.,* (1971) in press.
21. T.M. CHEN, R.H. BROWN, and C.C. BLACK, *Plant Physiol.,* 47 (1971) in press.
22. K.C. WOO, Jan M. ANDERSON, N.K. BOARDMAN, W.J.S. DOWNTON, C.B. OSMOND, and S.W. THORNE, *Proc. natn. Acad. Sci. U.S.A.,* 67 (1970) 18.
23. Y.S. KARPILOV, K.Y. BRILL, O.G. MALISHEV, and V.N. KARNAUKLOV, *Proc. Moldavian Res. Inst.,* II (1970) 25 (in Russian).
24. T. YAMASHITA and W.L. BUTLER, *Plant Physiol.,* 43 (1968) 1978.

LAMELLAR STRUCTURE AND COMPOSITION IN RELATION TO PHOTOCHEMICAL ACTIVITY

D.G. Bishop, Kirsten S. Andersen,[1] and Robert M. Smillie

Plant Physiology Unit, CSIRO, Division of Food Research, and
School of Biological Sciences, Macquarie University,
North Ryde, N.S.W., 2113, Australia

SUMMARY

The lipid composition and photochemical activity of preparations of mesophyll and bundle sheath chloroplasts of maize and Sorghum have been studied. In both species, the content of galactolipid on a chlorophyll basis is significantly higher in bundle sheath than in mesophyll chloroplast preparations.

Mesophyll chloroplasts are capable of the photoreduction of NADP but bundle sheath chloroplast preparations lack this capacity. However, the use of artificial electron acceptors showed that both photosystems are present in bundle sheath chloroplast preparations, but they are not linked for electron flow. Preparations of bundle sheath chloroplast fragments from secondary leaves of six-day old maize plants were capable of photoreducing NADP but this capacity was gradually lost as the leaves matured. Photosystem II activity did not decrease however, indicating that loss of the ability to photoreduce NADP was due to unlinking of the two photosystems.

INTRODUCTION

Direct comparisons of structure-function relationships between organelles from different sources is often complicated by species variation. Such situations are simplified by the existence of two distinct types of the same organelle within one tissue and the occurrence of two morphologically distinct chloroplasts within a single plant and provide an excellent opportunity for comparing the relationships between structure and function. The availability of methods for fractionating mesophyll and bundle sheath chloroplasts led us to

---

[1] *Visiting Scientist from Institute of Genetics, Copenhagen University, Oester Farimagsgade 2A, DK1353, Denmark.*

examine these organelles in the hope that the apparent structural differences between them would provide some information about their different functions in the plant.

The chloroplasts of higher plants contain a number of specific lipids, the major components being monogalactosyldiacylglycerol and digalactosyldiacylglycerol. The suggestions that P-700 particles of spinach (which are constituents of photosystem I [PS-I]) are devoid of monogalactosyldiacylglycerol (1), and that agranal bundle sheath chloroplasts are deficient in photosystem II (PS-II) (2) prompted us to investigate the galactolipid composition of mesophyll and bundle sheath chloroplasts in maize and *Sorghum*. At the same time, a study of the photochemical activities of the two types of chloroplast has been carried out.

EXPERIMENTAL

Maize (*Zea mays* var. DS 606A) and *Sorghum* (*Sorghum bicolor* var. Texas 610) were grown in a greenhouse under prevailing light conditions. Chloroplasts were isolated by the method of WOO *et al.* (2) and were washed once with suspension medium before assay. Lipid extraction and assay and photochemical measurements were carried out as described (3,4).

RESULTS AND DISCUSSION

Structure and lipid content

The ultrastructure of the two types of chloroplasts from 10-day old maize plants are shown in *Plate 1*. The bundle sheath chloroplast contains starch granules but only at 3 or 4 places can any suggestions of grana be detected. The mesophyll chloroplast is characterized by extensive granal stacks, but is devoid of starch. If the segments of lamellae in the granal stacks are considered to be in a hydrophobic environment (5), then it is apparent that more hydrophobic areas are present per unit length of membrane in the mesophyll chloroplast lamellae than in the bundle sheath chloroplast lamellae, in which essentially all the lamellae are in contact with the aqueous stroma.

A comparative analysis of the galactolipid composition of the two types of chloroplasts is shown in *Table 1*. In both maize and *Sorghum*, the bundle sheath preparations have a higher chlorophyll *a:b* ratio than the mesophyll chloroplasts. The galactolipid content of the maize bundle sheath cells is, on a chlorophyll basis, three times greater than that of the mesophyll chloroplasts, while the bundle sheath chloroplast fragments contain over four times the amount. The ratio of the two galactolipids in the various fractions does not vary significantly and the fatty acid composition of the monogalactosyldiacylglycerol and digalactosyldiacylglycerol from all four fractions is about the same, containing over 90% α-linolenic acid. A similar trend is evident in the *Sorghum* fractions, although the differences

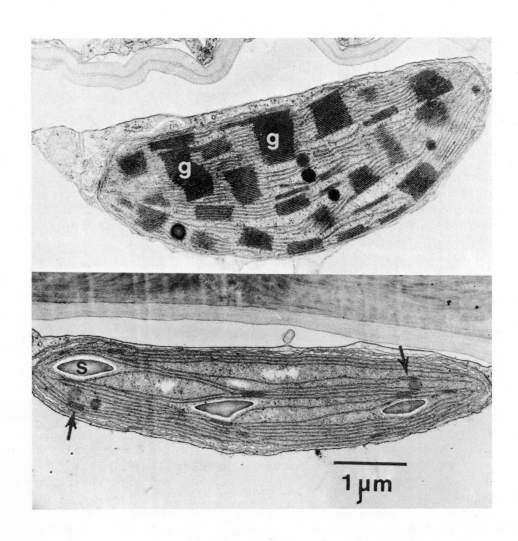

TABLE I

The galactolipid and chlorophyll composition of mesophyll and bundle sheath fractions (μmole μmole^{-1} chlorophyll).

Tissue	Fraction	Galacto-lipid	Ratio MGG:DGG*	Chlorophyll $a:b$
Maize	Whole leaves	3.4	2.0	3.15
(10 days)	Mesophyll chloroplasts	2.6	2.1	2.80
	Bundle sheath chloroplast fragments	11.5	2.4	4.86
	Bundle sheath cells	7.9	2.0	4.14
Sorghum	Whole leaves	6.3	1.6	3.31
(13 days)	Mesophyll chloroplasts	4.4	2.5	2.38
	Bundle sheath cells	6.7	1.8	5.26

* MGG, monogalactosyldiacylglycerol;
  DGG, digalactosyldiacylglycerol.

are not as great as in maize. Such variations highlight the difficulty of relating results obtained from different species.

We interpret the differences in the galactolipid content of the two types of chloroplasts as reflecting the lamellar structure of the chloroplast. In the model of chloroplast structure suggested by BENSON (5) the stroma lamellae (fret membranes) are envisaged as being surrounded by a hydrophilic environment. It therefore seems likely that such lamellae would require a higher content of amphiphilic lipids than the lamellae in the granal stacks, in which a hydrophobic environment is thought to exist (5). We feel our results are in accord with this hypothesis rather than that of ROSENBERG (6) who postulated that the fatty acid molecules of galactolipids serve as a binding site for the phytol side chains of chlorophyll molecules. Although the two theories are not mutually exclusive, the latter theory requires only 2 moles of galactolipid per mole of chlorophyll, and although such a ratio is present in some chloroplasts, the values for bundle sheath chloroplasts are much higher. It is possible, however, that the high molar ratio of galactolipid to chlorophyll in the bundle sheath chloroplasts is obtained because there is less chlorophyll per unit area of lamella.

Photoreduction of NADP

The photochemical activities of the chloroplast preparations are shown in *Figure 1*. The mesophyll chloroplasts of both maize and

Plate 1 Ultrastructure of mesophyll (upper) and bundle sheath chloroplasts (lower) of maize (g = grana; s = starch granules). Arrows indicate areas in the bundle sheath chloroplasts which appear to contain small grana.

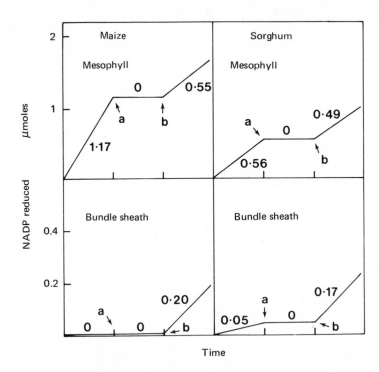

Figure 1 *Photoreduction of NADP by mesophyll and bundle sheath*
chloroplasts from 12-16 day old maize and Sorghum plants (4).
A, DCMU added; B, ascorbate and 2,6-dichlorophenolindophenol
added. The rates shown are in μmoles NADP reduced min^{-1} mg^{-1}
chlorophyll.

Sorghum were capable of carrying out the photoreduction of NADP. This
activity was inhibited by DCMU demonstrating that PS-II was present.
On the addition of 2,6-dichlorophenolindophenol (DCIP) and ascorbate,
photoreduction resumed, confirming the presence of PS-I. In contrast,
bundle sheath chloroplast fragments from maize were incapable of carrying out the photoreduction of NADP, although after the addition of
DCMU and DCIP-ascorbate, PS-I activity could be measured. The lack of
any photoreduction of NADP by the bundle sheath chloroplast fragments
indicated that the preparation was free of mesophyll chloroplast
fragments, as treatment of mesophyll chloroplasts under the conditions necessary to break bundle sheath cells, did not destroy any of
their activity.

Photosystem II activity in agranal chloroplasts
   The results above indicate that the isolated bundle sheath chloroplasts, which are unable to photoreduce NADP, either lack PS-II or
else contain a block or deficiency in the electron transfer pathway
linking PS-II with PS-I. To distinguish between these two possibilities, measurements were made of PS-II activity by a method which did
not involve PS-I, that is by use of artificial electron acceptors.
Bundle sheath chloroplasts of both maize and *Sorghum* were found to

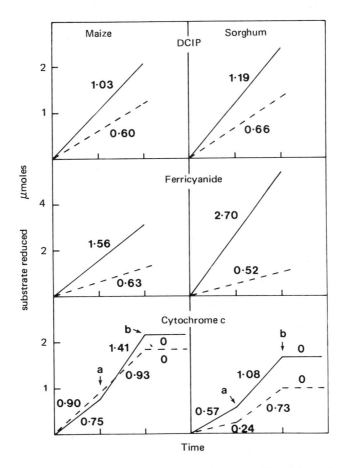

*Figure 2 Hill reaction by mesophyll and bundle sheath chloro-
plasts from 12-16 day old maize and Sorghum plants (4).
—— Mesophyll chloroplasts; ---- Bundle sheath chloroplasts.
A, ferredoxin added; B, DCMU added. The rates shown are in
μmoles substrate reduced min^{-1} mg^{-1} chlorophyll.*

contain Hill activity when measured with any of three commonly used
electron acceptors, DCIP, ferricyanide, or cytochrome c (*Figure 2*).
In other experiments we have shown that plastocyanin also serves as a
Hill oxidant for these chloroplasts. DCMU inhibited the photoreduction of all four oxidants. Treating the chloroplasts with 0.2 M Tris
buffer resulted in the loss of capacity to photoreduce DCIP, but activity was regained by adding semicarbazide as an electron donor.

The rate of substrate reduction by bundle sheath chloroplast fragments was generally lower than that of mesophyll chloroplasts. In the
case of cytochrome c reduction, the addition of ferredoxin, which
permits the reduction of cytochrome c by PS-I, resulted in an increase in the rate of reduction of cytochrome c by the maize mesophyll chloroplasts. Its failure to increase the rate of reduction of
cytochrome c in the maize bundle sheath chloroplast fragments suggests that the bundle sheath chloroplast fragments are free of

mesophyll contamination. The low activity of *Sorghum* bundle sheath chloroplast fragments for the photoreduction of NADP, and the 3-fold stimulation by ferredoxin of cytochrome *c* reduction, which is actually greater than the ferredoxin stimulation of mesophyll chloroplasts, might indicate that the two systems are not completely disconnected under the growth conditions employed, rather than that the bundle sheath chloroplast fragments are contaminated by mesophyll chloroplasts.

The effect of age and environmental conditions on photochemical activities

The results shown above indicate that PS-II is indeed present in bundle sheath chloroplasts of maize and *Sorghum* plants grown under our conditions, but that the two systems are not linked for electron flow, at least in the isolated preparations. In an attempt to produce bundle sheath chloroplasts which were deficient in PS-II, we have grown maize under a variety of conditions. The effect of leaf maturity on the photochemical activity of the bundle sheath chloroplast preparations from secondary leaves of maize is shown in *Table 2*.

TABLE 2

Photoreduction of NADP and Hill activity with cytochrome *c* in mesophyll (M) and bundle sheath (BS) chloroplasts of maize (μmoles substrate reduced min^{-1} mg^{-1} chlorophyll).

Days after sowing	Fresh weight g/100 leaves	Chlorophyll *a:b* ratio		NADP		Cytochrome *c*	
		M	BS	M	BS	M	BS
6	3.2	2.58	2.27	0.515	0.235	1.41	1.38
7	3.6	2.68	2.90	0.562	0.183	1.83	1.44
9	9.4	2.50	3.17	–	0.045	–	1.23
12	14.1	1.95	4.63	0.696	0.020	1.56	1.36
16	15.9	2.32	3.54	0.630	0.035	1.40	0.82

Growth is indicated by the increase in fresh weight of the leaves. It was found that 6 days after sowing, the bundle sheath chloroplast fragments were capable of photoreducing NADP, thus indicating that PS-I and PS-II were present and linked at least to some extent. This ability to photoreduce NADP was progressively lost until after twelve days growth the activity was only one-tenth of that after six days. However, measurement of PS-II by cytochrome *c* reduction shows that PS-II activity is retained during this period. Thus it is the link between PS-I and PS-II which is lost during growth. In contrast the capacity to photoreduce NADP is fully maintained by the mesophyll chloroplasts.

Electron microscopy of the secondary leaves at the various stages of growth suggests that at 6-7 days, when photoreduction of NADP takes place, the bundle sheath chloroplasts contain more granal stacks than in older leaves. While it is difficult to prepare bundle sheath cells free of mesophyll cells while the plants are young and the tissue very soft it is estimated that the contamination level was less than 10% in the 6-day sample, whilst the value for photoreduction of NADP by the bundle sheath preparation was 45% that of the mesophyll chloroplasts. As the secondary leaves develop, the linked activity between the two photosystems in the bundle sheath chloroplast fragments disappears, so that between 6 and 12 days growth there is a tenfold decrease in activity. There is a concomitant increase in the chlorophyll a:b ratio although it is not known if the two events are related. However, the activity of cytochrome c photoreduction remains essentially constant during this period, indicating strongly that PS-II is not disappearing, but rather that the two photosystems are becoming disconnected.

We have made other attempts to grow maize under conditions where PS-II is lacking. Plants have been grown under low light intensity, under high light intensity by supplementing normal daylight with horticultural lamps; under continuous light of varying intensities and at temperatures as high as 38°. Detached leaves have been allowed to senesce for 7 days in light. In all cases good PS-II activity, as measured by cytochrome c reduction, was obtained. Similar results were obtained from leaves taken from maize plants grown in the field for 7 weeks.

It was mentioned earlier that examination of the secondary leaves from 6-day old maize plants when PS-I and PS-II were linked indicated more granal stacks in the bundle sheath chloroplasts. Other workers have reported grana in partially developed bundle sheath chloroplasts of maize and sugarcane (7,8). Our results indicate that the presence of grana is not obligatory for PS-II activity. Although photoreduction of NADP is associated with granal stacking in most cases, exceptions have been reported. The photosynthetic mutant of *Chlamydomonas reinhardi*, ac-21, has chloroplasts with normal stacking and both PS-I and PS-II are present, but are not linked (9). Conversely, the ac-31 mutant is characterized by the absence of stacking but is capable of photosynthetic electron transport (10). Another group of *Chlamydomonas* mutants, which do not exhibit normal granal stacking have been found to be deficient in cytochrome b-559 (9), a cytochrome thought to be closely associated with PS-II. This cytochrome is absent from the bundle sheath cells examined by WOO *et al*. (2) and we also have been unable to detect it in our bundle sheath chloroplast preparations.

NADP photoreduction in intact bundle sheath cells

Our data confirm the results of WOO *et al*. (2) that agranal chloroplasts isolated from bundle sheath cells of maize and *Sorghum* are unable to photoreduce NADP. WOO *et al*. (2) consider that bundle sheath chloroplasts are deficient in the pigment-protein assembly of PS-II. If such is the case, bundle sheath cells would be incapable of

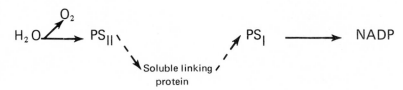

Figure 3 *Possible differences in the pathway of electron flow in granal mesophyll chloroplasts and agranal bundle sheath chloroplasts.*

photosynthetically reducing NADP coupled to oxygen evolution and this interpretation has been employed in some attempts at formulating the pathway of photosynthesis in C_4 plants (e.g. DOWNTON [11]). Our results, however, indicate that both photosystems are present but unlinked in the isolated bundle sheath chloroplast, and do not exclude the possibility of NADP photoreduction occurring in the intact bundle sheath cell. Soluble components present in the intact cell may act as an electron pathway bridge between the two photosystems. By following the photo-oxidation of cytochrome f (cytochrome c-554) in isolated intact bundle sheath cells, we have obtained evidence that the two photosystems are indeed linked in the intact cells, a conclusion which is supported by the observations of EDWARDS and BLACK (12) that isolated bundle sheath cells of *Digitaria sanguinalis* photosynthetically fix appreciable amounts of CO_2.

From these considerations we would suggest that a soluble linking protein(s) which mediates between PS-II and PS-I occurs in agranal bundle sheath cells (*Figure 3*, see also ref. 4). This component is presumably leached from the chloroplast during isolation resulting in a loss of the capacity to photoreduce NADP. In contrast, a complete membrane-bound electron transfer pathway exists in the grana-containing chloroplasts of the mesophyll cells.

Since bundle sheath chloroplasts apparently lack the membrane-bound cytochrome b-559 found in granal chloroplasts it will be of interest to see if soluble proteins possessing a potential similar to that of cytochrome b-559 [E_0' = +0.37, pH 7] (13) can act as a linking protein in isolated bundle sheath chloroplasts.

ACKNOWLEDGMENT

The authors wish to thank Dr. J. Bain and D. Gove for the electron micrographs and Ann Bartsch for capable technical assistance.

REFERENCES

1. C.F. ALLEN and P. GOOD, quoted by A.A. BENSON in Autonomy and Biogenesis of Mitochondria and Chloroplasts, eds. N.K. BOARD-MAN, A.W. LINNANE, and R.M. SMILLIE, North-Holland, Amsterdam, (1971) in press.
2. K.C. WOO, Jan M. ANDERSON, N.K. BOARDMAN, W.J.S. DOWNTON, C.B. OSMOND, and S.W. THORNE, *Proc. natn. Acad. Sci. U.S.A.*, 67 (1970) 18.
3. D.G. BISHOP, K.S. ANDERSEN, and R.M. SMILLIE, in preparation.
4. D.G. BISHOP, K.S. ANDERSEN, and R.M. SMILLIE, *Biochem. biophys. Res. Commun.*, in press.
5. T.E. WEIER and A.A. BENSON, in Biochemistry of Chloroplasts, ed. T.W. GOODWIN, Academic Press, New York, I (1966) 91.
6. A. ROSENBERG, *Science*, 157 (1967) 1191.
7. L.K. SHUMAY and T.E. WEIER, *Am. J. Bot.* 54 (1967) 773.
8. W.M. LAETSCH, *Am. J. Bot.*, 55 (1968) 875.
9. U.W. GOODENOUGH and R.P. LEVINE, *Plant Physiol.*, 44 (1969) 990.
10. U.W. GOODENOUGH, J.J. ARMSTRONG, and R.P. LEVINE, *Plant Physiol.*, 44 (1969) 1001.
11. W.J.S. DOWNTON, this volume, (1971).
12. G.E. EDWARDS and C.C. BLACK, this volume, (1971).
13. M. PLESNICAR and D.S. BENDALL, *Biochim. biophys. Acta*, 216 (1970) 192.

THE PROTEINS OF PHOTOSYSTEMS I AND II IN MESOPHYLL AND BUNDLE SHEATH CHLOROPLASTS OF *Sorghum bicolor*

J. Leggett Bailey, W.J.S. Downton, and Eva Mäsiar

Research School of Biological Sciences, Australian National University, Canberra City 2601, Australia.

SUMMARY

Mesophyll and bundle sheath chloroplasts of Sorghum bicolor were treated with sodium dodecylbenzenesulfate and the chlorophyll-protein complexes separated by electrophoresis in polyacrylamide gel. The protein ratios found for photosystems I and II were 1:2.2 for mesophyll chloroplasts (similar to spinach) and 1:0.7 for bundle sheath chloroplasts. Although a substantial amount of protein was present in the complex derived from photosystem II in bundle sheath chloroplasts, little pigment was associated with it.

INTRODUCTION

On the basis of their mode of carbon dioxide fixation, plants can be divided into two main categories, those that utilize the reactions of the Calvin cycle (C_3 plants) and those that utilize the β-carboxylation or C_4-dicarboxylic acid pathway (C_4 plants). The biochemical differences are well documented (1,2,3) and appear to be generally accepted. C_4 plants are further characterized by the occurrence of two distinctive cell layers surrounding the vascular bundles (4), the mesophyll and bundle sheath layers which contain varying proportions of chlorophyll. This morphological feature is exemplified in *Sorghum bicolor* which generally contains about 20% of the total leaf chlorophyll in the inner bundle sheath layer. The bundle sheath chloroplasts, which lack grana, also lack non-cyclic electron flow as measured by their capacity to reduce blue tetrazolium dye (5) and are apparently deficient in photosystem II units (6). This latter conclusion was based on photochemical and fluorescence measurements. A biochemical approach to the subject would first involve the simple step of physical separation of the two photosystems as their derived sodium dodecylbenzenesulfate(SDBS)-chlorophyll-protein complexes (7,8,9) or as complexes with sodium dodecylsulfate (10).

If bundle sheath chloroplasts are indeed deficient in photosystem II units then extraction with SDBS should afford mainly that complex related to photosystem I. We have carried out the experiment and wish to report that in spite of the lack of photosystem II activity, bundle sheath chloroplasts still contain a high proportion of associated protein although not as much as is contained in mesophyll cells.

EXPERIMENTAL

Seedlings of *Sorghum bicolor* L. (var. Texas 610) were grown in a glasshouse for 3-4 weeks. Mesophyll and bundle sheath chloroplasts were obtained from fully expanded leaves following the special grinding and fractionating procedures already described (6). The purity of the preparations was checked by examination under the light-microscope.

Chloroplasts were successively extracted ten times with 0.5% (w/v) SDBS-0.05 M sodium borate, pH 8.4 (detergent:chlorophyll, 5:1, w/w) and the crude chlorophyll-protein complexes precipitated with ammonium sulfate (9). The pooled extracts, neglecting the first extract, were dissolved in the minimum volume of 3% SDBS-5% sucrose in water and submitted to electrophoresis in 9% polyacrylamide gels. The procedure was similar to that described previously (7,9) except that for best results a 5 mm layer of a large pore stacking gel was also employed. Disks containing each zone were subsequently dissected from the gel columns and the SDBS complexes corresponding to the two photosystems were eluted, purified, and concentrated.

RESULTS AND DISCUSSION

Sorghum leaves (60 g) provided 8-10 mg of bundle sheath chlorophyll which had a chlorophyll *a:b* ratio from 6.0 to 7.0. This ratio is close to that (5.7) found for the light particles (enriched in photosystem I) obtained from spinach chloroplast lamellae split by digitonin (11) or by Triton X-100 (12) and to that (6.0) found for light particles obtained after mechanical breakage by sonication (13) or by passage through a French pressure cell (14).

All the SDBS extracts were blue-green in color and 8-10 treatments were required to completely solubilize the lamellar material of the bundle sheath chloroplasts. Components of the first SDBS extract were mainly free pigment together with small amounts of colorless proteins which failed to separate well in the gels. Electrophoresis of the combined extracts yielded two main green zones, a heavy slow-moving zone due to photosystem I and a diffuse banded area made up of pigment-detergent micelles. Above this pigment zone was a faint green zone exhibiting the properties of SDBS complex II a derivative of photosystem II. In *Figure 1* the colored components of bundle sheath chloroplasts (B) are compared with those obtained from mesophyll chloroplasts (M). Mesophyll chloroplasts provide SDBS

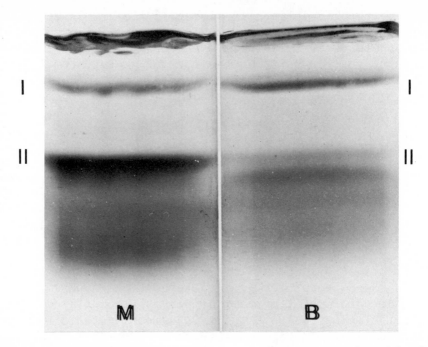

*Figure 1 Electrophoresis of SDBS extracts of Sorghum bicolor
mesophyll (M) and bundle sheath (B) chloroplasts in polyacrylamide gel (conc. 9%, cross-linking 1.66%). Zones corresponding
to complexes derived from photosystems I and II are indicated
by (I) and (II) respectively; the diffuse area below complex
II is due to pigment-detergent micelles.*

complexes I and II in roughly the same proportion (1:2.2) as given
by spinach, tobacco, and oat chloroplasts (1:2.0). The purified complexes from bundle sheath chloroplasts, however, yielded proteins related to photosystems I and II in the ratio 1:0.7. The photosystem I
band of the bundle sheath preparation contained six times as much
pigment as was found for the photosystem II band. Protein and chlorophyll $a:b$ ratios are presented in *Table 1*. The identity of the faint-
ly pigmented photosystem II zone is assumed from its electrophoretic
mobility under controled conditions and from the characteristic protein pattern obtained following electrophoresis in the absence of
SDBS.

When SDBS complex I is further subject to electrophoresis in the
absence of detergent, two colored protein zones are obtained containing different proportions of pigment. We have noticed that complex I
from bundle sheath chloroplasts usually gives a doublet, even in the
presence of SDBS, suggesting the existence of two very closely related structures in the original lamella. It is a matter of conjecture
at this stage whether this finding is any way relevant to the predictions (13,14) that there are two kinds of photosystem I, one occurring in grana lamellae and the other in inter-grana lamellae.

The main conclusion to be drawn from the above concerns the

Table I

Chlorophyll *a:b* ratios and protein ratios of SDBS complexes

	Chlorophyll *a:b*			Protein Complex I:Complex II
	Chloroplast	Complex I	Complex II	
Sorghum bundle sheath	6-7	34*	3.0	1:0.7
Sorghum mesophyll	3.2	20*	1.3	1:2.2
Spinach	3.2	12	1.2	1:2.0

*Determined by spectrofluorometric method (18)

question whether or not agranal chloroplasts contain the entity designated photosystem II. Certainly the protein moiety associated
with photosystem II occurs in an amount almost equal to the protein
moiety of photosystem I although in a proportion different from that
in granal chloroplasts. However, compared with photosystem I the
pigment complement is very low. Reports have appeared in the literature relating the morphology of the chloroplast (presence or absence
of grana) to changes in the environment (intensity of illumination)
or to the degree of maturation of the plant (15,16,17). It seems unlikely that gross changes in the basic chemical structure of chloroplast lamellae take place reversibly under changing light conditions.
We would rather believe that what might be termed the structural
protein of photosystem II remains largely intact in agranal chloroplasts. The lack of photosystem II activity may presumably be due to
some disorganization in the assembly of the pigment molecules within
the protein matrix.

REFERENCES

1. H.P. KORTSCHAK, C.E. HARTT, and G.O. BURR, *Plant Physiol.*, 40 (1965) 209.
2. M.D. HATCH and C.R. SLACK, *Biochem. J.*, 101 (1966) 103.
3. G.E. EDWARDS, S.S. LEE, J.M. CHEN, and C.C. BLACK, *Biochem. bio- phys. Res. Commun.*, 39 (1970) 389.
4. W.J.S. DOWNTON and E.B. TREGUNNA, *Can. J. Bot.*, 46 (1968) 207.
5. W.J.S. DOWNTON, J.A. BERRY, and E.B. TREGUNNA, *Z. PflPhysiol.* 63 (1970) 194.
6. K.C. WOO, Jan M. ANDERSON, N.K. BOARDMAN, W.J.S. DOWNTON, C.B. OSMOND, and S.W. THORNE, *Proc. natn. Acad. Sci. U.S.A.* 67 (1970) 18.

7. J.P. THORNBER, R.P.F. GREGORY, C.A. SMITH, and J.L. BAILEY, *Biochemistry*,   , 6 (1967) 391.
8. J.P. THORNBER, J.C. STEWART, M.W.C. HATTON, and J.L. BAILEY, *Biochemistry*, 6 (1967) 2006.
9. J.L. BAILEY and W. KREUTZ, in Progress in Photosynthesis Research, ed. H. METZNER, IUBS, Tübingen, I (1969) 149.
10. T. OGAWA, F. OBATA, and K. SHIBATA, *Biochim. biophys. Acta*, 112 (1966) 223.
11. Jan M. ANDERSON and N.K. BOARDMAN, *Biochim. biophys. Acta*, 112 (1966) 403.
12. L.P. VERNON, E. SHAW, and B. KE, *J. biol. Chem.*, 241 (1960) 4101.
13. G. JACOBI and H. LEHMANN, *Z. PflPhysiol.*, 59 (1968) 457.
14. P.V. SANE, D.J. GOODCHILD, and R.B. PARK, *Biochim. biophys. Acta*, 216 (1970) 162.
15. J.W. LYTTLETON, J.E.M. BALLANTINE, and B.J. FORDE, in Autonomy and Biogenesis of Mitochondria and Chloroplasts, ed. N.K. BOARDMAN, A.W. LINNANE, and R.M. SMILLIE, North-Holland, Amsterdam, (1971) in press.
16. W.J.S. DOWNTON and N.A. PYLIOTIS, *Can. J. Bot.*, (1971) in press.
17. W.M. LAETSCH and Ian PRICE, *Am. J. Bot.*, 56 (1969) 77.
18. H.R. HIGHKIN, N.K. BOARDMAN, and D.J. GOODCHILD, *Plant Physiol.*, 44 (1969) 1310.

EFFECT OF POTASSIUM CHLORIDE ON PHOTOSYSTEM II OF SPINACH CHLOROPLASTS[1]

Kozi Asada and Masa-aki Takahashi

The Research Institute for Food Science, Kyoto University,
Uji, Kyoto, 611, Japan

SUMMARY

The treatment of spinach chloroplasts with 0.5 M KCl resulted in the blocking of the water side of photosystem II. Hill reaction and fluorescence yield of KCl-treated chloroplasts were restored by the addition of hydroxylamine, diphenylcarbazide, or manganous ions, but not by the other electron donors which restore Hill reaction of Tris-washed chloroplasts. According to these observations, a scheme for the water side of photosystem II is proposed.

INTRODUCTION

Although it has been shown (1) that isolated bundle sheath chloroplasts of some C_4-pathway species lack photosystem II (PS-II), photosynthesis of green plants is characterized by oxygen evolution. The components participating in the splitting of water and the electron transport system between water and the reaction center of PS-II have not been disclosed, except for the involvement of manganese and, sometimes, of chloride ions. Several lines of evidence have been presented to show that oxidants which are supposed to be produced in the water side of PS-II play a role in the formation of glycolate and its metabolism (2-4).

In the present communication, the effect of KCl on spinach chloroplasts is reported. The observations presented here suggest that at least two steps occur on the water side of PS-II.

EXPERIMENTAL

Unless otherwise specified, "glycerinated chloroplasts" were used

---

[1] *This work was supported in part by a grant from the Ministry of Education.*

and were prepared as follows. Spinach chloroplasts were prepared as described previously (5), except that the grinding medium used contained 400 mM sucrose, 50 mM tricine-KOH, pH 8.0, and 10 mM NaCl. To the chloroplasts suspended in the grinding medium, the same volume of cold glycerol was added. After mixing, the suspension was stored at -20°. Under these conditions, less than 20% of the NADP, ferricyanide and 2,4 dichlorophenolindophenol (DCIP) photoreduction activity was lost during storage for 4 months (6).

KCl-washed chloroplasts were prepared from glycerinated chloroplasts. Water was added to the chloroplasts to bring the concentration of glycerol to 5% and the suspension was centrifuged. Chloroplasts were then suspended in 500 mM KCl to make the concentration of chlorophyll 150 μg ml^{-1}, and then stood for 30 min at 0°. The suspension was centrifuged and the sediment washed once with water. The KCl-washed chloroplasts were resuspended in the grinding medium with 50% glycerol. Tris-washed chloroplasts were prepared as described (7).

The rates of DCIP and NADP reduction were followed from the change of the absorbancy at 600 and 340 nm, respectively, using the following reaction mixture: 50 mM tricine-KOH, pH 8.0, 10 mM NaCl, and 0.2 ml of chloroplasts containing 30 μg of chlorophyll, in a total volume of 2 ml. To the above basal mixture, either 0.1 mM DCIP or 0.1 mM NADP plus 8.5 μM spinach ferredoxin was added. The reaction was carried out under the illumination (50,000 lux) from a projector lamp at 20°.

RESULTS AND DISCUSSION

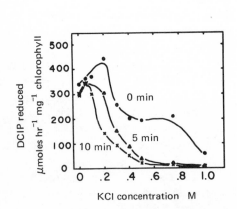 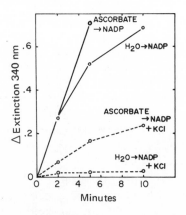

Figure 1 (left) Effect of KCl on DCIP reduction in glycerinated chloroplasts. KCl was added to the reaction mixture at 20° as indicated and after various periods reduction was determined.

Figure 2 (right) Effect of KCl on NADP reduction in glycerinated chloroplasts. KCl (200 mM) was added as indicated, other additions were 0.1 mM DCIP and 5 mM ascorbate.

The presence of KCl in the reaction mixture resulted in the decrease of the photoreduction of DCIP by glycerinated chloroplasts. As shown in *Figure 1*, about 10 min contact with 500 mM KCl at 20° brought about complete loss of the activity. Under these conditions O_2 evolution was not detectable. In the case of fresh chloroplasts a higher concentration of KCl was required to inhibit the photoreduction.

Figure 2 shows the effect of 200 mM KCl on NADP reduction. In the presence of KCl the activity was completely lost but was partially recovered by the addition of an electron donor for photosystem I (PS-I). These observations suggest that KCl mainly affected PS-II. As will be shown later, KCl-washed chloroplasts also lost the capacity to photoreduce NADP and DCIP.

In order to determine which side of PS-II was affected by KCl, fluorescence yield of the chloroplasts in the presence of KCl was measured. The addition of KCl at 100 mM to the glycerinated chloroplasts decreased the fluorescence yield at 680 nm on excitation by 436 nm actinic light to a level equivalent to that found for the KCl-washed chloroplasts (*Figure 3*). These observations support the conclusion that treatment of the chloroplasts with KCl brought about the limitation in electron drawing capacity to trapping center of PS-II, thus indicating that KCl inhibits the water side of PS-II.

Further evidence for the above conclusion was obtained by adding

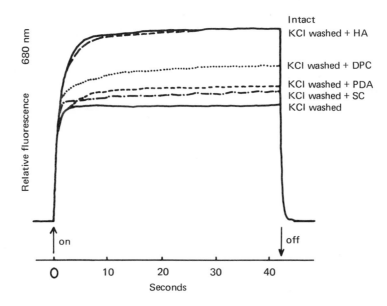

Figure 3 *Fluorescence yield of KCl-washed chloroplasts. The chloroplasts (7.5 µg chlorophyll ml⁻¹) were suspended in 50 mM tricine-KOH, pH 8.0 and 10 mM NaCl and irradiated with 436 nm actinic light, at room temperature. To the standard reaction mixture, 25 mM HA, 1 mM DPC, 0.1 mM PDA, and 1 mM SC were added as shown.*

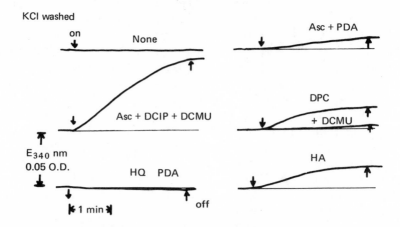

Figure 4 Photoreduction of NADP by KCl-washed chloroplasts. To the standard reaction mixture, 0.06 µM ferredoxin-NADP reductase was added. Ascorbate and DCIP were added as in Figure 2. Where indicated, 2 mM HQ, 0.1 mM PDA, 5 mM ascorbate, and 0.1 mM PDA, 1 mM DPC in the presence and absence of 5 µM DCMU and 25 mM HA were added.

electron donors for PS-II (*Figure 3*). The addition of hydroxylamine (HA) at 25 mM to the KCl-washed chloroplasts resulted in the restoration of the steady state level of fluorescence to that of the

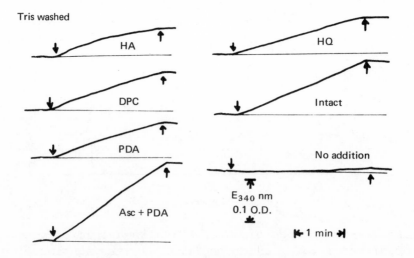

Figure 5 Photoreduction of NADP by Tris-washed chloroplasts. The electron donors were added as in Figure 4. Chloroplasts were intact before the Tris-treatment.

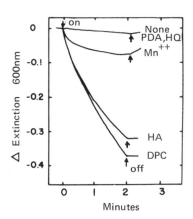

Figure 6 Photoreduction of DCIP by KCl-washed chloroplasts.
Addition of the electron donors was the same as in Figure 4.
MnCl₂ (0.5 mM).

unwashed chloroplasts. In addition to HA, diphenylcarbazide (DPC) also partially restored the fluorescence yield but p-phenylaminediamine
(PDA) and semicarbazide (SC) were not effective. Both HA and DPC are
known to donate electrons to PS-II of Tris-washed chloroplasts (7,8).

Because the steady state level of fluorescence by KCl-treated
chloroplasts was restored by the addition of the electron donors to
PS-II, it was expected that the Hill reaction of the treated chloroplasts would also be restored. In the case of KCl-washed chloroplasts, even in the presence of an electron donor to PS-I, the photoreduction of NADP was not observed without the addition of ferredoxin-NADP reductase. Therefore, in the following experiment, in addition to ferredoxin, the flavoprotein was added to the reaction mixture. NADP reduction by KCl-treated chloroplasts was restored by the
addition of DPC or HA (*Figure 4*). Since DPC-supported reduction was
inhibited by DCMU this restored activity was apparently not due to
electron donation to PS-I. However, PDA or hydroquinone (HQ) gave
little or no NADP reduction. When ascorbate was added to PDA, a low
activity was observed. On the other hand, in the case of Tris-washed
chloroplasts, all of the chemicals tested supported NADP reduction
(*Figure 5*).

As shown in *Figure 6*, DCIP reduction by KCl-washed chloroplasts
was also supported only by HA and DPC. The restored activity was inhibited by DCMU, suggesting that the two chemicals did not donate electrons to PS-I. The rate of the reduction in the presence of manganous ions was rapid upon turning on of light but decreased with time.
Upon turning off of light, rapid reoxidation of DCIP was observed.
These results suggest that manganous ions donated electrons to PS-II
of KCl-washed chloroplasts (9,10). Neither PDA and SC (shown in *Figure 6*), nor 1 mM dimethyl- and trimethyl-p-phenylenediamine, SC, ben-
zidine, and isonicotinic acid hydrazide supported DCIP reduction.

All of the evidence presented here suggests that KCl affected the

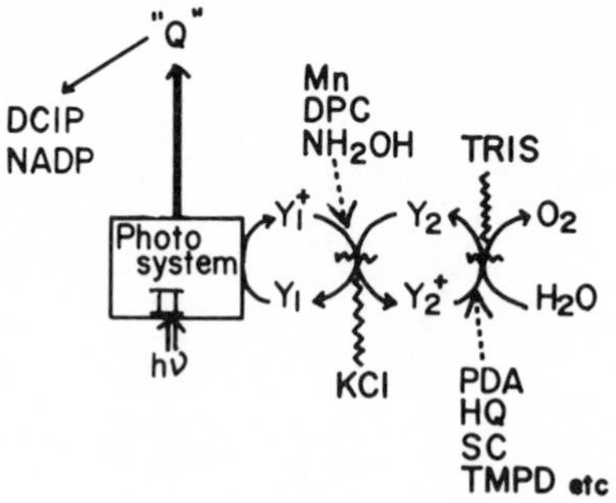

Figure 7 A proposed scheme for the water side of photosystem II.

water side of PS-II of spinach chloroplasts. However, compared with Tris-washed chloroplasts, we found a difference in the chemicals donating electrons to PS-II. In the case of Tris-washed chloroplasts, PDA, HQ, DPC, HA, SC, and others were good donors to PS-II, as evidenced by the recovery of fluorescence yield and Hill reaction, previously found by YAMASHITA and BUTLER (7,8), VERNON and SHAW (11) and ourselves with NADP reduction (*Figure 5*). On the other hand, with KCl-washed chloroplasts, only HA, DPC, and possibly manganous ions functioned as electron donors to PS-II.

To explain these observations, the following scheme is proposed (*Figure 7*). In the water side of PS-II, between the water and trapping center of PS-II, at least two reactants, Y_1 and Y_2, are involved. Tris-treatment of chloroplasts resulted in impairment between Y_2 and water. Thus, in the case of Tris-treatment, all chemicals were capable of donating electrons to PS-II. On the other hand, KCl acted either between Y_1 and Y_2, or possibly between both Y_1 and Y_2 and between Y_2 and water. Under these circumstances electron donors to PS-II were limited to HA, DPC, and manganous ions. Involvement of multisteps in the water side of PS-II has been also proposed by JOLIOT from flash light experiments on oxygen evolution and CHENIAE and MARTIN from their studies on PS-II activity and manganese contents of chloroplasts (12).

Irradiation of KCl-washed chloroplasts brought about the bleaching of the chloroplast pigments as in the case of Tris-washed chloroplasts (13,14). The addition of HA protected chloroplasts from the photobleaching. Thus it was supposed that the photobleaching was due to the oxidation of the pigments by oxidized Y_1 which was formed by the KCl-blockage of the water side of PS-II. In glycolate formation, oxidized Y_1 or Y_2 or both may play a role (2,3). The stimulation of glycolate oxidation by light (15) was also indicated by glyoxylate production. The possibility of the participation of oxidized Y_1 or Y_2 to this reaction is a future subject to be studied.

REFERENCES

1. K.C. WOO, Jan M. ANDERSON, N.K. BOARDMAN, W.J.S. DOWNTON, C.B. OSMOND, and S.W. THORNE, *Proc. natn. Acad. Sci. U.S.A.*, 67 (1970) 18.
2. Z. PLAUT and M. GIBBS, *Plant Physiol.*, 45 (1970) 470.
3. M. GIBBS, P.W. ELLYARD, Z. PLAUT, and Y. SHAIN, this volume, (1971).
4. H.A. TANNER, T.E. BROWN, C. EYSTER, and R.W. TREHARNE, *Biochem. biophys. Res. Commun.*, 3 (1960) 205.
5. K. ASADA, S. KITOH, R. DEURA, and Z. KASAI, *Plant Cell Physiol.*, 6 (1965) 615.
6. K. ASADA and M. TAKAHASHI, *Plant Cell Physiol.*, in press.
7. T. YAMASHITA and W.L. BUTLER, *Plant Physiol.*, 43 (1968) 1978.
8. T. YAMASHITA and W.L. BUTLER, *Plant Physiol.*, 44 (1969) 435.
9. S. IZAWA, *Biochim. biophys. Acta*, 197 (1970) 328.
10. G. BEN-HAYYIM and M. AVRON, *Biochim. biophys. Acta*, 205 (1970) 86.
11. L.P. VERNON and E.R. SHAW, *Biochem. biophys. Res. Commun.*, 36 (1969) 878.
12. G.M. CHENIAE, *A. Rev. Pl. Physiol.*, 21 (1970) 467.
13. K. YAMASHITA, K. KONISHI, M. ITOH, and K. SHIBATA, *Biochim. biophys. Acta*, 172 (1969) 511.
14. T. YAMASHITA and W.L. BUTLER, *Plant Physiol.*, 44 (1969) 1342.
15. L.A. DELAVAN and A.A. BENSON, in Brookhaven Symposia in Biology, 11 (1958) 259.

A COMPARATIVE STUDY OF THE COUPLING OF ION UPTAKE TO LIGHT REACTIONS IN THE LEAVES OF C_3 AND C_4 PLANTS

Ulrich Lüttge

Botanisches Institut der Technischen Hochschule,
6100 Darmstadt, Germany

SUMMARY

The C_4 pathway of photosynthesis requires several transport processes of different kinds to be integrated in the sequence of biochemical reactions. The model of salt uptake and excretion by the leaves of the C_4 plant *Atriplex spongiosa* is used to demonstrate light dependent symplasmatic transport and transport across membranes spatially distant from the chloroplasts with events occurring within these organelles. A comparative study of Cl^- uptake in relation to photosynthesis and ATP levels in the light suggests that in the C_3 plants spinach, *Atriplex hastata*, and *Oenothera*, light dependent Cl^- uptake is driven by ATP produced in photophosphorylation. In the C_4 plants *A. spongiosa*, *Zea mays*, and *Amaranthus caudatus* light dependent Cl^- uptake is independent of ATP or $\sim P$. In view of the effective symplasmatic transport of malate in C_4 plants it is concluded that reducing equivalents from the malate might provide energy in the cytoplasm for Cl^- transport across membranes distant from the thylakoids, where the reducing equivalents are generated.

INTRODUCTION

In the discussion of the many facets of the problem of C_4 photosynthesis, translocation processes appear to be much closer to the center of interest than one usually finds in the evaluation of biochemical reaction sequences. The co-operation of mesophyll and bundle sheath cells of C_4 plants can only work towards an efficient photosynthesis of carbohydrates if translocation processes are effectively integrated with the sequence of enzyme reactions. The scheme of C_4 photosynthesis requires transport of several intermediates carrying CO_2, $\sim P$, and reducing equivalents across the envelope of both mesophyll and bundle sheath chloroplasts. There must be symplastic transport of malate/aspartate and pyruvate in opposite directions and possibly also of triosephosphates and 3-phosphoglyceric acid. Furthermore, the long distance transport in the phloem and the deposition of starch in bundle sheath and mesophyll chloroplasts are related. The number of translocation arrows involved in the scheme of a photosynthesizing leaf becomes overwhelming when photorespiration is included, with transfer of intermediates between chloroplasts, mitochondria, and glyoxy/peroxisomes. (For references see the pertinent papers included in this volume).

Compared with this situation, a salt transport system which we began to investigate in leaves of the C_4 plant *Atriplex spongiosa* appears very simple, although transport physiologists would acknowledge its complexity. However, this system gives emphasis to the coupling of transport processes with events in the chloroplasts. This system involves ion transport from the green cells of the leaf blade, the epidermal cells, and the epidermal bladders supported by stalk cells, having a gland like dense cytoplasm [1,2]. The system may serve in the desalination of the leaf tissue of *A. spongiosa* growing in saline habitats.

EXPERIMENTAL

In some experiments 3 mm wide strips of leaf tissue with intact epidermal bladders were floated on labeled experimental solutions [1]. Uptake of $^{36}Cl^-$ by these tissues was sluggish due to slow equilibration of the tissue with experimental solutions. Thus in most experiments thin slices (0.5 mm) were used which could be manipulated much as an algal chain and in which equilibration times were very rapid [3]. Full experimental details are given elsewhere [4,5,6].

RESULTS AND DISCUSSION

Transport of Cl^- from a solution bathing the green cells of the leaf blade (in the apoplastic space) to the bladder vacuoles is highly stimulated by light. Light stimulation of uptake and transport became obvious after 20 hr. After 40 hr uptake by the blade cells was stimulated 1.5 times and translocation to the bladders 5 times [1]. This Cl^- transport is an active process. The Nernst potential calculated from chloride concentrations in the bladder vacuole and in the external solution, which contained 5 mM KCl, under the conditions of these experiments was +40 to +50 mV. By contrast, the membrane potential measured with a glass microelectrode in the bladder vacuole and a reference electrode in the external solution was -90 to -100 mV. Thus, chloride moves against an electro-chemical potential gradient [1].

In *Table 1* the effects of various treatments on light dependent $^{36}Cl^-$ entry from a 5 mM KCl solution into 0.5 mm wide leaf strips of *A. spongiosa* and on light dependent CO_2 fixation are depicted (c.f. 4). (DCMU = di-chlorophenyl-dimethyl-urea; FCCP = pCF_3O-carbonyl cyanide phenylhydrazone.)

TABLE I

The effect of far red light, DCMU, and FCCP on light dependent Cl⁻ uptake and CO_2 fixation in *A. spongiosa* leaf slices (percentage of control).

	Light >705 nm	DCMU $(5\times10^{-7}-2\times10^{-6}M)$	FCCP $(2\times10^{-7}-10^{-6}M)$
Cl⁻ uptake	0	50-0	100-150
CO_2 fixation	0	40-5	40- 10

It can be seen that the light dependent fraction of Cl⁻ entry into the leaf cells is entirely linked to photosynthesis. The experiments both with cut-off filters and DCMU suggest that the operation of both photosystems, i.e. of noncyclic electron flow, is required. However, the uncoupler FCCP leaves photosynthesis dependent chloride uptake unaffected or enhances it. Thus, we must conclude that ATP or a phosphorylated high energy intermediate produced in photophosphorylation is not the source of energy driving photosynthesis dependent Cl⁻ uptake in this system.

Since Cl⁻ transport into the bladder vacuole is strongly light dependent, one might expect that the stalk and bladder cells would exhibit extraordinary photosynthetic activity. However, chloroplasts are scarce in the bladder and stalk cell cytoplasm, and photosynthetic $^{14}CO_2$ fixation is very limited in these cells (7). Hence, energy must be derived from processes in the mesophyll, spatially separated from the site of ion accumulation in the bladder vacuole.

Electrical evidence shows effective symplasmatic contact between the mesophyll and the epidermal bladders (1,8). The problem remains, however, of *how* the metabolic dependence of these events might be linked. It was previously accepted that ATP is readily distributed from within the chloroplasts to other membranes. It appeared that the problem of linkage was compounded in that ATP was not the source of energy for active photosynthesis dependent Cl⁻ transport in *A. spongiosa*. In view of the discussion of HEBER'S recent results (9, and this volume) it is likely that transport of both ∿P and reducing equivalents across the chloroplast envelope must be mediated by "transport metabolites". It thus becomes more difficult to explain light driven, ATP dependent pumps than ATP independent pumps, since ATP can only move accompanied by reducing equivalents, whereas reducing potential can move in the absence of ∿P transport.

In principle there appear to be two basically different mechanisms for coupling ion transport at membranes distant from the chloroplasts with energy transfer at the thylakoids. First, there is a biophysical mechanism of linkage, because ion-imbalances and gradients of electrochemical potential are built up within the cells by photosynthetic

electron flow and the associated ion fluxes at the thylakoid membranes. There is evidence that this affects ion transport at other
membranes (e.g. 8,10).

Second, there are biochemical mechanisms of linkage. It has been
suggested that reducing equivalents formed through photosynthetic electron transport serve as the source of energy for ATP independent
ion uptake by green plant cells (HOPE and LILLEY, personal communication). Reducing equivalents are transported out of the chloroplasts
via transport metabolites (see above and 11). In C_4 plants symplasmatic transport of photosynthetic intermediates is obviously of particularly high efficiency. Among other compounds, malate is discussed
as an important intermediate transported symplasmatically in these
plants. It is clear that malate transport means a transport of reducing equivalents formed by photosynthetic electron flow. Some of the
malate in the symplast may be subject to reaction with NAD dependent
malate dehydrogenase, yielding reducing potential as NADH, which then
might be utilized to drive Cl^- uptake by a mechanism which has yet to
be explained. If this view is correct, one would expect to find more
active ATP independent, photosynthesis driven ion uptake in C_4 species. The above experiments with *A. spongiosa* encouraged a further
survey and we have compared three C_4 species (*A. spongiosa, Amaranthus caudatus*, and *Zea mays*) with three C_3 species (spinach, *Atriplex hastata, Oenothera*). In the C_4 plants light always stimulated Cl^-
uptake (between 1.3 and 2.2 times depending on the species). In the
C_3 plants light sometimes stimulated and sometimes inhibited Cl^- uptake, depending on the nutritional state of the tissue. In either
case, in C_3 plants, the effect of light was prevented by DCMU at a
concentration inhibiting CO_2 fixation in the light to less than 15%
of the controls. This provides additional evidence for our conclusion
that light only affects Cl^- uptake via the photosynthetic apparatus.
Light inhibition of Cl^- uptake may be explained by a competition of
photosynthetic CO_2 reduction for the total ATP available in a situation where total ATP production is limited by other factors. Light
stimulation of Cl^- uptake may be explained either by a higher availability of ATP due to photophosphorylation or by supply of photosynthetic energy to the ion pump independent of ATP.

It has been assumed that Cl^- uptake dependent on dark metabolism
continues in the light and dark uptake is subtracted from rates in
the light (4). *Table 2* shows the effects of FCCP on photosynthesis
dependent Cl^- uptake, on photosynthetic CO_2 fixation and O_2 evolution, on O_2 uptake in the dark, and on ATP levels in the light for C_3
and C_4 species (5,6). In the C_4 species FCCP concentrations between
10^{-7} M and 10^{-6} M enhance Cl^- uptake, in the C_3 plants concentrations
above 10^{-7} M are inhibitory. Photosynthesis and ATP levels in the
light are inhibited in both groups of plants. The increased O_2 uptake
in the dark is characteristic of uncoupling of respiration and shows
that FCCP is taken up by the cells and that it is present in the active form.

Thus it appears possible to explain the enhancement of Cl^- uptake

TABLE 2

Effect of FCCP on Cl⁻ uptake, photosynthesis, and respiration in leaf slices of C_3 and C_4 plants [Percentage of controls, showing range of values for three species. See (5,6) for details].

FCCP concentration (M)	Species	Cl⁻ uptake	Light CO_2 fixation	ATP level	Dark O_2 uptake
10^{-7}	C_4	125	85-100	–	120-145
	C_3	100	60-100	–	90-175
5×10^{-7}	C_4	140	50-95	80-95	135-170
	C_3	10-100	30-70	65-70	95-175
10^{-6}	C_4	100-115	25-75	45-90	120-170
	C_3	10-75	5-55	40-75	90-170
2×10^{-6}	C_4	50	10-60	30-85	120-150
	C_3	40	35-45	20-45	100

by FCCP in terms of uncoupling of phosphorylation which makes available an increased amount of reducing equivalents for active Cl⁻ uptake. Among the 6 species investigated, the occurrence of a photosynthesis dependent Cl⁻ pump enhanced by uncouplers is correlated with the occurrence of the C_4 pathway of photosynthesis. This suggests that the efficient translocation of energy carrying photosynthetic intermediates - a characteristic feature of C_4 plants - is also responsible for the coupling of ATP independent, photosynthesis driven Cl⁻ uptake with energy transfer within the chloroplasts.

However, we may not conclude that the C_4 pathway is an absolute prerequisite for this form of photosynthesis driven Cl⁻ pump. The classic examples of such pumps are giant algal coenocytes such as *Nitella* and *Hydrodictyon*(12,13). Dr. KLUGE has shown that amongst the initial products of light dependent $^{14}CO_2$ fixation by *Hydrodictyon*, neither malate nor aspartate contains considerable amounts of label after short term fixation.

In the C_4 species malate transport from the interior of chloroplasts into the symplast is assumed to occur mainly in the mesophyll but not in the bundle sheath parenchyma. It is therefore of particular interest in relation to the hypothesis developed above, that we have found isolated bundle sheath tissue of *Zea mays* to have a very low Cl⁻ accumulation capacity as compared with intact 0.5 mm wide leaf strips. Autoradiographs of leaf strips suggest that after uptake of ^{36}Cl⁻ label is equally distributed between mesophyll and bundle sheath cells. These recent results (SCHOCH, ENGEL, LÜTTGE unpublished) again emphasize co-operation between the two types of cells.

ACKNOWLEDGMENT

This work and attendance at the meeting, was supported by the Deutsche Forschungsgemeinschaft.

REFERENCES

1. C.B. OSMOND, U. LÜTTGE, K.R. WEST, C.K. PALLAGHY, and B. SHACHER-HILL, *Aust. J. biol. Sci.*, 22 (1969) 797.
2. U. LÜTTGE, *A. Rev. Pl. Physiol.*, (1971) in press.
3. C.B. OSMOND, *Aust. J. biol. Sci.*, 21 (1968) 1119.
4. U. LÜTTGE, C.K. PALLAGHY, and C.B. OSMOND, *J. Membrane Biol.*, 2 (1969) 17.
5. U. LÜTTGE, E. BALL, and K. von WILLERT, *Z. PflPhysiol.*, (1971) a in press.
6. U. LÜTTGE, E. BALL, and K. von WILLERT, *Z. PflPhysiol.*, (1971) b in press.
7. U. LÜTTGE and C.B. OSMOND, *Aust. J. biol. Sci.*, 23 (1970) 17.
8. U. LÜTTGE and C.K. PALLAGHY, *Z. PflPhysiol.*, 61 (1969) 58.
9. U. HEBER and K.A. SANTARIUS, *Z. Naturf.*, 25 b (1970) 718.
10. C.K. PALLAGHY and U. LÜTTGE, *Z. PflPhysiol.*, 62 (1970) 417.
11. U. HEBER and K.A. SANTARIUS, *Biochim. biophys. Acta*, 109 (1965) 390.
12. E.A.C. MacROBBIE, *Biochim. biophys. Acta*, 94 (1965) 64.
13. J.A. RAVEN, *New Phytol.*, 68 (1969) 45.

THE RELATIONSHIP BETWEEN GRANA AND STROMA LAMELLAE AND PHOTOSYSTEMS I AND II IN SPINACH CHLOROPLASTS

D.J. Goodchild

Division of Plant Industry, CSIRO, Canberra City, 2601, Australia

SUMMARY

Class II spinach chloroplasts were fragmented by passage through a French pressure cell and the fragments separated by fractional centrifugation. Sixty five percent of the chlorophyll appeared in the fraction sedimenting between 1000 x g and 10,000 x g (10 K) which had both photosystem I and II activity and a chlorophyll a/chlorophyll b ratio lower than 2.3. The fraction sedimenting between 40,000 x g and 160,000 x g (160 K) contained 7-10% of the chlorophyll with a chlorophyll a/chlorophyll b ratio of 6.0 and only photosystem I activity.

Thin sectioning, freeze fracturing, and deep etching showed the 160 K fraction to originate from stroma lamellae, and possibly end grana membranes of grana stacks, and to contain 110 Å particles. The 10 K fraction originated from the partition regions of grana stacks and contained both 110 Å and 175 Å particles. This distribution of particles on the fracture faces of stroma versus grana lamellae was also shown to exist for freeze fractured Class I chloroplasts.

Although it is tempting to correlate particle sizes with photosystem activities, recent evidence for the particle size distribution on freeze fractured faces of stacked and unstacked membranes suggests that such a correlation is not yet warranted.

INTRODUCTION

MICHEL and MICHEL (1) reported that the two photochemical systems of spinach chloroplasts could be partially separated on a density gradient following chloroplast breakage in a French pressure cell. The experiments reported here were undertaken to confirm this observation and to determine the biochemical characteristics, origin, and ultrastructure of the fragments from the French pressure cell.

EXPERIMENTAL

Class I spinach chloroplasts were isolated in buffer (0.05 M potassium phosphate, pH 7.4, 0.01 M KCl, 0.5 M sucrose) and resuspended in 0.15 M KCl in 0.05 M potassium phosphate buffer, pH 7.4. This suspension (Class II chloroplasts) was passed through an Aminco French pressure cell at 6,000 lb in^{-2} to yield a French press (FP) homogenate. The FP homogenate was fractionally separated by centrifugation at 1000 x g for 10 min, 10,000 x g for 30 min, 40,000 x g for 30 min, and 160,000 x g for 60 min (1 K, 10 K, 40 K, and 160 K fractions).

Chlorophyll a and b were determined by the spectrophotometric method of ARNON (2) and P 700 by the ferricyanide method (3). Photosystem I reduction of NADP was followed spectrophotometrically with added ferredoxin, plastocyanin, and NADP reductase (4). Photosystem II activity was also followed spectrophotometrically using water as reductant and dichlorophenolindophenol (DCIP) as oxidant (4).

For electron microscopy, fractions were resuspended in buffer containing 3% glutaraldehyde, pelleted, post-fixed with osmium tetroxide, embedded, sectioned, and then stained with uranyl acetate and lead citrate. Freeze fracturing and deep etching of the fractions were carried out as described by PARK and PFEIFHOFER (5) and Class I chloroplasts for freeze-fracturing were prepared by the method of JENSEN and BASSHAM (6).

RESULTS

Biochemical characteristics
Experiments to determine the maximum yield of the 160 K fraction

TABLE I

Distribution of chlorophylls, P 700, and photochemical activities (μmoles hr^{-1} mg^{-1} chlorophyll) of the fractions separated by differential centrifugation from a French Press homogenate (6,000 lb in^{-2}) of spinach chloroplasts

| Fraction | Chlorophyll | | | Activity | |
	$Chl\ a/$ chl b	% chl	$Chl/$ P 700	DCIP (pH 7.0)	NADP (pH 7.4)
Chloroplasts				174	172
FP	2.9	100	423	52	87
1 K				100	62
10 K				74	75
1 K + 10 K	2.4	67.1	650		
40 K	3.0	23.7	253	–	87
160 K	6.0	7.4	105	0	169
160 super.	4.7	1.8	0	–	–

from spinach chloroplasts showed that a pressure of 6,000 lb in^{-2} in the French pressure cell was optimal. The characteristics of the fractions obtained are shown in *Table 1*. The 160 K fraction was greatly enriched in *chl a* and P 700, was incapable of reducing DCIP but had a high Photosystem I (PS-I) activity. Thus the data suggested that the 160 K fraction was a PS-I fraction that had been separated from the French press homogenate. On the other hand the 1 K and 10 K fractions, which have both photosystems, have a low *chl a/chl b* ratio, are relatively depleted in P 700, are enriched in Photosystem II (PS-II) activity and depleted in PS-I activity compared with the FP homogenate. Further evidence for this distribution of the photosystems was obtained by the method of THORNBER *et al.* (7) when the 160 K fraction yielded predominantly complex I with a *chl a/chl b* ratio of 8 and the 10 K fraction yielded both complexes I and II.

Ultrastructural characteristics

Before passage through the French pressure cell Class II spinach chloroplasts, as seen in thin section, have well defined grana and interconnecting stroma lamellae (*Plate 1*). Treatment with the French pressure cell leaves intact grana stacks but the network of stroma lamellae is destroyed yielding vesicles of various sizes. When this FP homogenate is subjected to differential centrifugation the 10 K fraction (*Plate 2*) consists primarily of the grana stacks, while the 160 K fraction (*Plate 3*) is predominantly vesicles of various sizes. The 40 K fraction is intermediate in composition between these two fractions and consists of some large vesicles and small disrupted grana stacks. Freeze fracturing and deep etching of the 10 K and 160 K fractions showed that the 10 K fraction contained both large (175 Å) and small (110 Å) particles on the B and C fracture faces and had both A and D surfaces. This would be expected if the model of a gra-

---

Plate 1 Section through portion of a Class II chloroplast before French press treatment showing grana stacks with interconnecting stroma lamellae (St) x 47,500.

Plate 2 Section through a 10 K fraction showing remnants of grana (G) and some large vesicles with partitions x 29,000.

Plate 3 Section through a 160 K fraction showing vesicles of various sizes x 70,000.

Plate 4 Freeze-fractured and deep-etched 160 K fraction showing a vesicle with an A type surface and a fracture face, C, at either end with only small (110 Å) particles x 70,000.

Plate 5 Freeze-fractured Class I chloroplast showing a B fracture face with large (175 Å) particles in the grana region (G) and small (110 Å) particles in the stroma lamellar region (St). The end grana membrane (Em) also has only small particles x 100,000.

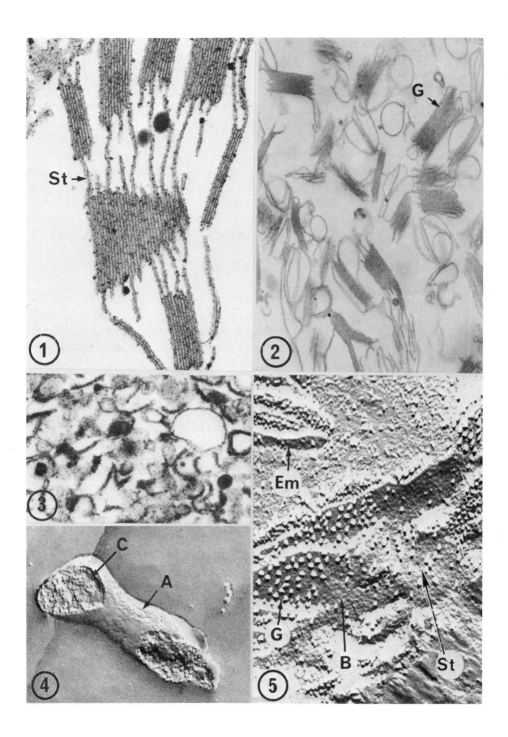

num proposed by PARK and PFEIFHOFER (5) is correct. The 160 K fraction possessed only small (110 Å) particles on its fracture faces and only A type surfaces (*Plate 4*). PARK and PFEIFHOFER (5) have shown that the B fracture face has large particles in the grana region. If the 160 K fraction, which has only small particles, is derived from stroma lamellae then the B fracture face in the stroma lamellar region of an intact chloroplast should have only small particles since stroma lamellae are an extension of grana lamellae. When Class I chloroplasts were freeze fractured, large particles were seen on the B fracture face in the region of a granum, (*Plate 5*), whereas, in the stroma lamellar region only a few small particles were seen. In addition, the B fracture face of an end grana membrane also appears to have only the small particles. This supports sectioning evidence from the FP fraction that end grana membranes may contribute to the vesicles of the 160 K fraction.

DISCUSSION

Since the 160 K membrane vesicles contain only small particles on their fracture faces it was concluded that they are unique and must have arisen from stroma lamellae and possibly end grana membranes. These vesicles have only PS-I activity whereas the grana fraction has both photosystem activities.

The differentiation of chloroplast membranes into regions having small and large particles on the same fracture face has been reported for broad bean (8) and for barley (9). In both cases the pictures can be interpreted as showing large particles on a B fracture face in a grana region with small particles on the same fracture face in the stroma lamellar region. Thus these data support the spinach data and suggest that stroma lamellae, with small particles, may also be the site of a PS-I in both broad bean and barley. GOODENOUGH and STAEHELIN (10) have investigated a wild-type and mutant strain of *Chlamydomonas reinhardi*. The mutant has no grana stacks but retains the ability to carry out photosynthesis at high light intensities. Where grana stacks are formed large particles are present on the B fracture faces but in the stroma lamellar regions of the wild-type and in all the lamellae of the mutant, which has no grana stacks, only a few large particles were observed. Thus it appears that certain particle size distributions are created as a consequence of the stacking process and this is independent of biochemical function. Therefore, at the present time it seems unwise to assign any biochemical function to particles of particular sizes.

ACKNOWLEDGMENTS

The author gratefully acknowledges the contribution of his coworkers Drs. P.V. Sane and R.B. Park of the Botany Department, University of California, Berkeley where this work was carried out.

REFERENCES

1. J. MICHEL and M. MICHEL-WOLWERTZ, in Progress in Photosynthesis Research, ed. H. METZNER, I.U.B.S., Tübingen, I (1969) 115.
2. D.I. ARNON, *Plant Physiol.*, 24 (1949) 1.
3. H.Y. YAMAMOTO and L.P. VERNON, *Biochemistry*, 8 (1969) 4131.
4. P.V. SANE, D.J. GOODCHILD, and R.B. PARK, *Biochim. biophys. Acta*, 216 (1970) 162.
5. R. PARK and A.O. PFEIFHOFER, *J. Cell Sci.*, 5 (1969) 299.
6. R.G. JENSEN and J.A. BASSHAM, *Proc. natn. Acad. Sci. U.S.A.*, 56 (1966) 1095.
7. J.P. THORNBER, R.P.F. GREGORY, C.A. SMITH, and J. LEGETT BAILEY, *Biochemistry*, 6 (1967) 391.
8. R. REMY, *C.R. Acad. Sci. (Paris)*, 268 (1969) 3057.
9. S. PHUNG NHU HUNG, A. LACOURLY, and C. SARDA, Z. *PflPhysiol.*, 62 (1970) 1.
10. U.W. GOODENOUGH and L.A. STAEHELIN, *J. Cell Biol.*, (in press).

THYLAKOID AGGREGATION CORRELATED WITH CHLOROPHYLL a/CHLOROPHYLL b RATIO IN SOME C$_4$ SPECIES

N.A. Pyliotis[1], K.C. Woo[2], and W.J.S. Downton[2].

Division of Plant Industry, CSIRO, Canberra City, 2601, Australia and Research School of Biological Sciences, Australian National University, Canberra City, 2601, Australia.

SUMMARY

The fine structure of chloroplasts in the mesophyll and bundle sheath cells of five species of C$_4$ plants was found to be related to the ratio of chlorophyll a to chlorophyll b of the respective chloroplasts. Within a given species, a distinct difference was found in the chl a/chl b ratio of the mesophyll and bundle sheath chloroplasts. A high chl a/chl b ratio was associated with a low value for total partition length per unit area of chloroplast.

Chl a/chl b ratio values were used to determine algebraically the contribution of the mesophyll and bundle sheath cell layers to total leaf chlorophyll. In the monocotyledons, the mesophyll cell layer contributed the higher proportion to total leaf chlorophyll, whereas, among the dicotyledons the bundle sheath cell layer generally contributed the higher proportion to total leaf chlorophyll.

INTRODUCTION

It has been suggested that it is necessary to have appressed thylakoids for photosystem II activity (1). When the artificial Hill oxidant tetranitro blue tetrazolium chloride was applied to sections of fresh leaf material (2) no dye reduction was found to occur in the essentially agranal bundle sheath chloroplasts of *Sorghum sudanense*. Some dye reduction was found in the bundle sheath chloroplasts of *Zea mays* (which have a few regions of appressed thylakoids) and extensive dye reduction in the granal bundle sheath chloroplasts of *Panicum miliaceum*. It was concluded that the agranal bundle sheath chloroplasts of *Sorghum sudanense* lack noncyclic electron flow. It has since been shown (3,4) that the essentially agranal bundle sheath chloroplasts of *Sorghum bicolor* and *Zea mays* are deficient in photosystem II activity. Determination of action spectra in photosynthesis (5), as well as chloroplast fractionation studies (6,7), indicate that photosystem II is enriched in chlorophyll b relative to photosystem I.

In view of the diversity in thylakoid aggregation found in the two types of chloroplast in C₄ species, it was of interest to determine whether thylakoid configuration could be correlated with *chl a/ chl b* ratio. We have also attempted to determine algebraically the respective contribution of the bundle sheath and mesophyll cell layers to total leaf chlorophyll.

EXPERIMENTAL

Seeds were grown in a glasshouse at 22°-27° under natural daylight. Only fully expanded leaves were used in this study. The mesophyll and bundle sheath chloroplasts were isolated as described previously (3) and *chl a/chl b* ratio values for each of the two types of chloroplast and the whole leaf determined spectrophotometrically (Cary 14R) in 80% acetone (8). Pieces of leaf tissue were processed for electron microscopy as previously described (9). A series of electron micrographs were quantitatively analyzed in order to obtain the average of total partition length per unit area of sectioned chloroplast (9). The term partition refers to the region of two appressed thylakoids. A granum is defined here as a stack of two or more tightly appressed thylakoids.

RESULTS AND DISCUSSION

Fine structure
Examination of the fine structure of the mesophyll and bundle sheath chloroplasts of the dicotyledons *Atriplex spongiosa* and *Gomphrena globosa* and the monocotyledons, *Chloris gayana*, *Zea mays* var. NES 1002, and *Sorghum bicolor* var. Texas 610, revealed a great diversity in the arrangement of their internal membranes (9). At the one extreme, *Sorghum bicolor* has mesophyll chloroplasts with numerous well-developed grana but bundle sheath chloroplasts with very few poorly-developed grana *(Plate 1)*. At the other extreme, *Chloris gayana* has numerous well-developed grana in both types of chloroplasts. In between these extremes, *Zea mays* has well-developed grana in the mesophyll chloroplasts but few, poorly-developed grana in the bundle sheath chloroplasts *(Plate 2)*. In the case of *Gomphrena*

Plate 1 *High magnification picture of part of a Sorghum bicolor bundle sheath chloroplast showing several small regions of two appressed thylakoids (arrows). Bar equals 0.5 μ in all plates.*

Plate 2 *High magnification picture of part of a Zea mays bundle sheath chloroplast showing several small regions of appressed thylakoids.*

Plate 3 *High magnification picture of part of a Gomphrena globosa bundle sheath chloroplast showing numerous regions of appressed thylakoids.*

globosa, the mesophyll chloroplasts have numerous well-developed grana, whereas, the bundle sheath chloroplasts have extremely numerous but small grana *(Plate 3).* In *Atriplex spongiosa,* both types of chloroplast have relatively well-developed grana but these were not as large as in *Chloris gayana.*

Correlation of structure with chl a/chl b ratio

The data obtained from a series of electron micrographs are summarized in *Table 1.* The main points to note are, (a) the large average number of partitions per chloroplast in both types of chloroplast of *Chloris gayana* and the mesophyll chloroplasts of *Gomphrena globosa, Zea mays,* and *Sorghum bicolor,* (b) the relatively small average partition length in the bundle sheath chloroplasts of *Zea mays* and in particular *Sorghum bicolor,* and (c) the variation within and between species in average chloroplast area.

TABLE I

Analysis of fine structure of bundle sheath (BS)
and mesophyll (M) chloroplasts

Species	(A) Partitions per chloroplast	(B) Partition length per granum (μ)	(C) Chloroplast area (μ^2)
Chloris gayana			
BS	577.0	0.4	25.4
M	488.7	0.4	15.6
Atriplex spongiosa			
BS	236.3	0.5	32.7
M	147.7	0.4	28.9
Gomphrena globosa			
BS	152.0	0.5	29.8
M	349.2	0.4	18.2
Zea mays			
BS	19.8	0.2	15.0
M	406.0	0.4	13.4
Sorghum bicolor			
BS	6.5	0.1	20.2
M	474.4	0.4	11.8

Figures for total partition length per unit area of chloroplast were obtained by multiplying the average number of partitions per chloroplast (A) by average partition length per chloroplast (B) and expressing this in terms of average area per chloroplast (C)

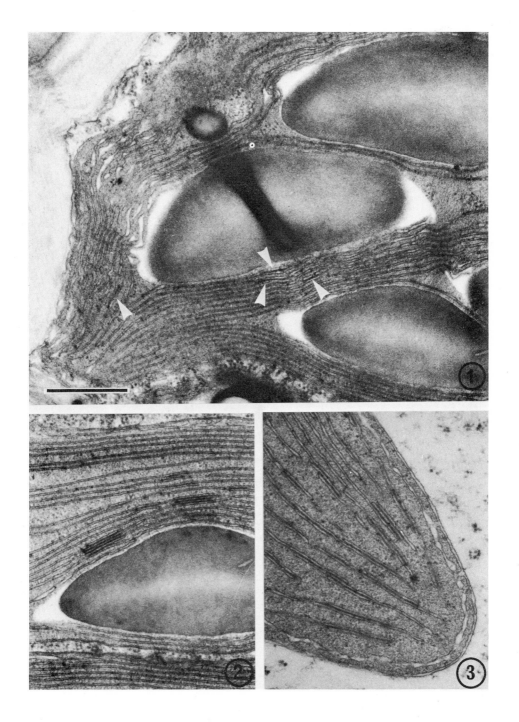

(*Table 2*). This was done to compensate for differences in the size
and shape of chloroplasts.

Correlation of total partition length per unit area of chloroplast with *chl a/chl b* ratio within species is shown in *Table 2*.
Points to note are, (a) the distinct difference in *chl a/chl b* ratio
between the two types of chloroplast, (b) the difference in total
partition length per unit area of chloroplast between the two types
of chloroplast (found to be highly significant), and (c) a higher
partition length per unit area of chloroplast is associated with a
lower *chl a/chl b* ratio value, irrespective of chloroplast type.

TABLE 2

Correlation between total partition length per unit
area of chloroplast (A × B/C) and *chl a/chl b* ratio

Species	$\dfrac{A \times B}{C}$	Chl a/Chl b
Chloris gayana		
BS	9.76	2.80
M	12.88	1.73
Atriplex spongiosa		
BS	3.62	3.29
M	1.89	4.63
Gomphrena globosa		
BS	2.30	4.57
M	7.30	3.40
Zea mays		
BS	0.24	4.96
M	13.29	3.30
Sorghum bicolor		
BS	0.04	5.62
M	16.84	3.12

The data shown in *Table 2*, which has been arranged by species and
chloroplast type, can be rearranged in increasing order of *chl a/
chl b* ratio. When this is done, with the exception of the bundle
sheath chloroplasts of *Atriplex spongiosa*, higher *chl a/chl b* ratio
values are generally associated with lower values for total partition length per unit area of chloroplast. These results suggest that
chloroplasts having a higher *chl a/chl b* ratio (indicative of lower
photosystem II activity) have fewer regions of closely appressed
thylakoids.

Chlorophyll distribution
    The amount of chlorophyll contributed by the two cell layers to total leaf chlorophyll was determined algebraically using *chl a/chl b* ratio values of the whole leaf (T), the mesophyll (M), and bundle sheath (BS) chloroplasts *(Table 3)*. The percentage contribution of the bundle sheath cell layer to total leaf chlorophyll is given by the following derived equation:

$$\frac{(BS + 1)}{(T + 1)} \frac{(T - M)}{(BS - M)} \times 100$$

TABLE 3

Chl a/chl b ratio of whole leaf (T), mesophyll (M), and bundle sheath (BS) chloroplasts, and percentage of total leaf chlorophyll in bundle sheath cell layer

Species	T	M	BS	% BS chlorophyll
Dicotyledon:				
Atriplex spongiosa	3.65	4.63	3.29	67
" *nummularia*	3.74	4.32	3.47	64
" *inflata*	3.76	4.06	3.36	39
" *lindleyii*	3.63	4.05	3.30	52
Gomphrena globosa	4.18	3.40	4.57	72
Monocotyledon:				
Chloris gayana	1.98	1.73	2.80	30
Zea mays	3.69	3.30	4.96	30
Sorghum bicolor	3.38	3.12	5.62	16
" *sudanense*	3.80	3.37	5.90	24

With the exception of *Atriplex inflata,* the greater proportion of total leaf chlorophyll in the dicotyledons is present in the bundle sheath cell layer. However, among the monocotyledons the greater proportion of total leaf chlorophyll is found in the mesophyll cell layer. The values derived by this algebraic method agreed very closely with those obtained quantitatively from isolated cell layers. The algebraic method is less tedious and is reliable provided the difference in *chl a/chl b* ratio values between the two types of chloroplast is not too small.

ACKNOWLEDGMENT

    We gratefully acknowledge the skilled technical assistance of Miss M.A. Paulin.

REFERENCES

1. P.H. HOMANN and G.H. SCHMID, *Plant Physiol.*, 42 (1967) 1619.
2. W.J.S. DOWNTON, J.A. BERRY, and E.B. TREGUNNA, *Z. PflPhysiol.*,
    63 (1970) 194.
3. K.C. WOO, J.M. ANDERSON, N.K. BOARDMAN, W.J.S. DOWNTON, C.B. OS-
    MOND, and S.W. THORNE, *Proc. natn. Acad. Sci. U.S.A.*, 67
    (1970) 18.
4. J.M. ANDERSON, K.C. WOO, and N.K. BOARDMAN, (this volume).
5. D.C. FORK and J. AMESZ, *A. Rev. Pl. Physiol.*, 20 (1969) 305.
6. N.K. BOARDMAN, *A. Rev. Pl. Physiol.*, 21 (1970) 115.
7. P.V. SANE, D.J. GOODCHILD, and R.B. PARK, *Biochim. biophys. Acta*
    216 (1970) 162.
8. D.I. ARNON, *Plant Physiol.*, 24 (1949) 1.
9. K.C. WOO, N.A. PYLIOTIS, and W.J.S. DOWNTON, *Z. PflPhysiol.*,

SOME ASPECTS OF ENVIRONMENTAL CONTROL OF THE PHOTOSYNTHETIC APPARATUS IN *Gomphrena globosa*

A.L. Hart and E.B. Tregunna[1]

Botany Department, University of British Columbia,
Vancouver, 8, British Columbia, Canada

SUMMARY

Leaves of Gomphrena globosa were allowed to develop at 2,000, 10,000, and 50,000 lux. Thylakoid overlaps occurred at all intensities; parallel, unappressed lamellae and some granal stacks also occurred at 2,000 lux. Mesophyll chloroplasts fluoresced at red wave lengths; bundle sheath chloroplasts fluoresced in the red and far-red. The bundle sheath of Panicum maximum fluoresced only in the red; that of Zea mays and Sorghum sudanense only in the far-red. Reduction of TNBT dye occurred readily except in the bundle sheath from leaves at the highest intensity. The amount of labeled aspartate after six seconds of photosynthesis increased as the light intensity decreased. Gomphrena appears to share the characteristics of high malate and high aspartate C_4 plants.

INTRODUCTION

Plants with the C_4 pathway of photosynthesis have been divided into malate formers and aspartate formers. *Gomphrena globosa* has been placed in the former category which suggests that the bundle sheath has a low capacity for generating reducing power (1).

A preliminary study in this laboratory with dye reduction indicated however that the bundle sheath of *Gomphrena* photoreduced this tetranitro blue tetrazolium (TNBT) dye. Based on this finding MELLOR and TREGUNNA (2) compared the distribution of enzymes for the nitrate reduction pathway in *Zea mays*, *Sorghum sudanense*, and *Gomphrena*. The bundle sheath of *Gomphrena* contained substantial amounts of nitrite reductase. The bundle sheath chloroplasts have also thylakoid overlaps which are sufficient for photosystem II (PS-II) activity.

One reason for the apparent discrepancy may be the environment

---

[1] *Paper read to meeting by Dr. W.J.S. Downton.*

under which the plants grow. The work reported here was intended to demonstrate whether the structure and function of the bundle sheath chloroplasts of *Gomphrena* were modified by the light intensity present during growth.

EXPERIMENTAL

Gomphrena globosa, var. globe amaranth, was grown in vermiculite with half strength Hoagland's solution. The plants were started in a growth chamber at 10,000 lux, 16 hour day, and 24°/18° day/night temperature. When approximately six leaves were present, plants were transferred to 2,000 lux and outdoors, during summer, to 50,000 lux. The leaves had grown more than 95% of their mature length in the environment indicated and had just stopped elongating. The parts of the monocotyledon leaves used as controls had also just stopped elongating. The leaves used for electron microscopy were removed after three hours illumination, fixed in 2.5% glutaraldehyde, pH 6.8, washed, postfixed in 1% OsO_4, dehydrated in ethanol and embedded in Spurr's medium (3). Sections were stained with uranyl acetate and lead citrate (4) and viewed with a Zeiss EM9A microscope.

The dye reduction studies were done with TNBT, using sections cut and maintained in 0.5 M sucrose containing Tris-HCl and EDTA at pH 7.0. As a control, ascorbate-dichlorophenol indophenol (DCIP) at $6x10^{-3}$ M and $1.5x10^{-4}$ M respectively was used as the alternate electron donor (5). Other controls were run in the dark or without TNBT. Bundle sheath cells were obtained by grinding the leaf pieces as described by BERRY *et al.* (6).

For the fluorescence microscopy, a Zeiss microscope with automatic exposure assembly and a mercury vapor lamp were used. A BG3 Zeiss exciter filter and a Kodak Wratten No. 12 (yellow) barrier filter were used. Kodak Ektachrome Infrared Aerofilm was used to detect the fluorescence and developed with the E3 process (7).

The malate:aspartate labeling ratio was determined by feeding $^{14}CO_2$ to detached leaves for 6 seconds at 20,000 lux. The leaves were transferred to liquid nitrogen, then extracted with boiling 80% ethanol followed by 20% ethanol containing 0.2 N formic acid. Samples of the combined extract were chromatogramed in one dimension and the chromatograms were scanned for qualitative and quantitative evaluation.

RESULTS AND DISCUSSION

Table 1 gives the results for comparison of the amounts of radioactivity in malate versus aspartate six seconds after feeding $^{14}CO_2$. The values shown are the mean and difference for replicate feedings. There was a progressive increase in the proportion of radioactivity in malate as the light intensity during growth decreased. That is, the *Gomphrena* grown at low intensity had a malate:aspartate ratio similar to that of *Zea* or *Sorghum*, while the *Gomphrena* grown at high

TABLE I

Effect of light intensity (lux) during growth on photosynthesis by *Gomphrena globosa*.

Parameter		Light intensity		
		50,000	10,000	2,000
Ratio of ^{14}C in malate to aspartate		0.9±0.2	2.4±0.9	4.5±0.5
TNBT reduction				
Control	Transverse section	Infrequent	Mosaic	Mosaic
	Isolated bundle sheath	Infrequent	+	+
DCIP-ascorbate	Transverse section	Mosaic	Mosaic	+
	Isolated bundle sheath	Mosaic	+	+

intensity had a much lower proportion of labeled malate. The differences in malate:aspartate ratio may, however, be a reflection of different pool sizes.

Table 1 also shows the results for dye reduction with TNBT. In all cases where TNBT was used, the mesophyll cells reduced the dye when they were illuminated. The bundle sheath cells showed a mosaic in transverse section. Some cells reduced the dye at a similar rate to that of the mesophyll cells, while others contained little or no reduced dye. This was particularly true without ascorbate-DCIP and for leaves grown at 50,000 lux. After isolating the bundle sheath, dye reduction occurred in it except for some cells from leaves grown at 50,000 lux. Since the addition of ascorbate-DCIP also did not cause reduction of dye by all of the cells, there must have been permeability problems or lack of photosynthetic activity by some of the cells. Leaves grown at 10,000 or 2,000 lux definitely had bundle sheaths which were capable of dye reduction without adding ascorbate-DCIP. This would indicate they have PS-II activity, unlike *Zea* or *Sorghum*.

Examination of the bundle sheath by fluorescence microscopy showed that *Gomphrena* bundle sheath fluoresced in both the red and far-red regions, unlike *Zea* and *Sorghum*. In photographs taken with infra-red film the film color changes from yellow to red as the fluorescence wavelength becomes longer than 700 nm. It has been postulated that chloroplasts showing a pure far-red fluorescence have only unappressed lamellae, and that those showing a pure red fluorescence have large grana (ELKIN, private communication). (The color photographs presented at the conference are held at the Botany Department, University of British Columbia.)

Some bundle sheaths of plants grown at 50,000 lux showed a slight tinge of red (far-red fluorescence) and also a yellow color (red fluorescence) in the film. In many cases however, only visible, red fluorescence was seen both in transverse sections and isolated bundle sheaths. Bundle sheaths of plants grown at 10,000 and 2,000 lux always showed red and far-red fluorescence, the far-red fluorescence being more apparent at these intensities than at 50,000 lux. In all cases far-red fluorescence coincided with the area of highest chloroplast density next to the vascular tissue. The mesophyll appears yellow as it fluoresces at visible red wavelengths. The bundle sheath of *Sorghum* showed only far-red fluorescence. The bundle sheath of *Panicum maximum* showed only visible fluorescence, appearing bright yellow on the film.

The fluorescence studies therefore seem to place *Gomphrena* in a rather intermediate position. The effect of decreasing light intensity during growth, if anything, was to increase the far-red fluorescence by the bundle sheath chloroplasts.

The ultrastructure of the *Gomphrena* chloroplasts was also examined. The mesophyll chloroplasts show extensive granal development. Bundle sheath cells from leaves grown at 50,000 lux contain many chloroplasts with numerous starch grains and a large number of plasmodesmata link adjacent bundle sheath cells. Pairs of appressed thylakoids are common (*Plate 1*); a few granal stacks and unappressed lamellae were also seen. Bundle sheath chloroplasts from leaves grown at 10,000 lux were characterized by massive starch deposition (*Plate 2*). Appressed thylakoids were common.

Leaves which developed at 2,000 lux show various degrees of granal development. The chloroplast in the center of *Plate 3* has only parallel lamellae. On both sides of it are chloroplasts which display a considerable number of appressed lamellae. Fusion of the bounding membrane of two chloroplasts is also apparent. *Plate 4* shows bundle sheath chloroplasts with fewer lamellae but there are numerous small grana stacks.

In summary, then, apart from the differences in the degree of starch formation, we also gained the impression that there were differences in lamellar structure between the three light intensities.

Plate 1 Bundle sheath chloroplast. 50,000 lux. Shows thylakoid overlaps. x 36,000.

Plate 2 Bundle sheath chloroplast. 10,000 lux. Large starch grains. x 12,000.

Plate 3 Bundle sheath chloroplasts. 2,000 lux. Chloroplast in center has only unappressed lamellae while those on either side have thylakoid overlaps. x 19,000.

Plate 4 Bundle sheath chloroplasts. 2,000 lux. Small grana are present (Arrows). x 12,000.

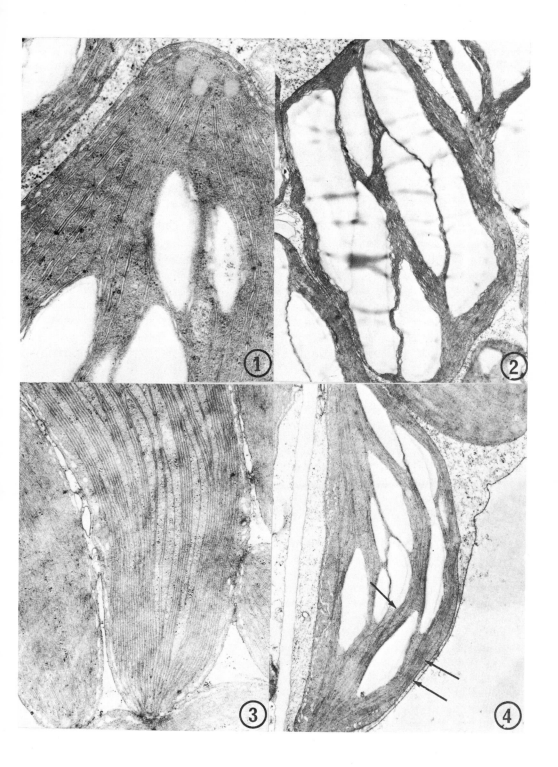

Unappressed lamellae seemed more common at 2,000 lux although grana occurred here as well. Thylakoid overlaps occurred at all intensities being particularly characteristic of bundle sheath chloroplasts at 50,000 lux.

These differences are such as to suggest that, with decreasing light intensity during growth of its leaves, *Gomphrena* makes less use of the bundle sheath chloroplasts for producing reductant. However, the qualitative microscopic techniques used make our conclusions quite tentative. If such a change does occur, it would be compatible with the developmental changes shown by *Sorghum* bundle sheath chloroplasts (7). As the leaves age, they may be shaded by younger leaves and so their loss of reductive capacity with aging could be a mechanism selected during evolution to contend with the lower light intensity available during ontogeny.

Referring back to the fluorescence microscopy, we feel unable, at present, to correlate, for *Gomphrena*, bundle sheath ultrastructure with a particular type of fluorescence, as detected by infra-red film, although the greater amount of visible fluorescence at 50,000 lux may be compatible with the seemingly greater incidence of thylakoid overlaps. The *Gomphrena* with which we have been working seems to be intermediate in type between the typical high malate and high aspartate plants.

REFERENCES

1. W.J.S. DOWNTON, *Can. J. Bot.*, 48 (1970) 1795.
2. G.E. MELLOR and E.B. TREGUNNA, *Can. J. Bot.*, in press.
3. A.R. SPURR, *J. Ultrastruct. Res.*, 26 (1969) 31.
4. E.S. REYNOLDS, *J. Cell Biol.*, 17 (1963) 208.
5. W.J.S. DOWNTON, J.A. BERRY, and E.B. TREGUNNA, Z. *PflPhysiol.*, 63 (1970) 194.
6. J.A. BERRY, W.J.S. DOWNTON, and E.B. TREGUNNA, *Can. J. Bot.*, 48 (1970) 778.
7. W.J.S. DOWNTON and N.A. PYLIOTIS, *Can. J. Bot.*, in press.

THE CHLOROPLASTS AND MITOCHONDRIA OF BUNDLE SHEATH CELLS IN RELATION TO C$_4$ PHOTOSYNTHESIS

W.J.S. Downton

Research School of Biological Sciences, Australian National University, Canberra City 2601, Australia.

SUMMARY

Chloroplasts often exhibit a definite orientation within bundle sheath cells. Monocotyledons with bundle sheath chloroplasts poor in grana development have chloroplasts positioned adjacent to the mesophyll wall (centrifugally arranged). The bundle sheath chloroplasts of all dicotyledons examined to date are centripetally arranged. C$_4$ plants synthesizing mainly aspartic acid have granal bundle sheath chloroplasts associated with great numbers of usually very large mitochondria. A possible role for mitochondria in C$_4$ photosynthesis is discussed.

The chloroplasts of the bundle sheath cells in C$_4$ plants usually assume a definite position within the cells. Their orientation seems to remain constant in both living and fixed material. An electron microscope study of the Gramineae led JOHNSON (1) to generalize that: (a) the bundle sheath chloroplasts of panicoid grasses were located centrifugally within the cell (were positioned adjacent to the mesophyll wall) and that (b) the bundle sheath chloroplasts of chloridoid-eragrostoid grasses assumed a centripetal position within the bundle sheath cell (were positioned adjacent to the vascular tissue). Plates 1 and 2 in "Adaptive and evolutionary aspects of C$_4$ photosynthesis" in this book illustrate the centrifugal and centripetal arrangements. A large number of C$_4$ species have now been scrutinized anatomically and biochemically. This report attempts to sort out some correlations between chloroplast position, ultrastructure, and the light and dark reactions of photosynthesis.

Recent studies on additional C$_4$ species invalidate JOHNSON'S generalizations (1). Both centrifugal and centripetal arrangements can be seen among panicoid grasses. For example, *Sorghum* spp., *Zea mays, Saccharum officinarum,* and *Panicum bulbosum* have centrifugally oriented bundle sheath chloroplasts, but *Panicum miliaceum, Panicum capillare* and other species of *Panicum* have centripetal orientation. The type genus of the Chlorideae, *Chloris gayana*, shows a centrifugal pattern (see Plate 1 in "Adaptive and evolutionary aspects of C_4 photosynthesis").

In the Gramineae, many C_4 species have agranal or poorly granal bundle sheath chloroplasts. *Saccharum officinarum*, *Sorghum sudanense*, *Sorghum bicolor*, *Anthephora pubescens*, *Anthephora cristata*, *Andropogon saccharoides*, *Sorghastrum mutans*, and *Eremochloa ophiuroides* are essentially agranal (1,2,3,4); *Zea mays*, *Euchlaenia mexicana*, *Coix* sp., and *Panicum bulbosum* have small stacks of grana very widely scattered throughout their bundle sheath chloroplasts (4, unpublished). Essentially agranal bundle sheath chloroplasts lack photosystem II and cannot generate reducing potential photochemically (5,6). Reducing potential is presumably provided by malate from the mesophyll layer which is decarboxylated to supply CO_2 and reduced pyridine nucleotide ($NADPH_2$) (5,7,8). C_4 species that preferentially synthesize malic acid have been termed "malate formers" (7). The bundle sheath chloroplasts of all the poorly granal grass species listed above are positioned centrifugally within the cell. This may ensure a tighter coupling of reactions between mesophyll and bundle sheath where the bundle sheath is dependent upon the mesophyll for photochemical products as well as CO_2.

None of the C_4 dicotyledons examined to date, which includes *Amaranthus* spp., *Atriplex* spp., *Froelichia gracilis*, *Bassia hyssopifolia*, *Tidestromia oblongifolia*, *Tribulus* spp., *Portulaca* spp., *Euphorbia* spp., and *Salsola kali*, have a centrifugal arrangement of bundle sheath chloroplasts. Although some C_4 dicotyledons *(Gomphrena globosa*, *Froelichia gracilis*, and *Tidestromia oblongifolia)* have undergone grana reduction and have high levels of "malic" enzyme to compensate for deficiencies in photosystem II (3,7,10), they have not evolved the close association found between the mesophyll and bundle sheath chloroplasts in poorly granal monocotyledons. This looser coupling in C_4 dicotyledons may be reflected in the time required for the label in C_4 acids to be transferred to phosphorylated compounds. For example, BERRY *et al*. (8) found that the half time for C_4-acid conversion to phosphorylated compounds during a pulse-chase experiment was about 20 seconds for *Zea mays* (centrifugal) but at least 60 seconds for *Gomphrena globosa* (centripetal). This possibility must be tested carefully, since differences in cell size and leaf thickness will also influence the rate of transfer. All of the "malate forming" C_4 species that have been examined so far show some grana reduction in their bundle sheath chloroplasts. This adds

---

Plate 1 Bundle sheath of Panicum miliaceum. Chloroplasts contain grana and a periferal reticulum. Numerous large mitochondria (M) are wedged between the chloroplasts. (Courtesy T.Bisalputra and H.Burton)

Plate 2 Bundle sheath of Atriplex spongiosa. Large mitochondria (M) surround the granal chloroplasts. (Courtesy D.J.Goodchild)

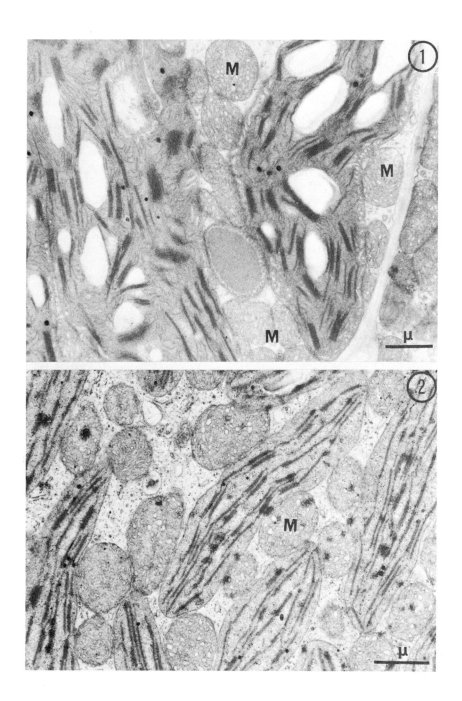

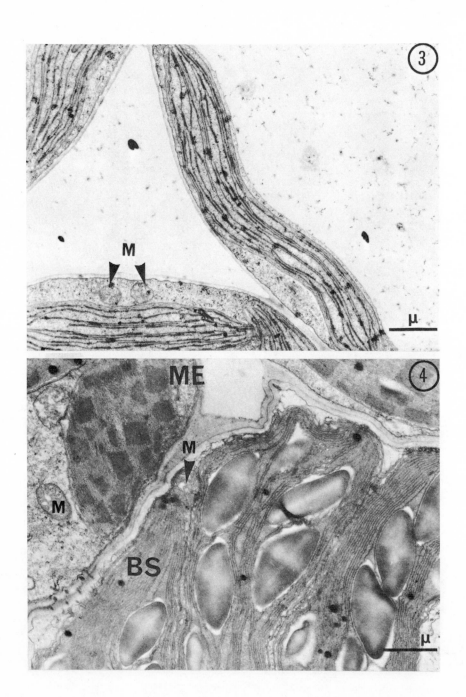

weight to the belief that malic enzyme activity compensates for deficient photosystem II activity in bundle sheath chloroplasts.

Bundle sheath chloroplasts with extensive thylakoid appression (good grana development and normal light reactions) may be distributed either centrifugally or centripetally within the cell. The chloroplasts of *Chloris gayana*, *Panicum maximum*, and *Panicum texanum* are centrifugally arranged but the granal bundle sheath chloroplasts of *Panicum miliaceum*, *Eragrostis intermedia*, *Sporobolus spicata*, *Cynodon dactylon*, *Hilaria mutica*, *Schedonnardus paniculatus*, and *Pappophorum bicolor* are centripetally distributed (1,3,unpublished). All of these species that have been tested synthesize mainly aspartic acid. Plants with this dominant pathway, which include a number of C$_4$ dicotyledons such as *Amaranthus* spp., *Atriplex* spp., and *Portulaca oleracea*, have been refered to as "aspartate formers" (7).

The high level of malic enzyme in "malate formers" provides the Calvin cycle with adequate CO_2 and some $NADPH_2$. "Aspartate formers" have low levels of malic enzyme (7) but contain high aspartate and alanine aminotransferase activity in both mesophyll and bundle sheath layers (11). This provides an enzymatic basis for preferential aspartate synthesis (7) and suggests that it is oxaloacetate that is decarboxylated to supply the bundle sheath with CO_2 (11). By analogy to the mechanism which transfers CO_2 and reducing potential to the bundle sheath in "malate formers", one would expect the decarboxylation of aspartate or oxaloacetate to occur enzymatically within the bundle sheath chloroplast.

The bundle sheath chloroplasts of all "aspartate formers" examined to date are associated with great numbers of usually very large mitochondria. Aggregations of large mitochondria have been seen in the bundle sheath cells of *Panicum miliacum (Plate 1)*, *Chloris gayana*, *Cynodon dactylon*, *Eragrostis intermedia*, *Sporobolus spicata*, *Schedonnardus paniculatus*, *Hilaria mutica*, and *Pappophorum bicolor* of the Gramineae and in *Atriplex spongiosa (Plate 2)*, *A. lentiformis*, *A. rosea*, *A. semibaccata*, *Amaranthus edulis*, *Amaranthus palmeri*, and *Portulaca oleracea* of the Dicotyledoneae (1,4,12,13). Mesophyll mitochondria are small and sparse *(Plate 3)*. A low frequency of small mitochondria are apparent in both mesophyll and bundle sheath cells of "malate formers" such as *Zea mays (Plate 4)* and *Sorghum* spp.

LAETSCH (4) has shown that bundle sheath chloroplasts of *Amaranthus edulis* and etioplasts of sugarcane have protrusions of periferal reticulum. The internal structure of protrusions seems similar to that of the large mitochondria found in bundle sheath cells [Figure 22 in (4)]. It is therefore difficult to ascertain if the orga-

---

Plate 3 Mesophyll cells of Atriplex spongiosa with granal chloroplasts and a few small mitochondria (M). (Courtesy D.J.Goodchild)

Plate 4 Mesophyll (ME) and bundle sheath (BS) of Zea mays. Both layers contain a few small mitochondria (M). (Courtesy N.A.Pyliotis)

nelles refered to here as mitochondria are discrete organelles or if they are merely a consequence of the plane of sectioning. However, serial sectioning indicates that most of the mitochondria-like organelles seen in electron micrographs are really separate from the chloroplast (LAETSCH, personal communication).

A large portion of alanine and aspartate aminotransferase activity in *Amaranthus palmeri* is not associated with the non-aqueously isolated chloroplast fraction (11). The loss of outer membranes when chloroplasts are exposed to non-aqueous solvent may disassociate the transaminases from the chloroplasts. On the other hand, the transaminases could be associated with mitochondria. It is possible that the bodies seen in the bundle sheath cells of "aspartate formers" and refered to as mitochondria may be specialized enzymatically for transamination and decarboxylation reactions. If so, these bodies would convert aspartate to oxaloacetate. Oxaloacetate would then be carboxylated yielding CO_2 (which can be detected as a postillumination CO_2 burst in "aspartate formers") and pyruvate (7,11). Pyruvate would acquire an amino group and return to the mesophyll to preserve a carbon and nitrogen balance between the two cell layers.

It should be a fairly simple task to isolate bundle sheath mitochondria and test this hypothesis.

ACKNOWLEDGMENTS

Mr. N.A. Pyliotis provided micrographs of *Zea mays* and examined *Chloris gayana, Gomphrena globosa,* and *Panicum bulbosum.* Dr. T. Bisalputra and Mr. H. Burton examined *Panicum miliaceum.* Dr. D.J. Goodchild provided photographs of *Atriplex spongiosa.* Dr. J. Berry examined *Froelichia gracilis* and *Tidestromia oblongifolia.*

REFERENCES

1. Sr. M.C. JOHNSON, Ph.D. Thesis (University of Texas, Austin) 1964.
2. T. BISALUPTRA, W.J.S. DOWNTON, and E.B. TREGUNNA, *Can. J. Bot.,* 47 (1969) 15.
3. K.C. WOO, N.A. PYLIOTIS, and W.J.S. DOWNTON, *Z. PflPhysiol.,* (1970) in press.
4. W.M. LAETSCH, *Sci. Prog. Oxf.,* 57 (1969) 323.
5. W.J.S. DOWNTON, J.A. BERRY, and E.B. TREGUNNA, *Z. PflPhysiol.,* 63 (1970) 194.
6. K.C. WOO, Jan M. ANDERSON, N.K. BOARDMAN, W.J.S. DOWNTON, C.B. OSMOND, and S.W. THORNE, *Proc. natn. Acad. Sci. U.S.A.,* 67 (1970) 18.
7. W.J.S. DOWNTON, *Can. J. Bot.,* 48 (1970) 1795.
8. J.A. BERRY, W.J.S. DOWNTON, and E.B. TREGUNNA, *Can. J. Bot.,* 48 (1970) 777.
9. J.A. BERRY, *Carnegie Inst. Wash. Year Book,* 68 (1970) in press.
10. W.M. LAETSCH, *Plant Physiol.,* 45 (1970) S-22.

11. T.J. ANDREWS, H.S. JOHNSON, C.R. SLACK, and M.D. HATCH, *Phytochemistry*, (1970) in press.
12. W.J.S. DOWNTON, T. BISALPUTRA, and E.B. TREGUNNA, *Can. J. Bot.*, 47 (1969) 915.
13. W.M. LAETSCH, *Am. J. Bot.*, , 55 (1968) 875.

CHLOROPLAST STRUCTURE: ASSESSMENT

D.J. Goodchild

Division of Plant Industry, CSIRO, Canberra City, 2601, Australia

Structurally, all C₄ pathway plants so far investigated have a
layer of bundle sheath cells surrounding the vascular tissue. This
bundle sheath layer contains chloroplasts and in some species the
chloroplasts lack grana. With one exception (*Atriplex rosea*) the only
common ultrastructural feature of these chloroplasts is the presence
of a well developed peripheral reticulum immediately inside the chloroplast envelope. This peripheral reticulum is always seen in the mesophyll cell chloroplasts but may be absent in bundle sheath cell
chloroplasts. It differs from the chloroplast envelope and other internal membranes in its sensitivity to $KMnO_4$ (LAETSCH) and while a
rudimentary type of reticulum has been seen in the chloroplasts of
some C₃ species its sensitivity to $KMnO_4$ has not been established.
Thus the peripheral reticulum appears to be a unique structural feature of C₄ plant chloroplasts. However this reticulum may not, as has
been suggested, be the site of the phosphoenolpyruvate carboxylase
since the enzyme is apparently absent from bundle sheath cells whose
chloroplasts have a peripheral reticulum (BLACK).
From work presented on a limited number of species it is clear
that light intensity can influence the number of grana to be found in
both bundle sheath and mesophyll cell chloroplasts although the influence appears to be greatest on the bundle sheath chloroplasts. It
was also shown (PYLIOTIS) that a reduction in the number of grana in
bundle sheath cells was correlated with the chlorophyll *a:b* ratio.
Fewer grana resulted in higher chlorophyll *a:b* ratios and when large
numbers of grana were present the chlorophyll *a:b* ratio resembled
that from spinach chloroplasts. If differences in chlorophyll *a:b*
ratios truly reflect the presence or absence of grana and this is related to the presence or absence of photosystem I the differences in
photosystem activities reported by BLACK and SMILLIE on the one hand,
and by BOARDMAN and co-workers on the other, may reflect the sensitivity of C₄ plants to their environment.
The mitochondria of bundle sheath cells, particularly in the dicotyledons which have relatively large numbers of grana in their bundle sheath chloroplasts, are large and appear to be highly

specialized in their ultrastructural relationship to the chloroplast. No biochemical studies have been attempted on these mitochondria but it should be noted that species having such specialized mitochondria form more aspartate than malate in their mesophyll cells. i.e. they have an aspartate bias (DOWNTON). It was generally concluded that such specialized mitochondria were not extensions of the peripheral reticulum.

Microbodies were found in both bundle sheath and mesophyll cells of C_4 plants and it was generally agreed that they are more prevalent in bundle sheath cells. In *Zea*, *Chloris*, and *Sorghum* which were studied quantitatively (NEWCOMB) there were from 2 to 12 times as many microbodies in bundle sheath as in mesophyll cells but such variation precludes any meaningful correlations being made between these figures and biochemical activity. There appear to be more microbodies per cell in the mesophyll cells of C_3 plants than in the bundle sheath cells of C_4 plants.

Two ultrastructural features that have so far been neglected should be mentioned. No clear evidence has been presented to show the frequency and distribution of plasmodesmata between bundle sheath and mesophyll cells yet the metabolic schemes proposed demand that malate and triose phosphates be transported in quantity between the mesophyll and bundle sheath and that 3-phosphoglycerate and pyruvate be transported in the opposite direction at the same time. In the Gramineae the presence of a suberin layer around the bundle sheath cells complicates the transport pathway. The second structure that demands study is the stomate, for the photosynthetic rate and gas exchange data suggest that it is the stomate that controls the rate of photosynthesis at high light intensities in C_4 plants. This is certainly not the case in C_3 plants and may reflect ultrastructural differences in the stomatal apparatus.

Note added, April 1971.

Since this conference was held *Atriplex rosea* has been grown during summer in Canberra from seed supplied by Dr. O. BJÖRKMAN. The characteristic peripheral reticulum of chloroplasts was seen when leaf material was examined in the electron microscope.

PHOTOPHOSPHORYLATION AND O_2 EVOLUTION: ASSESSMENT

M. Gibbs

Department of Biology, Brandeis University,
Waltham, Massachusetts, 02154, U.S.A.

AGREEMENT

Mesophyll chloroplasts and spinach chloroplasts function similarly in that both have a P/2e ratio of 1 or

$$TPN + ADP + Pi + H_2O \rightarrow TPNH + ATP + 1/2\ O_2 \tag{1}$$

Furthermore the chlorophyll *a:b* ratio in mesophyll chloroplasts is 3:1; a ratio described in spinach.

There is also considerable agreement that the chlorophyll *a:b* ratio in the bundle sheath cell chloroplasts is higher than 3:1. The significance of this altered ratio is not understood. The Hatch-Slack pathway is envisaged to be driven in the bundle sheath by reduced pyridine nucleotide delivered by malic enzyme and ATP generated by cyclic photophosphorylation. The high *a:b* ratio is interpreted by some in terms of cyclic photophosphorylation; however, BLACK'S bundle sheath cells of crab grass do not perform cyclic photophosphorylation unless fortified with PMS. Evidence for *in vivo* cyclic photophosphorylation is lacking.

DISAGREEMENT

Function of bundle sheath cell chloroplasts. Here we have three conflicting points of view:

(a) BLACK takes the traditional point of view. His evidence with crab grass bundle sheath cell chloroplasts establishes some photosystem II (PS-II) in this portion of the leaf. He has used benzoquinone as Hill oxidant and observed rates of O_2 evolution in the order of 50 to 100 µmoles hr^{-1} mg^{-1} chlorophyll. This rate is roughly half that of his mesophyll cells. This reaction is light dependent and is inhibited by DCMU.

(b) BOARDMAN and ANDERSON cannot demonstrate PS-II in isolated bundle sheath chloroplasts of *Sorghum*. Their preparations have

428

photosystem I (PS-I) as evidenced by TPN reduction with ascorbate as electron donor. The BOARDMAN and ANDERSON chloroplasts do not contain cytochrome b-559, and at liquid nitrogen temperature 95% of their fluorescence is emitted at the 735 nm band. There is little doubt that their preparations do not have detectable PS-II. The question remains whether the rather harsh methods used to prepare the *Sorghum* chloroplast fragments removed PS-II. Photophosphorylation data were not presented. These missing data are important since it is well known that the coupling of electron flow to phosphate esterification is the most sensitive step in the overall photochemical reaction.

(c) SMILLIE can detect PS-II and PS-I in maize bundle sheath chloroplast fragments. He employs the same procedure as BOARDMAN and ANDERSON in isolating chloroplasts. He has used cytochrome c reduction as the Hill oxidant to assay PS-II and the ascorbate-TPN reaction for PS-I. Unexpectedly, he reported that during maize development, the two photosystems when assayed independently seemed normal but the overall reaction (reaction 1) did not. He postulated that a coupling factor necessary to retain electron flow from H_2O to TPN was lost during development. This coupling factor was not characterized.

CONCLUSION

BLACK and SMILLIE report PS-II in bundle sheath chloroplast preparations of crab grass and maize. BOARDMAN and ANDERSON cannot detect this photosystem in *Sorghum* preparations. This assessor would have more confidence in the BOARDMAN and SMILLIE reports if they demonstrated ATP formation.

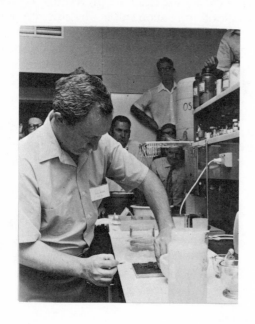

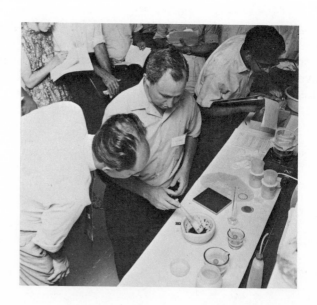

Separation of mesophyll and bundle-sheath cells from *Digitaria sanguanalis* (L.) Scop.

SECTION 4

Photorespiration and the Role of Microbodies

A. Review Papers

BIOSYNTHESIS OF GLYCOLIC ACID[1]

Martin Gibbs

Department of Biology, Brandeis University,
Waltham, Massachusetts, 02154, U.S.A.

SUMMARY

Glycolic acid is formed during photosynthesis as the result of an interaction between α, β-dihydroxyethylthiamine pyrophosphate (glyco-laldehyde-transketolase addition product) and an oxidant photogenerated either in photosystem I or II. This conclusion is based on experiments carried out with intact spinach chloroplasts and with a reconstituted system (spinach chloroplast fragments, fructose 6-phosphate, transketolase, ferredoxin and TPN).

INTRODUCTION

Glycolic acid is labeled early on during photosynthesis with $^{14}CO_2$ (1,2). It is now known that depending on the plant and on environmental conditions a considerable fraction of ^{14}C from newly assimilated $^{14}CO_2$ is metabolized through a metabolic pathway, termed the glycolic acid- or C_2-pathway (3). Since glycolic acid is produced and accumulates in chloroplast preparations (4-6), its subsequent metabolism must, therefore, occur elsewhere (7,8).

The formation of glycolic acid is increased by rate-limiting concentrations of carbon dioxide and by high partial pressures of oxygen (5). From the work of many laboratories (see review of TURNER and BRITTAIN (9)) it has become clear that the Warburg effect is also related to the levels of these gaseous substances. Indeed, a search of the literature indicates an interdependence among photosynthesis, photorespiration, and the Warburg effect, with glycolic acid as the common intermediate (10).

While glycolic acid is metabolized through a series of fairly-well established reactions, the biosynthesis of this α-hydroxyacid remains unknown. Glycolic acid is thought to derive from an intermediate of

---
[1] *This research was supported by grants from the National Science Foundation and the United States Atomic Energy Commission.*

the photosynthetic carbon reduction cycle by a reaction involving transketolase. The conclusion is based on the tracer study of WILSON and CALVIN (11) using the alga, *Scenedesmus* and on the demonstration of BRADBEER and RACKER (12) that the oxidation of the glycolaldehyde-transketolase addition product (α, β-dihydroxyethylthiamine pyrophosphate) to glycolic acid could be coupled to the reduction of ferricyanide. The natural oxidant of this two carbon-transketolase complex is most likely generated in the photochemical act since formation of glycolate from CO_2 (2) or glucose (13) is totally dependent upon light.

The purpose of this report is to describe some recent efforts of this laboratory to explain the coupling of the two photosystems of the photochemical act to glycolic acid formation in the spinach chloroplast.

EXPERIMENTAL

The general methods used to isolate spinach chloroplasts and to carry out these experiments are detailed elsewhere (5).

RESULTS AND DISCUSSION

It has been shown that inhibition of photosynthetic carbon dioxide assimilation by dichlorophenylmethyl urea (DCMU) can be relieved by ascorbate (14). Donation of electrons by ascorbate bypasses photosystem II (PS-II), thereby involving solely photosystem I (PS-I). When spinach chloroplasts were incubated anaerobically in the presence of an excess of carbon dioxide together with DCMU and 10 mM ascorbate, photosynthesis was restored but glycolic acid in contrast to the other products failed to accumulate (*Table 1*). When conditions were more favorable for glycolic acid formation, namely a rate-limiting level of carbon dioxide and air in place of N_2, the effect of ascorbate and DCMU on glycolic acid formation was diminished (*Table 2*).

These experiments carried out with intact spinach chloroplasts suggest two sites for oxidant formation in the electron transport chain. To account for the experimental findings recorded in *Tables 1* and *2*, the suggestion is offered that under conditions of saturating concentrations of carbon dioxide and low partial pressures of oxygen, glycolic acid is produced as the result of a reaction between a 2-carbon moiety and an oxidant photogenerated in PS-II. On the other hand, under conditions more favorable to glycolic acid biosynthesis such as low levels of carbon dioxide and high amounts of oxygen, an interaction between an oxidizing agent generated in PS-I and the 2-carbon moiety is favored.

In order to resolve the question, it became evident that it was necessary to prepare a reconstituted system using as substrate the 2-carbon precursor of glycolic acid and fragmented chloroplasts to produce the oxidant. The remainder of this paper reports on a reconstituted enzyme preparation capable of synthesizing glycolic acid at a

TABLE 1

Effect of ascorbate (10 mM) and DCMU (0.25 μM) at a saturating level of NaHCO$_3$ and under N$_2$ on the distribution of ^{14}C in photosynthesis products formed by spinach chloroplasts ($\times 10^2$ cpm in 1 ml reaction mixture). Chlorophyll concentration was 34 μg ml^{-1}, NaHCO$_3$ concentration 2 mM.

Experimental conditions	Total	3-PGA[*]	DHAP[**]	SMP[***]	SDP[#]	Glycolate	Insoluble[##]
Control							
5 min	439	64	217	35	22	56	45
10 min	785	146	282	68	62	133	94
Ascorbate							
5 min	1269	298	413	71	93	250	129
10 min	2926	488	840	144	98	895	430
DCMU							
5 min	151	30	55	23	10	12	21
10 min	291	79	86	28	26	40	32
DCMU + Ascorbate							
5 min	277	58	103	31	31	17	37
10 min	655	155	290	38	53	31	88

[*] 3-PGA = 3-phosphoglycerate
[**] DHAP = dihydroxyacetone phosphate
[***] SMP = sugar monophosphates (glucose fructose ribose)
[#] SDP = sugar diphosphates (fructose ribulose)
[##] polyglucan which yielded only glucose on heating in N HCl

TABLE 2

Effect of ascorbate (10 mM) and DCMU (0.25 μM) at a limiting level of NaHCO₃ and under air on fixation rates (μmoles hr^{-1} mg^{-1} chlorophyll) and distribution of ^{14}C in photosynthetic products (x10^2 cpm in 1 ml reaction mixture). Chlorophyll concentration 60 μg ml^{-1} and the NaH^{14}CO₃ concentration 0.1 mM. Products were sampled after 20 min photosynthesis.

Experimental conditions	CO₂ fixation rate	3-PGA	DHAP	SMP	SDP	Glycolate	Insoluble
Control	3.5	120	281	11	37	800	21
Ascorbate	8.2	204	300	18	70	1110	58
DCMU	1.3	77	151	4	21	198	18
Ascorbate + DCMU	1.9	172	265	14	48	487	48

rate approaching the rate characteristic of the intact plant. Formation of glycolic acid in this preparation which includes chloroplast
fragments, fructose 6-P, transketolase, ferredoxin, TPN and cofactors
is strictly light dependent and responds to the partial pressure of
oxygen.

TABLE 3

Glycolate formation by chloroplast fragments (μmoles hr^{-1} mg^{-1}
chlorophyll)

Contents	Glycolate formation
Complete*	2.92
-transketolase	0
-ADP, Pi	2.85
-ferredoxin	0.31
-ferredoxin, TPN	0.21

* The reaction mixture contained: 25 mM tris-HCl, pH 7.7, 1 mM ADP,
1 mM TPN, 7 mM MgCl$_2$, 1 mM potassium phosphate pH 7.7, 3 mM thiamine
pyrophosphate, 5 units ferredoxin, 0.5 μmolar units transketolase, 1
μM U-^{14}C-fructose 6-P (specific radioactivity of 0.4 mC/mmole), 20 μg
chlorophyll, in a total volume of 0.33 ml. Experiments were performed under air at a light intensity of 1200 ft-c for a period of 12
min.

The complete system is characterized in *Table 3*. Thiamine pyrophosphate was included in the reaction mixture but usually was not
limiting. Adenosine diphosphate and inorganic phosphate were not required but were routinely included. The stoichiometry of the reaction was approximately one mole of glycolic acid formed per mole of
fructose 6-P consumed. The glycolaldehyde-transketolase addition
product could replace fructose 6-P and transketolase but the procedure was time consuming and, in contrast to the system in which the
addition product is regenerated, resulted in lower glycolic acid
yields.

In their study, BRADBEER and RACKER (12) utilized ferricyanide as
the oxidant of the glycolaldehyde-transketolase addition product.
Other compounds such as benzoquinone and dichlorophenol indophenol
(DCPIP) can be substituted for ferricyanide. To be functional, the
compound needs a normal redox potential of about 120 mv.

Formation of glycolate was light dependent. Saturation was reached at roughly 1200 ft-c (*Figure 1*). A light intensity of 4,000 ft-c
depressed the glycolic acid yield by around 30%. Light above 620 nm,
produced by wrapping the reaction tubes with 2 layers of red cello-

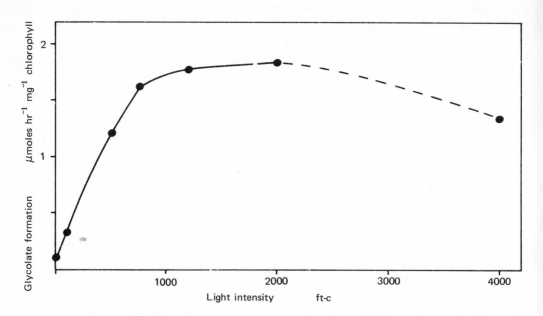

*Figure 1 The dependence of glycolic acid formation on light in
a reconstituted system.*

phane, can be substituted for white light.

The nature of the coupling between the photosystems and the 2-car-
bon precursor was determined by making use of the cysteine-DCPIP
electron donating system and the tris-washed chloroplast.

VERNON and ZAUGG (15) showed that the DCMU block of PS-II could be
by-passed and TPN reduction restored by supplying electrons from cysteine to the fragmented chloroplasts with the dye as mediator. Using
the cysteine-DCPIP electron donating system, TPNH formation but not
glycolate formation was restored in the presence of DCMU (*Table 4*).

TABLE 4

Effect of DCMU, its reversal by cysteine and DCPIP on TPN reduction and glycolate formation (μmoles hr^{-1} mg^{-1} chlorophyll)

Experimental conditions	TPN reduction	Glycolate formation
Complete	31	0.87
+ 50 μM DCMU	1.5	0.05
+ 50 μM DCMU, 5 mM cysteine, 100 μM DCPIP	27	0.1
+ 5 mM cysteine, 100 μM DCPIP	32	0.85
− fructose 6-P	32	----

TABLE 5

Glycolate formation in Tris-treated chloroplasts as mediated by p-phenylenediamine (µmoles hr^{-1} mg^{-1} chlorophyll)

Experimental Conditions	TPN reduction	Glycolate formation	TPN reduction as a result of p-phenylenediamine	Glycolate formation as a result of p-phenylenediamine
Complete, not tris-washed	56.0	2.54	–	–
Complete, tris-washed	3.62	0.12	–	–
+ 50 µM DCMU	0	0	–	–
+ 15 µM p-phenylene-diamine	6.98	2.66	3.36	2.54
+ 15 µM p-phenylene-diamine + 50 µM DCMU	0.85	0.85	0.85	0.85
+ 1.5 µM p-phenylene-diamine	4.93	1.62	1.31	1.50
+ 1.5 µM p-phenylene-diamine + 50 µM DCMU	0.60	0.63	0.60	0.63

Tris-washing of chloroplasts had been shown to block the photochemical act between water and PS-II as evidenced by the observation that
the electron donor, reduced P-phenylenediamine (PD), brought about
the photoreduction of TPN but that this photoreduction was inhibited
by DCMU (16).   As shown in *Table 5,* tris-washing of the spinach
chloroplast reduced TPN reduction together with glycolic acid formation to a low rate.   This low basal rate of TPN reduction and concomitant glycolate formation was increased on addition of PD.   The concentration of dye must be kept low since its oxidized form can bring
about the non-enzymic oxidation of the 2-carbon intermediate.   In
these preparations, a stoichiometry of one mole each of TPNH and glycolic acid was observed.

The data recorded in *Tables 4* and *5* would be in agreement with the
proposal that a site of formation of the oxidant is associated with
PS-II.

An increase in the partial pressure of oxygen resulted in a higher
rate of glycolic acid formation.   This effect of oxygen on glycolic
acid formation was eliminated by the addition of catalase to the reaction mixture.   The oxidant for glycolic acid formation under these
conditions is, therefore, most likely H_2O_2 formed as the result of
reaction between a reduced carrier of the photosynthetic electron
transport chain and molecular O_2.   The carriers of the chain would be
"over-reduced" under conditions of high light and low concentrations

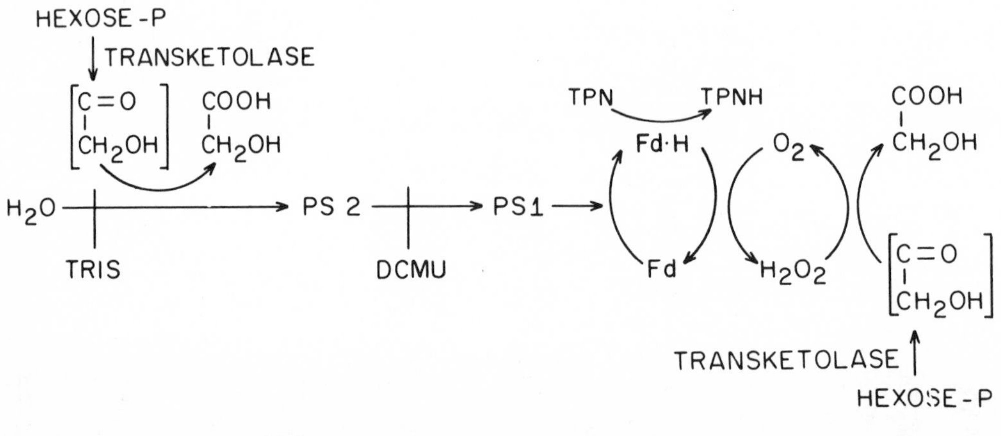

*Figure 2. Mechanism of glycolic acid formation.   Sugar phosph-
ate (fructose 6-P, xylulose 5-P, sedoheptulose 7-P) interacts
with transketolase to form α, β-dihydroxyethylthiamine pyrophosphate or glycolaldehyde-transketolase addition product
[$CH_2OH.C=O$].   Conversion of the addition product to glycolic
acid is mediated by p-phenylenediamine (PD) together with the
oxidant formed in photosystem 2 (PS II), or by H_2O_2 generated
during the re-oxidation of ferredoxin, or of TPNH catalyzed by
the flavoprotein, ferredoxin-TPN reductase.   The other product
of this peroxidatic reaction is given here as O_2 but H_2O may
be the more likely result.*

of CO_2. Under these conditions, and in the presence of elevated pressures of oxygen, H_2O_2 formation would result. Recently TELFER, CAMMACK and EVANS (17) demonstrated that H_2O_2 is a product in the autooxidation of reduced ferredoxin.

CONCLUSION

We would propose that in the biosynthesis of glycolic acid during photosynthesis coupling of the photosynthetic carbon reduction cycle to the light reactions may be associated either with PS-II or PS-I (*Figure 2*). Under physiological atmospheric conditions where glycolic acid formation is relatively low, PS-II would appear to be the likely site. However, under the extreme conditions of carbon starvation, high light and high partial pressure of oxygen where high glycolic acid yields prevail, both sites would be producing oxidant. Indeed, removal of oxidant through glycolic acid synthesis may be considered a detoxication mechanism since catalase activity is localized in the peroxisome (8).

Finally, the two carbon transketolase addition product can presumably be derived from fructose 6-P, xylulose 5-P or sedoheptulose 7-P, since each of these intermediates of the photosynthetic carbon reduction cycle can serve as two carbon donor for transketolase.

REFERENCES

1. L. SCHOU, A.A. BENSON, J.A. BASSHAM and M. CALVIN. *Physiologia Pl.*, 3, (1950) 487.
2. A.A. BENSON and M. CALVIN. *J. exp. Botany* 1 (1949) 63.
3. R. RABSON, N.E. TOLBERT and P.C. KEARNEY. *Archs. Biochem. Biophys.* 98 (1962) 154.
4. P.C. KEARNEY and N.E. TOLBERT. *Archs. Biochem. Biophys.* 98 (1962) 164.
5. P.W. ELLYARD and M. GIBBS. *Plant Physiol.* 44 (1969) 1115.
6. Z. PLAUT and M. GIBBS. *Plant Physiol.* 45 (1970) 470.
7. C.M. THOMPSON and C.P. WHITTINGHAM. *Biochim. biophys. Acta,* 153 (1963) 260.
8. N.E. TOLBERT, A. OSER, T. KISAKI, R.H. HAGEMAN and R.K. YAMAZAKI. *J. biol. Chem.* 243 (1968) 5179.
9. J.S. TURNER and E.G. BRITTAIN. *Biol. Rev.* 37 (1962) 130.
10. M. GIBBS, *Ann. N.Y. Acad. Sci.* 168 (1970) 356.
11. A.T. WILSON and M. CALVIN. *J. Amer. Chem. Soc.* 77 (1955) 5948.
12. J.W. BRADBEER and E. RACKER. *Fed. Proc.* 20 (1962) 88.
13. J. COOMBS and C.P. WHITTINGHAM. *Proc. R. Soc. (B)* 164 (1966) 511.
14. E.S. BAMBERGER, C.C. BLACK, C.A. FEWSON and M. GIBBS. *Plant Physiol.* 38 (1963) 483.
15. L.P. VERNON and W.S. ZAUGG. *J. biol. Chem.* 235 (1960) 2728.
16. T. YAMASHITA and W.L. BUTLER. *Plant Physiol.* 43 (1968) 1978.
17. A. TELFER, R. CAMMACK and M.C.W. EVANS. *FEBS Letters* 10 (1970) 21.

DISTRIBUTION AND STRUCTURE OF
PLANT MICROBODIES (PEROXISOMES)[1]

Eldon H. Newcomb and Sue Ellen Frederick

Department of Botany, University of Wisconsin, Madison,
Wisconsin, 53706, U.S.A.

SUMMARY

*The occurrence and distribution of microbodies throughout the
plant kingdom is reviewed. The fine structure of microbodies is described, and the application of cytochemical methods for the localisation of catalase is reported. Quantitative studies of the distribution of peroxisomes in leaf cells of different species, of the distribution between mesophyll and bundle sheath cells in leaves of C_4
species, and of the relationship between microbodies and other organelles are described.*

The term "microbody" was first used by RHODIN in 1954 to describe
a single membrane-bounded organelle observed in mouse kidney cells
(1). Subsequent work has established that microbodies occur consistently in certain cells of the kidney and liver, but are not found in
vertebrate cells generally. Biochemical studies carried out by DE
DUVE and co-workers (2,3) have shown that these organelles contain
catalase and certain flavin-linked oxidases that produce hydrogen
peroxide. DE DUVE termed particles with this enzyme complement
"peroxisomes" to emphasize their role in hydrogen peroxide metabolism (4). Thus identification of organelles termed "microbodies" is
on morphological grounds, and of those termed "peroxisomes" is on a
biochemical basis. Peroxisomes have been isolated and studies also
from the protozoan *Tetrahymena* (5).

Organelles now identifiable as plant microbodies were observed in-
cidentally in electron micrographs by a number of investigators in
the late 1950's and early 1960's. Their widespread occurrence in
plants and their morphological similarity to animal microbodies was
first pointed out by MOLLENHAUER et al (6) in 1966. The fine struct-

---

[1] *Preparation of this review was supported in part by Grant GB-15246
from the National Science Foundation (U.S.A.).*

ural characteristics of plant microbodies as preserved by glutaraldehyde fixation were described in detail by FREDERICK et al (7).

The organelles identified as plant microbodies are bounded by a
single membrane, are approximately 0.5 to 1.5 μ in the greatest diameter, exhibit spherical, elongate or irregularly shaped profiles in
section, have a granular to fibrillar matrix of variable but usually
moderate electron density, and at least in young stages usually show
a close physical association with one or more cisternae of the endoplasmic reticulum (*Plates 1, 2*). Dense amorphous nucleoids (*Plate 3*)
and crystalline inclusions (*Plates 6 and 9*) are commonly observed in
their matrix. In some plants the microbodies in the photosynthetic
tissues often show an intimate association with chloroplasts (*Plates
4, 6, 7*) and mitochondria (*Plate 5*).

Little is known about the nucleoids of plant microbodies. The
frequency with which they are encountered in tobacco leaf cells suggests that in this case there may be one in each microbody. Commonly
they are situated in a protuberance of the organelle. There is no
evidence that they give a positive reaction with diaminobenzidine in
the cytochemical test for catalase, or that they are physically associated with or give rise to the crystalline inclusions. Different
kinds of nucleoids of animal microbodies have recently been reviewed
in some detail (8).

Although it is now recognized that microbodies are widely distributed in the plant kingdom, there is no evidence as yet that this gen-
eralization can be applied to the algae. Among the few positive observations is that by BOUCK (9), who noted the presence of microbodies with crystalline inclusions in two species of brown algae. There
appears to be no published fine structural evidence for the occurrence of microbodies in the Chlorophyta. The first and only report of
their presence in the Euglenophyta which has been documented with
convincing electron micrographs may be that of GRAVES et al (10), who
have demonstrated that organelles resembling the microbodies of higher plants and animals are relatively abundant in cells of *Euglena
gracilis* grown on acetate or ethanol but are rarely observed in cells
grown on glucose. It is of particular interest that no catalase is
detectable in these microbodies when they are tested cytochemically
with the diaminobenzidine (DAB) reagent.

It appears that microbodies may be widespread in the fungi, judg-
ing from the diversity of the observations made thus far (11-16). In
yeast (15) and in *Botrytis cinerea* (13), catalase activity has been
localized in the microbodies by means of the DAB reaction and its
sensitivity to aminotriazole.

Microbodies with (17) and without (18) crystalline inclusions have
been reported in bryophytes. In our own laboratory, microbodies have
been seen in the photosynthetic tissues of mosses and a few lower
vascular plants, including *Psilotum nudum* and several species of
ferns. In flowering plants with high CO_2-photorespiration they are
particularly large and numerous in the photosynthetic tissues, and at
least in tobacco (19), are frequently appressed to the chloroplasts

Plate 1 Microbody in a meristematic cell of a root tip of bean (Phaseolus vulgaris). Rough endoplasmic reticulum is present on both sides of the microbody. The moderately electron-opaque matrix is not uniformly dispersed. Micrograph courtesy W.P. Wergin. x 53,000.

Plate 2 Microbody in a differentiating cell of a procambial strand in a bean root. The microbody bears a tail (arrow) and is closely associated with a cisterna of endoplasmic reticulum (ER). Whether the tail represents a connection with ER is not clear. Micrograph courtesy W. P. Wergin. x 59,000.

Plate 3 Microbody appressed to two chloroplasts (C) in a leaf cell of tobacco (Nicotiana tabacum). Note the amorphous electron-opaque nucleoid in a protuberance of the microbody. M, mitochondrion. x 47,000.

Plate 4 Amoeboid microbody in extensive contact with chloroplasts (C) and a mitochondrion (M). In a leaf cell of tobacco. x 42,000.

Plate 5 Microbody wrapped partly around a mitochondrion (M) in a leaf cell of tobacco. x 54,000.

Plate 6 Microbody (leaf peroxisome) with a large crystalline inclusion in a mesophyll cell of a tobacco leaf. At upper left the microbody is in contact with a chloroplast and conforms to its contour. x 130,000.

Plate 7 Microbody appressed to a chloroplast (C) in a mesophyll cell of a mature tobacco leaf. The close association does not appear to result from crowding by other organelles. x 41,000.

Plate 8 Microbody in a mesophyll cell from a segment of tobacco leaf incubated in the DAB medium. Dense reaction product (osmium black), attributable to the activity of catalase, has completely permeated the microbody (Mb). It is confined to the microbody and is not observed in the mitochondria (M) or the chloroplast (C). x 40,000.

Plate 9 A microbody with a large crystalline inclusion is adjacent to two chloroplasts in a mesophyll cell of a tobacco leaf. The material has been fixed and stained in the usual manner (i.e., without incubation in the DAB medium). A small mitochondrion lies to the right. x 55,000.

Plate 10 Microbody in a mesophyll cell from a segment of tobacco leaf incubated in the DAB medium. The microbody contains a crystalline inclusion which bears a heavy deposit of osmium black throughout. Some reaction product is also visible in the microbody matrix around the crystal and especially at the interfaces between microbody and chloroplasts. There is no evident product in the chloroplasts or the mitochondrion. x 45,000.

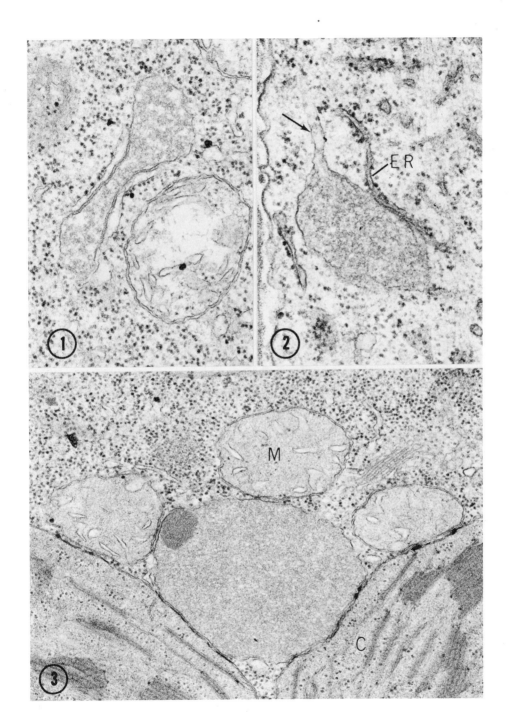

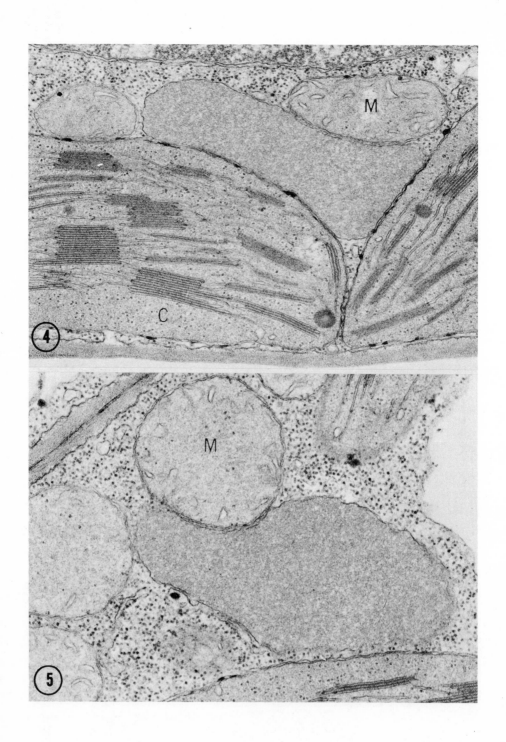

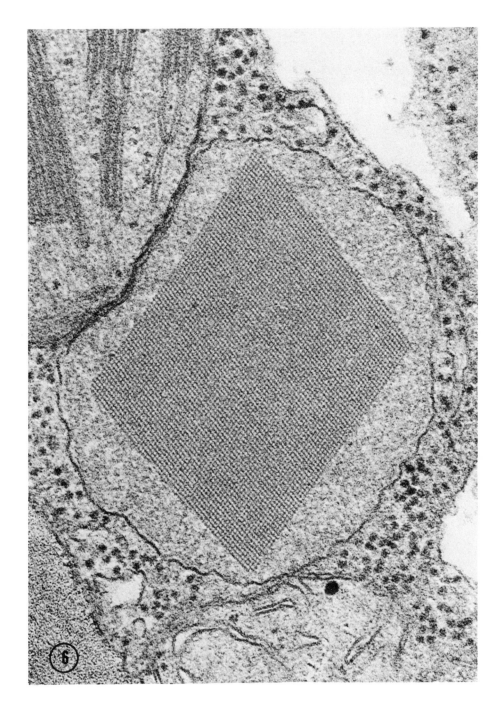

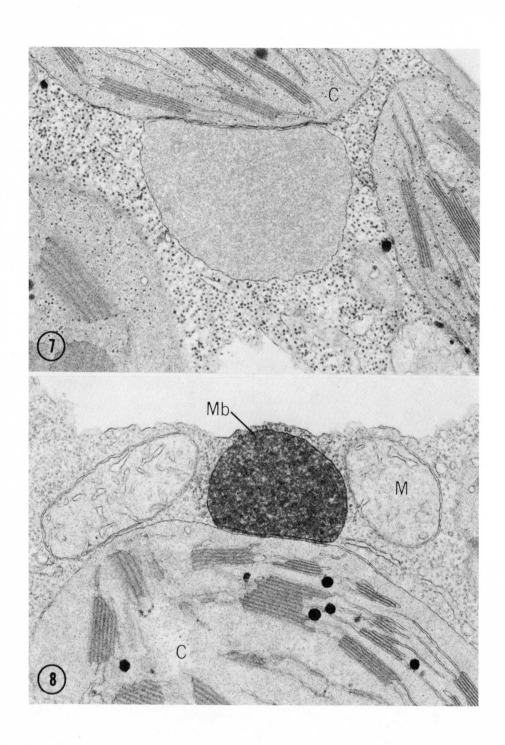

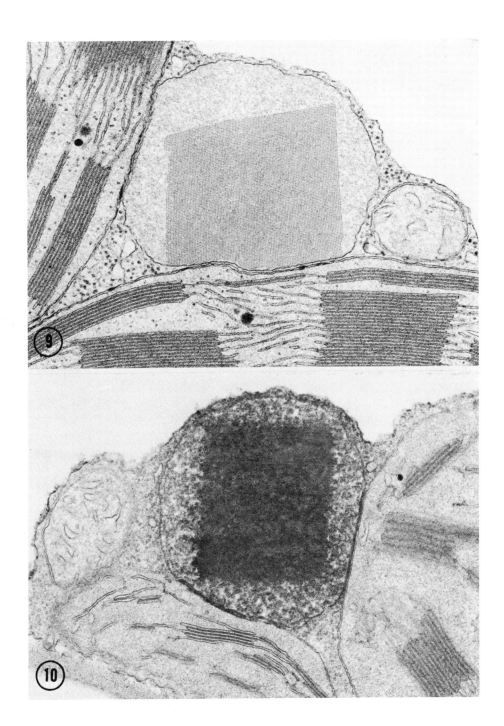

(*Plates 6, 7*). It is now known that they correspond to "leaf peroxi-somes", the biochemically distinct particles in which many of the steps in photorespiration take place, as shown by TOLBERT and co-workers (20). It appears that the photosynthetic organs of forest trees and other woody plants, e.g., conifers and oaks, have not yet been examined for the presence and abundance of microbodies, although they may have an especially high CO_2 compensation point (21).

Microbodies are by no means confined to the photosynthetic tissues of flowering plants, however. They are found in greatest concentrat-ion, in fact, in the fat-storing endosperm or cotyledons of fatty seeds during germination (22,23). They correspond to "glyoxysomes", particles isolated by BEEVERS and coworkers and named for their part-icipation in the conversion of fat to carbohydrate via the glyoxylate cycle (24,25).

Microbodies occur at least in small numbers in root tips (*Plates 1, 2*) and in a variety of other achlorophyllous tissues throughout the higher plant body. Before recognition of their similarity to an-imal microbodies, their presence in plant tissues was noted in pass-ing in a number of papers, particularly when they contained large crystalline inclusions. Some of the earlier references to these "crystal-containing bodies" were summarized by FREDERICK et al (7). Additional observations have been made in recent years on the presen-ce of microbodies in abscission cells (26), embryos (27), and else-where in the plant body (28-30).

In order to establish unequivocally that the enlarged forms of mi-crobodies common in leaf mesophyll cells are identical to the peroxi-somes isolated from leaf homogenates, it was necessary to show cyto-chemically that one or more of the enzymes localized in the isolated peroxisomes is also localized in the microbodies *in situ*. The equiv-alence between the two bodies was shown by employing the cytochemical reagent 3,3'-diaminobenzidine (DAB) on the electron microscopic level (31). This compound has been used in numerous recent studies to dem-onstrate peroxidatic activity of animal microbodies containing catal-ase (32-35) and to study the origin of new microbodies (36) and the turnover of catalase (37). The oxidation of DAB, often peroxidatic in nature, can be catalyzed by several hemoprotein compounds. The DAB is oxidized to a polymeric form that can interact with osmium te-troxide to yield electron-opaque osmium black (38).

Incubation of glutaraldehyde-fixed tobacco leaf segments in media containing DAB and hydrogen peroxide results in pronounced deposition of electron-opaque material within structures clearly recognizable as microbodies (31) (*Plate 8*). The reaction product is always sharply localized in the microbodies or in the microbodies and the immediate-ly adjacent cytoplasm. Thus this cytochemical method of localizing catalase shows clearly that only the microbodies of the leaf cells contain appreciable activity that can be attributed to this enzyme.

Tobacco leaf microbodies containing crystalline inclusions are highly reactive toward DAB, and much of the product is localized within the crystals (31) (*Plate 10*). It is clear that the reaction must take place throughout the crystal bodies during incubation, since the crystals show a uniformly heavy distribution of dense material when sectioned. Generally, the larger the crystal, the less reactive the granular or amorphous matrix, suggesting that the inclusions represent the progressive crystallization of catalase from the matrix. VIGIL has shown that the crystals in microbodies of the oat coleoptile (22) and in glyoxysomes of castor bean endosperm (23) are also heavily stained by DAB.

Evidence that accumulation of reaction product in leaf microbodies is mediated primarily by catalase rather than peroxidase is largely indirect and has been provided in part by experiments employing 3-amino-1,2,4-triazole (31). This compound, a potent inhibitor of catalase in several plant and animal tissues (39-41), completely prevents microbody staining when present at a final concentration of 0.02 M in the incubation medium. A similar sensitivity of the DAB reaction to aminotriazole has been shown also for the microbodies of castor bean endosperm (23) and the oat coleoptile (22).

Of particular current interest is a recent comparative study of the ultrastructure, distribution, and abundance of leaf microbodies in four species of "temperate" grasses with high and four "tropical" grasses with low CO_2-photorespiration (42). The temperate grasses investigated were all festucoid, while the tropical grasses included two panicoid species and two chloridoid. Profiles of microbodies, chloroplasts, and mitochondria were counted in thin sections in the electron microscope. Counts were made on five species of grasses, two temperate with high CO_2-photorespiration and three tropical with low CO_2-photorespiration.

The results are shown in *Table 1*. The values represent the mean numbers of organelles in 100-200 sectioned cells chosen at random from those completely visible between the grid bars. For the microbodies, standard errors of the mean were calculated. The approximate cell sizes given in the table represent the means of values obtained by using 1 µ thick sections of plastic-embedded material to measure 25-50 cells in the light microscope.

Examination of numerous sections of each species established that microbodies are considerably less numerous in the leaf cells of grasses with low photorespiration than they are in those with high photorespiration (see *Table 1*). Also, in the species with low photorespiration there are few microbodies in the mesophyll cells, most of them being concentrated in the bundle sheath cells. Finally, the microbodies in the grasses with low photorespiration are smaller than they are in those with high. The smaller size is particularly noticeable in the mesophyll cells, where the microbodies of the tropical grasses average less than 0.5 µ in diameter compared to approximately 1 µ for the festucoid grasses.

The data in *Table 1* also show that the numbers of microbody profiles relative to the number of mitochondrial and chloroplast profiles are generally higher in the species with high than in those with low CO_2-photorespiration. It is particularly noteworthy that in the

TABLE 1

Organelle distribution in leaf cells of grasses

Line	Species and cell type	Av. cell diameter (μ)	Number of profiles per cell section		
			chloro- plasts	mito- chon- dria	micro- bodies[α]
Festucoid	*Triticum aestivum* mesophyll	22	10.1	6.6	2.65 ±0.18
Festucoid	*Avena sativa* mesophyll	23	9.4	5.1	2.40 ±0.14
Panicoid	*Zea mays* bundle sheath	20	5.9	7.7	1.09 ±0.12
Panicoid	*Zea mays* mesophyll	21	4.9	2.8	0.29 ±0.06
Panicoid	*Sorghum sudanense* bundle sheath	11	2.6	3.4	0.68 ±0.10
Panicoid	*Sorghum sudanense* mesophyll	14	3.1	2.1	0.42 ±0.07
Chloridoid	*Chloris gayana* bundle sheath	19	5.0	16.1	1.07 ±0.13
Chloridoid	*Chloris gayana* mesophyll	12	1.6	0.7	0.08 ±0.04

[α] ±S.E.M. (Standard error of the mean)

chloridoid grass, *Chloris gayana*, the mitochondrial profiles are especially abundant in the bundle sheath cells, where they outnumber the microbody profiles by 16 to 1. In *Plate 15*, the abundance of mitochondria relative to microbodies can be seen in a bundle sheath cell of another chloridoid grass, *Cynodon dactylon*. In the mesophyll cells of these grasses, both mitochondrial and microbody profiles are encountered in far smaller numbers.

Morphologically the leaf microbodies of the eight grass species studied differ from one another primarily with respect to size and type of inclusion. In the festucoid species the microbodies contain numerous threads or fibrils with a distinct substructure (*Plate 11*). This holds true not only for *Triticum aestivum* and *Avena sativa* (42), but also for *Phleum pratense* (19). Cytochemical evidence from application of the DAB reaction (*Plate 12*) suggests that the fibrils may contain catalase.

Leaf microbodies of the two panicoid grasses, *Zea mays* and *Sorghum sudanense*, often contain a cluster of more or less radially distributed, small round bodies (*Plate 13*). These are more characteristic of mesophyll than of bundle sheath microbodies, and are not present in the bundle sheath microbody of *Sorghum* illustrated in *Plate 14*. Microbodies of the chloridoid grass, *Cynodon dactylon*, possess only faint structures suggestive of fibrillar inclusions (*Plate 15*).

It has been established (42) that the microbodies observed *in situ* in both types of grasses (i.e., in those with low CO_2-photorespiration as well as in those with high) are equivalent to peroxisomes as biochemically defined. This was determined by employing DAB to carry out the cytochemical test for catalase. In both temperate and tropical grasses, electron-opaque reaction product was deposited in all organelles recognizable as microbodies (*Plate 12*). In the two panicoid grasses tested, both mesophyll and bundle sheath microbodies reacted positively for catalase, although the reaction was less intense in the mesophyll microbodies, suggesting that they might be relatively poor in the enzyme. The specificity of the reaction was shown by incorporating aminotriazole at a final concentration of 0.02 M in the incubation medium. This relatively specific inhibitor of catalase completely prevented the darkening of microbodies by DAB in the four

---

Plate 11 *Microbody (Mb), mitochondrion (M) and chloroplasts (C) in a mesophyll cell of a festucoid grass, Avena sativa (oats). The microbody matrix contains fibrils bearing short lateral spurs. x 43,000.*

Plate 12 *Appearance of a microbody in a mesophyll cell of an oat leaf after incubation of the tissue in the DAB medium as a cytochemical test for catalase activity. Electron-opaque reaction product is confined to the microbody. Note that the fibrils in the microbody matrix show a positive reaction which may indicate the presence of catalase. x 29,000.*

Plate 13 *A microbody near chloroplasts in a bundle sheath cell of the panicoid grass Zea mays. Note the rows of electron dense structures in the microbody matrix. x 36,000.*

Plate 14 *A microbody and two small mitochondria near an agranal chloroplast in a bundle sheath cell of Sorghum sudanense (Sudan grass). The microbody lacks distinct inclusions in the matrix. x 34,000.*

Plate 15 *A microbody (Mb) among several mitochondria (M) in a bundle sheath cell of a chloridoid grass, Cynodon dactylon (Bermuda grass). The microbody matrix appears to have faint fibrillar structures. Note the presence of grana and starch grains (St) in the chloroplasts. x 35,000.*

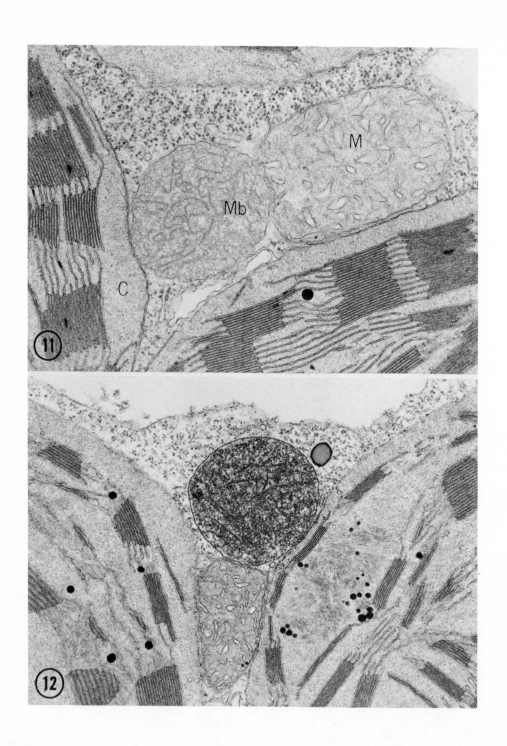

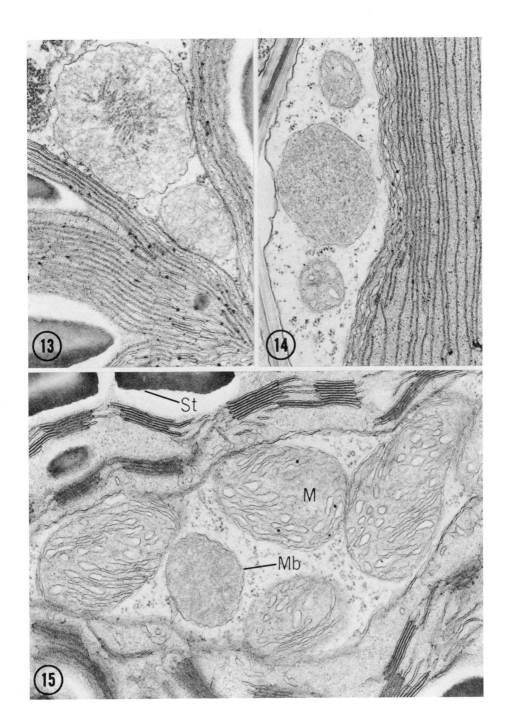

species tested.

The fine structural observations showing that microbodies (peroxisomes) in grasses with low CO_2-photorespiration are considerably smaller and less numerous than they are in grasses with high CO_2-photorespiration are consistent with published data on levels of peroxisomal enzymes in representatives of these two groups of plants (43). Furthermore, the observation that in the species with low CO_2-photorespiration most of the peroxisomes are located in the bundle sheath cells and that only small numbers of relatively small peroxisomes are present in the mesophyll cells correlates well with recent biochemical data on the levels of peroxisomal enzymes in these two cell types in leaves of *Zea mays* (43).

Since the role of photorespiration in the overall metabolism of the green leaf is still obscure, it is not clear why peroxisomes, which participate in the release of a large amount of fixed carbon from the leaf (20), should have been retained at such seemingly great expense to the cell. It is possible that in the tropical grasses, which show a high degree of evolutionary specialization in various anatomical features (44), the reduction in numbers of peroxisomes represents a stage in an evolutionary trend toward their complete loss from the leaf. It does not necessarily follow, however, that the same evolutionary course will be taken by higher green plants in general, since the (presumed) decline of the peroxisome in the tropical grasses would appear to have been accompanied by a variety of morphological, biochemical, and physiological specializations which may have consigned the peroxisomal metabolism to a role quite different from that in temperate plants with high photorespiration. After all, in those few plants with high photorespiration that have been examined thus far, the leaf peroxisomes are not only numerous, they are also the largest of any reported in either plant or animal cells. The reactions carried out by these leaf peroxisomes in their interaction with the chloroplasts and mitochondria may well be indispensable to these plants in some still unknown way.

REFERENCES

1. J. RHODIN, Akiebolaget Godvil, Stockholm (1954).
2. CH. DE DUVE, Proc. R. Soc. B, 173 (1969) 71.
3. CH. DE DUVE and P. BAUDHUIN, *Physiol. Rev.*, 46 (1966) 323.
4. CH. DE DUVE, *J. Cell Biol.*, 27 (1965) 25A.
5. M. MÜLLER, J.F. HOGG and CH. DE DUVE, *J. biol. Chem.*, 243 (1968) 5385.
6. H.H. MOLLENHAUER, D.J. MORRÉ and A.G. KELLEY, *Protoplasma*, 62 (1966) 44.
7. S.E. FREDERICK, E.H. NEWCOMB, E.L. VIGIL and W.P. WERGIN, *Planta*, 81 (1968) 229.
8. Z. HRUBAN and M. RECHCIGL, Jr., *Int. Rev. Cytol.*, 1 (Suppl.) (1969) 1.
9. G.B. BOUCK, *J. Cell Biol.*, 26 (1965) 523.

10. L.B. GRAVES, Jr., L. HANZELY and R.N. TRELEASE, *Protoplasma*, in the press.
11. C.J. AVERS and M. FEDERMAN, *J. Cell Biol.*, 37 (1968) 555.
12. H. BAUER and K. TANAKA, *J. Bact.*, 96 (1968) 2132.
13. D. PITT, *J. Histochem. Cytochem.*, 17 (1969) 613.
14. J.J. WOLKEN, *J. Cell Biol.*, 43 (1969) 354.
15. H.P. HOFFMANN, A. SZABO and C.J. AVERS, *J. Bact.*, 104 (1970) 581.
16. D.P. MAXWELL, D.H. WILLIAMS and M.D. MAXWELL, *Can. J. Bot.*, 48 (1970) 1689.
17. D.M.J. MUELLER, *J. Ultrastruct. Res.*, 30 (1970) 615.
18. J.H. MONROE, *Bot. Gaz.*, 129 (1968) 247.
19. S.E. FREDERICK and E.H. NEWCOMB, *Science*, 163 (1969) 1353.
20. N.E. TOLBERT, *A. Rev. Pl. Physiol.*, in the press.
21. D.N. MOSS, *Nature*, 4815 (1962) 587.
22. E.L. VIGIL, *J. Histochem. Cytochem.*, 17 (1969) 425.
23. E.L. VIGIL, *J. Cell Biol.*, 46 (1970) 435.
24. R.W. BREIDENBACH and H. BEEVERS, *Biochem. biophys. Res. Commun.*, 27 (1967) 462.
25. R.W. BREIDENBACH, A. KAHN and H. BEEVERS, *Plant Physiol.*, 43 (1968) 705.
26. T.E. JENSEN and J.G. VALDOVINOS, *Plant Physiol.*, 43 (1968) 2062.
27. T.A. VILLIERS, *Life Sci.*, 6 (1967) 2151.
28. F. MARTY, *Compt. Rend.*, 268 (1969) 1388.
29. T.P. O'BRIEN and K.V. THIMANN, *Protoplasma*, 63 (1967) 417.
30. H. PETZOLD, *Protoplasma*, 64 (1969) 120.
31. S.E. FREDERICK and E.H. NEWCOMB, *J. Cell Biol.*, 43 (1969) 343.
32. M.E. BEARD and A. NOVIKOFF, *J. Cell Biol.*, 42 (1969) 501.
33. E. ESSNER, *J. Histochem. Cytochem.*, 17 (1969) 454.
34. H.D. FAHIMI, *J. Cell Biol.*, 43 (1969) 275.
35. A. NOVIKOFF and S. GOLDFISCHER, *J. Histochem. Cytochem.*, 17 (1969) 675.
36. P.G. LEGG and R.L. WOOD, *J. Cell Biol.*, 45 (1970) 118.
37. R.L. WOOD and P.G. LEGG, *J. Cell Biol.*, 45 (1970) 576.
38. A.M. SELIGMAN, M.J. KARNOVSKY, H.L. WASSERKRUG and J.S. HANKER, *J. Cell Biol.*, 38 (1968) 1.
39. W.G. HEIM, D. APPLEMAN and H.T. PYFROM, *Am. J. Physiol.*, 186 (1956) 19.
40. E. MARGOLIASH and A. NOVOGRODSKY, *Biochem. J.*, 68 (1958) 468.
41. E. MARGOLIASH and A. NOVOGRODSKY, *Biochem. J.*, 74 (1960) 339.
42. S.E. FREDERICK and E.H. NEWCOMB, *Planta*, in the press.
43. D.W. REHFELD, D.D. RANDALL and N.E. TOLBERT, *Can. J. Bot.*, 48 (1970) 1219.
44. G.L. STEBBINS, *Am. J. Bot.*, 43 (1956) 890.

LEAF PEROXISOMES AND PHOTORESPIRATION[1]

N.E. Tolbert

Department of Biochemistry, Michigan State University,
East Lansing, Michigan, 48823, U.S.A.

Microbodies are subcellular organelles found widely distributed in various biological tissue. Two subclassifications are peroxisomes and glyoxysomes, as based upon different enzymatic composition, metabolic pathways, functions, and tissue distribution (1,2,3). Mammalian microbodies, particularly of the liver and kidney, were first investigated and masterfully reviewed by de DUVE (4,5). BEEVERS group (6,7) has investigated microbodies in germinating seeds and called them glyoxysomes to emphasize the functional role of the glyoxylate cycle in them. TOLBERT and associates (3, 8,9,10,11,12,13) have investigated microbodies in leaves and called them leaf peroxisomes because they appear similar in many respects to liver peroxisomes. Both types of plant microbodies have been carefully examined *in situ* by electron micrographs, particularly by NEWCOMB and associates (14). Microbodies range in size from 0.2 to 1.5 µ in diameter and are characterized by a single limiting membrane, a granular matrix and often a dense core or a crystalline inclusion. A feature of all microbodies isolated, is a specific density of 1.24 to 1.26 g cm^{-3} on a sucrose gradient, which places them, after isopycnic sucrose density gradient centrifugation, in a dense sucrose band well below the mitochondria or chloroplasts. Microbodies and mitochondria are two different types of subcellular particles for respiration, but they are readily distinguishable and entirely different in composition and function. Mitochondrial respiration conserves biochemical energy as ATP. Microbody respiration is characterized by flavin oxidases and H_2O_2 generation, which results in the loss of energy as heat during catalase destruction of the peroxide. A major enzyme, common to all microbodies is catalase, but the accompanying flavin oxidase will vary. There are numerous recent reviews on microbodies (3,4,5,7,8,9,15). This brief paper will focus only upon certain bio-

[1] *This work was supported in part by NSF Grant GB-17543, and is published as Journal Article No. 5355 of the Michigan Agricultural Experiment Station.*

458

chemical aspects of leaf peroxisomes and their relationship to photorespiration and photosynthesis. A more detailed examination of plant microbodies has been prepared (3).

Leaf peroxisomes have been found in photosynthetic tissue of all higher plants (3,14) and contain most of the enzymes for the glycolate pathway. This metabolic sequence from the photosynthetic carbon cycle is summarized in *Figure 1*. The total process of glycolate biosynthesis and metabolism is collectively manifested physiologically by the phenomenon of photorespiration and has been reviewed extensively (16,17,18). *In vivo* measurements of photorespiration may be considered as indicative of leaf peroxisomal activity, while dark respiration represents mitochondrial activity. Microbodies are detected *in vivo* by electron microscopy on the basis of their unique appearance and by cytochemical procedures for catalase, as discussed by NEWCOMB and LAETSCH (this volume). In tissue such as spinach and tobacco leaves, they are about 1/3 as numerous as mitochondria and contain 1-1.5% of the total protein. *In vitro* assays depend upon specific marker enzymes for leaf peroxisomes. Those that are exclusively in the peroxisome and that are relatively easy and rapid to assay are glycolate oxidase, catalase, and hydroxypyruvate reductase (8). The total activity of these enzymes in a completely homogenized leaf is believed to reflect total peroxisomal activity, even though the particles are nearly all broken by the complete grinding procedure. Assay for other peroxisomal enzymes in homogenates is unsatisfactory due to isoenzymes of similar activity or competing reactions for the same substrate, as for example the transaminases.

Because of their fragility, conditions of isolation for microbodies are critical. For this reason microbodies went undetected long after chloroplasts and mitochondria were isolated, and satisfactory isolation techniques are still a limiting factor. Initial grinding of the tissue must be done as gently as possible in buffered sucrose and consists of incomplete or partial chopping of the tissue. Separation of microbodies from other particles is dependent upon isopycnic ultracentrifugation on sucrose gradients. For details one must refer to the individual research papers and reviews on methods (8, 15). Markers for particles on sucrose gradients are chlorophyll for chloroplasts, cytochrome c oxidase for mitochondria, cytochrome reductase for endoplasmic reticulum, triose phosphate isomerase and phosphoglycolate phosphatase for proplastids or chloroplasts, and any of the marker enzymes cited for peroxisomes. The appearance of similar enzyme activity, such as NAD malic dehydrogenase (19), in more than one isolated particulate fraction is indicative of isoenzymes. From such results in detail, the total cellular activity and specific activity on a protein and chlorophyll basis can be calculated and used to evaluate the significance of the predicted metabolic pathway in a particle. These data on particulate location of enzymes have also shown that chloroplasts do not contain the peroxisomal enzymes, catalase, glycolate oxidase, and an isoenzyme of NAD malate dehydrogenase. Chloroplast preparations often contain these

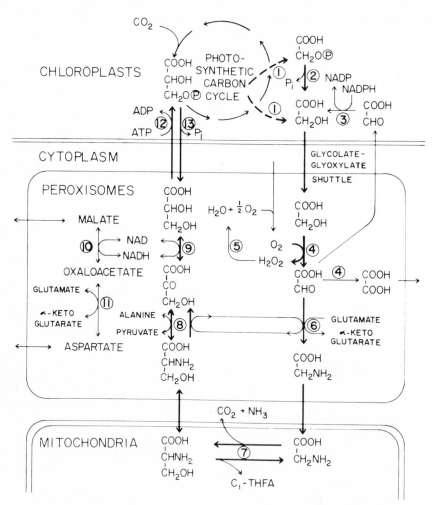

Figure 1 Reactions and site location of enzymes of the glyco-late pathway. 1. P-glycolate and glycolate biosynthesis from sugar phosphatases of the photosynthetic carbon cycle. 2. P-glycolate phosphatase. 3. NADPH glyoxylate reductase. 4. Gly-colate oxidase which catalyzes the oxidation of both glycolate and glyoxylate. 5. Catalase. 6. Glyoxylate-glutamate amino-transferase. A glyoxylate-serine aminotransferase, reaction 8, is also linked to glycine conversion to glycine. 7. A group of enzymatic reactions collectively referred to as serine hydr-oxymethyl transferase. 8. Glyoxylate-serine aminotransferase is not a reversible reaction, but as shown the enzyme also uses other amino donors to catalyze the reversed reaction, hydroxypyruvate-alanine transamination. 9. Hydroxypyruvate re-ductase. This enzyme has also been investigated as a glyoxy-late reductase and as D-glycerate dehydrogenase. 10. NAD ma-late dehydrogenase, the isoenzyme located in peroxisomes. 11. Aspartate-α-ketoglutarate aminotransferase, an isoenzyme in peroxisomes. 12. Glycerate kinase. 13. P-glycerate phosphatase.

enzymes either from contaminating microbodies or due to peroxisome breakage and absorption or binding of the peroxisomal enzymes to the chloroplasts.

Photorespiration and the glycolate pathway represents a metabolic sequence, parts of which are in the chloroplasts, peroxisomes and mitochondria, as depicted in *Figure 1*. This is a metabolic sequence in which non-phosphorylated compounds, glycolate and glycerate, derived from the photosynthetic carbon cycle are converted by two routes in the peroxisomes to glycine or serine. The pathway was first elucidated by ^{14}C tracer experiments on the rate of labeling and ^{14}C distribution in each carbon atom after [14]CO_2 fixation or after feeding labeled substrates (20,21). Enzymes for each step have been found, isolated and characterized. In fact four enzymes, catalase, glycolate oxidase, malate dehydrogenase, and NAD hydroxypyruvate reductase (glyoxylate reductase) have been crystallized. Most recent has been studies on the compartmentalization of various parts of this sequence, as described in this paper. Mechanisms of transport between these sites, if they exist, are poorly understood. The sequence of development and regulation of microbodies will also be subjects of much further investigation.

GLYCOLATE BIOSYNTHESIS

The first part of the glycolate pathway is in the chloroplast where the 2-carbon acid is formed in substantial amounts. Consideration of glycolate biosynthesis has been presented at these meetings by GIBBS, whose data supports its formation by a peroxidation of the thiamine pyrophosphate C_2-complex formed by transketolase from the top two carbon atoms of fructose-6-phosphate (22,23). This direct synthesis of glycolate does not utilize a specific phosphoglycolate phosphatase of the chloroplast, as discussed in another paper at these meetings. Since we theorize that this abundant phosphatase is involved in glycolate biosynthesis and excretion, this mechanism for glycolate formation is also shown in *Figure 1*. A third chloroplast enzyme for glycolate formation is a NADP glyoxylate reductase (12, 25), whose postulated function is discussed with respect to a glycolate-glyoxylate shuttle. All investigators seem to be in agreement that glycolate biosynthesis occurs in the chloroplasts, requires high light intensity, and is greatly enhanced by high O_2 and low CO_2 concentration.

GLYCOLATE CONVERSION TO GLYCINE

The glycolate excreted by the chloroplasts is oxidized in the peroxisomes to glyoxylate by glycolate oxidase (*Figure 1*). This flavoprotein has a relatively low affinity for oxygen, and requires 20% O_2 or air for rapid activity *in vivo*, and it is further stimulated by 60 to 100% oxygen. Oxygen uptake attributed to photorespiration certainly occurs during this peroxisomal respiration and probably

also during a Mehler reaction in the chloroplasts that produces the
peroxide for glycolate formation. For the glyoxylate formed from
glycolate there are three possible enzymatic reactions, so that glyoxylate is at a metabolic branch point, and all three reactions occur in limiting degrees. In the direct flow through of carbon along
the glycolate pathway the glyoxylate is converted in the peroxisome
to glycine by two different and essentially irreversible transaminases glyoxylate-glutamate (26), and glyoxylate-serine aminotransferases. The specific activity of the glyoxylate-glutamate aminotransferase is lower than that of other peroxisomal enzyme associated
with the glycolate pathway. Besides being present in limiting a-
mounts, the activity of this enzyme must be regulated by nitrogen
availability through the amino donors. Glyoxylate may also be slowly
oxidized to oxalate by the peroxisomal glycolate oxidase. This appears to be a non-functional overflow from the glycolate pathway.
Thus oxalate accumulation during nitrogen deficiency can be rationa-
lized by insufficiencies of glutamate to convert glyoxylate to glycine. A third metabolic fate for glyoxylate is its return to the
chloroplast where it is reduced by NADP glyoxylate reductase to glycolate to complete a glycolate-glyoxylate cycle between the two particles.

GLYCINE-SERINE INTERCONVERSION

These reactions have not been found in leaf peroxisomes or reported in any other type of microbody. Serine yields glycine and formyl-
tetrahydrofolic acid (C_1-THFA), and glycine can be converted into
CO_2, from the carboxyl group, and a C_1-THFA from the second carbon.
The C_1 may be transferred to another glycine for a net yield of CO_2
plus serine from two glycines. C_1-THFA represents the primary source
of all C_1 groups, and its biogenesis during photosynthesis in plants
is through the glycolate pathway. When particles from spinach leaves
were isolated on sucrose gradients (26), the loss of $^{14}CO_2$ from
($1-^{14}C$) glycine and formation of ^{14}C-formyl-THFA from ($2-^{14}C$) glycine was found in the particulate band represented by mitochondria
and some whole chloroplasts. These reactions and their cellular locations need to be more extensively investigated for plant tissue,
and our assignment of them to mitochondria, as in *Figure 1*, should
be subjected to further examination. This is an important reaction
for photorespiration, as it is a mechanism for CO_2 production. Consistent with this scheme is ZELITCH'S proposal (17,18) that CO_2 lost
during photorespiration arises from the carboxyl group of glycolate.
Likewise the carboxyl group of glycine is as effective as or better
than ($1-^{14}C$) glycolate as a substrate for CO_2 evolution during photorespiration (27). Glyoxylate decarboxylation as an alternate source
of this CO_2 loss has been considered, but enzymatic evidence for
this is lacking. ($1-^{14}C$) glyoxylate produced during ($1-^{14}C$) glycolate oxidation by isolated peroxisomes is not decarboxylated even in
the absence of glutamate for subsequent transamination to glycine

(26).

At present, we propose that O_2 uptake attributed to photorespiration occurs in the chloroplast during glycolate formation and in the peroxisomes during glycolate oxidation. CO_2 loss during photorespiration seems to occur in the mitochondria during the conversion of two glycines to one serine. Gas exchange during photorespiration thus involves three subcellular particles.

Carbon flow from phosphoglycolate to serine represents the first half of the glycolate pathway. All five enzyme, the phosphatase, oxidase, two transaminases, and glycine decarboxylase, are catalyzed by physiologically irreversible enzymatic reactions. This part of the glycolate pathway is extremely undirectional. It is strictly aerobic and increasing O_2 concentration stimulates both glycolate formation and the activity of glycolate oxidase with resultant carbon flow through to glycine and serine. For most plants the effect of increasing O_2 levels on the products of photosynthesis is an accumulation of glycine and serine. Glycolate accumulation does not occur to a significant extent unless the pathway is inhibited or broken up, as in isolated chloroplasts or algae which excrete the glycolate. The extensive use of sulfonate inhibitors *in vivo* to block glycolate oxidase and accumulate glycolate has proved however, that a massive flow of carbon normally moves through the glycolate pathway (17,18).

PHOSPHOGLYCERATE CONVERSION TO SERINE

The second part of the glycolate pathway in the left side of *Figure 1* interconverts carbon between serine and phosphoglycerate. This part is anaerobic and does not evolve CO_2, so it is not observed by measurements of photorespiration. In fact carbon flow from 3-phosphoglycerate (3-PGA) by this route produces serine without the losses suffered during photorespiration and glycolate conversion to serine. This is however not the major route to serine in higher plants in the light. When $(1-{}^{14}C)$ glycolate is fed to leaves, it is rapidly oxidized and converted into glycine, serine, and glycerate with labeling patterns consistent with the flow of carbon from glycolate to glycerate, as in *Figure 1* (20). In the light the glycerate is then phosphorylated and reduced to sugars, as predicted by the photosynthetic carbon cycle or by reversal of the Embdon Meyrhoff pathway (28). Thus the total flow of carbon from phosphoglycolate through the glycolate pathway to glycerate and then in the light to hexoses is gluconeogenic. That is to say, photorespiration is gluconeogenic. Two glycolates are converted to one phosphoglycerate and one CO_2, and there is the uptake of two atoms of oxygen. The missing link in our comprehension is why the leaf peroxisomal system exists at all. The glycolate pathway is a closed cycle starting from sugars of the photosynthetic carbon cycle and ending by sugar resynthesis in the chloroplasts. Glycine and serine and C_1-THFA can be formed directly from 3-PGA. The secret may hide in the essentiality of glycolate synthesis, in the first place. Glycolate seems to be the end product of an unavoidable or protective mechanism against excess photosynthetic capacity in high light and low CO_2. Glycolate conversion back to glycerate by photorespiration in leaves represents the salvage of 3/4 of this carbon for further sugar resynthesis.

Several enzymes for interconverting serine and glycerate are in leaf peroxisomes. The serine to hydroxypyruvate interconversion is catalyzed by a glyoxylate-serine aminotransferase, which is entirely located in these peroxisomes (D.W. REHFELD and N.E. TOLBERT, unpublished). This reaction conserves amino groups within the peroxisomes, but since it is a reaction with glyoxylate, it is irreversible (*Figure 1*). In the reverse direction other amino donors, particularly alanine and glutamate (29), can serve as amino donors for conversion of hydroxypyruvate to serine, apparently by the same enzyme. However, during the overall conversion of two glycolates to one glycerate, two glycines must be formed for each serine available as an amino donor. Thus the amount of serine conversion to hydroxypyruvate limits glyoxylate conversion to glycine, but this is alleviated by an additional peroxisomal transaminase, glyoxylate-glutamate in the particle.

The reduction of hydroxypyruvate is linked to NADH oxidation. This reductase is about four times more active for hydroxypyruvate reduction than glyoxylate reduction, and its K_m (hydroxypyruvate) is about 0.5 to 1.2 x 10^{-4} M as compared to a K_m (glyoxylate) of 1.6 to 5 x 10^{-2} M. From such considerations it has been concluded that the main function of this reductase of the peroxisome is for glycerate formation rather than glyoxylate reduction to glycolate (not shown in *Figure 1*) (12). For glycerate formation in the peroxisomes reducing equivalents must be carried in by a malate-aspartate shuttle. The peroxisomal isoenzyme of NAD malate dehydrogenase is the most active enzyme in the particle, besides catalase (19,29), and hydroxypyruvate reductase is nearly as active. An oxaloacetate-glutamate aminotransferase of the leaf peroxisome (29) should function in this shuttle for returning the oxidized C_4 carrier back to the chloroplast or to other cellular sites for reduction. During massive flow of carbon from glycolate to glycerate in the light, this reduction of hydroxypyruvate should represent another site for disposal of excess photosynthetic reducing capacity linked through the carriers, malate and oxaloacetate or aspartate.

A glycerate kinase of the chloroplast (30) would complete the glycolate metabolic cycle by utilizing photosynthetic ATP to reintroduce glycerate into the photosynthetic carbon cycle as 3-PGA. Under conditions, which promote large amounts of glycolate formation from the photosynthetic carbon cycle, the whole process could repeat so that a combination of glycolate formation from the photosynthetic carbon cycle and glycolate conversion to one CO_2 and 3-PGA would be a continuous process removing energy and losing CO_2 via photorespiration. This respiration can occur at the expense of the carbohydrate reserves of the leaf in the absence of external CO_2, and as shown by

MOSS (31) and others, C_3 plants cannot survive for long when this drain continues in the light.

The conversion of 3-PGA to serine is reversible and has several analogies with the conversion of phosphoglycolate to serine. Both start from phosphate esters formed photosynthetically and for both there is a specific phosphatase which is probably involved in the loss from the chlorplast of glycolate and glycerate (3,24). Whereas there is a glycerate kinase, no glycolate kinase has been found and indeed phosphorylation of glycolate seems not necessary, since all steps in its metabolism are irreversible.

The conversion of glycerate to serine, which occurs in the peroxisomes, appears limited by the pH optimum of hydroxypyruvate reductase which is at 6.3 for hydroxypyruvate reduction but above 9 for glycerate oxidation. In addition the reaction depends upon the availability of oxidized NAD. In the light with excess reducing capacity, NAD may be mostly reduced, and carbon flow from hydroxypyruvate to glycerate is favored. But in the dark with NAD in the oxidized form the reverse reaction can provide the essential serine, glycine, and C_1 derivatives. Thus movement of carbon from glycerate to serine primarily occurs in the dark, but the formation of carboxyl labeled serine during initial seconds of light $^{14}CO_2$ fixation in some cases, indicates that it does occur to a limited extent when serine synthesis from glycolate is blocked. In spite of the relative slow rate of serine formation from glycerate, in many plants and algae a large glycerate pool is labeled identically to the 3-PGA pools (32,33) and may reflect an additional transport function for the 3-PGA phosphatase and glycerate kinase.

PHOTORESPIRATION

In summarizing the biochemistry of the glycolate pathway there are seven sites at which photosynthetic assimilatory power could be consumed, which all together represents a considerable loss of energy by photorespiration.

(i)   During glycolate formation the 2-C of a ketosugar phosphate is oxidized to the carboxyl level.

(ii)   Phosphoglycolate phosphatase action would cost one phosphate bond.

(iii) The energy in glycolate oxidation in the peroxisomes is wasted through H_2O_2 formation and then destruction.

(iv)   Glyoxylate reduction by NADP glyoxylate reductase of the chloroplast is a potential cyclic shuttle system for utilization of excess reducing capacity in the plastids.

(v) CO_2 is lost during the oxidative decarboxylation of glycine to CO_2 and C_1-THFA.

(vi)   In the reduction of hydroxypyruvate to glycerate an additional reducing equivalent is required, and it is apparently transported from the chloroplasts to the peroxisomes by malate.

(vii) Phosphorylation of glycerate by a kinase in the chloroplasts requires ATP. Photorespiration can cost a varying amount of photosynthetic assimilatory capacity depending upon the magnitude of the glycolate-glyoxylate shuttle and upon whether the end products are glycine, serine, or 3-PGA. Further, the magnitude of photorespiration is dependent upon the environmental conditions regulating the amount of glycolate biosynthesis. Thus no constant respiratory quotient or rate can be predicted.

Photorespiration is a light stimulated rate of CO_2 release and O_2 uptake. Its magnitude necessitates the additional metabolism described as the glycolate pathway, for mitochondrial respiration rates are inadequate. The reason that photorespiration occurs only in the light is apparently due to glycolate formation only in the light. Oxidation of glycolate in the peroxisomes is not light dependent. If glyoxylate were reduced back to glycolate by the peroxisomal hydroxypyruvate reductase, then peroxisomal respiration should occur in the dark and consume reserves of reducing equivalents from the cell. This energy would be transported to the peroxisome as malate and linked through the peroxisomal malate dehydrogenase, to NAD reduction. Dark respiration in leaves is thought to be accounted for by mitochondrial activity. There is no evidence for peroxisomal respiration in the dark in leaves, but it probably does occur in liver (4, 5,13). Low glyoxylate concentrations may be the reason why a glycolate-glyoxylate terminal oxidation system does not function in leaf peroxisomes in the dark. Although the reductase is a major peroxisomal enzyme, its K_m (glyoxylate) at about 2×10^{-2} M (34) is probably much higher than cellular concentrations, particularly in the dark. Reactions in the dark for glyoxylate formation are not known in leaves. No glycolate is biosynthesized. Aminotransferase reactions between glyoxylate and other amino acids are nearly irreversible so that glycine conversion to glyoxylate is nil. In the light on the other hand, a large inflow of glycolate to glyoxylate may increase the glyoxylate pool until some additional O_2 uptake might occur by a direct glycolate-glyoxylate shuttle within the peroxisomes. However, *in vivo* ^{14}C-tracer experiments indicate that the glycolate is rapidly converted to glycine and serine and glyoxylate is not thought to accumulate in significant amounts.

GLYCOLATE-GLYOXYLATE SHUTTLE

Considerations against this terminal oxidation shuttle in peroxisomes were considered above. However, glycolate oxidase in the peroxisomes and a NADP glyoxylate reductase of the chloroplasts represents a possible cyclic process for linking NADPH oxidation to O_2 through these two-carbon carriers. The NADPH linked glyoxylate reductase in the chloroplast has a favorable K_m of about 2×10^{-4} M (25). As shown in *Figure 1*, a glycolate-glyoxylate shuttle between peroxisomes and chloroplasts would consume excess photosynthetic reducing power (NADPH) and transfer the hydrogen through the flavin glycolate oxidase of the peroxisomes to form H_2O_2, which is quickly

destroyed by the excess catalase. Catalytic amounts of either glycolate or glyoxylate could result in the oxidation of NADPH without
the release of any CO_2, and the oxygen uptake would be balanced by
oxygen production in the photosystem. It is not known to what extent
this occurs, if at all, but it should only occur in the light when
both NADPH and glyoxylate are formed photosynthetically. Its occurrence in the light need not be linked to CO_2 evolution and could go
unobserved by current assays for photorespiration.

There are theoretical reasons why this glycolate-glyoxylate shuttle is an attractive hypothesis. It is the same but more complex
concept as the Mehler reaction to dispense with excess reducing equivalents by transferring them to O_2. Since the catalase is all in the
peroxisomes and excess reducing capacity in the chloroplast, a glycolate-glyoxylate shuttle between the two particles would be an effective way to link the production of excess reducing capacity to
H_2O_2 production and catalase activity. H_2O_2 production in the chloroplasts in the absence of catalase should be undesirable, but may
be in part unavoidable and may account for glycolate formation. But
once some glycolate was formed, the glycolate-glyoxylate shuttle
could function. Regulation or balance between excess NADPH production and ATP formation could be achieved by consuming excess NADPH
in such a shuttle, linked to the peroxisomes without excess H_2O_2
production in the chloroplasts.

Although several reasons exist for postulating a glycolate-glyoxylate shuttle between chloroplasts and peroxisomes, *in vivo* demonstration of it has not been done. Of particular concern is the low
level of total NADP glyoxylate reductase activity in chloroplasts,
which is about 0.3 to 1 μmoles $min^{-1}mg^{-1}$ chlorophyll in spinach
leaves (J. ANDREWS and N.E. TOLBERT, unpublished). Compared to the
rates of CO_2 fixation of 3 to 4 μmoles min^{-1} mg^{-1} chlorophyll, the
activity of the reductase seems insufficient to alleviate excess
$NADPH_2$ production or to account for much of the glycolate biosynthesis. Also, the pH optimum of the chloroplast NADP glyoxylate reductase is between 5.8 to 6.5 which is lower by 2 pH units than most
chloroplast enzymes. Of course if the enzyme does function for disposal of excess NADPH its activity must be limited and regulated and
further investigations may reconcile these current discrepancies.

PHOTORESPIRATION IN C_4 PLANTS

Plants with the C_4 dicarboxylic acid cycle in mesophyll cells
have a near zero CO_2 compensation point. C_3 plants with only the
photosynthetic carbon cycle have compensation points in air and sunlight of 40 to 60 ppm CO_2 for wheat and beans to 150 ppm CO_2 for
leaves of trees. Several hypotheses have been considered to account
for this greater apparent efficiency in photosynthesis by the C_4
plant. Obviously first was the greater efficiency of the C_4 dicarboxylic acid cycle in the mesophyll cells to refix all CO_2, so that
even if photorespiration did occur in the bundle sheath cells, it

could be refixed before it could diffuse out through the mesophyll
cells. Data discussed by JACKSON and VOLK at this meeting indicate
that C_4 plants do have a substantial rate of O_2 exchange (16,35),
which is consistent with the occurrence of photorespiration in them.

It was also postulated that there might be lower levels of peroxisomal enzymes for glycolate metabolism as well as a lowered rate
of glycolate formation in C_4 plants. This seems to be partly correct.
Initial surveys found falsely low levels of peroxisomal enzymes in
C_4 plants, because grinding in the Waring blendor did not break the
bundle sheath cells (11,36). A re-evaluation by differential grinding procedures adequate to break both the mesophyll cells and bundle
sheath cells indicate that most of the peroxisomal enzymes and phosphoglycolate phosphatase are in the bundle sheath cells (37.38). OS-
MOND has further discussed these problems with C_4 plants at this
symposium. Our estimates are that most C_4 plants contain 1/3 to 1/2
as much peroxisomal enzyme activity as C_3 plants, and that this is
nearly all (70 to 90%) in the bundle sheath cells. Exact measurements of peroxisomal activity on the basis of bundle sheath chlorophyll has not been made, but could be as high as in C_3 plants. These
results have been confirmed by *in situ* electron micrographs of peroxisomes in C_4 leaves as reported by NEWCOMB and by LAETSCH at these
meetings. Large mitochondria and many peroxisomes are in the bundle
sheath cells, while the mesophyll cells contain fewer mitochondria
and very few microbodies. Thus the bundle sheath cells of most C_4
plants are like cells of a C_3 plant that contain the photosynthetic
carbon cycle, glycolate biosynthesis, and peroxisomes. When ZELITCH
(18) fed C_4 plants (1-^{14}C) glycolate, it was rapidly metabolized and
converted into the predicted products of the glycolate pathway, but
the $^{14}CO_2$ was all refixed into products of the photosynthetic carbon
cycle, as if it never escaped from the bundle sheath.

The few microbodies in the mesophyll cells have not been isolated
or their function explained. If we assume that these are peroxisomes
on the basis of enzyme activity in homogenates (37,38), then limited
glycolate synthesis and metabolism may also occur there, as does
perhaps some CO_2 fixation by the photosynthetic carbon cycle. It
will be interesting to learn about the level of glycolate biosynthesis and peroxisomal enzymes in the bundle sheath cells of *Sorghum
bicolor* grown in high light intensity. Agranal chloroplasts in this
tissue are reported to lack photosystem II (37) and to be unable to
produce photosynthetic reducing equivalents, in which case peroxide
formation and glycolate biosynthesis should also be prevented. Perhaps some correlation can be found in the extent of the agranal nature of bundle sheath chloroplasts, amount of photosystem II, and a-
mount of glycolate biosynthesis and metabolism.

At present photorespiration in C_4 plants is thought to occur in
the bundle sheath cells to a varying limited extent dependent upon
the amount of glycolate production. However the CO_2 produced is refixed before it can escape from the leaf and perhaps even from the
bundle sheath cells. It is correct to say that C_4 plants have a near

zero CO_2 compensation point because the CO_2 produced from respiration is not lost. These plants do have peroxisomes and photorespiration, but CO_2 loss is kept minimal by anatomical placement of most of the peroxisomes within the bundle sheath cells and by efficient refixation of all of the CO_2.

FUNCTION

Several things are accomplished by peroxisomal metabolism and photorespiration:

(i) Glycine and serine are formed and they are essential for protein and for C_1 biosynthesis.

(ii) Assuming that glycolate formation in the chloroplasts is an obligatory process, then the glycolate conversion back to glycerate and 3-PGA is a gluconeogenic process that salvages part of this carbon and energy.

(iii) Microbodies contain other enzymes for catabolism (3), which support the idea that they are involved in disposal of metabolites. This is more true for microbodies of other tissue, while leaf peroxisomes contain mainly the enzymes of the glycolate pathway.

(iv) Microbodies may regulate growth by disposing of excess energy. The diversion of part of the photosynthetic products through leaf peroxisomes and photorespiration may be such a regulation. The glycolate-glyoxylate shuttle between the peroxisomes and the chloroplasts is a further potential metabolic cycle for disposing of excess photosynthetic reducing power (NADPH). It is conceivable this process is essential because other mechanisms of regulating photosynthesis in the chloroplasts are insufficient. The glycolate glyoxylate cycle by using excess NADPH, may keep electrons flowing in the photosystems for additional ATP synthesis.

Additional functions can be postulated for leaf peroxisomes, and certainly the full complexity and role of these particles and the total process of photorespiration have not yet been completely elucidated. Such research involves the disciplines and participation of investigators in photosynthesis, plant biochemistry, physiology, and plant growth, as exemplified by this symposium.

REFERENCES

1. J.F. HOGG, *Ann. N.Y. Acad. Sci.*, 168 (1969) 209.
2. Z. HRUBAN and M. RECHCIGL, Jr., Microbodies and Related Particles, Academic Press, New York, (1969) 296 pp.
3. N.E. TOLBERT, *A. Rev. Pl. Physiol.*, 22 (1971) in press.
4. C. DE DUVE, *Physiol. Rev.*, 26 (1966) 323.
5. C. DE DUVE, *Proc. Roy. Soc. (B.)*, 173 (1969) 71.
6. R.W. BREIDENBACH and H. BEEVERS, *Biochem. biophys. Res. Commun.*, 27 (1967) 462.
7. H. BEEVERS, *Ann. N.Y. Acad. Sci.*, 168 (1969) 313.
8. N.E. TOLBERT, in Methods in Enzymology, Photosynthesis and Nitro-

gen Fixation, ed. A. SAN PIETRO, Academic Press, New York, (1971) in press.

9. N.E. TOLBERT, A. OESER, T. KISAKI, R.H. HAGEMAN, and R.K. YAMA-ZAKI, *J. biol. Chem.*, 243 (1968) 5179.

10. N.E. TOLBERT and R.K. YAMAZAKI, *Ann. N.Y. Acad. Sci.*, 168 (1969) 325.

11. N.E. TOLBERT, A. OESER, R.K. YAMAZAKI, R.H. HAGEMAN, and T. KIS-AKI, *Plant Physiol.*, 44 (1969) 135.

12. N.E. TOLBERT, R.K. YAMAZAKI, and A. OESER, *J. biol. Chem.*, 245 (1970) 5129.

13. S.L. VANDOR and N.E. TOLBERT, *Biochim. biophys. Acta*, 215 (1970) 449.

14. S.E. FREDERICK, E.H. NEWCOMB, E.L. VIGIL, and W.P. WERGIN, *Planta*, 81 (1968) 229.

15. F. LEIGHTON, B. POOLE, H. BEAUFAY, P. BAUDHUIN, J.W. COFFEY, S. FOWLER, and C. DE DUVE, *J. Cell Biol.*, 37 (1968) 482.

16. W.A. JACKSON and R.J. VOLK, *A. Rev. Pl. Physiol.*, 21 (1970) 385.

17. I. ZELITCH, *A. Rev. Pl. Physiol.*, 15 (1964) 121.

18. I. ZELITCH, in Harvesting the Sun, ed. A. SAN PIETRO, F.A. GREER, and T.S. ARMY, Academic Press, New York, (1969) 231.

19. R.K. YAMAZAKI and N.E. TOLBERT, *Biochim. biophys. Acta*, 178 (1969) 11.

20. R. RABSON, N.E. TOLBERT, and P.C. KEARNEY, *Archs. Biochem. Biophys.*, 98 (1962) 154.

21. N.E. TOLBERT, in Photosynthetic Mechanism in Green Plants, NAS, NRC Publication 1145 (1963) 648.

22. P.W. ELLYARD and M. GIBBS, *Plant Physiol.*, 44 (1969) 1115.

23. M. GIBBS, *Ann. N.Y. Acad. Sci.*, 168 (1969) 356.

24. D.D. RANDALL, N.E. TOLBERT, and D. GREMEL, *Plant Physiol.*, (1971) submitted.

25. I. ZELITCH and A.M. GOTTO, *Biochem. J.*, 84 (1962) 541.

26. T. KISAKI and N.E. TOLBERT, *Plant Physiol.*, 44 (1969) 242.

27. T. KISAKI and N.E. TOLBERT, *Plant Cell Physiol.*, 11 (1970) 247.

28. E. JIMENEZ, R.L. BALDWIN, N.E. TOLBERT, and W.A. WOOD, *Archs. Biochem. Biophys.*, 98 (1962) 172.

29. R.K. YAMAZAKI and N.E. TOLBERT, *J. biol. Chem.*, 245 (1970) 5137.

30. C.R. SLACK, M.D. HATCH, and D.J. GOODCHILD, *Biochem. J.*, 114 (1969) 489.

31. K.M. MENZ, D.N. MOSS, R.Q. CANNELL, and W.A. BRUN, *Crop Sci.*, 9 (1969) 692.

32. J.L. HESS and N.E. TOLBERT, *J. biol. Chem.*, 241 (1966) 5705.

33. J.L. HESS and N.E. TOLBERT, *Plant Physiol.*, 42 (1967) 371.

34. I. ZELITCH, *J. biol. Chem.*, 216 (1955) 553.

35. H. FOCK, H. SCHAUB, W. HILGENBERG, and K. EGLE, *Planta*, 86 (1969) 77.

36. C.B. OSMOND, *Biochem. biophys. Acta*, 172 (1969) 144.

37. D.W. REHFELD, D.D. RANDALL, and N.E. TOLBERT, *Can. J. Bot.*, 48 (1970) 1219.

38. C.B. OSMOND and B. HARRIS, *Biochim. biophys. Acta*, (1971) in

press.
39. K.C. WOO, Jan M. ANDERSON, N.K. BOARDMAN, W.J.S. DOWNTON, C.B. OSMOND, and S.W. THORNE, *Proc. natn. Acad. Sci. U.S.A.*, 67 (1970) 18.

THE ABSENCE OF PHOTORESPIRATION IN C_4 PLANTS: REAL OR APPARENT?

C.B. Osmond

Research School of Biological Sciences, Australian
National University, Canberra City 2601, Australia

SUMMARY

This paper reviews the evidence for glycolate production and its metabolism by the glycolate pathway in leaves of C_4 plants. Glycolate production during Calvin cycle photosynthesis occurs in the bundle sheath cells of many C_4 leaves, but may be defective in those lacking photosystem II in bundle sheath chloroplasts. Mesophyll chloroplasts lack the Calvin cycle and may be incapable of glycolate synthesis. Glycolate pathway activity and the potential for CO_2 release in some C_4 plants is considerable, but the site of CO_2 release is not clear. In these species, refixation of photorespiratory CO_2 during photosynthesis is implied. C_4 plants lacking photosystem II in bundle sheath chloroplasts have low glycolate pathway activity, and it is likely that photorespiration is absent in these species.

INTRODUCTION

The leaves of C_4 plants do not release CO_2 into CO_2 free air in the light, show no enhancement of net photosynthesis in O_2 free air, have CO_2 compensation values close to zero, and thus appear to lack photorespiration (1). To date there are no exceptions to this correlation, although leaves of C_4 plants may be induced to release CO_2 in low light, or in light after treatment with 3-(3,4-dichlorophenyl)-1, 1-dimethylurea (DCMU) (2,3). The CO_2 release so induced does not respond to O_2 concentration as expected of photorespiration (2). It is likely that both treatments are equivalent to darkness and that the CO_2 is of mitochondrial origin, (dark respiration).

The release of CO_2 during photorespiration is clearly associated with synthesis and metabolism of glycolate during photosynthesis (1, 4). The biochemistry of reactions leading to the evolution of CO_2 in photorespiration is not well understood. Both the oxidation of glyoxylate (4) and the condensation of two glycine to serine (5) have been proposed as the source of this CO_2. The overall process of photorespiration requires light, substrates derived from Calvin cycle intermediates and oxygen, and may involve chloroplasts, peroxisomes and mitochondria. Three simple alternatives may explain the apparent lack of photorespiration in leaves of C_4 plants. They may :

(i) not synthesize the C_2 substrates of photorespiration,

(ii) synthesize the C_2 substrates but be unable to metabolize them via the glycolate pathway at significant rates,

(iii) effectively synthesize and metabolize the C_2 compounds but the CO_2 evolved from the glycolate pathway may be refixed within the leaf.

These alternatives were reviewed briefly by JACKSON and VOLK (1) but recent developments in techniques for separating the biochemical activities of different cells in C_4 leaves have permitted several significant advances.

GLYCOLATE SYNTHESIS

When it was first recognised that C_4 plants did not release CO_2 in the light, several groups proposed that substrates for photorespiration might be otherwise involved in the carbon metabolism of C_4 photosynthesis (6,7). At that time "carboxyl-transfer" reactions were sought and the role of the Calvin cycle in C_4 photosynthesis was not clear. A C_2 compound may act as a CO_2 acceptor in a transcarboxylation reaction. As well, schemes could be constructed in which oxaloacetate and a C_2 compound condense and split yielding $2xC_3$ and, finally, 3-phosphoglycerate (3-PGA). These schemes side-stepped the apparently low ribulose-1,5-diphosphate (RuDP) carboxylase activity and suggested a diversion of glycolate precursors. However, careful studies failed to uncover significant label in glycolaldehyde or hydroxypyruvate, compounds likely to be involved as acceptors in the transfer of the C-4 carboxyl from oxaloacetate by this scheme (7).

It is now clear that all photosynthetic carbon eventually passes through the Calvin cycle during C_4 photosynthesis. There is no immediate reason to suspect that glycolate synthesis in C_4 plants should differ from that in the leaves of a C_3 plant.

Frequent literature references to lower rates of glycolate synthesis in C_4 plants than in C_3 plants (1,8,9) are based on relatively few data. Estimates of glycolate synthesis by colorimetric assay for glycolate after prolonged treatment with bisulfite compound inhibitors yield variable results. The data collected in *Table 1* show as much variability within as between species and, as is now recognised, are confounded by effects of bisulfite compounds on stomatal opening (10) and on CO_2 fixation via phosphoenolpyruvate (PEP) carboxylase (11). Furthermore, the rates measured in these experiments are generally less than 10% the rate of photosynthesis. MOSS (14) has demonstrated rates of glycolate formation approaching 20% of the tobacco photosynthetic rate (15). Yet, glycolate synthesis must greatly exceed these estimates if each 2 moles of glycolate gives rise to one

TABLE 1

Glycolate formation in the light (μmoles g^{-1} hr^{-1}) after treatment with 5 or 10 mM bisulfite compound. Data from (12,13) and OSMOND, unpublished.

C_3 plants		
	tobacco	3-30
	spinach	1.9-6.0
	tomato	3.2
	bean	4.3
	A. *hastata*	1.5
C_4 plants		
	maize	0.9-4.9
	A. *spongiosa*	0.6-10.0

mole of CO_2 and if CO_2 release approaches 50% of the photosynthetic rate.

More accurate data may be obtained by the use of $^{14}CO_2$ but the experiments suffer from the same technical criticisms. *Figure 1* compares the labelling of glycine and serine during $^{14}CO_2$ photosynthesis in *Atriplex spongiosa* (C_4), A. *hastata* (C_3) and tobacco (16). The appearance of label in these amino acids is much slower in the C_4 leaf and may constitute a lower proportion of the total fixed ^{14}C at radioactive saturation. *Figure 1* also shows that when leaves were

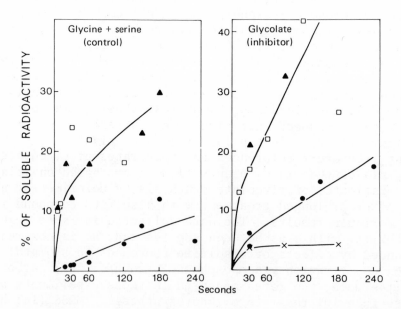

Figure 1 Labeling of glycine + serine and glycolate during $^{14}CO_2$ fixation in A. spongiosa (●), A. hastata (▲), S. bicolor (x) and and tobacco (□). The data for tobacco were taken from (16).

treated with glyoxal bisulfite, which among other effects, inhibits glycolate oxidation, glycolate label accumulated at about the same rate as glycine + serine label in the controls. In another C_4 species, *Sorghum bicolor*, the glycolate fraction contained little label and showed no increase with time. Labeled sugar phosphate precursors of glycolate were fed to leaves of *A. spongiosa* (C_4) and *A. hastata* (C_3) and were equally rapidly converted to labeled glycolate (17). However, as discussed later, this approach has very serious limitations.

If glycolate synthesis is deficient in C_4 plants, compared with C_3 plants, differences in the activity of enzymes involved in glycolate formation might be expected. Phosphoglycolate phosphatase and transketolase are thought to be involved in glycolate formation (18,19) yet in all C_4 species tested, the activity of these enzymes in the bundle sheath cells is equivalent to that in C_3 leaves (20). *Figure 2*, prepared from published data (17), shows that the activity of phosphoglycolate phosphatase and transketolase closely parallels the activity of RuDP carboxylase and aldolase in C_3 and C_4 species, and is equivalent when expressed in terms of the activity in the bundle sheath cells. The activity associated with the mesophyll fraction in *A. spongiosa* and *Sorghum* is probably contamination from the bundle sheath. When PEP carboxylase, which is almost exclusively associated with the mesophyll cells (21,22), is used as a marker, the enzymes of

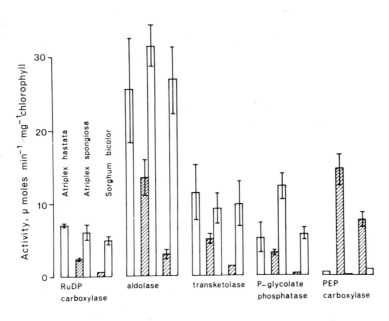

Figure 2 *The activity of photosynthetic carboxylases and enzymes of glycolate synthesis in extracts from A.* hastata, *A.* spongiosa *and S.* bicolor. *The extracts from* C_4 *leaves were prepared to give mesophyll (hatched bars) and bundle sheath (open bars) fractions. Constructed from data of (17).*

glycolate synthesis are clearly associated with the Calvin cycle reactions of the bundle sheath cells in C_4 leaves (17,20 and *Figure 3*).

A further requirement for glycolate synthesis, the production of
an oxidant generated by a photochemical act (19) cannot yet be evaluated in C_4 plants. This oxidant may be an intermediate of photosystem II and may be lacking in the bundle sheath cells of C_4 plants
such as *Sorghum* which are deficient in this photosystem (23). This
may explain the slow glycolate synthesis in *Sorghum* (*Figure 1*), and
the slow labelling of glycine and serine in this species (7). However, peroxides may also be involved in glycolate synthesis (19) and
may substitute for the photosystem II dependence.

Thus the available data do not make a strong case for deficient
glycolate synthesis in the bundle sheath cells of C_4 plants. Glycolate synthesis in the bundle sheath probably does not limit photorespiration in most C_4 plants. *Sorghum bicolor* may be an exception. The
synthesis of glycolate in mesophyll cells of C_4 plants which lack
Calvin cycle reactions (22) has yet to be examined.

GLYCOLATE METABOLISM

Cells of the leaves of C_4 plants contain microbodies which are
thought to be peroxisomes (24,25) although the frequency may be lower
than in C_3 plants. The yield of peroxisomes from many C_4 leaves is
low (26), but this may reflect extraction difficulties. Glycolate
oxidase, the first enzyme of the glycolate pathway is usually less
active in C_4 leaves (6) and some other enzymes of this pathway are
also less active (20). All C_4 plants tested will release $^{14}CO_2$ from
$(1-^{14}C)$ glycolate, glyoxylate or glycine when fed in the dark, but
compared with C_3 species, negligible $^{14}CO_2$ is released in the light
(4,17,25). These observations suggest a considerable potential for
glycolate pathway metabolism and photorespiration.

TABLE 2

Products of $(1-^{14}C)$ glycine metabolism in leaves in the light
or dark (Percentage of ^{14}C metabolized).

Species	$^{14}CO_2$	Serine	Sugars and Phosphates
A. *hastata* (C_3)			
light	21	22	26
dark	24	46	2
A. *spongiosa* (C_4)			
light	0.2	59	31
dark	22	57	2
S. *bicolor* (C_4)			
light	1	24	41
dark	29	50	0.3

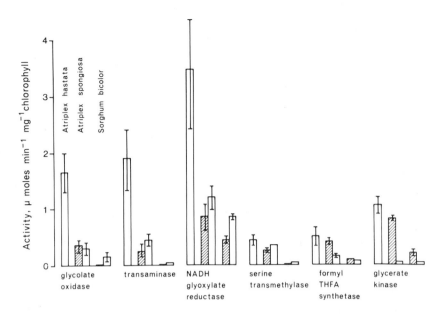

Figure 3 The activity of glycolate pathway enzymes in extracts of C_3 and C_4 leaves; notation as in Figure 2. Constructed from data of (17).

Only occasionally have these experiments shown, for example, that the $^{14}CO_2$ is released during the conversion of the labeled C_2 compound to serine (17) and there are few references to glycolate pathway reactions in C_4 plants (27). *Table 2* shows that the conversion of $(1-^{14}C)$ glycine to labeled serine and sugars is similar in leaves of C_3 and C_4 plants, although no $^{14}CO_2$ is released from the C_4 leaf. In fact, the proportion of label released as $^{14}CO_2$ in the dark and the effect of light on the labelling of phosphorylated compounds and sugars is identical in the species shown in *Table 2*. Before accepting these data as evidence for equal potential for photorespiration in plants of both types, it must be emphasised that labeled substrates fed to leaves, leaf discs or slices have to first cross a cell membrane. Compared with rates of photosynthesis or photorespiration, this transport is very slow; estimated at about 1% of the rate of photorespiration in the case of glycine (17). It follows that very large variations in the rate of glycolate pathway metabolism may go undetected by this technique.

The activity of a number of glycolate pathway enzymes was examined in mesophyll and bundle sheath fractions of the species listed in *Table 2* and *Figure 2*. In *A. hastata* these enzymes are sufficiently active to permit rates of glycolate metabolism approaching those required to give a rate of photorespiration 20-50% the rate of photosynthesis (*Figure 3*). The role of the C_1 activation enzymes in the glycolate pathway is not clear, nor is the enzymology of CO_2 evolution during serine synthesis. Unlike the Calvin cycle enzymes shown in *Figure 2*, those of the glycolate pathway are very much less active in the bundle sheath of C_4 leaves than in *A. hastata*. Comparison of data in *Table 2* and *Figure 3* emphasises the limiting role of transport in studies with labeled metabolites. There is a 10 to 100 fold difference in the activity of many glycolate pathway enzymes between *A. hastata* and *Sorghum*, yet *Table 2* suggests both are capable of equal rates of glycine metabolism.

The low glycolate pathway activity in *Sorghum*, coupled with the possibility of low glycolate synthesis, strongly suggests that this species may lack significant photorespiration i.e. that the potential for CO_2 evolution is about 1% of the photosynthesis rate. On the other hand, *A. spongiosa* resembles many other C_4 plants with considerable glycolate pathway activity (20) in which the potential for CO_2 evolution in the light may exceed 10% of the photosynthesis rate. In these species, considerable photosynthetic refixation of photorespiratory CO_2 is implied.

RECYCLING OF PHOTORESPIRATORY CO_2

The origin of the proposal that some plants are capable of refixation of photorespiratory CO_2 is difficult to place. It was discussed on the basis of correlation between "Kranz" anatomy and low CO_2 compensation values (28,29) and has been extended to biochemical phenomena as the further correlation with C_4 pathway activity has unfolded. In its present form, the "more efficient" carboxylation reaction (PEP carboxylase) of the mesophyll chloroplasts is invoked to prevent the release of CO_2 evolved in photorespiration. Three important questions must be answered to validate this hypothesis.

  (i) Is the species in question capable of glycolate pathway CO_2 evolution?
 (ii) In which tissue is the CO_2 evolved?
(iii) In which tissue and by which carboxylation reactions is the CO_2 refixed?

These questions cannot yet be answered satisfactorily for any one species. *A. spongiosa*, as shown in *Figure 3* and in common with several other C_4 plants (20), has the capacity for significant CO_2 production from the glycolate pathway. Recent experiments with this species have provided hints as to the sites and processes of CO_2 evolution and refixation.

Figure 4 shows the distribution of glycolate pathway and photosynthetic enzymes in *A. spongiosa* leaves calculated from the specific activity on a chlorophyll basis and chlorophyll distribution (17). It is clear that the glycolate pathway enzymes of the peroxisome and serine transmethylase are largely restricted to the bundle sheath cells which also contain 90% of the Calvin cycle activity. This is in agreement with comments on the distribution of peroxisomes (24, 25). Were it not for the strong association of the C_1 activation enzymes and glycerate kinase with the mesophyll cells there would be little question of CO_2 evolution in the bundle sheath cells.

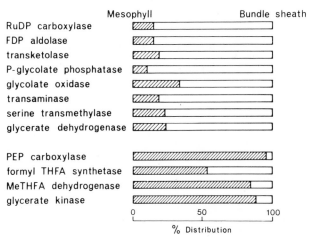

Figure 4 *Calculated distribution of enzymes between mesophyll and bundle sheath cells in leaves of A. spongiosa. Constructed from data of (17).*

If CO_2 was evolved in the bundle sheath cells it would presumably be fixed by RuDP carboxylase in this tissue and could enter sugars via the Calvin cycle. In that external CO_2 does not seem to have extensive direct access to the bundle sheath chloroplasts during C_4 photosynthesis, it is unreasonable to propose that CO_2 generated by bundle sheath photorespiration could leak to the mesophyll for refixation. Furthermore, glycolate pathway activity in the bundle sheath cells must stop short of 3-PGA, for glycerate kinase is a mesophyll enzyme (22).

If the C_1 activation enzymes and glycerate kinase are associated with the glycolate pathway and CO_2 evolution, then an alternative explanation is possible. The mesophyll chloroplasts of C_4 plants may be incapable of glycolate synthesis, for they lack Calvin cycle intermediates. If however, glycine moved from the bundle sheath peroxisomes to mesophyll cells, the activity of serine hydroxy transmethylase and subsequent glycolate pathway enzymes in these cells is sufficient to permit considerable CO_2 evolution, serine formation and metabolism to 3-PGA. The 3-PGA, following conversion to PEP, could provide the acceptor for refixation of the photorespiratory CO_2. For each 2 moles of glycine moving to the mesophyll, 1 mole of C_4 acid would be returned to the bundle sheath.

This hypothesis accounts for the distribution of some enzymes, but is there evidence that PEP carboxylase is involved in the refixation of CO_2 derived from glycolate pathway intermediates? Feeding experiments, although they show rapid labelling of C_4 acids from labeled glycolate pathway intermediates, are too lengthy (minutes) to deter-

TABLE 3

$^{14}CO_2$ release from glycolate pathway intermediates fed to leaves in light or dark. (Percentage of ^{14}C metabolized.)

Treatment	A. hastata (C_3)	A. spongiosa (C_4)
$(1-^{14}C)$glyoxylate - 1 hr		
light	6.8	1.4
" 10 mM glyoxal bisulfite	12.4	4.8
dark	14.0	17.0
$(1-^{14}C)$glycine - 30 min		
light	20.8	0.2
" + 10 mM glyoxal bisulfite	25.6	3.7
dark	24.2	22.1

mine the initial products of refixation. After a few minutes, the labelling pattern of C_4 photosynthesis does not differ from that of the Calvin cycle. However, bisulfite compounds, inhibitors of PEP carboxylase (11), stimulate the release of $^{14}CO_2$ from $(1-^{14}C)$ glyoxylate and glycine fed to the leaves of C_4 plants in the light (Table 3). The inhibitor does not influence the rate of serine formation but inhibits the labelling of C_4 acids. The stimulation of $^{14}CO_2$ release observed in all experiments, was small and seldom exceeded 20% of the dark level. Perhaps higher concentrations of inhibitor are required to control the high activity of PEP carboxylase. Even when 90% inhibited the activity would be sufficient to refix CO_2 at a rate equal to that of CO_2 evolution during photorespiration. Although the distribution of enzymes and the inhibitor studies tend to favor the view that photorespiratory CO_2 evolution and refixation occurs in mesophyll cells of some C_4 plants, other alternatives cannot be excluded at present.

CONCLUSIONS

The data available suggest that there are two explanations for the apparent absence of photorespiration in C_4 leaves. Both are intimately associated with the photosynthetic carbon metabolism of these leaves. In the first, species such as Sorghum bicolor seem to have evolved in the direction of reduced glycolate synthesis and reduced glycolate pathway activity. The reduction in glycolate synthesis during bundle sheath photosynthetic carbon metabolism may be associated with high CO_2 (9), low O_2 and low light intensity in the agranal chloroplasts. In the second, in species such as Atriplex spongiosa, the CO_2 evolved during normal photorespiratory metabolism may be refixed. The location of this refixation and the carboxylation processes involved have yet to be specified. Of the two alternatives, the latter is the most expensive in terms of NADPH and ATP utilisation per mole net CO_2 fixed. These speculations provide interesting test systems for many problems of photorespiratory metabolism, over and above the role of photorespiration in the photosynthetic efficiency of C_4 plants.

REFERENCES

1. W.A. JACKSON and R.J. VOLK, *A. Rev. Pl. Physiol.*, 21 (1970) 385.
2. W.J.S. DOWNTON and E.B. TREGUNNA, *Plant Physiol.*, 43 (1968) 923.
3. M.A. EL-SHARKAWAY, R.S. LOOMIS and W.A. WILLIAMS, *Physiologia Pl.* 20 (1967) 171.
4. I. ZELITCH, in "Harvesting the Sun, Photosynthesis in Plant Life" A. San Pietro, Academic Press, N.Y., (1967) 231.
5. N.E. TOLBERT and R.K. YAMAZAKI, *Ann. N.Y. Acad. Sci.*, 168 (1969) 325.
6. C.B. OSMOND, *Biochim. biophys. Acta*, 172 (1969) 144.
7. H.S. JOHNSON and M.D. HATCH, *Biochem. J.*, 114 (1969) 127.
8. C.B. OSMOND, J.H. TROUGHTON and D.J. GOODCHILD, *Z. Pfl Physiol.*, 61 (1969) 171.
9. A. GOLDSWORTHY and P.R. DAY, *Nature*, 228 (1970) 687.
10. I. ZELITCH, *Proc. natn. Acad. Sci. U.S.A.*, 47 (1961) 1423.
11. C.B. OSMOND and P.N. AVADHANI, *Plant Physiol.*, 45 (1970) 228.
12. I. ZELITCH, *J. biol. Chem.*, 233 (1958) 1299.
13. I. ZELITCH, *Plant Physiol.*, 41 (1966) 1623.
14. D.N. MOSS, *Crop Sci.*, 8 (1968) 71.
15. C.P. WHITTINGHAM and G.G. PRITCHARD, *Proc. R. Soc. (B)*, 157 (1963) 366.
16. J.L. HESS and N.E. TOLBERT, *J. biol. Chem.*, 241 (1965) 5705.
17. C.B. OSMOND and B. HARRIS, *Biochim. biophys. Acta* (1971) in press.
18. K.E. RICHARDSON and N.E. TOLBERT, *J. biol. Chem.*, 236 (1961) 1285.
19. Z. PLAUT and M. GIBBS, *Plant Physiol.*, 45 (1970) 470.
20. D.W. REHFELD, D.D. RANDALL and N.E. TOLBERT, *Can. J. Bot.*, 48 (1970) 1219.
21. O. BJÖRKMAN and E. GAUHL, *Planta*, 88 (1969) 197.
22. C.R. SLACK, M.D. HATCH and D.J. GOODCHILD, *Biochem. J.*, 114 (1969) 489.
23. K.C. WOO, J.M. ANDERSON, N.K. BOARDMAN, W.J.S. DOWNTON, C.B. OSMOND and S.W. THORNE, *Proc. natn. Acad. Sci. U.S.A.*, 67 (1970) 18.
24. W.M. LAETSCH, *Sci. Prog., Oxf.*, 57 (1969) 323.
25. T. KISAKI and N.E. TOLBERT, *Pl. Cell Physiol.*, 11 (1970) 247.
26. N.E. TOLBERT, A. OESER, R.K. YAMAZAKI, R.H. HAGEMAN and T. KISAKI, *Plant Physiol.*, 44 (1969) 135.
27. E. JIMINEZ, R.L. BALDWIN, N.E. TOLBERT and W.A. WOOD, *Archs. Biochem. Biophys.*, 98 (1962) 172.

28.   G. HOFSTRA and J.D. HESKETH, *Planta*, 85 (1969) 228.
29.   J. DOWNTON, J. BERRY and E.B. TREGUNNA, *Science*, 163 (1969) 78.

COMPARATIVE BIOCHEMISTRY OF MICROBODIES (GLYOXYSOMES, PEROXISOMES)[1]

Harry Beevers

Division of Natural Sciences, University of California,
Santa Cruz, California 95060, U.S.A.

SUMMARY

Microbodies are seen in electron microscope pictures of plant cells as spherical or oblate organelles of roughly 1 micron in diameter with a finely granular matrix and single bounding membrane. They are particularly numerous in fatty seedling tissue and are readily noticeable constituents of green cells of leaves but are recognized as minor constituents of most plant cells which have been examined. By suitable grinding procedures followed by differential and sucrose density centrifugation these organelles have now been isolated from a variety of plant tissues as a protein band of specific density 1.23-1.25. The proportion of the particulate protein represented by the microbody fraction is some 20% in castor bean endosperm but in leaves and other tissues the value is less than 1%. All of the microbody fractions which have been isolated have catalase as a major enzyme constituent and most contain glycolate oxidase and urate oxidase as well. This relatively restricted enzyme constitution, which was that used functionally to characterize microbodies from liver as peroxisomes, is typical of microbody fractions from many plant tissues. Preparations from leaves contain in addition, glyoxylate reductase, malate dehydrogenase and various transaminases. They are referred to as leaf peroxisomes and they are believed to function in the metabolism of glycolate in photorespiration. In fatty seedling tissue the microbodies play a central role in the conversion of fat to carbohydrate. They contain each of the enzymes of the glyoxylate cycle as well as a system for the activation and β-oxidation of long chain fatty acids. They are known functionally as glyoxysomes.

INTRODUCTION

The term microbody was introduced to describe a distinctive type

---

[1] *Supported by NSF Grant GB24961 and AEC Contract AT-1 04/3*

of organelle observed in electron microscope pictures of kidney and liver cells (1,2). These spherical organelles, 0.5-1.0 microns in diameter, had a granular matrix with a single bounding membrane, and showed occasional connections with the endoplasmic reticulum. A possible functional role was suggested when it was shown cytochemically that the microbodies were the major intracellular site of catalase and uricase activity (3).

By a combination of differential and density gradient centrifugation, DE DUVE and his colleagues (4,5,6) were able to separate a fraction from liver homogenates which was enriched in such organelles and they have intensively studied their enzymology and physical characteristics. The isolated organelles represented 2-5% of the cell protein and had an equilibrium density of 1.25 on sucrose gradients. They were similar in appearance to the microbodies observed *in situ* and they were shown by direct assay to contain most of the cellular catalase. Certain other enzymes, notably α-OH acid oxidase, uricase and some other flavin-linked oxidases generating H_2O_2 were shown to be specifically associated with the isolated organelles. Their functional role thus appeared to center on reactions producing H_2O_2 on the one hand and decomposing it (by either the catalatic or peroxidatic action of catalase) on the other. For this reason the term peroxisome, descriptive of function, was introduced (4) and it is by this term that microbodies isolated from the organs (mostly liver and kidney) of a variety of animals have come to be known. In spite of intensive investigations, however, it is still not clear whether these organelles do in fact play a central and essential role in the metabolism of such organs (6). To be sure, they do offer an extramitochondrial pathway of oxidation of $NADH_2$ (with lactate or ethanol carrying reducing equivalents into the organelles). Such a pathway would be wasteful in the sense that no coupled ATP synthesis occurs and indeed there seems to be no estimate of what part (if any) of the normal O_2 uptake of such cells is accounted for by the microbodies.

In the meantime, microbodies with the same morphological characteristics as those described from liver were found to be present in many kinds of plant cells (7). In contrast to their limited distribution in animal cells, microbodies have been seen in electron microscope pictures of most higher plant cells which have been examined. In the past few years progress has been made in isolating these organelles and in defining their enzymological makeup and metabolic functions. It is now clear that microbodies in general have a certain minimal enzyme complement (glycolate oxidase, catalase) but in some tissues the presence of other enzymes fit these organelles for more specialized metabolic roles.

MICROBODIES IN FATTY SEEDLING TISSUES

The first such organelles to be characterized from plants were those from the endosperm of germinating castor bean (8). In this and other fatty seedling tissues the dominant metabolic event is the con-

version of fat to carbohydrate and in this process the glyoxylate cycle plays a central role (9). The distinctive enzymes of this cycle,
isocitrate lyase and malate synthetase, are found together in plants
only in such fatty tissues: they are produced *de novo* during germination and disappear when fat utilization is complete (9,10).

From our earlier work it had appeared that only a relatively small
part of the activity of these key enzymes was particulate but when
BREIDENBACH reexamined this question in my laboratory he was able to
show, using appropriate preparative conditions, that most of the activity of both enzymes could be recovered in the crude particulate
fraction sedimenting at 10,000 g. When this material was layered on
a continuous sucrose density gradient and centrifuged at high speed,
three distinct protein bands were obtained. The uppermost band
(equilibrium density 1.19) was shown to be the exclusive location of
enzymes of the TCA cycle, and examination in the electron microscope
showed it to be comprised almost exclusively of intact mitochondria.
Isocitrate lyase and malate synthetase were not present in this fraction and were confined to the lowest band (equilibrium density 1.25).
Each of the enzymes of the complete glyoxylate cycle was eventually
recovered in this band (11) and the name glyoxysome was therefore
given as a functional description of the organelles contained within
it (8,12). Electron microscope pictures showed that these organelles
had the morphological features of microbodies (8,12). Subsequently
we recognized that these isolated organelles also contained most of
the cellular catalase and glycolate oxidase (12) and also uricase
(13), marker enzymes for the microbodies isolated by DE DUVE'S group,
and, surprisingly, the enzyme system concerned with activation and β-
oxidation of long chain fatty acids (14). Glyoxysomes have now been
isolated from the fatty organs of many germinating seedlings. Electron microscope pictures of such organs show that these are very rich
in microbodies and their appearance is remarkably similar to the isolated glyoxysomes (15,16). The function of the glyoxysome as revealed by its distinctive enzyme complement is shown in *Figure 1*.

No major role is at present assigned to the glycolate oxidase in
the glyoxysomes, but the fact that the enzyme was present in a special organelle raised the possibility that this may also be true in
leaves where this enzyme is highly active and apparently plays a
major metabolic role in the process of photorespiration. An examination of this possibility led TOLBERT to the discovery of a second
major functional group of plant microbodies. These organelles are
the subject of separate contributions to this volume from TOLBERT and
from NEWCOMB who respectively have contributed most to the present
understanding of their biochemistry and morphology.

MICROBODIES IN LEAVES

By sequential differential and stepped sucrose density gradient
centrifugation a fraction was isolated from a variety of leaves which
was highly enriched in glycolate oxidase and catalase (17,18). Draw-

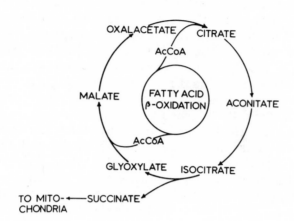

Figure 1 Reactions in the glyoxysome from castor bean endosperm (14). Succinate, the product of the reactions of the glyoxylate cycle, and also NADH$_2$ generated in the glyoxysome do not undergo further reactions there and are presumably metabolized in the mitochondria.

ing on the analogy with the preparations that had been isolated from liver TOLBERT refers to these organelles as peroxisomes. However, it was clear from the original (17) and later reports (18,19) that the enzymatic constitution of the microbodies from leaves was considerably more extensive than that of the liver peroxisomes and, consequently, that their metabolic role might be more important.

The presence of hydroxypyruvate (glyoxylate) reductase and specific transaminases was of particular interest since these enzymes can be assigned roles in the pathway by which glycolate is apparently metabolized in the light in green leaves (20). This pathway, whereby carbon from glycolate is converted to sugars, includes glyozylate, glycine, serine, hydroxypyruvate and glycerate as intermediates. As shown in *Figure 2* each of the reactions of this pathway (with the important exception of that in which 2 moles of glycine are converted to serine and CO$_2$) can occur in the microbodies (peroxisomes) isolated from leaves. Together, the reactions in which glycolate is produced in the light in the chloroplast and subsequently oxidized and converted into glycerate and CO$_2$ can be regarded as photorespiration. On this view glycolate is regarded as an incidental by-product of photosynthesis and a potential loss to the system. By the reactions outlined, including steps absorbing O$_2$ and yielding CO$_2$, three fourths of the glycolate carbon is recovered and reintroduced into the photosynthetic flow at the level of phosphoglycerate. The function

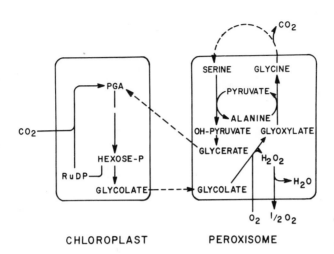

Figure 2 Representation of the major reactions in leaf peroxi-somes. Based on the work of TOLBERT (17,19).

of photorespiration in this view is the recovery of some fixed carbon which would otherwise be lost; clearly if glycolate were not produced in quantity and lost from the chloroplast in the first place the recovery would not be necessary and the reactions in the leaf peroxisome would then assume a proportionately greater significance in the production of glycine and serine for protein synthesis.

Regardless of whether such an interpretation prevails it is clear that the microbodies in leaves are the site of important reactions which are an adjunct to photosynthesis; the striking pictures provided by NEWCOMB'S group of microbodies in leaf cells suggest an intimate relationship with reactions in the chloroplast (21,22).

A general problem in the separation of organelles from cells is to find a grinding procedure which breaks open the cells but leaves the organelles intact. This problem is more aggravated when, as for microbodies, the organelles are particularly susceptible to mechanical and osmotic shocks. Moreover the recovery of intact and functional organelles by the present methods requires that the organelles and their enzymes remain stable during prolonged centrifugation through gradients. Clearly the chances of recovery of functional organelles are increased when the numbers are relatively large. This is so for glyoxysomes; they account for some 20% of the protein in the crude particulate fraction from 5 day castor bean endosperm. However, the particulate preparations from leaves are dominated by chloroplasts and the microbodies account for less than 1% of the particulate protein. This increases the difficulty of making pure preparations in good yield from leaves, and although visible bands of microbodies,

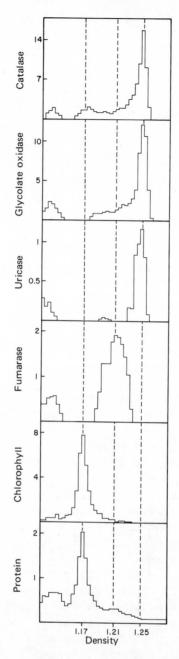

*Figure 3 Separation of the particulate fraction (2,000-10,000
g) of spinach leaves into chloroplast, (density 1.17) mitochondrial (1.21) and peroxisome (1.25) components on a linear sucrose gradient. Note that the chloroplast region contains most
of particulate protein. The mitochondrial and peroxisome regions
are defined by specific enzyme constituents. (HUANG and BEEVERS
unpublished).*

free of chlorophyll, can be obtained in linear gradients, the amount
of protein recovered is very small and a more satisfactory way of de-

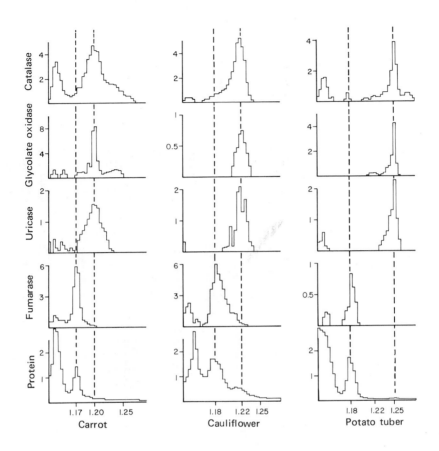

*Figure 4 Separation of microbodies from the particulate fract-
ions from potato tuber, carrot root and cauliflower buds on
linear sucrose gradients. A major protein band in each preparation is that due to mitochondria (fumarase marker). The microbody region is defined by the association of marker enzymes
(uricase, catalase, glycolate oxidase) at a lower point in the
gradient than the mitochondria. (HUANG and BEEVERS unpublished)*

fining the microbody fraction and estimating contamination by other
organelles is to plot the distribution of marker enzymes across the
linear gradient (*Figure 3*).

MICROBODIES IN OTHER PLANT TISSUES

HUANG and I have now applied this method to twelve different non-
green plant tissues, using catalase as the marker enzyme for microbodies and fumarase for mitochondria (*Figure 4*). From each of these
tissues particulate catalase was obtained and on centrifuging on a
linear sucrose gradient it was found to come to equilibrium at a density of 1.22-1.25. The distribution of glycolate oxidase and urate
oxidase over the gradient followed a closely similar pattern and from

TABLE 1

Enzymatic constituents of microbodies from plant and animal cells. In each instance the organelle fraction was obtained as a region of mean density about 1.25 on sucrose gradients. Note that catalase and glycolate oxidase are common constituents of each preparation and that uricase is usually present. *Tetrahymena*, a protozoon with a functional glyoxylate cycle, yields microbodies with only some of the enzymes of glyoxylate cycle. The specialized nature of the organelles from castor bean endosperm and green leaves is indicated. The relatively unspecialized microbodies from potato tuber are typical of those from ten other plant tissues examined. (Some of the information in this table is based on articles in the Annals of the New York Academy of Sciences, Volume 168, which is a report of a conference on microbodies, 1969.)

ENZYME	RAT LIVER	CHICKEN KIDNEY	TETRA-HYMENA	SPINACH LEAVES	CASTOR BEAN ENDOSPERM	POTATO TUBER
Catalase	■	■	■	■	■	■
α-OH Acid oxidase	■	■	■	■	■	■
Urate oxidase	■	■	■	■	■	■
D-Amino acid oxidase	■	■				
L-Amino acid oxidase						
Glyoxylate reductase	■			■		
Allantoinase			■			
Isocitrate dehydrogenase				■	■	
Transaminases						
Isocitrate lyase			■		■	
Malate synthetase			■		■	
Aconitase					■	
Malate dehydrogenase				■	■	
Citrate synthetase					■	
Acetothiokinase					■	
Fatty acyl thiokinase					■	
Acyl CoA oxidase					■	
β-OH Acyl CoA dehydrogenase					■	
Enoyl-CoA hydratase					■	
Thiolase					■	

most of the tissues these three enzymes were recovered together in a single band. The distribution of fumarase in the gradients coincided with the major visible protein band of density 1.17-1.20. The amounts of protein recovered in the region of the microbody marker enzymes were at best only 1-2% of the total applied as crude particulate suspension, but from potato and cauliflower visible and measurable protein peaks coincided with the peaks of enzyme activity. It thus appears that organelles with a density greater than that of mitochondria and containing enzymes distinctive to microbodies can be isolated from a wide variety of plant tissues. These organelles are clearly minor cell constituents and their function is unknown. It is of interest that their enzymic constitution is reminiscent of that of the microbodies from liver for which the name peroxisome was originally introduced.

    As indicated earlier, organelles with the morphological features of microbodies have been seen in sections of many kinds of plant cells (7,22). The present biochemical evidence allows the conclusion that the organelles can be separated from other cellular particles by virtue of their specific density, and that they have a few enzymes (catalase, glycolate oxidase) in common. In leaves the microbodies have other enzymes in addition and this fits them for an apparently important role in photorespiration. In fatty seedlings a different kind of specialization has occurred; the microbodies are present in large numbers and house the enzymes of the glyoxylate cycle and of β-oxidation. A summary of the enzymic constitution of microbodies from a variety of tissues is shown in *Table 1*.

DEVELOPMENT OF MICROBODIES

    The striking increase in activity of the enzymes of the glyoxylate cycle in the endosperm of the germinating castor bean was shown to be accompanied by a corresponding increase in the number of glyoxysomes, and the numbers declined in parallel with enzyme activity after fat utilization was complete (23). The precise timing of fat utilization and the confinement of this process to a particular tissue during a few days of the germination period is closely related to the genesis and destruction of the glyoxysomes. The present evidence suggests that the development of these organelles is not autonomous. No ribosomes have been observed in these or other microbodies and DNA was not detected in purified glyoxysomes (24). RNA was present but this may be ascribed to the membranes of the endoplasmic reticulum from which the microbodies originate (24).

    An interesting problem is encountered in the cotyledons of germinating cucurbits. In the early stages of germination fat is converted to sugar; the enzymes of the glyoxylate cycle are produced and glyoxysomes are present (25). If the seedlings are exposed to light in early growth, the cotyledons become green and the enzymes typical of leaf microbodies (glycolate oxidase, glyoxylate reductase) increase strikingly at the time the activities of the enzymes of the glyoxylate cycle are declining (25,26). This raises the possibility that in these tissues one kind of microbody, the glyoxysome, is converted to another, the leaf peroxisome. One piece of evidence against this possibility is the following. If light is withheld during growth at 30°C the declining phase of activity of the enzymes of the glyoxylate cycle is virtually completed by 10 days. Exposure to light after this stage nevertheless allows the development of chlorophyll and of the enzymes of the leaf peroxisome (25,26). Kagawa and I have now shown (26) that the decline of enzyme activity in the dark is accompanied by a marked loss of microbody protein (that recovered from a sucrose gradient at a density of 1.25). We interpret this to mean that, as in the castor bean endosperm, glyoxysomes are destroyed after fat utilization is completed. The provision of light results in some acceleration of the rate of loss of microbody protein. After 2 days of illumination, when the activity of enzymes of the leaf peroxisomes has been fully developed, the amount of microbody protein is only a small fraction of that present at the peak of glyoxysome development. The weight of this evidence is clearly against the possibility that glyoxysomes are transformed into leaf peroxisomes and is consistent with the view that the glyoxysomes are dispensed with and a new and smaller microbody population with different enzymic constitution is produced during greening. NEWCOMB and his colleagues have made an investigation of this problem using the electron microscope, and have come to a different conclusion on what I consider to be less definitive evidence (27).

Whatever the final resolution of this particular aspect of microbody ontogeny, it is quite clear that in leaves generally, peroxisomes are present and that they do not arise from glyoxysomes. FEIERABEND and I are examining the development of microbodies in the leaves of germinating wheat, using catalase, glycolate oxidase and glyoxylate reductase as marker enzymes. Low levels of these enzymes are produced in young leaves in darkness; most of the activity is particulate but occupies a position on the sucrose gradient corresponding to a density of about 1.2. When light is provided the enzyme activities increase rapidly and they are then recovered together in a sharp band of density 1.25. However, the development of these organelles is not directly related to that of the chloroplasts, since an amino triazole treatment which prevents chloroplast morphogenesis has little effect on the development of glycolate oxidase and glyoxylate reductase or on their association with the microbodies.

REFERENCES

1. J. RHODIN, *Aktiebolaget Godvil*, Stockholm, 1954.
2. C. ROUILLET and W. BERNHARD, *J. Biophys. Biochem. Cytol.*, 2 (1956) 355.
3. Z. HRUBAN and M. RECHCIGL, *Int. Rev. Cytol. Suppl.*, 1 (1969).
4. P. BAUDHUIN, H. BEAUFAY and C.E. DE DUVE, *J. Cell Biol.*, 26 (1965) 219.

5. C.E. DE DUVE and P. BAUDHUIN, *Physiol. Rev.*, 46 (1966) 323.
6. P. BAUDHUIN, *Ann. N.Y. Acad. Sci.*, 168 (1969) 214.
7. H.H. MOLLENHAUER, D.J. MORRÉ and A.G. KELLEY, *Protoplasma*, 62 (1966) 44.
8. R.W. BREIDENBACH and H. BEEVERS, *Biochem. biophys. Res. Commun.*, 27 (1967) 462.
9. H. BEEVERS, *Nature*, 191 (1961) 433.
10. G. LONGO, *Plant Physiol.*, 43 (1968) 660.
11. T.G. COOPER and H. BEEVERS, *J. biol. Chem.*, 244 (1969) 3507.
12. R.W. BREIDENBACH and H. BEEVERS, *Plant Physiol.*, 43 (1968) 705.
13. R. THEIMER and H. BEEVERS, *Plant Physiol.*, in press.
14. T.G. COOPER and H. BEEVERS, *J. biol. Chem.*, 244 (1969) 3507.
15. H.T. HORNER and J.J. ARNOTT, *Bot. Gaz.*, 127 (1966) 48.
16. E.L. VIGIL, *J. Cell Biol.*, 46 (1970) 435.
17. N.E. TOLBERT, A. OESER, T. KISAKI, R.H. HAGEMAN and R.K. YAMA-ZAKI, *J. biol. Chem.*, 243 (1968) 5179.
18. N.E. TOLBERT, A. OESER, R.K. YAMAZAKI, R.H. HAGEMAN and T. KIS-AKI, *Plant Physiol.*, 44 (1969) 135.
19. N.E. TOLBERT and R.K. YAMAZAKI, *Ann. N.Y. Acad. Sci.*, 168 (1969) 325.
20. N.E. TOLBERT, in Photosynthetic Mechanisms in Green Plants, NAS, NRC Publication, 1145 (1963) 648.
21. S.E. FREDERICK and E.H. NEWCOMB, *Science*, 163 (1969) 1353.
22. S.E. FREDERICK, E.H. NEWCOMB, E.L. VIGIL and W.P. WERGIN, *Planta*, 81 (1968) 229.
23. B.P. GERHARDT and H. BEEVERS, *J. Cell Biol.*, 44 (1970) 94.
24. B.P. GERHARDT and H. BEEVERS, *Plant Physiol.*, 44 (1969) 1475.
25. D.I. McGREGOR and H. BEEVERS, *Plant Physiol.*, 44s (1969) 33.
26. T. KAGAWA and H. BEEVERS, *Plant Physiol.*, 46s (1970) 38.
27. R.N. TRELEASE, W.N. BECKER and E.H. NEWCOMB, *Plant Physiol.*, 46s (1970) 38.

SECTION 4

Photorespiration and the Role of Microbodies

B. Research Papers and Assessments

INTERMEDIARY METABOLISM OF PHOTOSYNTHESIS IN RELATION TO CARBON DIOXIDE EVOLUTION IN SUNFLOWER[1]

C.A. Atkins[2], D.T. Canvin and H. Fock[3]

Department of Biology, Queen's University,
Kingston, Ontario, Canada.

SUMMARY

During steady rate photosynthesis by leaf discs of sunflower the substrate(s) of 'photorespiration' reach isotopic equilibrium at the same time and show similar turnover as the intermediates of the Calvin cycle, glycine and serine. The effects of O_2 concentration on the metabolism of sugar diphosphates and serine suggest that the CO_2 evolved in the light might be derived from these compounds. At least 75% of the total fixed carbon may pass through serine indicating that in this tissue it is a major intermediate of CO_2 fixation. Effects of O_2 and light suggest that serine might also be an intermediate of CO_2 evolution.

INTRODUCTION

Carbon dioxide evolution from green leaves in light (photorespiration) has been investigated from many different aspects (1) but neither the substrate(s) which yields the CO_2 nor the mechanism have been clearly established (1). Recent experiments with feeding $^{14}CO_2$ and gas analysis under steady state conditions (2,3) have delineated several properties of the substrate(s) which must be satisfied by any proposed compound(s). For example, the substrate(s) for CO_2 evolution in light is closely linked to photosynthesis (2,3,4,5). It becomes rapidly labeled when $^{14}CO_2$ is supplied to the leaf and is rap-

[1] *Supported in part by the National Research Council of Canada and by Deutsche Forschungsgemeinschaft.*
[2] *Present address: Plant Physiology Unit, CSIRO School of Biological Sciences, Macquarie University, North Ryde, Sydney, 2113, N.S.W., Australia.*
[3] *Present address: Botanisches Institut, Universität Frankfurt, Frankfurt a.M., Germany.*

idly turned over when CO_2 is subsequently supplied (4). The pool size of this immediate substrate(s) is small and, when labeled with ^{14}C, it is rapidly depleted and evolved as [14]CO_2 in a CO_2-free gas stream containing 21% O_2, but [14]CO_2 is not evolved in a CO_2-free gas stream containing 1% O_2 (5). We studied the metabolism of the carbon compounds of photosynthesis under conditions similar to those of gas exchange measurement to find which ones had properties consistent with those deduced from gas exchange experiments.

EXPERIMENTAL

Discs cut from sunflower leaves (*Helianthus annuus* L. 'Mennonite') were used (6,7). Measurement of gas exchange and [14]CO_2 feeding in the 'open system' employed have been described previously (2,3,6,7). After 15 min in [14]CO_2 the turnover of the labeled products was studied in a 10 min flushing period with [12]CO_2 in air, or CO_2-free gas containing 1% or 21% O_2. Extraction and fractionation of the water soluble products of photosynthesis from discs will be described elsewhere (7). The specific activity of organic acids was determined by radio-gas-liquid chromatography of their trimethylsilyl derivatives (6).

RESULTS AND DISCUSSION

During steady state photosynthesis in 300 ppm CO_2 air labeled with [14]CO_2, the CO_2 evolved from the leaf was saturated with [14]C 10 min after [14]CO_2 was supplied to the leaf and it was no longer radioactive 10 min after [12]CO_2 was subsequently applied (5). Any compound that is proposed as the substrate must exhibit kinetics consistent with this pattern.

The labeling of some of the early products of [14]CO_2 fixation are shown in *Figure 1*. Since the feedings were done under steady state conditions with constant specific activity [14]CO_2 it is reasonable to

TABLE 1

Specific activity of PGA, malate and succinate ($\times 10^4$ dpm μg^{-1} carbon) after photosynthesis in [14]CO_2. Conditions as specified for *Figure 1*.

	Specific Activity
[14]CO_2 supplied during photosynthesis	9.20
PGA after 10 min photosynthesis	9.50
Malate after 15 min photosynthesis	0.15
Succinate after 30 min photosynthesis	0.007

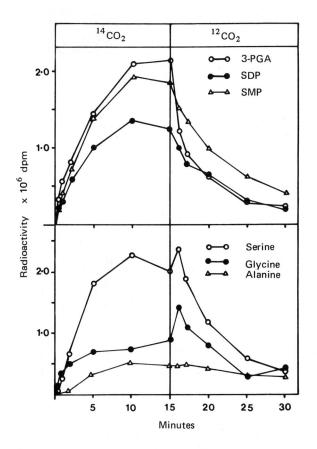

Figure 1 Labelling kinetics and turnover of some of the inter-mediates of photosynthesis in leaf discs of sunflower. 0.55 dm^2 of tissue was used in a 40.5 cm^3 cuvette in an open gas exchange system. The $^{14}CO_2$ gas stream contained 270 ppm CO_2, 21% O_2 and $^{14}CO_2$ at a specific activity of 9.2 x 10^4 dpm μg^{-1} carbon. The $^{12}CO_2$ gas stream was identical but contained no $^{14}CO_2$. All gases were supplied at 500 ml min^{-1} and 52% relative humidity. The temperature was maintained at 21° and the light intensity was 2,200 ft-c. Before the labeled gas stream was introduced the tissue was held for 10 min under the same conditions used in the feeding period. The steady rate of net photosynthesis was 10.2 ± 0.2 mg CO_2 fixed dm^{-2} hr^{-1}. 3-PGA = phosphoglyceric acid, SDP = sugar diphosphates and SMP = sugar monophosphates.

assume that the leveling off in ^{14}C content indicates that these compounds were saturated with ^{14}C. The saturation of 3-phosphoglycerate (3-PGA), however, was confirmed by a direct specific activity determination (*Table 1*). Malate and succinate from the same tissue are included for comparison. The results for 3-PGA show quite clearly that it was equal in specific activity to the supplied $^{14}CO_2$ and

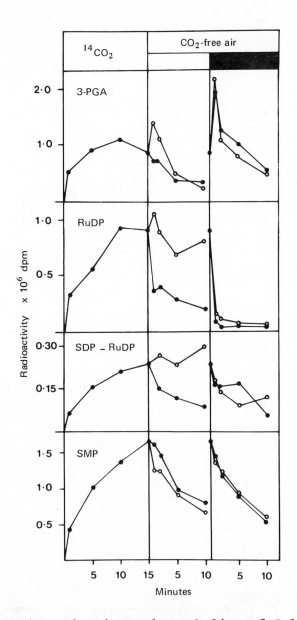

Figure 2 Labeling kinetics and metabolism of Calvin cycle intermediates in leaf discs of sunflower. CO_2-free gas contained
less than 0.5 ppm CO_2 with 21% O_2 (closed circles) or 1% O_2
(open circles). All other conditions as in Figure 1. RuDP =
ribulose diphosphate with other abbreviations as for Figure 1.

that there were no significant unlabeled pools of 3-PGA in this tissue. In the subsequent flushing period with $^{12}CO_2$ air, the intermediates except alanine, were depleted of label. Thus, intermediates of
the Calvin cycle and glycine and serine satisfied the predicted characteristics of the 'photorespiratory substrate': namely, isotopic
equilibrium with $^{14}CO_2$ after 10 min photosynthesis and rapid carbon

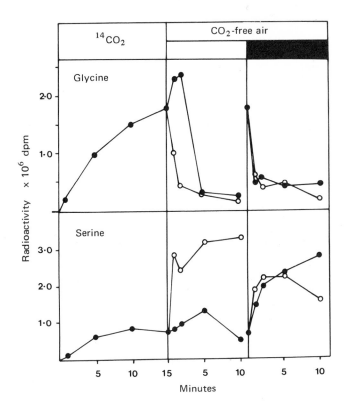

Figure 3 Labeling kinetics and metabolism of glycine and ser-
ine in CO$_2$-free gas with 21% O$_2$ (closed circles), or 1% O$_2$ (open
circles). All other conditions of the experiment as in Figure 1.

turnover.

When discs were placed in light in CO$_2$-free air, after 15 min in
^{14}CO$_2$, ^{14}CO$_2$ was evolved and this evolution was markedly inhibited in
low O$_2$ (6,7). For example, in a typical experiment 3.13 x 10^6 dpm
^{14}CO$_2$ was evolved in 10 min in 21% O$_2$ but only 0.11 x 10^6 dpm in 1%
O$_2$. The difference in ^{14}C between the amount of ^{14}CO$_2$ evolved in 21%
and 1% O$_2$ should be in the substrate(s) for 'photorespiration' and we
have examined several compounds for this difference. The only ones
which contained more ^{14}C after flushing in 1% O$_2$ in contrast to 21%
O$_2$ were the sugar diphosphates, including ribulose-1,5-diphosphate
(RuDP) (*Figure 2*), and serine (*Figure 3*). Also they contained more
than enough additional ^{14}C to account for the different amounts of
^{14}CO$_2$ evolved in the two gas streams.

BASSHAM and KIRK (8) also found increased levels of ^{14}C-sugar diphosphates in N$_2$ compared to O$_2$. Glycolate production was depressed
and they suggested that the rise in diphosphates was due to an absence of competition for carbon by glycolate synthesis. We did not
isolate glycolate quantitatively or completely free of a number of
other labeled acids, so a comparison with the observed changes in the
sugar phosphates could not be made. However, if the reduced carbon

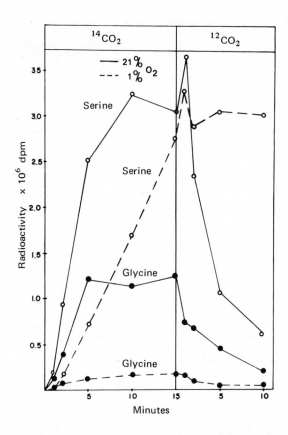

Figure 4 Labeling kinetics and metabolism of glycine and ser-
ine in leaf discs of sunflower with the gas containing 21% O_2
(unbroken lines) or 1% O_2 (broken lines). The CO_2 concentrat-
ion was 386 ppm in 21% O_2 and 404 ppm in 1% O_2 and the lab-
eled gas stream contained 6.5 x 10^4 dpm.$\mu g^{-1}C$. All other ex-
perimental conditions as in Figure 1. The steady rates of net
CO_2 uptake were 20.9 mg.$dm^{-2}.hr^{-1}$ in 21% O_2 and 26 mg.$dm^{-2}.hr^{-1}$
in 1% O_2.

flow through the diphosphates in 1% O_2 led to lower glycolate production then this might have resulted in reduced $^{14}CO_2$ evolution, but
only if glycolate metabolism is the source of CO_2 in the light (9-
11). The accumulation of ^{14}C-serine suggests that oxygen may not only affect the formation of serine (see later) but also the subsequent
metabolism of serine. TOLBERT (10,11) has suggested that the conversion of 2 glycines to serine and CO_2 is the reaction yielding the CO_2
in the light and MARKER and WHITTINGHAM (12) have indeed shown that
when (1-^{14}C) glycolate and glycine were supplied to illuminated pea
leaves about 25% of the label was recovered in CO_2. We show (*Figure*
3) that the ^{14}C accumulated in serine in 1% O_2 was about equal to
that which was lost from glycine but the expected amount of $^{14}CO_2$ was
not observed. Thus, either $^{14}CO_2$ was produced and not evolved or no
$^{14}CO_2$ was produced and serine was formed by a pathway separate to

that of glycine.

When the discs were flushed in the dark with 1% or 21% O_2 there was no effect on the metabolism of any of the compounds examined (*Figures 2,3*). Apart from a post-illumination CO_2 burst in 21% O_2 there was also little effect of O_2 on $^{14}CO_2$ evolution (2.24 x 10^6 dpm $^{14}CO_2$ evolved in 10 min in 21% O_2 and 1.27 x 10^6 dpm in 1% O_2). The accumulation of ^{14}C-serine in the dark (*Figure 3*) is consistent with previous studies (13-15) which have shown that while the reactions from glycolate to serine proceeded in the dark, subsequent metabolism of serine required light.

In another investigation of glycine-serine metabolism the tissue was given $^{14}CO_2$ in a gas stream containing either 1 or 21% O_2 for 15 min. This was followed by a flushing period of 10 min in $^{12}CO_2$ with the same O_2 concentrations used in each case (*Figure 4*). The total water soluble pool sizes of glycine and serine and their ^{14}C content in the tissue after photosynthesis in the two O_2 concentrations were measured. In 21% O_2 there was 50 µg of glycine per sample (16 µg carbon); in 1% O_2 this was reduced to 10 µg (3.2 µg carbon). The opposite occurred for serine where the pool size increased from 120-160 µg (41-55 µg carbon) in 21% O_2 to 400-500 µg (137-171 µg carbon) in 1% O_2. The specific activity of the $^{14}CO_2$ supplied was 0.65 x 10^5 dpm/µg carbon. Thus if the water soluble glycine in the tissue was saturated it would be expected to contain about 1 x 10^6 dpm in 21% and about 0.2 x 10^6 dpm in 1% O_2. *Figure 4* shows that values similar to these were obtained suggesting that glycine was saturated, or nearly so, and if there was an unlabeled soluble pool it was relatively small. In 21% O_2 about 3 x 10^6 dpm would be expected in serine at saturation and this was found (*Figure 4*). In 1% O_2, however, around 1 x 10^7 dpm might be expected. About 30% of this value was found (*Figure 4*) indicating, as the labeling kinetics show, that the carbon flow into serine was markedly reduced.

These results support our earlier assumptions that glycine and serine were saturated with ^{14}C after photosynthesis in $^{14}CO_2$ with 21% O_2 and that both compounds turn over rapidly at this O_2 concentration. The results in 1% O_2 support the idea that both serine biosynthesis and serine utilization (see also *Figure 3*) are suppressed at low O_2 concentrations. If the CO_2 that is evolved in light is produced by a pathway involving the metabolism of serine (10,11) this would suggest that there is no CO_2 production at low O_2 concentrations rather than a production and refixation of CO_2.

Using whole leaves of sunflower, LUDWIG (2,3) has measured rates of CO_2 evolution in light into normal air equal to 20% of the rate of total carbon fixation. Bearing in mind that this CO_2 is equilibrated with the supplied CO_2 within 10 min and allowing for some refixation it seems reasonable to conclude that at least 25% of the CO_2 fixed in photosynthesis is concurrently evolved. Thus carbon flow through substrate(s) must be adequate to allow for such rates. For example, if the CO_2 is derived from the synthesis of serine by the glycolate pathway as proposed by TOLBERT (11) all the carbon fixed would be ex-

pected to pass through glycine and 75% through serine with 25% released as CO_2. While it has been previously estimated that 50% of the
total carbon fixed in photosynthesis may flow through the glycolate
pathway (11,16) this is not sufficient to produce the measured amount
of CO_2 evolved if CO_2 is produced as suggested by TOLBERT (11). If
all the carbon fixed in photosynthesis flows through the glycolate
pathway then we can no longer view this pathway as an adjunct to the
Calvin cycle but must incorporate it completely into the carbon fixation scheme for photosynthesis.

Although we have tried to determine carbon flow through glycine
and serine we have not yet been successful. Firstly, we do not know
the specific activity of the precursor of these compounds and secondly our kinetic data on glycine and serine labeling gives an unsatisfactory and confused picture. The ^{14}C content (and presumably pool
size) of serine usually exceeds the ^{14}C content of glycine (*Figures
1,4*) but in some cases the opposite is found (*Figure 3*). The early
labeling patterns are, in some cases, consistent with glycine being
the precursor of serine (*Figures 1,3* and 1% O_2 *Figure 4*) but in one
instance the opposite is found (21% O_2, *Figure 4*). In spite of these
difficulties the data can be used to derive some estimate of carbon
flow through serine. Assuming that the precursor is at least equal
to, or lower in, specific activity than 3-PGA at 1 min, there is in
the second minute a net flow of 20 µg carbon into serine (*Figure 1*).
This is nearly 80% of the net fixation rate in that experiment
(52 µg carbon min^{-1} dm^{-2}). *Figure 4*, using the same assumptions,
shows a net flow of 36 µg carbon into serine in the same time period
and this rate is 60% of the net rate of CO_2 fixation (58 µg carbon/
minute). Thus carbon flow through serine is indeed in the region of
75% of the rate of CO_2 fixation and, coupled with the effects of
light and oxygen, makes the metabolism of serine a likely prospect as
a CO_2-yielding pathway in 'photorespiration'. It is not yet clear if
serine is totally derived from glycine via the glycolate pathway (11)
or from other pathways (14,17). If it is derived from the glycolate
pathway then relatively little carbon can be lost by the nonenzymic
oxidation of glyoxylate which was suggested by ZELITCH (9) as then
the rate of carbon flow through glycolate would have to exceed the
rate of carbon fixation in photosynthesis. Our results are consistent with a carbon flow through serine nearly equal to the carbon fixation rate of photosynthesis. Thus serine is a major intermediate in
photosynthetic CO_2 fixation and may be a major intermediate in CO_2
evolution in light.

REFERENCES

1. W.A. JACKSON and R.J. VOLK, *A. Rev. Pl. Physiol.*, 21 (1970) 385.
2. L.J. LUDWIG, Ph.D thesis, Queen's University, Kingston, Ontario
   (1968).
3. L.J. LUDWIG and D.T. CANVIN, *Can. J. Bot.*, submitted, 1971.
4. A.L. D'AOUST, Ph D thesis, Queen's University, Kingston, Ontario
   (1970).

5. L.J. LUDWIG, G. KROTKOV and D.T. CANVIN, in Progress in Photosynthesis Research, ed. H. METZNER, IUBS, Tübingen, I (1969) 494.
6. C.A. ATKINS, Ph D thesis, Queen's University, Kingston, Ontario
   (1969).
7. C.A. ATKINS and D.T. CANVIN, *Can. J. Bot.*, submitted, 1971.
8. J.A. BASSHAM and M. KIRK, *Biochim. biophys. Acta*, 9 (1962) 375.
9. I. ZELITCH, *Plant Physiol.*, 41 (1966) 1623.
10. T. KISAKI and N.E. TOLBERT, *Plant Physiol.*, 44 (1969) 242.
11. N.E. TOLBERT and R.K. YAMAZAKI, *Ann. N.Y. Acad. Sci.*, 168 (1969)
    325.
12. A.F.H. MARKER and C.P. WHITTINGHAM, *J. exp. Bot.*, 18 (1969) 732.
13. B.J. MIFLIN, A.F.H. MARKER and C.P. WHITTINGHAM, *Biochim. bio-
    phys. Acta*, 120 (1966) 266.
14. R. RABSON, N.E. TOLBERT and P.C. KEARNEY, *Archs. Biochem. Bio-
    phys.*, 98 (1962) 823.
15. A. ONGUN and C.R. STOCKING, *Plant Physiol.*, 40 (1965) 823.
16. I. ZELITCH, *J. biol. Chem.*, 240 (1965) 1869.
17. D.C. MORTIMER, *Can. J. Bot.*, 38 (1960) 623.

GLYCOLATE PATHWAY IN ALGAE[1]

N.E. Tolbert , E.B. Nelson[2] and W.J. Bruin[3]

Department of Biochemistry, Michigan State University,
East Lansing, Michigan 48823, U.S.A.

SUMMARY

The status of glycolate metabolism in algae is reviewed. Some information on the activity of glycolate pathway enzymes and the labeling of glycolate from $^{14}CO_2$ in algae is provided.

In algae, as in higher plants, uniformly labeled glycolate is formed photosynthetically in large amounts by unknown mechanisms and for unknown reasons (1,2). In fact most of the earlier photosynthetic investigations were done with algae and thus included work on glycolate formation and metabolism (3). Although the mechanism of glycolate biosynthesis is unknown, no differences in physiological conditions for its formation are apparent for algae or higher plants. By 1956 it was learned that much of the glycolate produced by laboratory cultures of algae was excreted (4), and to some extent by unicellular algae in nature (5,6). Later a very active P-glycolate phosphatase was found in both higher plants (7,8) and algae (1,7). This specific phosphatase in chloroplasts is postulated as being involved in glycolate formation and excretion and is discussed in more detail elsewhere in this symposium. It thus appears that chloroplasts and algae are *symbolically* similar in that both form and excrete glycolate during photosynthesis at high O_2 concentration, in high light intensity, at low CO_2 concentration, and at high pH (9.10). Our failure to detect glycolate oxidase in some unicellular green algae (2) seemed consistent with these scheme, and it has been proposed that the ex-

[1] *This work was supported in part by National Science Foundation Grant GB-17543, and published as Journal Article No. 5356 of the Michigan Agricultural Experiment Station.*
[2] *Present address: Department of Biochemistry, Southwestern Medical School at Dallas, Dallas, Texas 75235, U.S.A.*
[3] *Present address: Department of Physiological Chemistry, University of Wisconsin Medical School, Madison, Wisconsin 53706, U.S.A.*

creted glycolate is an end product of photosynthesis.

Other data are interpreted to mean that part of the glycolate formed in algae is metabolized within the cell. Labeling patterns for glycine and serine formation during CO_2 fixation are consistent with metabolism by the glycolate pathway in an analogous manner to that in higher plants. Although added glycolate is not readily absorbed by algae, some ^{14}C-labeled glycolate is converted to labeled glycine and serine in a manner predicted by the glycolate pathway (1, 3). Synchronized cultures of *Scenedesmus* will take up and metabolize exogenous glycolate-^{14}C during the dark phases of cell division but not during the light stages of growth. Glycolate excretion, on the other hand, occurs during the light stages but not during the dark stages of growth (1,11,12). Further, other investigators reported glycolate oxidation by algal extracts, although the enzyme was not fully characterized (13,14,15). It is now realized that the algal enzyme in most cases is not an oxidase but a glycolate dehydrogenase of relatively low specific activity, which is regulated by CO_2 availability in the growth media (1,16,17).

Present concepts are summarized as follows. Algae biosynthesize glycolate as do chloroplasts. The glycolate is mostly excreted if the algae had been grown in a media aerated with high CO_2 (0.2% CO_2 or more), as is the case with most laboratory cultures. Algae can also metabolize part of the glycolate by the same metabolic pathway as in higher plants. If grown on air (low CO_2), the algae metabolize all the glycolate and do not excrete it. This has been proven by [14]CO_2 fixation products, by feeding glycolate-^{14}C, by degrading the products glycine, serine and glycerate and by isolating the enzymes for each step (1,2,18). The algae contain P-glycolate phosphatase, glycolate dehydrogenase, L-glutamate:glyoxylate aminotransferase, serine hydroxymethylase, hydroxypyruvate reductase, glycerate kinase and 3-phosphoglycerate (3-PGA) kinase, all enzymes catalyzing consecutive steps of the glycolate pathway (10) as described in *Figure 1* of my paper on leaf peroxisomes.

Several significant differences in glycolate metabolism between algae and higher plants exist. The major difference is in the enzymatic oxidation of glycolate as summarized in *Table 1*. In leaves glycolate oxidase is a flavoprotein catalyzing O_2 uptake and generating H_2O_2, and its activity in leaf homogenates of C_3-plants is 20 to 60 nmoles min^{-1} mg^{-1} protein. This oxidase is located in the leaf peroxisomes along with a great excess of catalase to destroy the H_2O_2. The activity of the algal enzyme in crude extracts is as low as 1/10 that found in leaves. In contrast to the leaf oxidase, the algal enzyme is a dehydrogenase, which couples to the reduction of the dye, dichlorophenolindophenol (DCPIP), during the oxidation of glycolate to glyoxylate, but this oxidation does not couple directly to oxygen and no H_2O_2 is produced. As an apparent correlation, no peroxisomes have been observed in electron micrographs or isolated on sucrose gradients from phototropically grown unicellular algae. In algae, catalase levels are also about 1/10 those found in higher plants (1).

TABLE 1

Comparison of glycolate oxidation by leaves and algae

	Leaves	Unicellular green algae
Enzyme	glycolate oxidase (glycolate:O_2 oxidoreductase)	glycolate dehydrogenase (glycolate:DCPIP oxidoreductase)
Electron acceptor	oxygen	DCPIP
Products	glyoxylate+H_2O_2	glyoxylate+DCPIPH_2
Rate[*]	20 - 60	3 - 11
Lactate oxidation	L-lactate (66%)	D-lactate (66%)
Inhibition by KCN	none at 10^{-2} M	by 10^{-3} M
Inhibition by PCMB[**]	partial at 10^{-2} M	by 10^{-5} M
Cellular location	peroxisomes	unknown

[*]nmoles x min^{-1} x mg^{-1} protein in extracts
[**] p-chloromercurobenzoate

The absence or presence of only low rates of photorespiration in algae (19,20,21) seem consistent with the absence of glycolate oxidation by O_2. O_2 inhibition of photosynthesis by algae is also less
pronounced (22). Thus glycolate excreted by chloroplasts in higher
plants is oxidized in leaf peroxisomes by the glycolate pathway (10,
see TOLBERT this volume) to account in large part for photorespiration, while the glycolate excreted by algae is lost in the media or
utilized by other organisms. Only respiration associated with glycolate formation should be observed with algae. It is also possible
that the ultimate electron acceptor from the glycolate dehydrogenase
may be oxygen.

Glycolate oxidase of leaf peroxisomes is characterized by O_2 uptake during the oxidation of glycolate or L-lactate, but not D-lact-
ate; it is insensitive to cyanide and sulfhydryl reagents. In algal
oxidation, glycolate dehydrogenase catalysis does not transfer the e-
lectrons to O_2 but only to the dye. It oxidizes glycolate or D-lact-
ate, but L-lactate is oxidized only slowly. It is inhibited by 10^{-4}
KCN and by sulfhydryl reagents. Consequently by these several differences one can differentiate between the two enzymes in plant and
algae homogenates (17). So far we have found only glycolate oxidase
in higher vascular plants including liverwort, aquatic plants and
green leaves of all land plants. In the few unicellular green algae
that we examined, only glycolate dehydrogenase has been found. CODD,
LORD and MERRETT (23) have confirmed that their *Chlorella* enzyme is a
glycolate dehydrogenase and not an oxidase, as first reported (14).
Already one exception in the algae seems to be a yellow *Chlorella* mu-

TABLE 2

Specific activity (nmoles min^{-1} mg^{-1} protein) of enzymes of the glycolate pathway in algal extracts*

Enzyme	Specific activity
Hydroxypyruvate reductase	150–500
L-Glutamate-glyoxylate aminotransferase	50–100
P-Glycolate phosphatase	30–60
Glycolate dehydrogenase	3–11
Serine hydroxymethylase	3–6
Catalase	2.4×10^4

* In most cases *Chlorella*, *Scenedesmus* and *Chlamydomonas* were examined.

tant which contains a glycolate oxidase that links directly to O_2 (24).

Other major differences between the glycolate pathway in algae and leaves of higher plants are in the specific activities of the enzymes (*Table 2*). The absence of the H_2O_2 producing glycolate oxidase is consistent with the absence of peroxisomes in algae and about 1/10 as much catalase as in leaves. In the green leaf peroxisomal pathway, glyoxylate:glutamate aminotransferase has a lower specific activity than other enzymes involved in glycolate metabolism (10). Among the algal enzymes of the glycolate pathway, glycolate dehydrogenase and serine hydroxymethylase have the lowest activity (*Table 2*), although they are very active in the leaf. The low glycolate dehydrogenase activity seems to be correlated with regulation of how much of the glycolate is metabolized or excreted. The low serine hydroxymethylase, which interconverts serine and glycine, is consistent with the labeling patterns found after short exposures to $^{14}CO_2$ in these two amino acids (*Table 3*). In higher C_3-plants serine and glycine are uniformly labeled at short time periods, when 3-PGA is still carboxyl labeled (25). This has been interpreted to show the rapidity of carbon flow from uniformly labeled glycolate through the pathway to uniformly labeled serine. On the other hand, serine and glycine are formed also in algae during photosynthesis, but the serine pool appears to be labeled first and it is labeled predominantly in the carboxyl group with nearly the same distribution as in 3-PGA (1,2). Glycine is nearly uniformly labeled, and must have come mainly from glycolate. A detailed examination of labeling patterns in glycolate and glycine after 2 to 10 sec of $^{14}CO_2$ fixation indicates that some carboxyl-labeled serine is also contributing to the initial formation of carboxyl-labeled glycine (1). The initial label in glycolate, as

TABLE 3

^{14}C Labeling patterns after 4 to 10 sec of $^{14}CO_2$ fixation in light (percentage distribution)

| | | Higher Plants | | Algae | |
| | | * | ** | *** | *** |
		Tobacco	Soybean	Chlorella	Chlamydomonas
Glycolate	$\overset{\bullet}{C}OOH$	52		35	30
	CH_2OH	48		65	70
Glycine	$\overset{\bullet}{C}OOH$	50		50	49
	CH_2NH_2	50		50	51
Serine	$\overset{\bullet}{C}OOH$	40	28	78	69
	$\overset{\bullet}{C}HNH_2$	40	38	6	7
	CH_2OH	20	34	16	24
3-PGA or	$\overset{\bullet}{C}OOH$	65	74	82	79
glycerate	$\overset{\bullet}{C}HOH$	16	11	3	4
	$CH_2O\,\textcircled{P}$	19	15	15	17

  * From (28)
 ** From (25) after 20 sec of $^{14}CO_2$ fixation
*** From (1), (11)

shown in *Table 3*, is similar to the top two carbon atoms of the sugar phosphates, which along with glycolate, quickly become equally label- ed. These data are consistent with a flow of carbon from the photo- synthetic carbon cycle into glycine and serine by two routes depicted earlier (TOLBERT, this volume). In algae uniformly labeled glycine is formed from glycolate, but its conversion to serine is slow. Car- bon also flows from carboxyl-labeled 3-PGA, to glycerate, to hydroxy- pyruvate and to carboxyl-labeled serine. In the higher plant both of these pathways are in the peroxisomes and carbon flows rapidly from glycolate all the way to serine and even on to glycerate. The exact significance of these differences reflected by the labeling patterns and enzyme activity needs to be further examined.

Glycolate excretion by green algae has been an intriguing problem because so many questions about it have not been answered. How sign- ificant is it in nature and how much does glycolate contribute to the growth of other organisms? In laboratory cultures of algae, glycol- ate excretion is so enormous that it represents a serious drain on net photosynthesis. In fact, cultures in the light in the absence of added CO_2, continue to excrete glycolate at the expense of their stored reserves (26). This is indeed also true for photorespiration in higher plants, but in this case only one CO_2 is lost in converting two glycolates to one serine or glycerate. Since both phenomena ap- pear wasteful to the photosynthetic cell, their regulation seems nec- essary. Glycolate formation is regulated by the environmental conditions, being highest in the presence of high light, high O_2 and low CO_2. Another method is to prevent glycolate excretion, and this occurs if algae are grown on the low CO_2 levels supplied by aeration only with air.

For algae, CO_2 concentrations effect the glycolate pathway in two ways, which can occur together and cause some confusion in interpretation of data. Low levels of CO_2 stimulates glycolate biosynthesis during photosynthesis. During continued growth on low CO_2 (air) glycolate dehydrogenase activity increases, glycolate metabolism by the cells increases, and glycolate excretion stops over a time period of hours. A low CO_2 concentration, at the time of the photosynthetic measurement of $^{14}CO_2$ fixation, increases glycolate biosynthesis and the percentage ^{14}C in glycolate, glycine and serine, while a high CO_2 concentration decreases the percentage ^{14}C in these products. This latter effect is reversed as rapidly as the CO_2 concentration changes, and apparently is related to glycolate formation.

The effect of CO_2 level during growth of *Chlamydomonas reinhardii* was investigated by NELSON and TOLBERT (16). When grown on air supplemented with 1% CO_2, the algae secreted 0.7 to 1.0 mmoles glycolate min^{-1} mg^{-1} chlorophyll, but when transferred to air without CO_2 the excretion rate dropped to zero in about 15 hrs. Thereafter the *Chlamydomonas* grew much slower, but they did not excrete any glycolate. Glycolate metabolism by the algae was occurring, for if this were inhibited by isonicotinyl hydrazide, the air grown culture also excreted glycolate. Concurrent with changes in glycolate excretion rates, the activity of two enzymes, glycolate dehydrogenase and carbonic anhydrase also changed. Activity of P-glycolate phosphatase, involved in glycolate synthesis, did not change. When the algae grown on 1% CO_2 were being adapted to growth on air, the glycolate dehydrogenase activity increased about 3-fold in 10 to 12 hours. Levels of carbonic anhydrase activity were 20-fold higher, when the algae were grown on air rather than on 1% CO_2 in air (27). The changes in glycolate dehydrogenase were inhibited by cycloheximide, indicative of protein synthesis. When air grown algae were given air supplemented with 1% CO_2, glycolate dehydrogenase activity was repressed. Thus, CO_2 in the growth medium had a reversible effect upon the level of the dehydrogenase activity. Glycolate excretion paralleled these changes. In the absence of adequate CO_2, glycolate production occurred but it was all internally metabolized by increased levels of glycolate dehydrogenase. In the presence of excess CO_2, the dehydrogenase was repressed and much of the glycolate was excreted. No similar effect has been reported on the regulation of leaf photorespiration and the peroxisomal glycolate oxidase. The low levels of CO_2 in air and in natural waters suggest that in nature glycolate excretion is not nearly as great as in the algal cultures growing in the laboratory on excess CO_2. However, the ecologists must evaluate the effect of increased CO_2 in waters on glycolate excretion and glycolate feeding of the phytoplankton.

The differences described here between glycolate metabolism in algae and higher plants bears on the speculative question of the origin and function of peroxisomes. Since algae do not have glycolate oxidase and peroxisomes, this particle may have developed in land plants in response to the altered environment, such as the high level of oxygen encountered in the air. That glycolate formation and metabolism occurs in both algae and plants during photosynthesis, indicates that it is an obligated process, even if not understood. Further surveys of the algae and plants for differences in glycolate metabolism may offer more clues in this evolutionary problem. Enzymatic differences between glycolate oxidase and glycolate dehydrogenase make these surveys feasible through relatively simple enzyme assays.

REFERENCES

1. W.J. BRUIN, E.B. NELSON and N.E. TOLBERT, *Plant Physiol.*, 46 (1970) 386.
2. J.L. HESS and N.E. TOLBERT, *Plant Physiol.*, 42 (1967) 371.
3. L. SCHOU, A.A. BENSON, J.A. BASSHAM and M. CALVIN, *Physiologia Pl.*, 3 (1950) 487.
4. N.E. TOLBERT and L.P. ZILL, *J. biol. Chem.*, 222 (1956) 895.
5. G.E. FOGG, *Br. phycol. Bull.*, 2 (1963) 195.
6. G.E. FOGG, D.J. EAGLE and M.E. KINSON, *Vert. int. Verein Liminol*, 17 (1969) 480.
7. D.D. RANDALL, N.E. TOLBERT and D. GREMEL, *Plant Physiol.*, (1971), submitted.
8. K.E. RICHARDSON and N.E. TOLBERT, *J. biol. Chem.*, 236 (1961) 1285.
9. N.E. TOLBERT, in "Photosynthetic mechanism in green plants", NAS, NRC Publication 1145 (1963) 648.
10. N.E. TOLBERT, *A. Rev. Pl. Physiol.*, 22 (1971) in press.
11. W.H. CHANG and N.E. TOLBERT, *Plant Physiol.*, 46 (1970) 377.
12. H. GIMMLER, W. ULLRICH, J. DOMANSKI-KADEN and W. URBACH, *Plant Cell Physiol.*, 10 (1969) 103.
13. W.J.S. DOWNTON and E.B. TREGUNNA, *Plant Physiol.*, 43 (1968) 923.
14. M.J. LORD and M.J. MERRETT, *Biochim. Biophys. Acta*, 159 (1968) 543.
15. I. ZELITCH and P.R. DAY, *Plant Physiol.*, 43 (1968) 289.
16. E.B. NELSON and N.E. TOLBERT, *Biochim. biophys. Acta*, 184 (1969) 263.
17. E.B. NELSON and N.E. TOLBERT, *Archs. Biochem. Biophys.*, 141 (1970) 102.
18. M.J. LORD and M.J. MERRETT, *Biochem. J.*, 117 (1970) 929.
19. D.L. BROWN and E.B. TREGUNNA, *Can. J. Bot.*, 45 (1967) 1135.
20. G. HOCH, O.v.H. OWENS and B. KOK, *Archs. Biochem. Biophys.*, 101 (1963) 171.
21. I. ZELITCH and P.R. DAY, *Plant Physiol.*, 43 (1968) 1838.
22. O. BJÖRKMAN, *Carnegie Inst. Wash. Yearbook*, 65 (1967) 446.
23. G.A. CODD, M.J. LORD and M.J. MERRETT, *FEBS Letters*, 5 (1969) 341.

24. G.H. SCHMID and P. SCHWARZE, Hoppe Seylers Z. *Physiol. Chem.*, 350
    (1969) 1513.
25. R. RABSON, N.E. TOLBERT and P.C. KEARNEY, *Archs. Biochem. Bio-
    phys.*, 98 (1962) 154.
26. J.L. HESS, N.E. TOLBERT and L.M. PIKE, *Planta*, 74 (1967) 278.
27. E.B. NELSON, A. CENEDELLA and N.E. TOLBERT, *Phytochemistry*, 8
    (1969) 2305.
28. J.L. HESS and N.E. TOLBERT, *J. biol. Chem.*, 241 (1966) 5705.

LIGHT STIMULATED RESPIRATORY GAS EXCHANGE IN ALGAE AND ITS RELATION TO PHOTORESPIRATION

Wolfgang Kowallik

Botanisches Institut der Universität Köln,
5000 Köln-Lindenthal, Germany

SUMMARY

Several observations suggest that the reduced rate of O_2 evolution in algae during photosynthesis in blue light is due to increased respiratory O_2 uptake. The blue light response is most clearly seen in a yellow, non-photosynthetic mutant of Chlorella. Under anaerobic conditions this organism also shows a blue light stimulation of acid excretion. Glycolate is among the acids and glycolate oxidase is active in this alga. However, a number of observations indicate that there are pronounced differences between light-enhanced respiration in higher plants and that in unicellular algae.

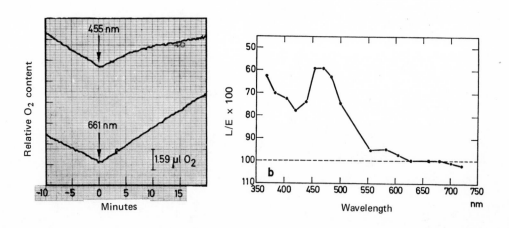

Figure 1 (left) Oxygen exchange of a suspension of Chlorella pyrenoidosa in darkness and in blue or red light.

Figure 2 (right) Action spectrum for a drop in apparent O_2 release during the first minutes of photosynthesis of Chlorella; expressed as ratio of [O_2 release(15-20 min)/O_2 release(0-5 min)] x 100 = L/E x 100.

Under photosynthetic conditions the O_2 output of *Chlorella pyre-noidosa* is constant for 20 min in red light. In blue light of equal quantum flux it is almost as high initially but decreases to a smaller steady rate in the course of 5-10 min *(Figure 1)*. This drop in O_2 release during photosynthesis in blue light is most pronoun-ced at wavelengths around 460-470 nm and in near ultraviolet *(Fig-ure 2)* (1).

Such a decrease in apparent O_2 liberation could depend on some unspecific photoxidation, on a gradual decrease in photosynthesis, or on increasing respiratory O_2 uptake in blue light. The latter assumption is supported by several observations. Firstly, if res-piration is stimulated to a maximal value by addition of exogenous glucose, the gradual drop in O_2 output in blue light is not observ-ed. Secondly, if photosynthesis is inhibited by 10^{-5} M DCMU the en-dogenous dark O_2 uptake is enhanced on exposure to short wavelength light; it does not, however, increase in red light. *Figure 3* is an original recording of this effect. On illumination there is an ini-tial decrease in both red and blue light. Although this low rate stays constant over the time of observation in red light, it is ov-ercome by a gradual increase in O_2 uptake in blue light. The time

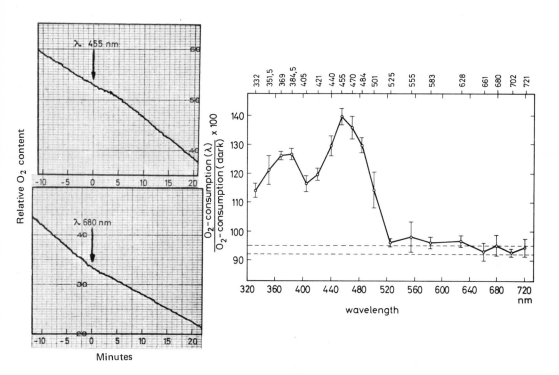

Figure 3 (left) Oxygen exchange of a suspension of DCMU-pois-oned (10^{-5} M) Chlorella in darkness and in blue or red light.

Figure 4 (right) Action spectrum for an enhancement in O_2 up-take of DCMU-poisoned (10^{-5} M) Chlorella in the light.

period needed to establish a steady enhanced O_2 uptake in blue
light is 5-10 min. (The initial decrease in rate in blue and in red
light will be left aside for the moment.) An action spectrum (Figure 4) for the rather slowly established light-enhanced O_2 uptake
in DCMU-poisoned green *Chlorella* cells quite closely resembles that
for the wavelength-dependent drop in apparent photosynthetic O_2 release (2). Thirdly, the O_2 uptake of a completely chlorophyll-free,
carotenoid-containing, non-photosynthetic mutant of *Chlorella* is
also strongly enhanced by light. Again, only blue and near ultraviolet wavelengths are effective. Again, a period of 5-10 min elapses
before the O_2 uptake is fully enhanced in short wavelength light.
The action spectrum shows the same maxima at 460 and 370 nm and the
minimum around 400 nm (3) as that for the enhanced O_2 uptake of
DCMU-poisoned green cells and as that of the drop in apparent photosynthesis.

Because of the conformity of the induction periods and of the action spectra, and since there are comparable low energy requirements
for all three effects, it is felt that all three observations - the
drop in O_2 release under photosynthetic conditions, the enhanced O_2
uptake on DCMU-poisoning, and the enhanced O_2 uptake of a non-photosynthetic mutant - depend on the same reaction. This then means that
the drop in apparent photosynthetic O_2 release in blue light is due
to an increased O_2 uptake.

Four observations with the yellow mutant suggest that enhanced O_2
uptake in blue light is not due to an unspecific photoxidation (4).
First , very small amounts of blue light ($\sim$400 ergs $cm^{-2}sec^{-1}$) are
sufficient to saturate the effect. Second ·, the light-enhanced O_2
uptake is reversible. Third , an increase in temperature leads to
an increase in both the blue-enhanced and the dark rate. Fourthly,
the release of CO_2 is equally enhanced in blue light as is the O_2
uptake.

From the data presented thus far it is concluded, that in unicellular green algae such as *Chlorella*, respiration is enhanced by
light, in the presence and absence of photosynthesis. Since from
this statement one might automatically expect that endogenous dark
respiration is enhanced by blue light, the preferred view is that
the respiratory gas exchange of green algae is enhanced by short
wavelength illumination.

In addition to *C. pyrenoidosa*, *C. vulgaris*, and their mutants,
this effect has also been seen in *Scenedesmus quadricauda* and one
of its green mutants. It was also observed in the unicellular red
alga, *Porphyridium cruentum*.

A first approach to the answer of the question if dark respiration or some other catabolic reaction is enhanced in the light comes
from experiments with the yellow *Chlorella* mutant. In an atmosphere
of nitrogen this alga does not produce CO_2 and alcohol, but organic
acids as fermentation end products. These are excreted into the medium. This acid production is enhanced by short wavelength light.
Quantitatively this can be followed by titration of the acids pro-

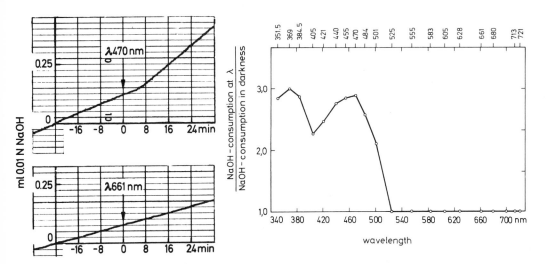

Figure 5 (left) Amount of NaOH needed to neutralize the acids excreted under N$_2$ in a suspension of the yellow Chlorella mutant 211-11h/20 in darkness and in blue or red light. From (5) with permission.

Figure 6 (right) Action spectrum for the light stimulation of organic acid excretion under N$_2$ by the yellow Chlorella mutant 211-11h/20.

duced and by recording of the base needed. *Figure 5* gives the 0.01 N NaOH-amount needed to keep constant pH in a suspension of the yellow mutant under nitrogen over a period of 64 min. The steady rate of base addition in the dark does not vary on transfer to red light. It does, however, increase considerably on exposure to blue radiation. An action spectrum for the light-enhanced acid production obtained by this method shows greatest effect of wavelengths around 460 nm and again of those in near ultraviolet. There is no effect whatsoever of yellow, red, and far red light *(Figure 6)*. [The extent of the influence of ultraviolet light shown here is not absolutely correct. It is partly this great because of an effect of these wavelengths on the glass pH-electrode used (5).] Essentially, this spectrum resembles those shown for the enhanced respiratory gas exchange in the same alga and in its green parent. One might conclude, therefore, that the light effect is the same for both processes. It might consist in the production of a respirable substrate which under anaerobiosis would be excreted into the medium, and which, in the presence of oxygen, would be oxidized.

The observation that the acids produced under nitrogen can be respired on addition of oxygen fits this assumption. In the experiment of *Figure 7*, 3 of 4 parallel samples of the yellow mutant under nitrogen have been illuminated with blue light for different periods of time, whereas the fourth one was kept in the dark. On addition of

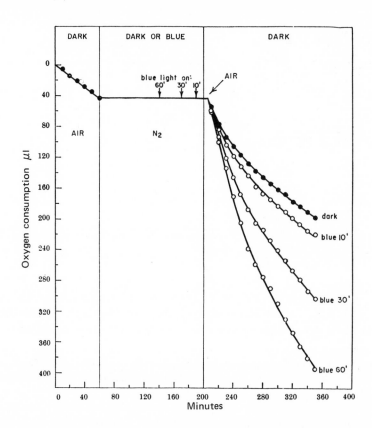

*Figure 7 O₂ uptake by the yellow Chlorella mutant 211-11h/20
in darkness after different periods of anaerobiosis in darkness or blue light. From (6) with permission.*

oxygen the respiratory rate of the previously illuminated cells is
greater, the longer the exposure to blue light (6). That this depends on a substance excreted into the medium during the anaerobic
phase of the experiment, and not on some long-lived light stimulation of cell metabolism, is proved by the fact that "fresh" cells
show greatest O_2 uptake when suspended in the medium isolated from
an anaerobic, blue-light culture, and least O_2 uptake in fresh buffer.

Color tests and chromatography provide evidence, that glycolic
acid is among the acids produced under nitrogen and excreted into
the medium by the yellow *Chlorella* mutant. Glycolate can be oxidized
by our *Chlorella* strains. On addition of exogenous glycolate the endogenous O_2 uptake increases by a factor of 4-5, while the corres-
pondent CO_2 release increases by a factor of 5-6 (7). In an enzyme
preparation glycolate oxidase activity can be demonstrated by the
uptake of oxygen on addition of glycolate, as well as by the reduction of the artificial electron acceptor dichlorophenolindophenol
(DCPIP), shown in *Figure 8.*

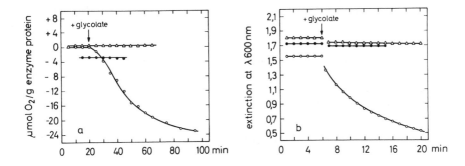

Figure 8 a) O_2 uptake and b) DCPIP reduction on addition of glycolate to 3 fractions of an enzyme preparation of *Chlorella vulgaris* ($o—o = 0-30\%$, $●—● = 30-60\%$, $▲—▲ = 60-100\%$ $(NH_4)_2SO_4$ precipitate).

From these results it seems probable that the light-enhanced respiratory gas exchange in unicellular green algae during photosynthesis corresponds to photorespiration of higher plants and is, in addition, blue light dependent. Six criteria are provided to test this assumption. First , the Q_{10} of photorespiration of higher plants is greater than that of dark respiration of these organisms (8). If we calculate the O_2 uptake per 10 min for an illuminated culture and for a dark culture of the yellow *Chlorella* mutant at different temperatures during a period of 90 min, this is also true for the O_2 uptake. At 30° the rate in blue light is 2.4 times greater than the dark rate, while at 15 and 22.6° it is only 1.7-1.8 times greater. If however, we calculate the same rate from the first 30 min only under each temperature, the blue rate is only slightly

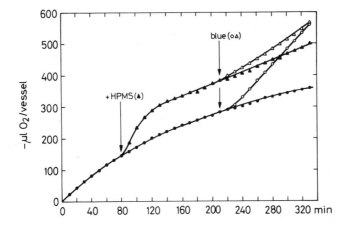

Figure 9 O_2 uptake by the yellow *Chlorella* mutant 211-11h/20 in the dark and in blue light, with and without 2×10^{-3} M HMS .

greater than the dark rate under all three temperatures. This discrepancy is brought about by the fact, that the rate of O_2 uptake in
blue light stays constant over several hours, while that in darkness
decreases gradually with time.

Second , α-hydroxymethanesulfonates (HMS) inhibit photorespiration severely but leave dark respiration uneffected (9). The effect
of HMS on respiratory gas exchange of the yellow *Chlorella* mutant is
quite inconsistent with higher plant studies. First of all the dark
O_2 uptake of that alga is usually - at least temporarily - stimulated. The blue-enhanced O_2 uptake is sometimes depressed *(Figure 9)*,
and sometimes uneffected by HMS. That our mutant might respond to
HMS differently than higher plants is shown by the complete ineffec-
tiveness of 2×10^{-3} M HMS on the oxidation of exogenous glycolate
(Figure 10).

Third , photorespiration is greatest at high light intensities
(10,11). However the blue-light-enhanced respiratory gas exchange in
Chlorella is already saturated at 400 to 500 ergs cm^{-2}sec^{-1}. If one
assumes that the effective wavelengths for photorespiration are also
those of the short end of the visible spectrum this might be only a
technical difference.

Fourth , the action spectrum of photorespiration resembles that
for photosynthesis (12) which is completely different from that of
light-enhanced respiratory gas exchange in algae. A publication of
POSKUTA (13) shows that photorespiration of spruce twigs is twice as
high in blue compared with red light; an observation which suggests
more than a close connection between photorespiration and photosynthesis.

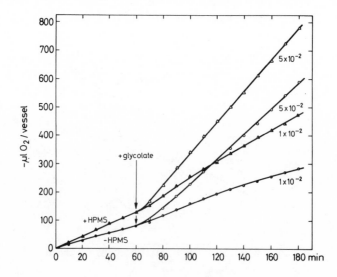

*Figure 10 O_2 uptake by the yellow Chlorella mutant 211-11h/20
in the presence and absence of Na glycolate, with and without
2×10^{-3} M HMS.*

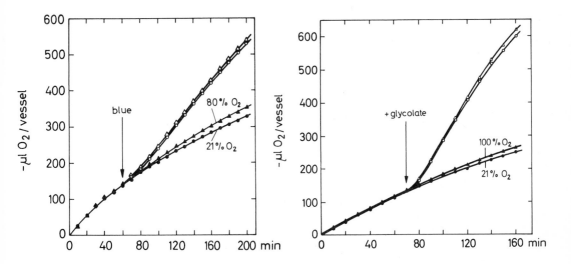

Figure 11 (left) O_2 uptake by the yellow *Chlorella* mutant
211-11h/20 in darkness and blue light under 21 and 80% O_2.

Figure 12 (right) O_2 uptake by the yellow *Chlorella* mutant
211-11h/20 in the presence and absence of 10^{-1} M Na glycolate
under 21% and 100% O_2

Fifth , photorespiration in higher plants is stimulated under
atmospheres richer in oxygen than air (10). This is not so with the
blue-enhanced O_2 uptake of the yellow *Chlorella* mutant. The dark O_2
uptake and the enhanced light O_2 uptake are equal under 21% O_2 and
80% O_2 (Figure 11). In contrast to higher plants the consumption of
exogenous glycolate is not O_2 dependent (Figure 12).

None of the five criteria mentioned thus far shows good agreement
between the behavior of photorespiration and of light-enhanced respiratory gas exchange in unicellular green algae. However, none
clearly excludes the possibility that both reactions are essentially
of the same kind. No conformity between both reactions can be seen,
however, in the last criterion. Photorespiration is said to depend
on continuing photosynthesis. Blue-light-enhanced respiratory gas
exchange in algae definitely does not. If this difference proves unequivocally correct it seems, that in spite of some similarities,
there are profound differences between the light-enhanced respiration in higher plants and that in unicellular algae - at least in
the provision of the substrate.

ACKNOWLEDGMENT

Thanks are due to Deutsche Forschungsgemeinschaft for financial
support and a travel grant.

REFERENCES

1. U. KOWALLIK and W. KOWALLIK, *Planta*, 84 (1969) 141.
2. W. KOWALLIK, *Planta*, 86 (1969) 50.
3. W. KOWALLIK, *Plant Physiol.*, 42 (1967) 672.
4. W. KOWALLIK, *Brookhaven Symp. Biol.*, 19 (1966) 467.
5. W. KOWALLIK, *Planta*, 87 (1969) 372.
6. W. KOWALLIK and H. GAFFRON, *Nature*, 215 (1967) 1038.
7. W. KOWALLIK and G.H. SCHMID, *Planta*, (1970) in press.
8. I. ZELITCH, *Plant Physiol.*, 41 (1966) 1623.
9. I. ZELITCH, *J. biol. Chem.*, 233 (1958) 1299.
10. M.L. FORRESTER, G. KROTKOV, and C.D. NELSON, *Plant Physiol.*, 41 (1966) 422.
11. E.B. TREGUNNA, G. KROTKOV, and C.D. NELSON, *Physiologia Pl.*, 19 (1966) 723.
12. N.R. BULLEY, C.D. NELSON, and E.B. TREGUNNA, *Plant Physiol.*, 44 (1969) 678.
13. J. POSKUTA, *Experientia*, 24 (1968) 796.

MICROBODIES IN FAT-STORING COTYLEDONS: ULTRASTRUCTURAL AND ENZYMATIC CHANGES DURING GREENING[1]

Richard N. Trelease, Peter J. Gruber,
Wayne M. Becker and Eldon H. Newcomb

Department of Botany, University of Wisconsin,
Madison, Wisconsin 53706, U.S.A.

SUMMARY

The marked changes in activities of microbody enzymes detected during greening of sunflower, tomato, and cucumber cotyledons have been correlated with changes in the fine structure of microbodies and in their association with other cellular components. Microbodies are interspersed among lipid bodies during early stages of greening but become associated preferentially with chloroplasts as the cotyledons turn deep green; it appears that they are capable of associating with other structures in the cell in receiving and metabolizing particular substrates as they change in metabolic activity at different stages of development. In light-grown cotyledons, cytoplasmic invaginations containing clustered ribosomes are seen in microbodies only during the greening process. In dark-grown cucumber cotyledons, similar invaginations are observed in microbodies throughout the dark period, but then disappear quickly (within 12 hr) after exposure of these cotyledons to light. The evidence suggests that the changeover from a glyoxysomal to a peroxisomal function involves a single ongoing population of microbodies rather than two distinct populations. Although the role of the cytoplasmic invaginations is unknown, their restriction largely to the period when the enzyme activities are undergoing rapid change suggests that they may participate in the conversion of microbodies from one functional state to another during development.

INTRODUCTION

Cotyledons of many seedlings convert stored lipid into carbohyd-

[1] *This work was supported in part by Grant GB-15246 from the National Science Foundation (U.S.A.)*

rate during the early period of postgerminative growth, then expand
and differentiate into green photosynthetic organs. During the transition from the one stage to the other, the decline in the activities
of the lipid-metabolizing enzymes, which are housed in glyoxysomes
(1,2,3,4), is accompanied by a rapid rise in the activity of glycolate oxidase, which is associated, as in green leaves, with peroxisomes (4,5). Hence, in these greening cotyledons a succession exists
between glyoxysomes and peroxisomes, two organelles which are enzymatically distinct but both morphologically characterized as microbodies (6). This report correlates changes in activities of glyoxysomal
and peroxisomal enzymes with changes in the fine structure and cellular disposition of microbodies during this transitional period in
three species of fatty cotyledons.

MATERIALS AND METHODS

Cucumber seeds (*Cucumis sativus* L., cv. "Improved Long Green") and
sunflower seeds (*Helianthus annuus* L., cv. "Russian Sunflower") were
planted in moistened vermiculite overlying soil and grown at 20-28°
under a 12-12 hr light-dark cycle. Tomato seeds (*Lycopersicum escul-
entum* Mill., cv. "Marglobe") were grown on gauze moistened with 1/10-
strength Hoagland solution. Some cucumber seeds were grown in complete darkness at 20-24° before exposure to the light-dark cycle described above. Germination was considered to begin when the seeds were
first moistened.

Crude homogenates of sunflower and cucumber cotyledons prepared as
previously described (4,7) were used in determining microbody enzyme
activities. Homogenates were assayed for catalase (8), glycolate oxidase (9), glyoxylate reductase (9), malate synthetase (2), and isocitrate lyase (2) activity, and for protein (10). Segments of cotyledons at various stages of greening were prepared for electron microscopic examination by fixing first in a buffered glutaraldehyde
solution, then in buffered osmium tetroxide (4,7).

RESULTS

The enzyme profiles for homogenates of cucumber cotyledons show
that a transitional period occurs between the third and seventh days
of post-germinative growth in the light (*Figure 1*). The activities
of isocitrate lyase and malate synthetase (glyoxysomal enzymes) undergo the characteristic rise and fall reported for other fatty seedlings. Of particular interest is the sharp decline in these activities from near peak levels at day 4 to low levels only 24 hours later.
The activity of glycolate oxidase (characteristic of leaf peroxisomes) rises markedly during the greening period, then remains at a relatively constant level through day 10 (*Figure 1*). The increase in
glycolate oxidase activity occurs initially during a period of high
isocitrate lyase and malate synthetase activity, indicating substantial overlap of glyoxysomal and peroxisomal enzyme complements during

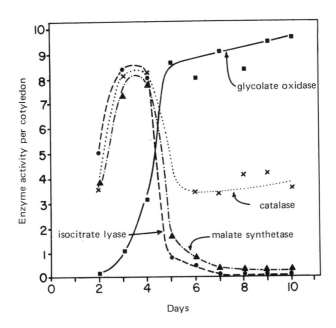

*Figure 1 Changes in glyoxysomal and peroxisomal enzyme activit-
ies during greening determined from crude homogenates of cucumber cotyledons. Enzyme activity: isocitrate lyase and glycolate
oxidase, nmoles substrate consumed min^{-1} per cotyledon; malate
synthetase, nmoles x 0.04 substrate consumed min^{1} per cotyledon;
catalase, units x 0.2 (a unit of catalase is that amount of enzyme required to catalyze the decomposition of 50% of the H_2O_2
present per 100 sec at 25°).*

the early greening phase. Catalase, an enzyme common to both types
of microbodies, follows the glyoxylate cycle enzymes closely until
day 5, then declines to about 40% of its peak activity and maintains
this plateau level for several days (*Figure 1*). A similar pattern of
changing enzyme activities was observed during greening of sunflower
cotyledons (7).

In early stages of greening, the cotyledonary cells of all three
species contain numerous microbodies interspersed among lipid bodies
(*Plates 1-3*). A high proportion of the microbody profiles have pockets of cytoplasm containing ribosomes and occasionally membranous
components. The pockets appear to represent invaginations of cytoplasm (*Plates 2,3,8*). The ribosomes in the pockets usually appear
clustered and may or may not represent polysomes.

As greening progresses and the activities of the glyoxylate cycle
enzymes drop markedly, the granal stacks of the plastids become more
prominent and a large central vacuole forms (*Plate 4*). After most of
the storage lipid has disappeared, the microbodies, are found near
both the plastids and the remaining lipid bodies (*Plate 4*).

The final stage of greening in the light-grown cotyledons is characterized by vacuolate cells containing well developed chloroplasts

(*Plates 5,6,7*). Activities of the glyoxylate cycle enzymes are low while glycolate oxidase activity is near its peak (*Figure 1*). The close association of the microbodies with the chloroplasts is their most striking feature; in most cases they are squeezed between adjacent chloroplasts (*Plate 5*) or are in surface contact with one or more of them (*Plates 5,6*). Cytoplasmic inclusions are not observed in the microbodies at this time in any of the three species.

Dark-grown seedlings

     In an attempt to separate glyoxysomal and peroxisomal enzymes temporally, greening was delayed by growing cucumber cotyledons in the dark for 10 days before exposing them to the 12-12 light-dark cycle. Again, ultrastructural observations were correlated with the results of enzyme assays (11). The activities of the glyoxylate cycle en-

---

Plate 1,2,3 Electron micrographs of microbodies (Mb) dispersed among lipid bodies (L) in tomato, cucumber, and sunflower cotyledon cells, respectively, at the early stage of greening (day 4). Portions of cytoplasm containing clustered ribosomes and membranous components are invaginated into microbodies (arrows). Plastid (P). Plate 1, x 14,700; Plate 2, x 11,900; Plate 3, x 30,600.

Plate 4 Portion of a cucumber cell at day 5 with large vacuoles (V) forming in the interior of the cells. Microbodies (Mb), some of which still exhibit cytoplasmic inclusions, are located near both the developing chloroplasts (P) and the few remaining lipid bodies (L). x 17,200.

Plates 5,6,7 Cells in sunflower, tomato, and cucumber cotyledons, respectively, with microbodies (Mb) closely appressed to well developed chloroplasts (P). Mitochondria (M) are also near microbodies and plastids. Plate 5, x 24,200; Plate 6, x 35,800; Plate 7, x 29,600.

Plates 8-12 Electron micrographs of cells in etiolated cucumber cotyledons, some of which have been exposed to light. Plate 8. Microbody (Mb) with a homogeneous matrix located among lipid bodies (L) at day 4. Note the small cytoplasmic invaginations (arrow). x 25,600. Plate 9. Microbody with a non-homogeneous matrix contains a large cytoplasmic invagination (day 7). The plastid (P) has a prolamellar body (Pb) indicative of etiolation. x 27,900. Plate 10. Several microbodies (Mb) from a cotyledon grown in the dark for 10 days. Some contain cytoplasmic inclusions and all have matrices which are flocculent in texture. Plastid (P). x 25,900. Plate 11. Portion of a cell from a cotyledon grown in the dark for 10 days, then exposed to 12 hr light. Some microbodies (Mb) contain cytoplasmic invaginations. x 20,400. Plate 12. A cell with microbodies (Mb) closely appressed to well developed chloroplasts (P). The cotyledon was grown in the dark for 10 days, then exposed to 51 hr of the light-dark cycle. x 24,300.

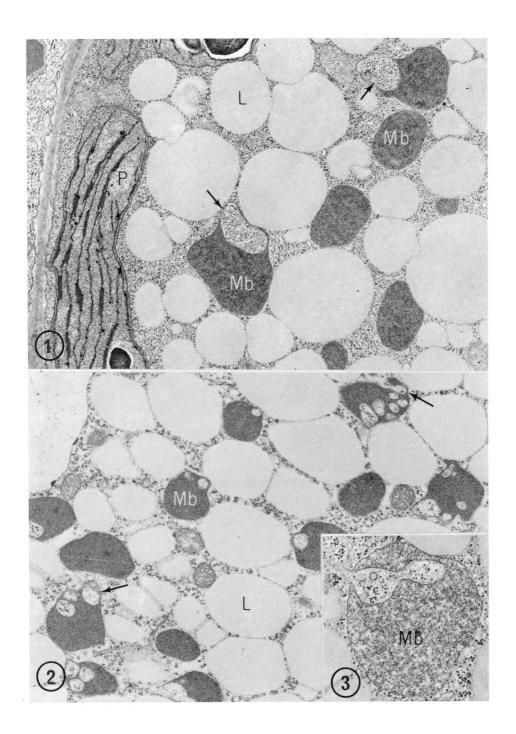

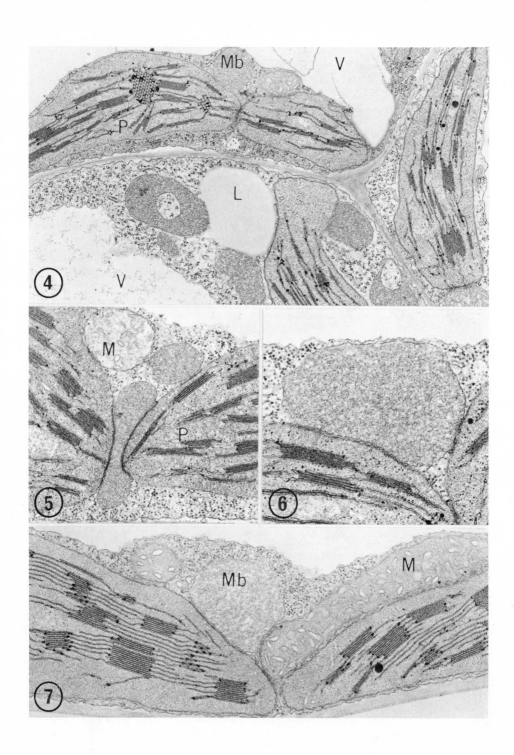

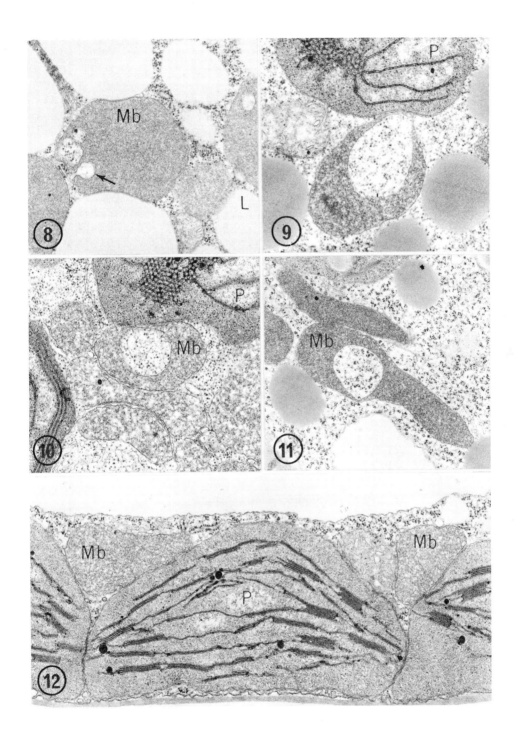

zymes and of catalase undergo an initial increase similar to that
seen in the light, but then decline gradually rather than sharply.
Glycolate oxidase activity remains low but detectable throughout the
dark period.

Microbodies are prominent in the etiolated cotyledons at all
stages and contain cytoplasmic inclusions similar to those described
in light-grown seedlings. At day 4 the cytoplasmic pockets in microbodies are small but numerous (*Plate 8*); at later stages, a single
large pocket is more characteristic (*Plates 9,10*). Lipid bodies persist in the dark far longer than in the light, but appear to decrease
in number per cell.

Upon illumination, the etiolated cotyledons turn from yellow to
light green within 12 hr. During this time, glycolate oxidase and
glyoxylate reductase activities increase substantially, while the activities of the glyoxylate cycle enzymes decrease (*Table 1*). During
the remaining 51 hr exposure to light, glycolate oxidase and glyoxylate reductase activities continue to increase while catalase activity reaches a plateau and isocitrate lyase activity becomes nearly
undetectable (*Table 1*).

TABLE 1

Changes in peroxisomal and glyoxysomal enzyme activities measured in homogenates of cucumber cotyledons grown in the dark
for ten days then exposed to light.

	Isoci- trate Lyase*	Malate Synthe- tase*	Glycolic acid oxidase*	Glyoxy- late Reduc- tase*	Cata- lase**
Dark (10 day)	4.6	79.1	0.73	31.2	11.4
Dark + 12 hr	4.3	47.3	1.39	55.0	7.2
Dark + 27 hr	2.0	-	4.29	63.9	6.9
Dark + 51 hr	0.6	-	6.04	120.3	8.4

 * nmoles of substrate utilized/min per cotyledon
** catalase units - that amount of enzyme required to catalyze
   the decomposition of 50% of the H_2O_2 present per 100 sec
   per cotyledon at 25°.

Microbodies tend to be distributed among lipid bodies during the
dark period, but after 51 hr exposure to light, commonly show a close
association with chloroplasts (*Plate 12*). These microbody-chloro-
plast associations show the same characteristics as are observed in
the final stages of greening in light grown seedlings. Cytoplasmic
inclusions are still observed in some microbodies after 12 hr exposure to light, but not in cotyledons exposed 51 hr (*Plates 11,12*).

DISCUSSION

The changing appearance and cellular disposition of microbodies can be correlated with the changing activities of glyoxysomal and peroxisomal enzymes during the greening process in the fatty cotyledons of the three species. The rapid rise and fall in the activities of key glyoxysomal enzymes in light-grown cotyledons and the slow decrease in these activities in etiolated cotyledons correlate well with the close association observed between microbodies and lipid bodies as long as the latter persist in the cells (11). The close associations lend further support to the suggestion that lipid bodies and glyoxysomes are mutually involved in the net conversion of lipid to carbohydrate (4,7,11,12). On the other hand, when the cotyledons are green and peroxisomal enzyme activity is near its peak, microbodies are preferentially associated with chloroplasts, presumably to receive and metabolize glycolate (7,13). Thus it appears that microbodies in these cotyledons have the capacity to associate with at least two different organelles.

It seems worthwhile to ask whether peroxisomes are derived from pre-existing glyoxysomes or constitute a separate population of microbodies. In the latter case, one population might replace the other during development, or the two might coexist, each being active in cellular metabolism at a different stage. If peroxisomes replace glyoxysomes, one would expect to see evidence in the electron micrographs for a degradation of glyoxysomes and a synthesis of new peroxisomes. KAGAWA and BEEVERS (14), using etiolated watermelon cotyledons, found a progressive loss of activity of glyoxysomal enzymes with a concomitant 75% decrease in the protein of isolated glyoxysomes. On illuminating the seedlings they observed an increase in peroxisomal enzyme activity and concluded that peroxisomes present after greening are not derived from pre-existing glyoxysomes. In our study, no ultrastructural evidence for the destruction of glyoxysomes or synthesis of new particles either arising from endoplasmic reticulum (15) or budding from pre-existing particles (16) was obtained in light-grown cotyledons or in etiolated cucumber cotyledons exposed to light. Under the latter conditions, synthesis of peroxisomes should be rapid and evident in the electron microscope if indeed new particles were being made. Furthermore, counts of microbodies per cell section in cucumber cotyledons showed only a 25 to 35% decrease during the dark period (11). The flocculent texture and less electron opaque appearance of microbody matrices seen in our dark-grown seedlings after day 4 suggests that the microbodies may become more fragile and therefore more difficult to isolate intact. This could account for the progressive decrease in isolatable particles observed by KAGAWA and BEEVERS without necessitating the destruction of the particles in the cell. Thus, our evidence does not support the concept that one population of microbodies replaces another during the greening transition.

No definitive data in this study have been obtained that allow us to determine whether the changeover in microbody function involves a single population or the coexistence of two populations. However, detailed fine structural observations of both light- and dark-grown cotyledons have produced no evidence suggesting a dichotomy in the microbody population, either in morphological appearance or in association with other organelles. For example, distinctions could not be made either between inactive peroxisomes and functional glyoxysomes or between enzyme-depleted glyoxysomes and active peroxisomes. Furthermore, no population of microbodies failed to take up associations with chloroplasts in the late stages of greening. Our observations, therefore, seem most consistent with the concept that a single ongoing population of microbodies is involved in the succession of glyoxysomes to peroxisomes during greening.

Whether there is a single kind of microbody capable of changing its enzyme complement or two kinds of particles with different non-overlapping metabolic roles, some sort of mechanism must exist to degrade or inactivate glyoxysomal enzymes and to synthesize or activate peroxisomal enzymes. The most consistent and unique feature of microbodies in the transitional stage is the presence of cytoplasmic inclusions. The clusters of ribosomes within these invaginations might provide the necessary protein synthesis either for the acquisition of peroxisomal enzyme activity or for the proteolytic activity responsible for the loss of specific enzymes such as those of the glyoxylate cycle. It seems unlikely from our ultrastructural observations that these invaginations are involved in the destruction of a major part of the microbody population. The presence of invaginations during the transition from glyoxysomal to peroxisomal function in light-grown cotyledons and their persistence in etiolated cotyledons strongly suggest that they may participate in the alteration of microbody function.

REFERENCES

1. H. BEEVERS, *Ann. N.Y. Acad. Sci.*, 168 (1969) 313.
2. T.G. COOPER and H. BEEVERS, *J. biol. Chem.*, 244 (1969) 3507.
3. D.I. McGREGOR and H. BEEVERS, *Plant Physiol.*, 44 (1969) S-33.
4. R.N. TRELEASE, P.J. GRUBER, W.M. BECKER and E.H. NEWCOMB, *Plant Physiol.*, in the press.
5. N.E. TOLBERT, this volume (1971).

6. S.E. FREDERICK, E.H. NEWCOMB, E.L. VIGIL and W.P. WERGIN, *Planta*, 81 (1968) 229.
7. P.J. GRUBER, R.N. TRELEASE, W.M. BECKER and E.H. NEWCOMB, *Planta*, 93 (1970) 269.
8. H. LÜCK, *Methods Enzymat. Anal.*, (1963) 885.
9. N.E. TOLBERT, in Methods in Enzymology (1971) in press.
10. O.H. LOWRY, N.J. ROSENBROUGH, A.L. FARR and R.J. RANDALL, *J.*

biol. Chem., 193 (1951) 265.

11. R.N. TRELEASE, W.M. BECKER and E.H. NEWCOMB, *Plant Physiol.*, in the press.
12. E.L. VIGIL, *J. Cell Biol.*, 46 (1970) 435.
13. S.E. FREDERICK and E.H. NEWCOMB, *Science*, 163 (1969) 1353.
14. T. KAGAWA and H. BEEVERS, *Plant Physiol.*, 46 (1970) S-38.
15. Z. HRUBAN and M. RECHCIGL, Jr., *Int. Rev. Cytol.*, 1 (1969) 1.
16. J.L. RIGATUSO, P.G. LEGG and R.L. WOOD, *J. Histochem, Cytochem.*, 18 (1970) 893.

ON THE LOCALIZATION OF PLANT CELL ORGANELLES

Irwin P. Ting, Victor Rocha, S.K. Mukerji and Russell Curry

Department of Life Sciences, University of California,
Riverside 92502, U.S.A.

SUMMARY

A method is outlined in which intact organelles can be prepared from plant tissues in high purity with the use of aqueous solvents. The method is recommended for subcellular enzyme localization studies.

INTRODUCTION

Because subcellular compartmentation of enzymes most likely provides for cellular regulation of metabolite flow, it is important to accurately determine enzymic location. In our laboratory we have been concerned with the cellular localization of multiple forms of enzymes (isoenzymes) important in CO_2 metabolism. Recently, we described a rate sedimentation-equilibrium density (S-ρ) method to prepare organelles from spinach leaves (1). The method was successfully applied to the study of tissue and subcellular distribution of malate dehydrogenase isoenzymes in spinach leaf and seedling tissue (2).

Both aqueous and nonaqueous procedures are available for the isolation and purification of subcellular organelles. In the nonaqueous procedure, organic solvents are used to prevent leaching of water soluble components. Because of extraction of lipoidal components by the organic solvents, recovery of microbodies and mitochondria is not satisfactory. Chloroplasts are isolated with stroma present, but lack outer envelopes and usually are associated with adsorbed cytoplasmic protein. The latter frequently represents the major problem in enzymic localization using nonaqueously prepared chloroplasts. Aqueously prepared organelles are frequently ruptured or disturbed such that significant leaching of water soluble protein occurs. In many instances, however, aqueously prepared organelles can be obtained such that the outer membranes are intact and loss of protein is minimal.

In this paper, we present work concerned with the localization of

534

some plant cell enzymes in spinach leaf tissue and the application of the S-ρ organelle isolation procedure to plant tissue other than spinach.

EXPERIMENTAL

Preparation of cellular organelles

Fresh tissue is washed thoroughly with distilled water and deribbed. The washed tissue should be destarched by storage overnight in the cold. 100 g of the prepared tissue is thoroughly minced with a food chopper or sharp knife in 250 ml of grinding medium (0.5 M sucrose, 1 mM EDTA, 1 mM dithiothreitol, 0.1% bovine serum albumin) all in 0.05 M tris buffer, pH 7.5. After mincing, the preparation is sheared in a blendor for 2 or 3 seconds at low speed. The blendor switch should be rapidly turned on and off 3 times. After each second, the tissue should be stirred and remixed by hand while in the blendor. The initial mince and short, low speed shearing is absolutely necessary for the isolation of intact chloroplasts in high yield. The homogenate is gently filtered through eight layers of cheese cloth without pressure. The filtrate containing organelles is centrifuged at 250 g for 90 seconds to remove debris, intact cells, and unbroken tissue. The supernatant fluid is centrifuged at 1000 g for 5 minutes in a swing-out rotor to obtain a 1000 g pellet (250 to 1000 g). The latter supernatant fluid is recentrifuged in the swing-out rotor for 15 minutes at 3000 g. Pellets should be gently suspended in 40% sucrose (w/v) by agitation. All operations must be conducted in a cold room or over ice (0-4°).

Linear sucrose density gradients of 40-80% (w/v) are prepared in 50 ml cellulose nitrate centrifuge tubes with the use of a double chamber, gravity flow apparatus or other suitable gradient maker. The sucrose is prepared in 0.05 M tris buffer (pH 7.5) containing 1 mM EDTA and 1 mM dithiothreitol. Linearity of the gradient should be confirmed. Prior to use, the gradients must be chilled to about 0°.

The 1000 and 3000 g resuspended pellets are layered over separate linear gradients by pipetting down the side of the centrifuge tubes. Gradients are centrifuged for 3 hours in a swinging bucket rotor at 20,000 rpm ($\omega^2 t = 74 \times 10^9$ rad^2 sec^{-1}). Centrifugation at lower speeds in swinging bucket rotors results in a percentage of the particles coming to their equilibrium positions, but a high degree of contamination of low density components by high density components. The three hour centrifugation is sufficient for spinach organelles; other tissues, however, may require extended centrifugation times.

After centrifugation to equilibrium, the gradients are fractionated through a hole in the bottom of the centrifuge tube into approximately 30 equal fractions of about 1.2 ml. Prior to enzymic assay, each fraction should be treated by sonication or with detergents to insure rupture of organelles with subsequent liberation of protein.

Location of organelles within the gradient

Organelles can be located within the gradient fractions with the use of enzymic markers. Catalase appears to be the best marker for microbodies. Glycolate oxidase can be used to detect microbodies in green tissues. Mitochondria can be located by the presence of cytochrome oxidase. Since the latter is associated with membranes, broken as well as intact mitochondria will be detected. Fumarase could be used to detect the presence of intact mitochondria. NADP glyceraldehyde 3-phosphate dehydrogenase and chlorophyll can be used to locate intact chloroplasts. Chlorophyll without the latter enzyme locates broken or stripped chloroplasts.

Enzyme assays

(a) Cytochrome oxidase Cytochrome oxidase was assayed spectrophotometrically by following the oxidation of reduced cytochrome c at 550 mμ (3). The reaction mixture contained 0.03 mM reduced cytochrome c, enzyme (treated with Brij 58) and 0.05 M phosphate buffer (pH 7.0) in a total volume of 1.25 ml. Cytochrome c was reduced chemically with dithionite.

(b) Glycolate oxidase The activity of glycolate oxidase was determined using the redox dye, o-dianisidine. The standard assay contained 1.0 mM FMN, 1.58 mM o-dianisidine:diHCl, 1 mM glycolate, 0.04 ml peroxidase (1 mg/ml) (treated with Brij 58), and 0.05 M tris (pH 8.0) in a total volume of 1.0 ml. Initial velocities were determined by following the increase in absorption of oxidized o-dianisidine at 435 mu. NADP glyceraldehyde 3-phosphate dehydrogenase was assayed by coupling with phosphoglyceric acid kinase (4).

(c) Other assays Chlorophyll was determined by the method of ARNON (5). Chlorophyll was extracted to a final concentration of 80% acetone and read at 663 mμ and 645 mμ. Catalase was assayed spectrophotometrically by measuring the disappearance of hydrogen peroxide at 240 mμ (6). Aspartate aminotransferase activity was measured by coupling with malate dehydrogenase (7) while NADH and NADPH glycolate reductase were assayed by following the oxidation of the reduced

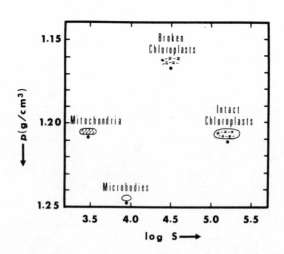

Figure 1 Sedimentation velocity (S) - equilibrium density (ρ) plot of spinach leaf organelles.

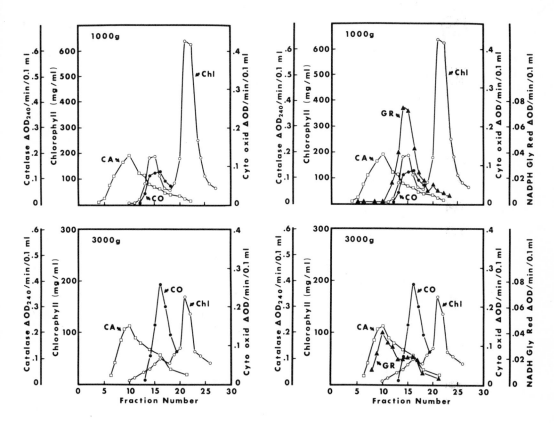

Figure 2 (left) Organelle profile of S-ρ separation of spinach leaf organelles. Upper = 1000 g pellet. Lower = 3000 g pellet. CA = catalase, CO = cytochrome oxidase, Chl = chlorophyll.

Figure 3 (right) Organelle profile and localization of glyoxylate reductases in spinach leaf tissue (see Figure 2). Upper GR = NADP glyoxylate reductase; lower GR = NAD glyoxylate reductase.

pyridine nucleotide at 340 mμ (8).

RESULTS AND DISCUSSION

Figure 1 shows an S-ρ diagram for spinach leaf organelles. Diagrams such as the one shown can be used to predict proper sedimentation velocity centrifugation times and forces, and proper density gradients to obtain maximum separation and purification of organelles.

The results of a typical organelle isolation procedure applied to spinach leaf tissue are presented in *Figure 2*. After centrifugation to equilibrium, the 1000 g organelle pellet (*Figure 2:* upper) resolved into intact chloroplasts (fractions 14 and 15) and broken chloro-

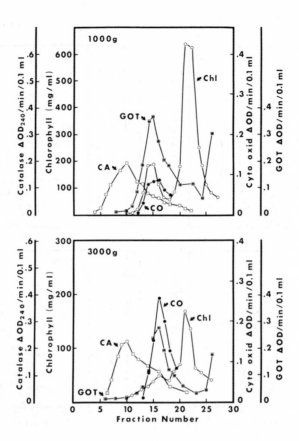

Figure 4 Organelle profile and localization of aspartate amino-transferase (GOT) in spinach leaf tissue. See Figure 2.

plasts (fractions 21 and 22). Some microbodies were also resolved from the 1000 g pellet as indicated by the marker enzyme catalase (fractions 8 and 9). The 3000 g organelle pellet (*Figure 1*: lower) resolved into broken chloroplasts (fractions 21 and 22), mitochondria as indicated by the cytochrome oxidase peak (fractions 15, 16 and 17) and a microbody peak identified by catalase (fractions 9 and 10).

Using this technique, the compartmentation of NADH and NADPH glyoxylate reductase and aspartate aminotransferase were investigated in spinach leaf tissue. It appears that NADPH glyoxylate reductase is an enzyme of intact chloroplasts (*Figure 3*: upper). The fact that broken chloroplasts are devoid of this enzyme suggests that the enzyme resides in the stroma of intact chloroplasts. In contrast, NADH glyoxylate reductase appears to be associated with the microbody fraction (*Figure 3*: lower).

Aspartate aminotransferase (GOT) activity may be associated with intact chloroplasts and mitochondria (*Figure 4*). The ratio of GOT activity to cytochrome oxidase activity in the mitochondrial region of the 3000 g gradient is approximately 0.7 while in the intact chloroplast region of the 1000 g gradient it is approximately 4.2.

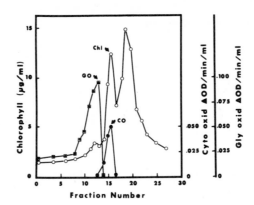

Figure 5 Organelle profile of 600 g pellet from cotton leaf tissue. See Figure 2. 25-60% gradient.

If the GOT activity in the intact chloroplast fraction was present because of mitochondrial contamination, the ratio would be on the order of 0.7. The most reasonable interpretation of these data is that GOT activity is associated with both mitochondria and intact chloroplasts (stroma). No evidence was found for activity in the microbody region.

Using a linear sucrose gradient (25-60%), the 600 g organelle pellet from cotton leaves was resolved into two distinct chlorophyll containing bands probably representing broken and intact chloroplasts (*Figure 5*). In addition, a distinct glycolate oxidase peak (fractions 12 and 13) indicated microbodies.

The spinach organelle isolation procedure applied to corn mesophyll cells resulted in the separation of two distinct chlorophyll containing bands (*Figure 6*). The lower band (fractions 10 and 11) was composed of intact chloroplasts while the upper band contained broken chloroplasts. The mitochondria that sediment in the 1000 g pellet were located intermediate between the two chlorophyll bands. NAD-malate dehydrogenase activity (not shown) paralleled the cytochrome oxidase activity in the gradient. Significantly, no glycolate oxidase or catalase activity was detected in the gradient.

Several *Brassica* spp. were used for the separation of organelles. In order to obtain good separation of the organelles, centrifugation of the loaded sucrose gradients was extended (12-16 hours). *Figure 7* shows the organelle marker components. Chlorophyll was distributed in a double peak indicative of broken and intact chloroplasts. The cytochrome oxidase activity indicated that the mitochondrial peak was slightly more dense than the intact chloroplasts. Glycolate oxidase activity indicated that the microbody fraction was separated from the other organelles.

The S-ρ centrifugation technique for the isolation of intact organelles of high purity seems to be generally applicable to a wide variety of plant species. We have successfully used spinach tissues,

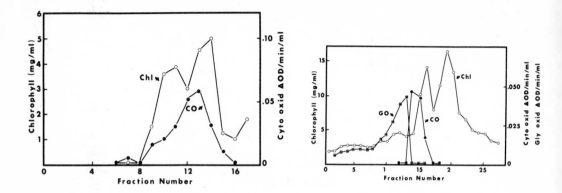

*Figure 6 (left) Organelle profile of 1000 g pellet from corn
leaf mesophyll cells. See Figure 2.*

*Figure 7 (right) Organelle profile of 1000 g pellet of Brassica
oleracea. See Figure 2.*

Brassica oleracea leaf tissue, *Atriplex* mesophyll cells, *Zea mays*
mesophyll cells, and cotton leaf tissue. Ideally, sedimentation coefficients and equilibrium densities of the organelles should be
known prior to designing separation procedures. A single standard
separation using the method outlined here should allow prediction of
proper modification such that a second run would result in the desir
ed separation and purification.

REFERENCES

1. V. ROCHA and I.P. TING, *Archs. Biochem. Biophys.*, 140 (1970) 398.
2. V. ROCHA and I.P. TING, *Plant Physiol.*, 46 (1970) 754.
3. E.W. SIMON, *Biochem. J.*, 69 (1958) 67.
4. U. HEBER, N.G. PON and M. HEBER, *Plant Physiol.*, 38 (1963) 355.
5. D.I. ARNON, *Plant Physiol.*, 24 (1942) 1.
6. B.C. MALHLY and B. CHANCE, *Meth. Biochem. Anal.*, 1 (1954) 357.
7. I.W. SIZER and W.T. JENKINS, in Methods in Enzymology, eds. S.P.
     COLOWICK and N.O. KAPLAN, 5 (1962) 677.
8. I. ZELITCH, in Methods in Enzymology, eds. S.P. COLOWICK and N.O.
     KAPLAN, 1 (1955) 528.

PHOTORESPIRATION : ASSESSMENT

Harry Beevers

Division of Natural Sciences, University of California,
Santa Cruz, California 95060, U.S.A.

Although no final answers were forthcoming and formidable problems remain, some agreement was reached and questions defined as follows.

DEFINITION AND MEASUREMENT

Light dependent O_2 uptake and/or CO_2 output constitute photorespiration. Its recognition and measurement are obviously made difficult by concurrent photosynthesis and problems of CO_2 diffusion and biochemical recycling within the leaf. In C_3 plants estimates of rates of photorespiration of over 50% of the photosynthetic CO_2 fixation rate have been made by indirect methods. These are based on the observed inhibitory effect of 21% O_2 on apparent photosynthesis and the assumption that the major effect of removing O_2 is to prevent photorespiration which is strongly dependent on the O_2 concentration. Measurements of light stimulated uptake of O^{18} constitute unequivocal evidence for photorespiration and provide the best, though minimal estimates of the rate. JACKSON applied this method and concluded that light also stimulates O_2 consumption by 2-3 fold, in corn, a C_4 species. Agreement was not reached on whether light suppresses normal respiration completely; it was assumed that photorespiration is quantitatively of much greater significance in its effect on gas exchange. The occurrence of photorespiration in a leaf in air can be reasonably inferred if

(a) the rate of apparent photosynthesis is increased by lowering the O_2 concentration;

(b) the concentration of CO_2 increases when a CO_2-free air stream is passed over an illuminated leaf;

(c) the CO_2 compensation point (at $25°C$) is some 40 ppm rather than nearly 0.

Negative results in (a) and (b) and (c) do not rule out the participation of photorespiration since peculiar leaf anatomy and biochemistry in the C_4 plants probably allow more efficient refixation of any CO_2 respired. Methods based on the ability of leaves to convert labeled glycolate to $^{14}CO_2$ are not reliable, since they measure only the capacity of the enzyme systems and not their operation under normal conditions.

MECHANISM

Although a variety of ways can be envisaged whereby intermediates produced in photosynthesis are oxidized in ancillary respiratory reactions almost all of the discussion was focussed on the oxidation of glycolate as the essential photorespiratory event.

GIBBS has shown one mechanism whereby glycolate can originate as a by-product of photosynthesis. The C_2 TPP-addition product generated in transketolase reactions is readily oxidized by oxidants generated in photosystem II or by H_2O_2. H_2O_2 could be generated in chloroplasts by a Mehler reaction, dependent on O_2 tension. Such a mechanism would make the production of glycolate an O_2 consuming event. Other mechanisms of glycolate production are not ruled out. BJÖRKMAN leans towards a mechanism involving interaction of O_2 with the carboxylation reaction which would account for the competitive effects of O_2 and CO_2 on net photosynthesis and apparently also on glycolate production. TOLBERT favors phosphoglycolate as a glycolate precursor. Whatever the precise mechanism or extent of glycolate production in different plants there is agreement that its oxidation must occur in the microbodies (leaf peroxisomes), which house the glycolate oxidase. Conversion to glyoxylate in an O_2-consuming reaction would constitute an unambiguous respiratory event; this reaction is not itself light dependent but depends on concurrent photosynthesis for a supply of its substrate glycolate. H_2O_2 generated in this reaction would be subject to breakdown by catalase, which is a major enzymatic component of the microbodies. Other enzymes in the microbodies would allow conversion of glyoxylate to glycine and also the conversion of serine to glycerate. These compounds have been implicated previously in the glycolate pathway by which carbon from glycolate is converted to carbohydrate. The reaction: 2 glycine → serine + CO_2 apparently occurs in mitochondria and not in microbodies. The suggestion by ZELITCH that glyoxylate by further oxidation could give rise to CO_2 and the C_1 donor for serine production from glycine would give the same result but there is no evidence that this or any other reaction yielding CO_2 from glyoxylate occurs in the isolated microbodies. Potentially 75% of the carbon in glycolate could be recovered as glycerate and reintroduced into the chloroplast by the reactions outlined. Whether such a stoichiometry is in fact achieved is not known.

Photorespiration on this view is an O_2-consuming and CO_2-yielding sequence which depends on light only for the supply of its substrate, glycolate. The respiratory aspects of the above reaction sequence from glycolate to glycerate are the uptake of O_2 and the production of CO_2. The "photo" aspect is restricted to the generation of the substrate, glycolate. It should be emphasized that no useful form of energy is generated in the sequence outlined; if indeed the function of the process is the conservation of glycolate carbon, two additional energy inputs are required, one as NADH in the reduction of hydroxypyruvate and one as ATP to convert the glycerate to phosphoglycerate. The suggestion (TOLBERT) that a glycolate-glyoxylate shuttle between chloroplasts and microbodies offers a means of disposing of excess reducing equivalents generated during photosynthesis did not meet with unbridled enthusiasm.

The question of how C_4 plants circumvent the loss of carbon in photorespiration is of great interest. The leaves of at least some C_4 plants contain adequate amounts of glycolate oxidase and the EM pictures show that microbodies are present both in mesophyll and bundle sheath cells. Any general explanation of the *apparent* lack of photorespiration in C_4 plants (as judged from CO_2 exchange) has to accomodate these observations and additionally cannot depend on the fact that in some C_4 plants the bundle sheath chloroplasts are agranal. The high affinity primary CO_2 fixing mechanisms in such plants might of course completely recycle CO_2 released during photorespiration. However, such a recycling would require photosynthetically generated reducing power and ATP and can therefore not adequately explain the absence of an apparent photorespiration in C_4 plants at low light intensities where the rate of photosynthesis is light limited. Hence it seems tempting to suppose that it is the rate of glycolate *production* as a photosynthetic by-product which is lower in C_4 than in C_3 plants.

PHOTORESPIRATION : ASSESSMENT

C.B. Osmond

Research School of Biological Sciences, Australian
National University, Canberra City 2601, Australia.

Photorespiration is a major component of the carbon budget of
plants in the light. It appears to be maximal under conditions limiting for carbon incorporation but nonlimiting for photosynthetic energy production (low ambient CO_2 concentration, high O_2 concentration, high light and high temperature). This emphasis on photorespiration in terms of carbon metabolism rather than energy metabolism
is further conditioned in that CO_2 evolution in the light is the most
accessible parameter of photorespiration, and in that the process
consumes, but does not produce, useful energy.

As presently understood, photorespiration hinges largely on the
production and excretion of glycolate from chloroplasts capable of
Calvin cycle metabolism. Subsequent metabolism of glycolate to glycerate and the re-entry of glycerate to the chloroplast provide for a
maximum recovery of 75% of glycolate carbon in C_3 plants and up to
100% in C_4 plants. The actual losses of carbon due to photorespiration in C_3 plants are difficult to assess accurately but depend
first, on the flow of carbon into glycolate and second, on several
alternative paths for glycolate metabolism.

Problems of measuring photorespiration, reviewed recently by LUD-
LOW and JARVIS (1), are additionally confounded by the likelihood of
mitochondrial respiration in the light. The O_2 exchange measurements
(JACKSON, this volume) provide clear evidence for photorespiration
but are difficult to apportion between the several sites of O_2 uptake
in photorespiration. The stoichiometry of CO_2 evolution and O_2 uptake during photorespiration is likely to be highly variable.

"LIGHT" REACTIONS OF PHOTORESPIRATION

The "light" reactions of photorespiration are probably restricted
to the synthesis of glycolate, concomitant with Calvin cycle metabolism. The close association between glycolate synthesis and photosynthetic electron transport and the interaction of O_2 and light was emphasised by GIBBS (this volume). These studies provide mechanisms

for O_2 and CO_2 responses similar to those observed in vivo. Alternative paths of glycolate synthesis are less satisfactory from these standpoints. In addition, these studies underline the consequences for carbon metabolism of peroxide generating reactions in the chloroplast. With catalase located more or less exclusively in the microbodies it is surprising that other direct peroxidative interactions between carbon metabolism and light have not been described. Perhaps the transketolase "addition complex" is particularly sensitive to peroxidation. It will be of interest to examine these interactions in mesophyll chloroplasts of C_4 plants which appear to lack Calvin cycle metabolism. The capacity for glycolate synthesis in these chloroplasts is unknown.

Light may interact at other sites of carbon metabolism involved in photorespiration. The labeling of glycine and serine during photosynthesis is sensitive to light quality (2,3) which may be linked to the photosensitivity of glycolate oxidase (4).

"DARK" REACTIONS OF PHOTORESPIRATION

The "dark", energy consuming reactions of photorespiration appear to be those of the glycolate pathway. The central role of glycolate oxidase, the inducible initial reaction of the pathway, is widely acknowledged but the fate of its product, glyoxylate, is disputed. It is clear that a large proportion of glycolate carbon is converted to glycine in the peroxisome as described by TOLBERT (this volume). Both TOLBERT and ZELITCH (5) agree that glyoxylate may return to the chloroplast and participate in an energy consuming shuttle between chloroplast and peroxisome.

The key reactions leading to CO_2 evolution and the localisation of these reactions are not clear. TOLBERT'S view, that CO_2 is released during the conversion of glycine to serine, is supported by semi quantitative, in vivo studies which emphasise the relationship between these events during photosynthesis (see ATKINS *et al.* this volume), and by studies of the metabolism of labeled glycolate, glycine and glyoxylate. Yet the enzymic reactions involved in a direct decarboxylation of glycine and activation of the remaining C_1 unit are not known. It is ironical that this crucial event of photorespiration is currently relegated to the mitochondrion (TOLBERT, this volume).

ZELITCH denies this association between CO_2 release and serine formation, proposing the decarboxylation of glyoxylate as a source of photorespiratory CO_2. It is very likely, however, that formate from glyoxylate decarboxylation could be activated and participate in serine synthesis. Enzymes of formate activation involving ATP and NADP are well known in leaves and seem to be related to glycolate pathway activity (OSMOND, this volume). *Acetabularia* chloroplasts show O_2 sensitive photorespiration (6) and the synthesis of glycine and serine has been demonstrated in many chloroplast preparations (7). Accepting that glycolate oxidase is restricted to microbodies, part of of the glyoxylate re-entering the chloroplast may be metabolised in this way, yielding photorespiratory CO_2 (8).

The overall result of these two photorespiratory pathways is similar. CO_2 is released and serine is formed from 2 x C_2 compounds. Whichever proposal proves correct, the regulated flow of metabolites between chloroplasts and other organelles is likely to emerge as a key feature of photorespiration. NEWCOMB's elegant micrographs (this volume) show that the proximity of other organelles to the chloroplast is not a problem but the role of membrane transport processes is not clear. As pointed out by TOLBERT, the coupled serine-hydroxypyruvate, glyoxylate-glycine transaminase of the peroxisome may provide internal regulation of carbon flow in the leaf peroxisome and determine the proportion of glyoxylate returned to the chloroplast.

These problems aside, it is clear that the leaf peroxisome is instrumental in ensuring the conservation of glycolate carbon during Calvin cycle photosynthesis. It may be compared to the glyoxysome of germinating fatty seeds in that both microbodies conserve the carbon of specific C_2 substrates, carbon which would otherwise be lost as CO_2 during peroxidative or mitochondrial metabolism.

REFERENCES

1. M.M. LUDLOW and P.G. JARVIS, in Plant Photosynthetic Production : Manual of Methods, eds. Z. SESTAK, J. CATSKY and P.G. JARVIS, Dr. Junk, The Hague, (1971) in press.
2. N.P. VOSKRESENSKAYA and G.S. GRISHINA, *Fiziol. Rast.*, 5 (1958) 147.
3. J.L. HESS and N.E. TOLBERT, *Plant Physiol.*, 42 (1967) 1123.
4. G.H. SCHMID, *Hoppe-Seylers Z. Physiol. Chem.*, 350 (1969) 1035.
5. I. ZELITCH, in Harvesting the Sun, eds. A. SAN PIETRO, F.A. GREER and T.J. ARMY, Academic Press, N.Y. (1967) 231.
6. R.G.S. BIDWELL, W.B. LEVIN and D.C. SHEPHARD, *Plant Physiol.*, 44 (1969) 946.
7. S.P.J. SHAH and E.A. COSSINS, *Phytochemistry*, 9 (1970) 1545.
8. T. KISAKI and N.E. TOLBERT, *Plant Physiol.*, 44 (1969) 242.

Appendix

POSTSCRIPT (MUSICAL)[1]

THE C-TWO THREE THROUGH FOUR PATHWAY

To TANNENBAUM: with feeling

The spinach is a C_3 plant
It has a Nobel pathway:
Its compensation point is high
From early morn till noon-day.
It has carboxydismutase
But has no malate transferase;
The spinach is a regal plant
But lacks the Hatch-Slack pathway.

The sorghum is a C_4 plant
It keeps the malate flowing,
From mesophyll to bundle sheath,
That's how it keeps on growing.
There's System I and System II,
So Smillie says and Clanton too;
Yet Boardman says it isn't true:
But still it keeps on growing.

Now Slack he wants a C-through plant
To solve the distrobution:[2]
He'd look down through his microscope
With splendid resolution.
O, has it CAM or has it not,
Or is it just a Björkman plot?
The Atriplex confusion's due
to C-4 evolution.

---

[1] *This is a revised version of a contribution written by F.A. Smith, H. Beevers and twelve others, and sung at the conference by a choir conducted by Beevers.*
[2] *distrobution (sic): Slack is primarily interested in enzyme distrobution.*

There is a pathway for C-2
Says Tolbert with elation.
Gibbs says it is most likely through
Peroxide generation.
The microbodies then dictate
Conversion into glycerate
Which shuttles back to chloroplasts -
That's photorespiration.

LIGHT MEASUREMENT IN PHOTOSYNTHETIC RESEARCH

Margaret C. Anderson

CSIRO Division of Plant Industry, Canberra, 2601, Australia

Figure 1 is a conversion table for many of the great variety of units used for energy measurement in the course of photosynthetic research, and shows both radiometric (based on the total energy content in a specified waveband) and photometric units (based on the spectral sensitivity of the human eye, with a peak at 555 nm). Far too many plant physiologists still fail to appreciate that photometric units, such as foot-candles are unsuitable in their work. The common foot candle meters use a photo cell, of different spectral sensitivity to the human eye, calibrated against a source of a particular spectral composition. The readings are meaningless when the instrument is used with other sources (1,2).

'Noon sunlight' is a nebulous term, since the solar energy from a clear sky depends on solar altitude and the particle content of the atmosphere (3,4). In temperate and tropical regions the range would be about 700 - 1000 W m^{-2}. At high altitudes, or with a few thin highly reflective clouds near the sun far higher figures may be achieved, and with multiple forward scattering, the solar constant may be exceeded.

For many purposes, photon rather than energy flux is required. To convert, the spectral distribution of the source must be known, since the conversion is wavelength dependent:

$$1 \text{ erg} = 8.37 \times 10^{-23}\lambda \text{ Einsteins}$$

$$1 \text{ Einstein} = 1.19 \times 10^{22}/\lambda \text{ ergs cm}^{-1}$$

where wavelength (λ) is measured in cm^{-1} (5).

An interesting point is that the photon/energy ratio for natural sky and sun in the 400-700 nm waveband is largely independent of weather conditions, and very low compared with other energy sources in that waveband. ANDERSON (6) suggests that a figure of 1.2×10^{19} photons cal^{-1} (1.03×10^{23} photons watt hr^{-1} sec^{-1}) can be adopted for radiation from sun and sky. From this, given precise knowledge of reaction times, it is possible to decide whether or not photon flux is rate limiting.

551

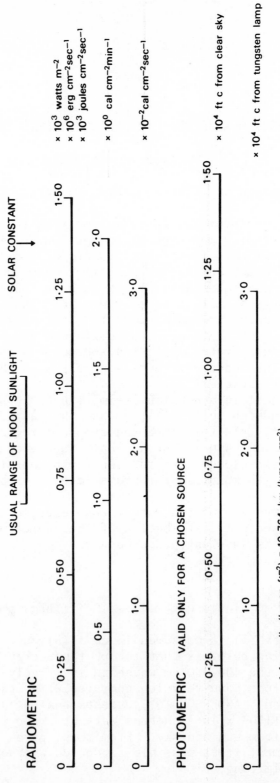

Figure 1 Conversion table for radiometric and photometric energy units

REFERENCES

1. M.C. ANDERSON, in IBP Manual on Photosynthetic Production, Dr.
   Junk, Hague (1971) in press.
2. K.H. NORRIS, A. *Rev. Pl. Physiol.*, 19 (1968) 490.
3. N. ROBINSON, ed. Solar radiation, Elsevier, Amsterdam (1966)
   pp. 347.
4. K.Ya. KONDRATYEV, Radiation in the Atmosphere. International
   Geophysics Series, Vol. 12, Academic Press, New York, (1969)
   pp. 912.
5. P. GAASTRA, in UNESCO Natural Resources Research, ed. F.E.
   ECKHARDT Vol. V. (1968) 467.
6. M.C. ANDERSON, *Ecology*, 48, (1967) 1049.

CHECK LIST OF C$_4$ SPECIES

Compiled by W.J.S. Downton

Research School of Biological Sciences, Australian National
University, Canberra City 2601, Australia

The C$_4$ species cited in this list have been selected from works
dealing with various aspects of C$_4$ photosynthesis. Since a number
of characteristics are invariably associated with the C$_4$ pathway,
any of them can be used as indicators of a C$_4$ species. The criteria
used to designate a given species are indicated, along with the reference at the end of the check list. Only one reference is given for
each species even though some plants have been studied by several
workers. In these instances the reference given generally has hist-
orical priority.

I MONOCOTYLEDONS

CYPERACEAE

Cyperus albomarginatus	(2)	*Aristida uniplumis*	(10)
Cyperus bowmannii	(18)	*Arundinella hirta*	(10)
Cyperus eragrostis	(2)	*Astrebla pectinata*	(15)
Cyperus esculentus	(4)	*Axonopus argentinus*	(3)
Cyperus polystachyos	(18)	*Axonopus compressus*	(10)
Cyperus rotundus	(4)	*Bouteloua curtipendula*	(15)
Kyllinga monocephala	(18)	*Buchloe sp.*	(6)
		Cenchrus ciliaris	(4)
### GRAMINEAE		*Cenchrus echinatus*	(4)
		Cenchrus myosuroides	(13)
Andropogon gayanus	(15)	*Chloris gayana*	(2)
Andropogon saccharoides	(13)	*Coix lacryma-jobi*	(15)
Andropogon scoparius	(4)	*Cynodon dactylon*	(2)
Andropogon virginicus	(4)	*Dactyloctenium aegyptium*	(4)
Anthephora cristata	(13)	*Dichanthium aristatum*	(15)
Anthephora pubescens	(6)	*Digitaria argyrograpta*	(15)
Aristida adscendsionis	(11)	*Digitaria sanguinalis*	(2)
Aristida longiseta	(13)	*Digitaria smutsii*	(15)
Aristida ternipes	(13)	*Digitaria decumbens*	(23)

Digitaria pentzii	(4)	*Paspalum distichum*	(2)
Distichlis spicata	(12)	*Paspalum hartwegianum*	(13)
Echinochloa colonum	(4)	*Paspalum notatum*	(4)
Echinochloa crus-galli	(2)	*Pennisetum glaucum*	(2)
Echinochloa stagnina	(15)	*Pennisetum pedicellatum*	(15)
Eleusine coracana	(15)	*Pennisetum purpureum*	(4)
Eleusine indica	(4)	*Saccharum officinarum*	(3)
Eragrostis brownei	(3)	*Saccharum robustum*	(26)
Eragrostis chloromelas	(4)	*Saccharum sinense*	(3)
Eragrostis curvula	(17)	*Saccharum spontaneum*	(3)
Eragrostis intermedia	(13)	*Schedonnardus paniculatus*	(13)
Eragrostis mexicana	(2)	*Setaria italica*	(2)
Eremochloa ophiuroides	(13)	*Setaria lutescens*	(2)
Eragrostis pilosa	(2)	*Setaria sphacelata*	(15)
Eragrostis rigidior	(15)	*Setaria viridis*	(4)
Erianthus maximus	(3)	*Sorghastrum nutans*	(13)
Euchlaena mexicana	(14)	*Sorghum almum*	(2)
Heteropogon contortus	(4)	*Sorghum bicolor*	(4)
Hilaria mutica	(13)	*Sorghum halepense*	(3)
Hyparrhenia hirta	(17)	*Sorghum propinquum*	(16)
Imperata arundinacea	(22)	*Sorghum sudanense*	(2)
Leptochloa dubia	(4)	*Sorghum vulgare*	(2)
Leptochloa fusca	(4)	*Spartina foliosa*	(26)
Melinis minutiflora	(15)	*Sporobolus caroli*	(24)
Miscanthus sacchariflorus	(26)	*Sporobolus cryptandrus*	(6)
Muhlenbergia racemosa	(6)	*Sporobolus spicatus*	(13)
Panicum anceps	(5)	*Themeda australis*	(15)
Panicum antidotale	(5)	*Tragus australianus*	(15)
Panicum bulbosum	(7)	*Trichachne californica*	(4)
Panicum capillare	(2)	*Trichachne insularis*	(4)
Panicum coloratum	(5)	*Trichloris sp*	(6)
Panicum dichotomiflorum	(5)	*Tripsacum dactyloides*	(4)
Panicum filipes	(13)	*Urochloa mosambicensis*	(13)
Panicum hallii	(5)	*Vaseyochloa multinervosa*	(6)
Panicum laevifolium	(5)	*Zea mays*	(3)
Panicum maximum	(4)	*Zoysia japonica*	(17)
Panicum miliaceum	(2)		
Panicum milioides	(24)		
Panicum minus	(5)		
Panicum obtusum	(5)		
Panicum plenum	(5)		
Panicum polygonatum	(6)		
Panicum prolutum	(5)		
Panicum stapfianum	(5)		
Panicum texanum	(6)		
Panicum turgidum	(5)		
Panicum virgatum	(4)		
Pappophorum bicolor	(6)		
Paspalum dilatatum	(3)		

II DICOTYLEDONS

AZIOACEAE

Mollugo cerviana (26)
Trianthema portulacastrum (25)

AMARANTHACEAE

Amaranthus albus (1)
Amaranthus blitoides (1)
Amaranthus caudatus (1)
Amaranthus edulis (1)
Amaranthus hypochondriacus
   nanus (1)
Amaranthus palmeri (16)
Amaranthus patulus (1)
Amaranthus retroflexus (1)
Amaranthus melancholicus
   ruber (1)
Amaranthus tricolor (1)
Amaranthus viridis (1)
Froelichia gracilis (1)
Gomphrena celosioides (18)
Gomphrena globosa (1)
Gomphrena haageana (1)
Tidestromia oblongifolia (8)

CHENOPODIACEAE

Atriplex angulata (19)
Atriplex argentea (20)
Atriplex canescens (20)
Atriplex halimus (9)
Atriplex holocarpa (9)
Atriplex inflata (19)
Atriplex lentiformis (20)
Atriplex lindleyi (19)
Atriplex nummularia (10)
Atriplex polycarpa (20)
Atriplex rhagodioides (9)

Atriplex rosea (1)
Atriplex semibaccata (12)
Atriplex sibirica (9)
Atriplex spongiosa (9)
Atriplex suberecta (10)
Atriplex tatarica (10)
Atriplex vesicaria (19)
Bassia hyssopifolia (9)
Kochia childsii (1)
Kochia scoparia (1)
Salsola kali (1)

COMPOSITEAE

Pectis leptocephala (21)

EUPHORBIACEAE

Euphorbia drummondii (24)
Euphorbia glyptosperma (5)
Euphorbia maculata (5)
Euphorbia missurica (5)
Euphorbia serpyllifolia (5)
Euphorbia supina (5)
Euphorbia wheeleri (24)

NYCTAGINACEAE

Boerhaavia paniculata (25)

PORTULACACEAE

Portulaca grandiflora (1)
Portulaca oleracea (1)

ZYGOPHYLLACEAE

Tribulus hystrix (24)
Tribulus terrestris (24)

REFERENCES

1. E.B. TREGUNNA and J. DOWNTON, *Can. J. Bot.*, 45 (1967) 2385. (Low compensation point.)
2. W.J.S. DOWNTON and E.B. TREGUNNA, *Can. J. Bot.*, 46 (1968) 207. (Low compensation point and "Kranz" type leaf anatomy.)

3. M.D. HATCH, C.R. SLACK and H.S. JOHNSON, *Biochem. J.*, 102 (1967) 417. (C_4 acids as initial products of $^{14}CO_2$ fixation.)
4. T.M. CHEN, R.H. BROWN and C.C. BLACK, *Weed Sci.*, 18 (1970) 399. (Low compensation point).
5. D.N. MOSS, E.G. KRENZER and W.A. BRUIN, *Science*, 164 (1969) 187. (Low compensation point).
6. E.G. KRENZER and D.N. MOSS, *Crop Sci.*, 9 (1969) 619. (Low compensation point).
7. J. DOWNTON, J. BERRY and E.B. TREGUNNA, *Science*, 163 (1969) 78. (C_4 acids as initial products of $^{14}CO_2$ fixation and low compensation point).
8. J. BERRY, unpublished ("Kranz" type leaf anatomy).
9. E.B. TREGUNNA, J. DOWNTON and P.A. JOLLIFFE, in Progress in Photosynthesis Research, ed. H. METZNER, IUBS Tubingen, I (1969) 488. (Low compensation point and "Kranz" type leaf anatomy).
10. E.B. TREGUNNA, B.N. SMITH, J.A. BERRY, and W.J.S. DOWNTON, *Can. J. Bot.*, 48 (1970) 1209. ("Kranz" type leaf anatomy, low compensation point and ^{13}C discrimination).
11. W.J.S. DOWNTON, Ph.D. thesis, University of British Columbia, Vancouver (1969). ("Kranz" type leaf anatomy and low compensation point).
12. O. BJÖRKMAN and E. GAUHL, *Planta* 88 (1969) 197. ("Kranz" type leaf anatomy and no low O_2 enhancement of photosynthesis).
13. Sr. M.C. JOHNSON, Ph.D. thesis, University of Texas, Austin (1964). ("Kranz" type leaf anatomy).
14. W.M. LAETSCH, *Sci. Prog., Oxf.*, 57 (1969) 323. ("Kranz" type leaf anatomy).
15. R.W. DOWNES and J.D. HESKETH, *Planta* 78 (1968) 79. (No low O_2 enhancement of photosynthesis).
16. J. HESKETH, *Planta* 76 (1967) 371. (No low O_2 enhancement of photosynthesis).
17. J.P. COOPER and N.M. TAINTON, *Herb. Abst.*, 38 (1968) 167.
18. H.S. JOHNSON and M.D. HATCH, *Phytochemistry*, 7 (1968) 375. (C_4 acids as initial products of photosynthetic $^{14}CO_2$ fixation).
19. C.B. OSMOND, *Z. PflPhysiol.*, 62 (1970) 129. (C_4 acids as initial products of photosynthetic $^{14}CO_2$ fixation).
20. W.M. LAETSCH, *Am. J. Bot.*, 55 (1968) 875. ("Kranz" type leaf anatomy).
21. E. SCHÖCH, *Z. PflPhysiol.* (1971) in press. ("Kranz" type leaf anatomy and C_4 acids as initial products of photosynthetic $^{14}CO_2$ fixation).
22. E. SCHÖCH and U. LÜTTGE, unpublished. ("Kranz" type leaf anatomy and C_4 acids as initial products of photosynthetic $^{14}CO_2$ fixation).
23. J.H. HILLIARD and S.H. WEST, *Science* 168 (1970) 494. ("Kranz" type leaf anatomy).
24. W.J.S. DOWNTON, unpublished. ("Kranz" type leaf anatomy).

25. R.K. CROOKSTEN and D.N. MOSS, *Plant Physiol.*, 46 (1970) 564.
    ("Kranz" type leaf anatomy).
26. W.M. LAETSCH, this volume (1971). ("Kranz" type leaf anatomy).

Author Index

Subject Index